Integrative Sexual Health

Integrative Medicine Library

Published and Forthcoming Volumes

SERIES EDITOR

Andrew Weil, MD

Timothy P. Culbert and Karen Olness: *Integrative Pediatrics*

Randy Horwitz and Daniel Muller: *Integrative Rheumatology, Allergy, and Immunology*

Stephen DeVries and James Dalen: *Integrative Cardiology*

Mary Jo Kreitzer and Mary Koithan: *Integrative Nursing*

Myles D. Spar and George E. Munoz: *Integrative Men's Health*

Robert Norman, Philip D. Shenefelt, and Reena N. Rupani: *Integrative Dermatology*

Donald I. Abrams and Andrew Weil: *Integrative Oncology 2nd edition*

Victoria Maizes and Tieraona Low Dog: *Integrative Women's Health 2nd edition*

Robert Bonakdar and Andrew W. Sukiennik: *Integrative Pain Management*

Aly Cohen and Frederick S. vom Saal: *Integrative Environmental Medicine*

Richard Carmona and Mark Liponis: *Integrative Preventive Medicine*

Mikhael Kogan: *Integrative Geriatric Medicine*

Daniel A. Monti and Andrew B. Newberg: *Integrative Psychiatry and Brain Health 2nd edition*

Barbara Bartlik, Geovanni Espinosa, and Janet Mindes: *Integrative Sexual Health*

Shahla Modir and George Munoz: *Integrative Addiction and Recovery*

Gerard Mullin: *Integrative Gastroenterology*

Integrative Sexual Health

EDITED BY

Barbara Bartlik, MD, ABOIM
Private Practice, Integrative Medicine and Psychiatry
Voluntary Faculty
NewYork-Presbyterian/Weill Cornell Medical Center
Lenox Hill Hospital (Northwell Health)
New York, NY
Psychiatrist, Yonkers Service Center of Rockland Psychiatric Center
Orangeburg, NY

Geovanni Espinosa, ND, LAc, IFMCP, CNS
Integrative and Functional Urology
Clinical Assistant Professor
New York University Langone Health
New York, NY

Janet Mindes, PhD
Faculty, Research Instruction and Development
Tri-State College of Acupuncture
New York, NY

OXFORD
UNIVERSITY PRESS

OXFORD
UNIVERSITY PRESS

Oxford University Press is a department of the University of Oxford. It furthers the University's objective of excellence in research, scholarship, and education by publishing worldwide. Oxford is a registered trade mark of Oxford University Press in the UK and certain other countries.

Published in the United States of America by Oxford University Press
198 Madison Avenue, New York, NY 10016, United States of America.

Library of Congress Cataloging-in-Publication Data
Names: Bartlik, Barbara, editor. | Espinosa, Geovanni, editor. | Mindes, Janet, editor.
Title: Integrative sexual health / edited by Barbara Bartlik, Geovanni Espinosa, Janet Mindes.
Description: New York, NY : Oxford University Press, [2018] | Includes bibliographical references.
Identifiers: LCCN 2017030817 | ISBN 9780190225889 (pbk.)
Subjects: | MESH: Reproductive Health | Integrative Medicine—methods |
Sexual Dysfunction, Physiological—therapy |
Sexual Dysfunctions, Psychological—therapy | Sexuality
Classification: LCC RC556 | NLM WQ 200.1 | DDC 616.6/9—dc23
LC record available at https://lccn.loc.gov/2017030817

1 3 5 7 9 8 6 4 2
Printed by Webcom Inc., Canada

I want to express my gratitude to my husband, David Carroll, whose words of encouragement and endless support have made this book possible. Thank you for always being there for me.

—Barbara Bartlik

A special dedication to my children, Mia, Gianna, and Leonardo, and to my amazing wife, Johanna, for your unwavering and unconditional love and support.

—Geovanni Espinosa

To my late, beloved mother, psychotherapist Dr. Selma Mindes, who to age 96 was a living example of what a mindfully healthful life can yield in vitality and longevity. You understood the worth and originality of this project from the start.

—Janet Mindes

We all offer and dedicate this book to advocate for an integrative sexual health model, to help clinicians better understand more varied causes of sexual difficulties and to offer a sound basis for new integrative treatment strategies, toward improved vitality and enhanced emotional and life satisfaction for all.

—Barbara Bartlik, MD
Geovanni Espinosa, ND, LAc, IFMCP, CNS
Janet Mindes, PhD

CONTENTS

SERIES FOREWORD

ANDREW WEIL, MD

Series Editor

Despite its importance, the topic of sexual health is often slighted in the training of physicians and allied health professionals. So many factors influence sexual health, and sexual behavior has so many implications for general health that the prevailing biomedical paradigm is of very limited utility. Only a comprehensive model that includes the psychosocial, behavioral, and spiritual dimensions of human life can serve.

Integrative medicine provides such a model. It pays attention to the nonphysical aspects of patients and looks at all elements of lifestyle that influence health. This volume is the first effort to apply it to the field of sexual health.

Drs. Bartlik, Espinosa, and Mindes draw on the philosophy of integrative medicine to help practitioners understand the complex causes of sexual dysfunction and treat sexual problems with diverse, safe, effective, evidence-based strategies rather than solely with pharmaceuticals. Many of the approaches they recommend are not yet included in the standard training of physicians.

Integrative Sexual Health is a ground-breaking, welcome addition to the Oxford University Press growing *Integrative Medicine Library* series. It is a milestone in the advancement of the field and will greatly benefit both practitioners and patients.

PREFACE: AN OVERVIEW OF *INTEGRATIVE SEXUAL HEALTH*

Integrative Sexual Health is an up-to-date, evidence-based, multifaceted examination of sexual problems as integrative medicine would treat them. Promoting optimum health is a fundamental goal and also a path to mitigating or even resolving many common sexual difficulties. We believe this book will be useful to clinicians who practice integrative and functional medicine, as well as to primary and specialty care clinicians, to researchers, and ultimately to patients. We intend that *Integrative Sexual Health* will extend the benefits of integrative medicine into sexual medicine and into individuals' sexual health understanding and behavior.

Fundamentally, this book itself is integrative—it draws on very interdisciplinary content and synthesizes a wide range of biological, psychological, and behavioral knowledge and information that we believe has not been synthesized in this way previously, in the service of sexual health.

The authors of individual chapters in this volume are authorities in integrative medicine across a range of disciplines and detail theory and practice of their treatment of sexual dysfunctions in men and women. Research is extensively cited, and clinical examples and practical strategies are offered. Problem-solving for sexual life is frequently linked to strategies for improving physical and mental health.

Hundreds of millions of dollars are spent each year on diverse products and supplements thought to alleviate sexual symptoms. By and large, these products have not been extensively studied and proven to be safe and effective. Nor are the proposed mechanisms by which they are said to work understood by the people who use them, and often not even by their clinicians. Pharmaceuticals used to treat male and female sexual dysfunction often are not fully effective or are undesirable due to adverse effects, expense, or inconvenience. Approximately half of the prescriptions written for erectile-enhancing medications, such as sildenafil (Viagra), are not refilled, and the new female sexual desire enhancing medicine flibanserin (Addyi) has been slow to gain popularity. There is a clear need for new thinking and better, more holistic clinical strategies.

Integrative Sexual Health has accessed and converged evidence from very diverse, relevant, and reputable scientific and medical sources. The many randomized clinical trials cited in chapters in this book sometimes address a narrowly sexual question but more often address closely related topics in biology, physiology, psychology, and behavior. The reader should be aware that clinical and basic science research in sexual dysfunction is a relatively underfunded area. Most sexual dysfunction—distressing as it is to those afflicted—does not involve grave medical problems, although types of sexual dysfunction now are seen as often very intertwined with a range of chronic health conditions, some of which may become serious. The paucity of randomized clinical trials in sexual medicine, and the even greater absence of clinical trials in sexual medicine incorporating any integrative aspect, made it impossible to approach writing this book based only on that so-called gold standard of evidence. We view it as a strength of this volume that very diverse sources of evidence were consulted and from a range of conventional Western, integrative, and traditional medicine disciplines.

There is a large and ever-growing literature on the diverse health impacts of unfortunately very common systemic conditions such as obesity, diabetes, depression, and poor gut health. Most clinicians and patients are not aware that, via complex hormonal, inflammatory, psychological, and other pathways affected in states of chronic ill health, sexual function often is impacted. Gathering data on sexual function in studies of chronic health conditions also can shed new light on biological pathways underpinning the primary health concern. Mitigating obesity, diabetes, depression, and other chronic ills, and the sexual problems they often cause, increasingly involves recommendations for the kinds of lifestyle changes and nutritional improvements that integrative medicine has long advocated. These recommendations are now in the mainstream of medicine.

Sexual problems also are caused by commonly prescribed medications for a very wide range of conditions. Many Americans currently live with this reality. Medications often deplete micronutrients, and this also can contribute to sexual dysfunction. This connection is infrequently made. Rather than ignore the sexual problem or rush to treat it with pharmaceuticals, targeted micronutrient therapy and other integrative strategies may be used to improve sexual function.

This book emphasizes a multitiered approach: (a) improving lifestyle and health habits to minimize the need for many common medications that cause sexual side effects, (b) preserving sexual function using integrative strategies and mitigating the adverse effects of needed medications, and (c) therefore also minimizing the need for many mainstream pharmaceuticals and other treatments for sexual dysfunction. This holistic strategy designed to promote long-term health and sexual viability also promotes an integrated view of the self and of one's own health status, choices, and behaviors.

Many people now use integrative medicine for a wide range of health and even sexual difficulties, but their physicians often cannot advise them; they mostly are not aware of the integrative medicine research and clinical literature, nor often the sexual medicine literature. The intersection of these two domains is so new it is mostly invisible to clinicians. Use of integrative medicine for any reason is infrequently discussed between doctors and patients, although this is beginning to change. Effective integrative and functional medicine interventions and approaches still are not part of the standard of care in medical schools and in most subspecialties and practices. Even trained sex therapists rarely address the use of integrative medicine for sexual problems.

The overarching aim of this book is to offer its professional readers an *integrative sexual health model.* We seek to provide clinicians with a comprehensive, well-supported overview of safe integrative approaches to treating sexual problems, to be used alone or with conventional approaches. Physicians, even those trained in integrative medicine, are primarily concerned with general health and the management of illness and infrequently address sexuality. They often encounter sexual difficulties in their patients and may feel they have little to offer, beyond suggesting a pharmaceutical, a device, or psychosexual therapy. All these certainly can be relevant and effective, but they no longer constitute a complete enough picture. An integrative sexual medicine approach is needed, one that can offer the varied complementary and traditional medicine approaches described in this volume, such as nutritional, botanical, aromatherapy, behavioral, physical, and other integrative modalities and targeted interventions.

Integrative Sexual Health is divided into four sections. Section I sets out an integrative medicine framework for understanding sexual dysfunction, sexual biochemistry, menopause, and andropause. Section II discusses lifestyle issues that impact sexual function, including macronutrition, micronutrient deficiencies, effects of hormonal contraception and environmental toxins on health and sexuality, the importance of physical exercise to sexuality, and the threats posed to sexuality by the obesity epidemic. Section III describes physiological and psychiatric disturbances that affect sexual function, such as inflammation and the gut-brain-mood connection, depression and anxiety, and sexual addiction, as well as psychological and behavioral perspectives on sexuality in the context of integrative medicine. Section IV presents traditional Eastern and Western approaches to healing sexual problems, including the Ayurvedic tradition, traditional East Asian medicine such as acupuncture, homeopathy, chiropractic medicine, aromatherapy, European water therapy, and botanical medicine.

Human beings now live far longer than our remote ancestors and even our immediate forebears. With longer lifespans, and the health and lifestyle issues of our times, treating and managing chronic medical and psychological problems is more likely to be necessary over the lifespan; often these conditions may affect sexual function and expression. In general, broader awareness of the many factors that may impact sexual function is needed, and physicians and allied health professionals should be updated. This volume alerts physicians, for example, to the direct and indirect adverse effects on sexuality of obesity, hormonal contraception, macro- and micronutrition status, sexual side effects and micronutrient depletion associated with common medications, poor gut health, habits and lifestyle factors such as poor sleep quality, lack of exercise, smoking, the growing burden of chemicals in the environment, and other risk factors. Sexuality also may be compromised by psychological factors and stressors, social phenomena such as Internet pornography addiction and its impact on sexual response and relationships, dealing with the aftermath of prostate and breast cancer surgeries as they affect sexuality, and other medical, psychological, and psychiatric contexts.

Those interested in maintaining an active sex life may confront some of these diverse impediments to satisfying relationships and sexual expression; they may also be unaware that their less than satisfying sexual response may be due to some of the aforementioned factors. Those who do want a sexual life and are prevented from having one by medical problems and treatment side effects often become depressed, anxious, or more socially isolated as a result, adding to their burden. More solutions—preferably also addressing a person's overall health status—are needed for men and women, perhaps particularly for female patients, given relatively few approved pharmacological treatments for female sexual dysfunction.

To be blocked from satisfying sexual expression is a distinct loss to health and well-being. Intimate contact and orgasm can be very gratifying and destressing, in the context of loving physical closeness and release of oxytocin and other neurochemicals. Orgasm enhances sleep, reduces blood pressure, and improves mood, yet too many men and women are deprived of these benefits, at every age. Most physicians are neither inclined nor typically in a position to offer opinions on how couples may improve their health by improving their sexual connection and vice versa. Sex therapists and psychological professionals may be able to suggest useful behavioral and psychological strategies to improve sexual communication and satisfaction but may lack specific knowledge of the relevant biology and awareness of potentially useful integrative medicine interventions.

Physicians and psychological professionals alike need more knowledge and expanded clinical strategies to help patients with sexual problems. Some patients may not actively seek specific sexual solutions because their sexual life is not currently active or may be less important to them. Yet they may still wish to be alive to sexual feelings and to their bodies. Addressing vitality, medical problems and lifestyle concerns may help those individuals experience more aliveness, more interest in intimacy, and at a minimum achieve better and more enduring health and increased physical and emotional self-esteem.

Some professionals and many of the lay public increasingly question the medicalization of all aspects of health. Many now question the almost obligatory use of pharmaceuticals, while still insufficient emphasis is placed on the benefits of proven lifestyle changes as genuine healthcare solutions. For example, not long ago the American Heart Association issued new guidelines for the use of statins, leading to increased use of these drugs. Some physicians commented in the media that many who might be prescribed statins would be better served by changes in diet and exercise and that, of people prescribed statins who do not need them to guard against high-risk heart disease, up to approximately 20% may experience sexual side effects. Compliance with healthful regimens of course is a concern, compared to the ease of taking a pill. However, we must at the same time support a healthcare culture that makes it increasingly attainable for patients to be motivated and educated to act mindfully in the service of their own bodies and their own health.

A growing number of Americans are not only resisting overuse of pharmaceuticals and procedures but are seeking less invasive and more "natural" treatment options. They prefer more holistic approaches and are less willing to accept side effects caused by medications, including those disrupting desire and sexual function. Physicians who wish to improve patient compliance with needed treatments can be helped by increased knowledge of integrative clinical strategies and treatment options.

Our modern culture may well rely somewhat too often on pharmaceuticals and quick fixes for health and sexual problems. In addition, we are surrounded by continual superficial sexual stimulation via the media and Internet. It can be argued that cultural norms of sexual behavior have become more superficial, more aggressive, and more sympathetic nervous system driven, with diminished sensitivity to deeper human needs for emotional and sensual connection. These behavioral trends in themselves are the source of certain kinds of sexual problems. By contrast, being more psychologically and emotionally receptive, and achieving more balance between the sympathetically and parasympathetically driven aspects of sexuality, more fully engaging the senses and emotions in sexual connection, will support healthy sexuality. Traditional medicine systems, such as Ayurveda, have always advocated this understanding of sexual health and enjoyment.

With respect to all the issues and concerns mentioned here, integrative medicine has something to contribute, as we hope the reader will discover in our diverse chapters. The integrative medicine movement seeks to marshal lifestyle changes, greatly increased attention to nutritional quality, reducing the chemical burden on the body, use of effective treatments and perspectives from world traditional medicine, responsible self-care, stress reduction, emotional awareness and expression, and promotion of health and vitality into one system. When medical treatment is needed, both evidence-based conventional and alternative interventions may be considered. The integrative medicine framework can and, we contend, should incorporate sexuality into the broader discussion of health in a new way which could be beneficial for individuals, for medicine, and for society.

In conclusion, we believe *Integrative Sexual Health* will make a useful contribution, by gathering the clinical and research knowledge of diverse experts on uses of integrative medicine for sexuality and by covering a wide range of topics. The primary professional aim of this book is to offer an innovative *integrative sexual health model* to help clinicians better understand the varied causes of sexual difficulties and to better address a range of sexual problems, a model that offers a sound basis for new integrative strategies. An integrative approach to sexual medicine can help in guiding patients to achieve and maintain better physical and psychological health. The rewards of gratifying intimacy could do much to help motivate patients to stay on track toward living healthier lifestyles. Vitality and quality of life could be increased, management of sexual sequelae of illness and needed treatment also could be improved, medical management costs possibly could be reduced, and emotional and life satisfaction might be increased. We offer this book in the sincere hope that we are contributing to these worthy goals.

Barbara Bartlik, MD, ABOIM
Geovanni Espinosa, ND, LAc, IFMCP, CNS
Janet Mindes, PhD

ACKNOWLEDGMENTS

We all wish to thank the many colleagues, family and friends who have been interested and supportive as we developed this book. With special thanks for early encouragement to Victoria Maizes, MD, of the University of Arizona College of Medicine, and to Dr. Andrew Weil for accepting our book proposal to be a volume in the Weil Integrative Medicine Library. With thanks in warm memory of our late friend and colleague, integrative nutrition pioneer Dr. Annemarie Colbin, who helped the first ideas for this book take form. We offer much sincere appreciation to our very patient editor at Oxford University Press, Andrea Knobloch, and her team. Dr. Bartlik also thanks her assistants, Denisa Music, BS, Vittoria Turano, BA, and Sabeen Arif, for their invaluable help with this book, and Joseph Nazareth, MD for his wisdom and advice.

To all our wonderful chapter authors: it has been a pleasure and an education working with all of you.

Barbara Bartlik, MD, ABOIM
Geovanni Espinosa, ND, LAc, IFMCP, CNS
Janet Mindes, PhD

CONTRIBUTORS

Jeffrey A. Albaugh, PhD, APRN, CUCNS
Director of Sexual Health
John & Carol Walter Center for Urological Health
North Shore University Health System
Glenview, IL

Karen Allen, CCH
Faculty, American Medical College of Homeopathy
Phoenix Institute of Herbal Medicine and Acupuncture
Phoenix, AZ
Private Practice
San Francisco, CA

Barbara Bartlik, MD, ABOIM
Private Practice, Integrative Medicine and Psychiatry
Diplomate, American College of Psychiatrists
Diplomate, American Board of Psychiatry and Neurology (in psychiatry)
Diplomate, American Board of Sexology
Voluntary Faculty, NewYork-Presbyterian/Weill Cornell Medical Center
Lenox Hill Hospital (Northwell Health)
Chair, Complementary and Integrative Medicine Committee
New York County Psychiatric Society
New York, NY
Psychiatrist, Yonkers Service Center of Rockland Psychiatric Center
Orangeburg, NY

Jonathan Bekisz
MD/MSci Candidate, Class of 2018
New York University School of Medicine
New York, NY

Janine A Blackman, MD, PhD
Epidemiology & Preventive Medicine
Private Practice: Functional Medicine and Bio-Identical Hormone Therapy for Women and Men
Potomac, MD

Jillian Capodice, MS, LAc
Assistant Professor
Director, Integrative Urology and Wellness
Icahn School of Medicine
Mount Sinai Health System
New York, NY

Madeleine M Castellanos, MD, IFMCP
Functional Medicine Psychiatrist, Sex Therapist
Private Practice in New York, NY and Boca Raton, FL
Medical Director—Restoration Women's Health
Hauppauge, NY

Susan R. Eisen, DC
Holistic Chiropractic
New York, NY

Geovanni Espinosa, ND, LAc, IFMCP, CNS
Integrative and Functional Urology
Clinical Assistant Professor
New York University Langone Health
New York, NY

Ralph Esposito, ND, LAc
Naturopathic Physician, Licensed Acupuncturist
Adjunct Professor, New York University
Department of Nutrition and Food Studies
New York, NY
Adjunct Clinical Professor
Center of Excellence in Generative Medicine
University of Bridgeport College of Naturopathic Medicine
Bridgeport, CT

Keesha Ewers, PhD, ARNP, AAP
Doctor of Sexology
Advanced Ayurvedic Medicine Practitioner
Certified Psychotherapist
Medical Director and Founder, Fern Life Integrative Medical Center
Issaquah, WA

Katherine Falk, MD, ABIHM
Private Practice, Integrative Psychiatry
Diplomate, American Board of Psychiatry and Neurology
Distinguished Life Fellow, the American Psychiatric Association (2007)
Adjunct Assistant Clinical Professor of Psychiatry
Mount Sinai Medical Center
New York, NY

Jennifer Yonaitis Fariello, MSN, MEd, WHNP-BC
Smith Institute for Urology
Pelvic Pain Treatment Center
Garden City, NY

Alexa Fleckenstein, MD, MEd
Author, European Natural Medicine
Brookline, MA

Mindy Green, MS, RA, RH (AHG)
Private Practice, Green Scentsations
Boulder, CO

Sheila Sperber Haas, PhD
Science Writer
New York, NY

Ronald L. Hoffman, MD, CNS
Medical Nutritionist
New York, NY

Amandeep Kaur, MD
Manhattan Psychiatric Center
New York, NY

Julie Kolzet, PhD
Licensed Psychologist
Ferkauf Graduate School of Psychology
New York, NY

Steven Lamm, MD
Professor of Clinical Medicine
Medical Director, Preston Robert Tisch Center for Men's Health
New York University Langone Health
New York, NY

Monica Meyer, PhD, CSAT-S
Clinical Director
Gentle Path at The Meadows
Wickenburg, AZ

Janet Mindes, PhD
Faculty, Research Instruction and Development
Tri-State College of Acupuncture
Consulting Research Psychologist, Independent Writer and Editor
New York, NY

Leyla Muedin, MS, RD, CDN
Nutritionist
New York, NY

Jacquelyn M. Paykel, MD, FACOG
Director of THRIVE
Chief of Whole Health Services
James A. Hailey Veterans' Hospital and Clinics
Tampa, FL

Chelsea Schoen, PhD
Post-Doctoral Fellow
New York University Langone Health—Rusk Rehabilitation
New York, NY

Susan Kellogg Spadt, PhD, CRNP, IF, FCST, CSC
Director, Female Sexual Medicine
Center for Pelvic Medicine
Bryn Mawr, PA
Professor, Human Sexuality
Widener University
Chester, PA
Associate Faculty
University of Pennsylvania Colleges of Medicine and Nursing
Rutgers University College of Medicine
Thomas Jefferson University
Philadelphia, PA

Adrienne Sprouse, MD, MEd
Medical Director
Manhattan Health Consultants
New York, NY

Jade Teta, ND
Naturopathic Health Clinic of North Carolina
Metabolic Effect
Winston-Salem, NC

Keoni (John) Teta, ND, LAc
Naturopathic Health Clinic of North Carolina
Metabolic Effect
Winston-Salem, NC

Dana Ullman, MPH, CCH
Homeopathic Educational Services
Foundation for Homeopathic Education and Research
Berkeley, CA

Eric Yarnell, ND, RH (AHG)
Associate Professor
Department of Botanical Medicine
Bastyr University
Kenmore, WA

Integrative Sexual Health

SECTION I

An Integrative Medicine Framework for Understanding Sexual Dysfunction

1

Female Sexual Biochemistry

MADELEINE M. CASTELLANOS

The biology of being female is both beautiful in its intricacies and complex in its interactions. Although we have come far in our understanding of sexual biochemistry, there are volumes yet to discover about how that biochemistry affects not just the reproductive organs but also the arteries and cardiovascular system, connective tissues and bones, regulation of muscle and fat tissue, as well as the entire nervous system and brain just to name a few. Here we dive into an overview of female sexual biochemistry and function.

When considering female sexual biochemistry, one must appreciate that there is a constant interplay between what is happening physically and what is happening psychologically. Considering the psychological is of the utmost importance because thoughts and emotions result in physiological responses in both the brain and the body, both producing and affecting sexual response. Not only are we created as sexual beings but we are also constantly receiving messages about sex and sexuality that continue to shape us throughout our lives. Hormones work together in harmony; not only do they affect the development and the functioning of the body but they are affected by diet, stress, moods, and even thoughts. It is not a static system with absolute levels. Instead, there is constant variation, which works as a balance of perpetual adjustments to internal states and environmental (external) forces.

Understanding female sexual chemistry is about recognizing the multitude of interactions that take place between several hormones at any given time and the scores of tissues that rely on them. The interplay between hormones in females is complicated by the many different changes that occur not just over the whole life cycle but also the cycles within the reproductive years and the changes that those bring.

Common Pathway of Hormone Production

Because nature's design is inherently economical, it uses cholesterol as the main building block of all the sex steroid molecules. Cholesterol is important for the production of phospholipids in cell membranes as well as conversion to cortisol, progesterone, estrogen, dehydroepiandrosterone (DHEA), and testosterone (Figure 1.1). It is transported to the cells via a molecule of LDL cholesterol that is taken up by the cell through receptor-mediated endocytosis. In order for cholesterol to be converted to a sex steroid molecule, it must first be transported to the inner mitochondrial membrane. This movement of cholesterol from the outer mitochondrial membrane to the inner mitochondrial membrane constitutes the rate-limiting step in the formation of all corticosteroids and all sex steroid molecules (Rone, 2009).

Once cholesterol is inside the mitochondria, it can be converted to pregnenolone, the precursor to all sex steroids. The enzyme responsible for this conversion is CYP11A1 (previously known as P450scc), which is found on the inner mitochondrial membrane and is controlled by the tropic hormones of the anterior pituitary (adrenocorticotropic hormone [ACTH], luteinizing hormone [LH],

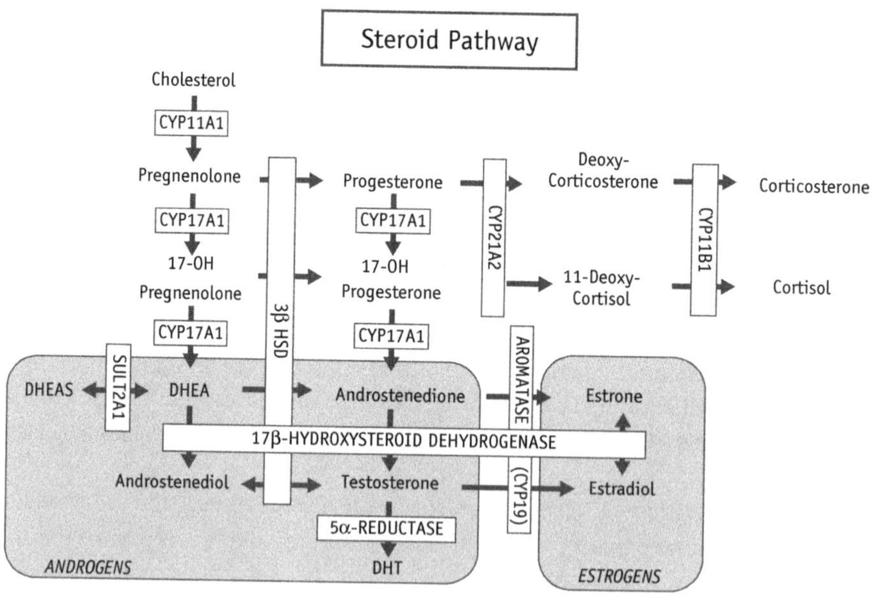

FIGURE 1.1. Steroidogenesis for all sex steroids begins with a cholesterol molecule that is converted to pregnenolone. The final steroid molecule produced is under gonadotropin influence and the presence of particular enzymes in each cell type that carries out the enzymatic conversion.

follicule-stimulating hormone [FSH]). Pregnenolone is then moved back out into the cell to the endoplasmic reticulum where it is further converted to downstream steroids (like progesterone, 17-OH pregnenolone—the precursor to cortisol, DHEA, testosterone, or estrogen) depending on the type of enzymes available in that particular cell. Studies have shown that, under experimental conditions, CYP11A1-mediated production of pregnenolone may be inhibited by ketoconazole, posaconazole, carbenoxolone, and selegiline by at least 67%. These studies also demonstrate increased CYP11A1 enzyme activity by up to 70% by the drugs pemirolast, clobenpropit, desogestrel, dexmedetomidine, and tizanidine (Mast, 2013).

Estrogens

Estrogen is the principle hormone associated with female sexuality. It is not a single hormone but a class of hormones with a similar structure that have a multitude of effects on a wide range of different tissues in the body, many of which are still not fully understood. Initially, it was thought that estrogens exerted their effects on tissues only by regulating DNA transcription. This process could take from hours to days for its effects to be seen. Now we know that estrogens can exert a more immediate effect on cells and tissues as well.

The types of estrogen receptors (ER) that affect DNA transcription are known as ERα and ERβ, each with their subtypes. ERα and ERβ seem to have opposite effects with ERα being more growth-promoting and ERβ more inhibiting of growth. Estrogens can readily pass in and out through the cell membrane and then bind the ERs found on the membrane of the cell nucleus. ERs are structured as dimers that are either homodimers (αα or ββ) or heterodimers (αβ). Two molecules of an estrogen or estrogen-activator must be present to activate the receptor. The entire complex then translocates into the nucleus where it binds to DNA in order to control transcription of different genes.

Estrogens have also been found to affect cells through cellular membrane receptors (mER), which include mERα, mERβ, ER-X, and G protein-coupled estrogen receptor 1 (GPER), which allow for rapid signaling within a cell. Some of the mERs are complexes that contain signaling proteins acting as second messengers in the cell. GPER seems to mediate the action of estrogen in a cell, but it is not clear if it does this alone as well as with the presence of an ER. It seems that cells may have ERs, both ERs and GPERs, or just GPERs. It is interesting to note that in cancer cells there is often a higher expression of GPERs when compared to the cells from the originating cell line, but the significance of their role in the development of cancer is still unclear (Soltysik, 2013). The ER-X receptor is known to be present in neonates, as well as in adult brain tissue after a stroke, indicating that it may play an active role in regeneration of neural tissues (Toran-Allerand, 2002).

There are three primary estrogens produced in women—estradiol (E2), the main estrogen of adulthood and the most active estrogen; estriol (E3), the weakest of the natural estrogens and the main estrogen of pregnancy; and estrone (E1), the primary estrogen circulating in menopause. Besides supporting endometrial tissue, ovarian tissue, vaginal tissue, tissues of the external female genitals, and breast tissue, estrogens also have action in the brain; in bone; in blood vessels; and in the heart, liver, kidney, lung, bladder, skin, and gastrointestinal tract.

Estradiol (E2) is produced in the granulosa cells of the ovary from aromatization of either testosterone or androstenedione via estrone (E1). The theca cells of the ovary are first stimulated by LH to produce testosterone, which can then diffuse into the granulosa cells. FSH then regulates the conversion of testosterone to estradiol throughout a woman's cycle by controlling the levels of aromatase enzyme in the granulosa cells of the ovary. This two-cell model indicates that both theca cells and granulosa cells are needed for estrogen production in the ovaries.

The adrenal gland is also responsible for production of estrogens by synthesizing the precursor androstenedione, which then circulates to target tissues, such as the ovary, for conversion via aromatase enzyme to estrone (E1) and then via 17-beta-HSD to estradiol (E2). Aromatase enzyme is also naturally found in adipose tissue, skin, bone, blood vessels, brain tissue, and the placenta, so conversion to estrogen can take place in these tissues as well.

Interestingly, tissues that are considered abnormal in nature such as endometriosis, uterine fibroids, breast cancer, and endometrial cancer all have elevated levels of aromatase and therefore are capable of producing higher levels of estradiol that stimulate their further growth. These are thought to be inflammatory in nature, and it has been found that PGE2 is the most potent stimulator of aromatase in breast cancer (Bulun, 2009) and endometriosis (Noble, 1997). This implies that reducing chronic inflammation in the body could impact the development of such conditions. Aromatase activity in human granulosa cells is also increased by atrazine, a common pesticide in the United States that persists in the water supply (Holloway, 2008). This increases estrogen production, thereby increasing stimulation of tissue growth and increasing cancer risk.

PUBERTY

Although newborns have very high levels of estradiol, these fall to low levels within a few days of delivery and stay relatively low until puberty. The average estradiol level for girls before puberty is 0.6+/−0.6 pg/mL (Klein, 1994) and does not appear to be correlated with body mass index or age. Estrogen is a growth hormone that stimulates growth of tissues. It is probably hormetic, as it is stimulating at lower doses and inhibitory or toxic at higher doses. In girls, estrogen is responsible for the secondary sex characteristics that appear in puberty such as breast development, greater distribution of body fat over hips, and the commencement of menstrual cycles. In puberty, estradiol stimulates changes in both internal reproductive organs and external genitalia. In addition to the growth of pubic hair, lengthening and darkening of the labia minora and majora occur during puberty. This is due to the effects of estrogen, which increases the production of pheomelanin, giving the labia a darker color than the surrounding skin.

With prolonged stimulation of estrogen, the uterus elongates and the ovaries also grow and mature. Estradiol also causes thickening of the myoepithelial layer as well as the smooth muscle of the vagina. The vaginal epithelium that is rather thin and flat throughout childhood becomes thicker and starts to produce glycogen-containing cells. This glycogen is fermented to lactic acid by the natural vaginal flora, which is what makes the vaginal pH mildly acidic. Estrogen also stimulates the production of a watery cervical mucus that is favorable for the passage of sperm. Progesterone, on the other hand, reduces the amount of cervical mucus, making it more challenging for sperm that may be present during the luteal phase of the menstrual cycle.

Estrogens (and androgens) are controlled by the tropic hormones LH and FSH released from the anterior pituitary. These are both stimulated by gonadotropin-releasing hormone (GnRH), which is secreted in pulses from the hypothalamus to the anterior pituitary. There is a negative feedback loop back to both the pituitary and the hypothalamus in which androgens and/or estrogens inhibit the secretion of LH. Interestingly, pure androgens such as dihydrotestosterone (DHT) have shown to reduce LH pulse frequency, and estrogen has been shown to reduce overall LH pulse amplitude. It is believed that the inhibition taking place at the level of the hypothalamus is primarily from estrogen derived from local conversion of testosterone by aromatase.

Estrogen and testosterone are transported through the circulation primarily bound to sex hormone-binding globulin (SHBG), which has a stronger affinity for testosterone. There is also some very weak binding to albumin. SHBG production in the liver is increased by estrogen and decreased by testosterone. Consequently, it is usually twice as high in women as it is in men. Interestingly, the proportion of free and bound estradiol does not vary much during the menstrual cycle. An increased SHBG, however—such as that seen with use of synthetic estrogens like birth control pills—can significantly impact the amount of free testosterone available to bind to androgen receptors.

SEXUAL AROUSAL

Estrogen helps facilitate the production of nitric oxide and vasoactive intestinal peptide. In sexual arousal, this allows for smooth muscle dilation and increased blood flow to the vagina and the erectile tissue of the clitourethrovaginal complex.

In this way, it greatly increases vaginal lubrication in response to sexual arousal (Ottesen, 1987). By modulating the blood flow to the tissues of the vagina, clitoris, and external genitals, estrogen has a key role in engorgement of the clitoris and tenting of the upper portion of the vagina in sexual arousal as well (Park, 1997).

Estrogen has also been found to be a significant factor in driving sexual desire for women through centrally acting mechanisms. As estrogen increases during the follicular phase, there is a positive correlation with increases in sexual desire with a two-day lag (Roney, 2013). This results in peak sexual desire coinciding with ovulation and women engaging in more flirtation at this time. At mid-cycle when estrogen is at its highest, women have also been found to tend to prefer men with more masculine features, indicative of higher testosterone levels (Gangestad, 2005).

METABOLISM OF ESTROGENS

Estrogens are metabolized via Phase I (cytochrome P450 system) and Phase II detoxification for elimination. In Phase I, estrogens can follow three different paths. The first is via CYP1A1 to the 2-OH estrogens, which are considered weak estrogens and are protective—possibly blocking the effects of more toxic estrogens. Next, CYP1B1 metabolizes estrogen to a 4-OH estrogen form, which is very potent and possibly carcinogenic. Last, the CYP2C and CYP3A4 enzymes convert estrogen to a 16-OH estrogen form, which is also considered a toxic form of estrogen. Knowledge of genetic variants in the CYP families of patients may help determine which pathway is favored. In addition, indole-3-carbinol and its dimer diindolylmethane, which are found naturally in cruciferous vegetables, induce CYP1A1 and favor the production of 2-OH-estrogens (Auborn, 2003).

Phase II detoxification involves conjugation with another molecule to make it water soluble for excretion in urine or feces. In the liver, the primary reaction is *methylation* via the enzyme catechol-O-methyltransferase (COMT). This process attaches a methyl group to the estrogen molecule to form either 2-Me-OH-estrogen (induces apoptosis and is protective) or 4-Me-OH-estrogen (considered neutral). 4-OH-estrogens are metabolized by COMT faster than 2-OH-estrogens (which is favorable) unless there is a genetic variant in COMT at 108Met, which is three times less active in methylating the 4-OH type (Dawling, 2001). A person's risk for buildup of toxic metabolites can be assessed by knowledge of the genetic picture. Adequate intake of B6, B12, and folic acid are needed for proper methylation, and persons with a genetic variant of the MTHFR gene will need to supplement with methyl-cobalamin and methyl-folate.

The methylated forms of estrogens, also known as catechol estrogens, can also be converted to quinones. This occurs from exposure to oxidative fats and heavy metals such as mercury or arsenic. Estrogen O-quinones can then go on to damage

DNA directly, which is a major mechanism of estrogen carcinogenesis. This effect is enhanced in tissues that have ERs, suggesting that the catechol estrogen or O-quinone form may reach the interior of the nucleus by riding on the ER and then directly damaging DNA (Bolton, 2008).

Sulfation is another way that estrogen is converted to a more water-soluble and inactive form, which allows it to be transported through the circulation. Because of its long half-life, sulfonated estrogen can be used as a way for the body to store estrogen until needed in order to maintain high levels. This reaction is carried out by sulfotransferase and utilizes the cofactor 3'-phosphoadenosine-5'-phosphosulfate (which has both an adenosine and a sulfate group). Sufficient amounts of sulfur from the diet is important for this process and can be obtained from garlic, onions, egg yolks, Brussels sprouts, and animal protein.

Glucuronidation is yet another way that estrogens are metabolized in the body. Estrogen glucuronides are cleared faster, are excreted both in urine and in bile, and are less abundant in the body. Estrogen glucuronides may serve as a source for resorption of estrogen back into the circulation from the GI tract if there are abundant "bad" bacteria present in the gut or if transit time in the gut is greatly increased. The enzyme beta-glucuronidase found in these bacteria can remove the glucuronic acid side group, allowing the estrogen to be resorbed through the gut. This can be one source of increased levels of estrogens in circulation. Calcium-D-glucarate, found in fruits and vegetables, as well as available as a supplement, inhibits this enzyme in gut bacteria, thereby preventing resorption of estrogens through the gut.

Metabolism of estrogens can be influenced by genetic differences in CYP enzyme profile as well as dietary factors and/or deficiencies. Encouraging metabolism down neutral pathways helps eliminate the burden of more harmful estrogens. Also, timely and efficient intestinal elimination helps prevent resorption of estrogens from the gut, which can increase the buildup of toxic metabolites.

Androgens

Testosterone is made in theca cells of the ovaries as well as in the adrenal and other peripheral tissues. Androgen precursors including androstenedione, androstenediol, and DHEA are produced in the adrenal glands and are released into the circulation for further conversion to testosterone and DHT by target cells. Testosterone production is controlled by the pituitary hormones LH and FSH. These are in turn controlled by the hypothalamus via secretion of corticotropin-releasing hormone, which signals the pituitary to produce LH and FSH. The principle androgens in women are (in decreasing order of concentration) dehydroepiandrosterone sulfate, DHEA, androstenedione, testosterone, and DHT (Panjari, 2007).

Androstenediol is another androgen produced by the adrenal gland that can be converted to testosterone but also has very weak estrogenic properties itself. Levels of circulation androstenediol are about two to three times that of E2, but it has only about 0.1% of the potency of E2 (Lasley, 2011).

As girls go through puberty, they experience a rise in testosterone levels along with estrogen, but their levels of androgens are always substantially less than in boys. In fact, women's testosterone levels will remain about one-tenth that found in men. Besides being used for the production of estrogen, androgens themselves have specific functions in the female body. They help maintain muscle mass, increase red cell production, as well as maintain the erectile tissue of the clitoris, urethral sponge, Bartholin's glands (vestibular glands), and vaginal vestibule. These have been found to have an abundance of androgen receptors (Goldstein, 2009). Women with untreated vulvar lichen sclerosus have been found to have lower levels of androstenedione, testosterone, and DHT compared to other women their age (Friedrich, 1984).

There is no consistent correlation of androgen levels with libido in women, although androgens do improve mood, energy, and motivation. Some have speculated that any increase in libido with androgen administration is probably due to these factors, with specific effects on libido due to estrogen instead. Studies have shown, however, that when women imagine a positive sexual encounter with a partner who they find attractive, they experience significant increases in testosterone levels within 15 minutes of having the thought (Goldey, 2011). Interestingly, testosterone levels in women seem to increase more by cuddling than by sexual intercourse but are overall higher after intercourse (van Anders, 2007), possibly due to the increase caused by anticipatory erotic thoughts. Of note, orgasmic dysfunction has been found in postmenopausal women with low testosterone and DHEA levels and improves with replacement of these androgens (Guay, 2002). Studies looking at the effect of DHEA on sexual functioning and libido have had mixed results and therefore are inconclusive.

Progesterone

The effects of progesterone (P4) in the body are quite varied. Like estrogen, its actions stretch far beyond the reproductive system alone. Progesterone has a role in metabolism of carbohydrates, fats, and proteins. It can compete with aldosterone at the level of the renal tubule to cause a decrease in reabsorption of sodium. This in turn causes an increase in aldosterone secretion by the adrenal cortex (as is seen in pregnancy when progesterone levels are high). Progesterone can lead to increase in insulin levels and increased insulin resistance, which is a mechanism contributing to gestational diabetes.

Progesterone also increases body temperature, but the mechanism is not known. It works on respiratory centers to increase the ventilator response to CO_2, which results in lower arterial PCO_2 seen in pregnancy as well as the luteal phase of the menstrual cycle. Progesterone and its metabolites also exert a hypnotic effect on the brain. Note that synthetic progestins do not have the same effects, and in some cases have the opposite effects of progesterone (e.g., vasoconstriction with progestins versus vasodilation with progesterone). Studies demonstrate that progesterone in the luteal phase appears to have a suppressive effect on sexual desire (Roney, 2013). Although progesterone can be calming and hypnotic, it can cause dysphoria and moodiness in some women.

Thyroid

There is a connection between female gonadal functioning and the thyroid gland that has been documented but is still not completely understood. Hypothyroidism presenting in infancy results in delayed puberty and anovulatory cycles if left untreated. Not only do increased levels of estrogen increase blood flow to the thyroid (as in the follicular phase), but there are also thyroid hormone receptors expressed in ovarian tissue suggesting the regulation of ovarian function in some way by the thyroid (Doufas, 2000). Women with clinical and subclinical hypothyroidism have been found to have significantly higher incidence of sexual dysfunction (Atis, 2010). Hypothyroid can drive mild elevation of prolactin (PRL) (Wolfgang, 2003), which also has a negative effect on libido and arousal.

Sexual Activity, Responsiveness, and the Dual-Control Model

The classic model of sexual responsiveness was put forth by Masters and Johnson in the 1960s (Masters, 1966). They described a linear progression beginning at excitation (arousal) and proceeding to orgasm, plateau, and resolution. They did note that women often have a continued plateau after orgasm as well as the capacity for multiple orgasms without a refractory period as is found in men. Although the model was limited, it did provide a very important starting point for the examination of sexual responsiveness in humans. Helen Singer Kaplan (1995) later added sexual desire to the Masters and Johnson model to emphasize motivation and willingness for sexual activity. Although this seemed to work well for men and many women, in a considerable number of women desire did not seem to happen spontaneously.

In the 1990s, Rosemary Basson observed that women's sexual desire did not always precede sexual arousal. This led her to propose that women might need some level of arousal before they were able to recognize any sexual desire at all (Basson, 2000). It was then that people started to understand that sexual desire in women is mostly responsive and not always spontaneous—a concept that is now becoming widely accepted.

To understand how sexual arousal and desire proceed, it is important to think of the dual-control model of sexual arousal (Figure 1.2). This model clearly demonstrates that there is a balance between increasing excitatory factors and decreasing inhibitory factors that leads to continued sexual arousal at any given point in time. Factors that can be considered excitatory or inhibitory can be physical, psychological, cultural, behavioral, and relational. They work either in concert or in conflict to cause a Sexual Tipping Point™ either favoring or preventing sexual response (Perelman, 2009). How thoughts are processed in the brain can either add to arousal by enhancing transmission or can add to inhibition by short-circuiting

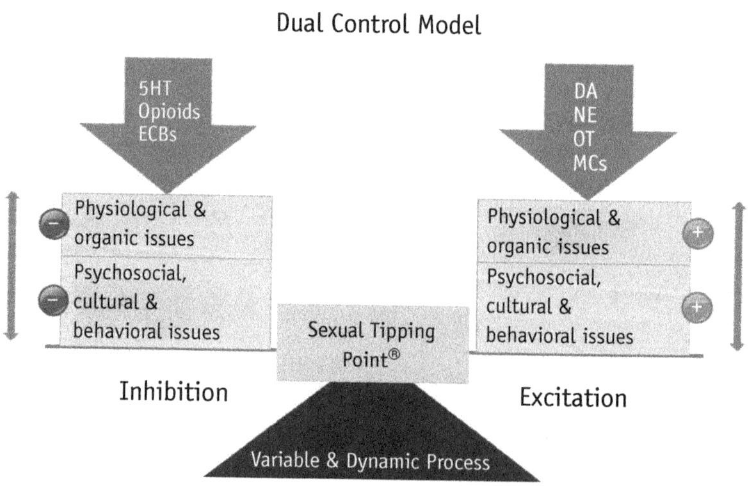

FIGURE 1.2. Dual Control Model of sexual responsiveness showing excitatory and inhibitory factors including neurotransmitters working that work to increase arousal or inhibit it.

Sources:

Bancroft J, Graham CA, Janssen E, Sanders SA (2009). The Dual Control Model: Current status and future directions. *J Sex Res.* 46(2–3):121–142.

Pfaus JG (2009). Reviews: Pathways of sexual desire. *J Sex Med.* 6(6):1506–1533.

The registered trademark and copyright for the Sexual Tipping Point Model images is owned by the MAP Educational Fund, a 510(3)(c) public charity and is used with permission of Michael A. Perelman, PhD. (Perelman MA. 2009). The Sexual Tipping Point®: A mind/body model for sexual medicine. *J Sex Med.* 6(3):629–632 transmission. These are expressed in different pathways in the brain, causing sexual responsiveness to be robust versus truncated or sluggish. This dual-control model recognizes individual variation in all dimensions, which contributes to different thresholds of excitation or inhibition from person to person (Bancroft, 2009).

BIOCHEMICAL MECHANISM OF SEXUAL AROUSAL

In the central nervous system (CNS), there are neurotransmitters that are more excitatory and those that are more inhibitory for sexual responsiveness. Sex steroids play an important role in the sexual response cycle by encouraging the synthesis and release of excitatory neurotransmitters so that they respond to learned sexual cues. In this way, estrogens, androgens, and progesterone prime the brain to respond to particular sexual stimuli. When levels of sex steroid hormones are greatly reduced, such as in menopause, there is a decreased response of excitatory neurotransmitters to sexual stimuli, as well as a decrease in sexual desire. In women, hormone replacement with estrogen helps restore the necessary neurochemical state. The excitatory neurotransmitters are dopamine (DA), norepinephrine (NE), oxytocin (OT), and melanocortins (MCs), which act in the hypothalamus and the limbic system to focus attention and create arousal in response to sexual cues (Pfaus, 2009).

Dopamine

DA is an excitatory neurotransmitter that plays a major role in sexual arousal. It helps focus attention to a sexual cue, triggers the sympathetic nervous system for increased blood flow to the heart, controls the parasympathetic nervous system to increase blood flow to the genitals, and encourages behavior in the direction of said cue. Estradiol appears to facilitate the release of DA in some areas of the brain while decreasing it in others with the overall effect being heightened sexual arousal. This helps explain why desire and arousal seem to be higher during periods of high estrogen, such as in the follicular phase and especially pronounced around the time of ovulation.

Norepinephrine

NE also works as an excitatory neurotransmitter by controlling aspects of motivation, controlling autonomic outflow, and balancing behavior toward a sexual cue with that of a fear response. It is triggered by DA in sexual arousal to help focus concentration, increase blood flow to the heart, and drive behavior in the direction of more arousal. Estradiol has also been found to increase NE synthesis and release, again showing its role in priming the brain for good sexual responsiveness. NE appears to have a positive role in vaginal response to erotic stimuli as well as in sexual desire. Too much activation of the sympathetic nervous system, however, such as in anger or fear, triggers a fight-or-flight response that can work against arousal by shunting blood flow away from the genitals and focusing on avoiding stimuli perceived as dangerous.

Oxytocin

OT has the reputation of being the bonding hormone, but it has various functions, both in the body as a hormone and in the brain as a neuromodulator. It is produced in the hypothalamus and then stored in the posterior pituitary gland until it is released into circulation. During childbirth, stretching of the uterus and cervix close to delivery induces release of oxytocin, which helps facilitates the uterine contractions of labor. Oxytocin also has an important role in breastfeeding by producing the letdown reflex of milk after stimulation of the nipples by the feeding baby.

Oxytocin works as an excitatory neurotransmitter for sexual responsiveness. In sexual arousal, there appears to be bidirectional activation between dopamine and oxytocin. OT increases levels of DA in the nucleus accumbens, which appears to have an important role in eliciting sexual arousal in response to sexual stimuli. In women, oxytocin levels rise significantly with stimulation, especially over the breasts and belly. At the moment of orgasm, oxytocin is released in levels up to five times what is usually seen in the body, which helps a woman feel closer to her partner. Of note, the strength of orgasm is correlated with the levels of oxytocin (Carmichael, 1994). There is a short window of action for oxytocin however, since its half-life is between three to five minutes.

Melanocortins

MCs are neuropeptides that are derived from propiomelanocortin and are involved in a variety of different physiological processes. Examples of melanocortins are β-endorphins, ACTH, and α-melanocyte stimulating hormone (α-MSH). Both ACTH and α-MSH bind to many different melanocortin receptors, and both of these have been found to induce physiological arousal in animal studies (Argiolas, 2000). MC agonists have been found to stimulate sexual desire and arousal (Pfaus, 2009) and are currently being studied as formulations for treatment of sexual dysfunction. Here again, there appears to be a priming effect by sex steroids on this system as estradiol increases α-MSH levels in the hypothalamus (Medina, 1998).

BIOCHEMICAL MECHANISM OF INHIBITION OF SEXUAL AROUSAL

On the inhibitory side of sexual arousal are the neurotransmitters serotonin (5HT), opioids, and endocannabinoids (ECBs). The mechanisms of inhibition are not as well known as those for excitation but appear to involve satiety, reward, and sedation.

Serotonin

5HT has long been identified with satiety in appetite, as well as functions related to sleep, mood, cognition, and motor functioning. It has also been found to have an inhibitory effect on sexual arousal and orgasm. The serotonin system has been found to interact with the melanocortin system as well as the endocannabinoid system (Feijó, 2011). It is well known that there is significant connection between 5HT and DA systems in the brain and that they are mutually inhibitory for sexual responsiveness (Pfaus, 2009). Selective serotonin reuptake inhibitors (SSRIs) prescribed for depression or anxiety increase levels of 5HT in the synaptic space, causing desensitization/downregulation of 5HT1A receptors in areas of the hypothalamus. For some people, this may result in sexual side effects such as decreased libido, difficulty with sexual arousal, and/or orgasmic dysfunction. SSRIs also appear to indirectly inhibit mesolimbic DA release via effects of serotonin on the prefrontal cortex (PFC) (Bortolozzi, 2005).

Opioids

Opioids are well known for acting in the body to signal pleasure and decrease pain. They also have significant roles in sexual responsiveness. "Endorphin" is the term given to endogenous substances that have activity on opioid receptors. A critical component of sexual reward is the release of endorphins at the moment of orgasm. They complete the reward part of the attention-reward loop so that, once the opioid is released, the brain is rewarded with pleasure, thereby reinforcing the cycle. Brain opioid systems that are activated at orgasm decrease activity in hypothalamic areas that are involved in sexual arousal and desire, contributing to the feelings of bliss after orgasm. It also appears that the level of arousal prior to orgasm is related to the level of reward derived from the experience.

Opioids work at the level of the CNS to inhibit LH and testosterone production, so that long-term opioid users will have decreased levels of circulating testosterone. Exogenous opioids are well known for decreasing sexual arousal (Tanco,

1993), and chronic opioid abusers have decreased frequency of sexual intercourse as well as decreased intensity of orgasm (Saso, 2002). Opioid release with orgasm in women is responsible for increased pain tolerance precisely at the moment of orgasm without a decrease in discrimination of stimulus.

Endocannabinoids

ECBs are a group of molecules produced in the body that act on cannabinoid receptors and have both neuromodulary and immunomodulary roles in the body. The cannabinoid receptors currently known are CB1 and CB2 and were named after it was discovered that THC was a strong agonist to these receptors. They bind the ECBs anandamide and 2-arachidonylglycerol as well as exogenous cannabinoids like THC. CB2 is found mostly in the periphery and on immune cells, whereas CB1 is found in the CNS in high concentrations in the cerebral cortex, limbic system, and cerebellum. CB1 receptors have also been found at the peripheral terminals of primary sensory neurons and pain pathways of the brain and spinal cord.

Although the exact role of CB2 is not currently known, agonists acting at CB1 have been found to be inhibitory for sexual responsiveness. Since agonists at CB1 decrease sensation over the body by blunting somatosensory awareness, they inhibit arousal and delay or impair orgasm and ejaculation. They also create a state of sedation, slow cognition, and motor functioning and decrease sexual motivation. It is believed that both anandamide and 2-arachidonylglycerol are synthesized on demand at low levels (rather than being stored in vesicles) and are broken down quickly to arachidonic acid and ethanolamine. The analgesic action of acetaminophen is thought to occur because its primary metabolite is a weak agonist at both CB1 and CB2 as well as an inhibitor of anandamide uptake, thereby increasing its availability.

There appears to be a bidirectional relationship between ECBs and sex steroids. Whereas gonadal hormones seem to modulate protein expression in the ECB system, ECBs appear to downregulate hypothalamic–pituitary–gonadalaxis activity. Anandamide levels seem to be positively correlated with levels of estrogen and FSH, and peak at the time of ovulation when estrogen and FSH are at their highest and appear to be the lowest at the optimal time for implantation of a fertilized egg. High levels of anandamide or other cannabinoids may impair implantation and have been associated with higher incidence of miscarriage (Habayeb, 2008). Anandamide levels are also high at the time of labor, possibly assisting in delivery.

Some women have attempted to use cannabis as a way of enhancing sexual desire and arousal, but results may depend on the balance between THC concentrations and cannabidiol (CBD) concentrations. THC is a CB1 agonist with psychotomimetic effects, whereas CBD is an indirect CB1 antagonist and also has anxiolytic properties through its 5HT1A agonist effects. This could be helpful for women who are overly anxious or sexually inhibited. The moderate to strong antihistaminic effects of cannabis, however, can result in severe vaginal dryness and difficulty lubricating. This paired with the distractibility of decreased cognition, blunted peripheral sensation, and reduced motor functioning may not outweigh any mild anxiolytic effects that cannabis may have for sexual responsiveness.

Sexual Response Cycle

DESIRE

Sexual desire is hardly a spontaneous event. There is constant communication between different body states, external stimuli, and thought processes so that not only are there multiple contributors to sexual desire, but they each influence each other as well. Sexual desire is also actively molded by expectations, past experiences, mood states, energy level, overall nutritional status, hormone balance, stress levels, and perceived consequences of sexual activity. Taking the dual-control model into consideration, the goal is to increase excitatory factors while decreasing inhibitory factors in order to enhance sexual responsiveness.

AROUSAL

Arousal is technically defined as the stage in which blood starts to move into the erectile tissues, which for a woman includes the entire structure of the clitoris, the vestibular bulbs, the urethral sponge, and the perineal sponge. When engorged, they surround the opening of the vagina and place outward pressure on the labia and glans clitoris. This arousal heightens sensation in the genitals during sexual stimulation, which in turn drives more arousal. This cycle of arousal depends on the systems of attention and reward that rely on excitatory neurotransmitters mentioned earlier. Adequate levels of estrogen in the brain ensure proper priming and release of DA and NE.

Higher levels of perceived sexual arousal are correlated with increased blood flow to the ventral midbrain and the right caudate nucleus. During clitoral stimulation, there is increased activity in the left secondary somatosensory cortex (SII) and the dorsal convexity of the right primary somatosensory cortex (SI). Activation of the SII was also seen with stimulation of the vaginal vestibule (Georgiadis, 2006). Women's brains have increased connection and communication between many areas of the brain in comparison to men, which results in a greater distractibility and judgment, which can be significant obstacles in building arousal.

Good arousal for a woman depends on her perceiving the stimuli around her as positive and in harmony with the sexual stimulation she receives. If a woman's thoughts are positively correlated with excitement and sexual arousal, this will add to the excitatory side of the dual-control model. If instead the thoughts are negative, punitive, guilty, or anxious, this will add to inhibitory factors, causing the balance to tip away from arousal. Messages learned over a lifetime are stored and can affect cognitions about sexuality in the moment.

ORGASM

In orgasm, there are rhythmic one-second contractions of the muscles of the pelvic floor for an average of eight seconds. These are accompanied by slower contractions of the vaginal wall and the uterine smooth muscle. The intensity of orgasm can vary greatly from just a localized pleasurable sensation to waves of pleasure that are felt in the entire body. The degree of involvement seems to be related to levels of psychological arousal as well as physical arousal, as well as hormone status and build-up of oxytocin during arousal. Interestingly, when women do reach orgasm, they may experience it more intensely than their male counterparts. Orgasms last longer for women, and many report feeling that their whole bodies are involved rather than just the genitals. Women also experience a wide range of physical and emotional feelings during orgasm, from a gentle wave or shudder over their body, to powerful pelvic floor contractions, to involvement of their entire bodies tensing and relaxing rhythmically.

Orgasm is different for women's brains too. With the experience of orgasm, a woman's brain can relax for those few seconds to the same level as one can achieve in transcendental meditation. Studies have shown that when women do reach orgasm, their brain activity seems to silence (especially in the dorsal medial PFC and the left lateral orbitofrontal cortex), which allows them to feel completely in their bodies without thinking or processing emotion. During orgasm, women experience a decrease in blood flow to the temporal regions (especially the left temporal lobe) and neocortex and an increase in blood flow to the deep cerebellar nuclei (which are involved in muscle contractions during orgasm). There is significant decrease in activity and blood flow to the left lateral orbitofrontal cortex, which signifies a loss of inhibition and self-control at that moment. Another area showing decreased activity during orgasm in women is the dorsomedial PFC, which is typically involved in judgment and moral reasoning (Georgiadis, 2006). Women have a larger PFC in proportion to men, predisposing them to more judgment and less impulsivity, which can be an obstacle to orgasm.

Orgasm brings with it a surge of oxytocin, prolactin, and endorphins (opioids). Oxytocin and opioids have a significant role in sexual arousal and orgasm for women, as discussed earlier. PRL surges at orgasm for both men and women, producing a feeling of sleepiness. This is counterbalanced in women by a continued plateau of arousal and desire for continued sexual stimulation. PRL also increases activity in the stem cells of the olfactory bulb, possibly enhancing the sense of smell to one's partner. During the last trimester of pregnancy and continuing throughout breastfeeding, a woman's body starts to secrete higher levels of PRL to produce milk for the infant. With these higher levels of PRL, there can be inhibition of libido, which is often seen postpartum.

Women who are orgasmic can typically reach orgasm with self stimulation and no distractions in about 1 to 1½ minutes. Unfortunately, this is not typical for women in their sexual encounters, where it takes most women between 10 and 20 minutes to reach orgasm. There are several things to consider when it comes to a woman's ability to reach orgasm. First, it is not uncommon for a woman's arousal to proceed more slowly than a man's. She does not receive sensations from her erectile tissue as easily as a man does because hers is almost entirely internal. Next, because her brain is wired to give attention to a variety of details simultaneously, it is more difficult for her to quiet her thoughts and focus just on the sexual experience.

RESOLUTION

Women also have the ability to have multiple orgasms without a significant refractory period as men experience. Although it is not yet known what causes a refractory period in men and why it is different for women, we do know that women's arousal stays at a plateau after orgasm, making it easier to achieve another orgasm relatively quickly. Although the neurotransmitter mechanisms responsible for a refractory and resolution phase in men is still not known, there is a very similar increase in PRL levels in both men and women after orgasm, making PRL alone unlikely to be responsible for the refractory period.

Pregnancy

In pregnancy, there is coordination between the hormones produced in the mother, those produced in the fetus, and metabolism of these in the placenta. Starting at the eight-cell stage, the blastocyst begins secreting HCG and other growth factors and cytokines. HCG can be detected in maternal serum after three days of implantation and continue to increase until the 12th week gestation. One function of HCG early in pregnancy is to support the function of the corpus luteum. HCG binds to the LHCGR on the corpus luteum and stimulates it to continue to produce estrogen and progesterone. It may also stimulate production of relaxin from the corpus luteum. HCG can also bind to the TSH receptor in the thyroid so that high levels of HCG in early pregnancy will increase the production of thyroid hormones in the pregnant mother (Glinoer, 1997).

After nine weeks of pregnancy, the placenta begins to produce large amounts of estriol as a result of conversion of androgens. Estradiol and estrone are also produced as a result of aromatization of androgens in the placenta. After this conversion takes place, it appears that estriol and estradiol are preferentially passed to the maternal circulation, with the weaker estrone passing to the fetus (Walsh, 1981). Many women find that the increase in estrogens translates into increased sensitivity of their breasts and genital tissues. Some women have also described increased sexual thoughts and heightened ability to orgasm during pregnancy, possibly due to increased vasocongestion of the genitals.

Generally, there is a decrease in sexual desire and sexual activity during the first trimester, usually due to the discomfort of nausea that accompanies the beginning of pregnancy. Usually by the second trimester, there is resolution of the physical symptoms and an increase in sexual desire and sexual activity. Not until late in the third trimester, with significant increase of the size of the fetus as well as increased levels of progesterone, is there a noticeable decline in sexual desire. Interestingly, preterm labor is found to be reduced in women who have sex late in pregnancy (Sayle, 2001).

Childbirth

In pregnancy, sex steroids are produced in both the fetus and the placenta, causing the levels of sex steroids to rise significantly in the mother's circulation, especially estrogens and progesterone. After delivery, there is a sudden drop in the circulation of these hormones so that, by the fifth day after delivery, hormones levels are similar to what they were prior to conception. These changes have a significant impact on sexual functioning for the new mother.

When estrogens decline, there can be physical symptoms of vaginal burning and dryness that cause pain and discomfort during sexual activity. Women also experience a transient change in thyroid hormones as TBG drops after delivery and the thyroid hormones adjust accordingly. These hormonal changes can have a negative impact on mood as well as sexual desire. Of course, when considering the dual-control model, others factors such as fatigue, physical recuperation of lacerations or surgical wound, stress of a new baby, demanding feeding schedule with or without breastfeeding, and expectations of a partner's cooperation all can work to inhibit sexual responsiveness.

PRL rises during late pregnancy and drops off by the three weeks after delivery. When women breastfeed, the breast stimulation causes release of oxytocin, which in turn keeps PRL levels elevated. For these women, PRL levels may be 20 times those prior to pregnancy and may contribute significantly to inhibition of sexual desire and arousal.

Often relational factors, adaptation to the role of mother, and other responsibilities coupled with the factors already mentioned present a significant challenge to many women in regaining their sexual responsiveness.

Perimenopause

Perimenopause is a period of time that can span several years before the onset of menopause. It is characterized by changes in the length and/or quality of the menstrual period such that women may have shortened or lengthened periods, heavier or lighter periods, or skipped periods altogether. These variations in the menstrual cycle occur because the ovaries are no longer reliably producing a mature follicle with each cycle. The result is that there is fluctuation of hormones with resultant physical symptoms.

As the ovaries age, they are no longer as responsive to the gonadotropin FSH. Consequently, FSH levels keep rising in an effort to produce a follicle. Without production of a follicle, there is no rise in progesterone or inhibin B in the second half of the cycle, which both normally create a negative feedback loop to the pituitary to decrease levels of FSH (Pezzani, 2000). As FSH levels remain elevated, the ovary can only respond by producing increased levels of estrogen. Without a normal balance between estrogen and progesterone, a state of estrogen dominance is created that results in symptoms of breast tenderness, moodiness, weight gain, headaches, bloating, and heavy vaginal bleeding.

Eventually, the increase in estrogen levels or production of a follicle that secretes progesterone will inhibit FSH sufficiently for estrogen levels to drop. For many women, this can result in hot flashes, insomnia, low libido, and decreased concentration similar to menopause. With these fluctuations between estrogen dominance and estrogen withdrawal, a woman may have a significant negative impact on her sexual activities. This combination of physical and mood symptoms can decrease sexual responsiveness and often leads to an avoidance of sexual activity during this time.

Menopause

Menopause is defined as the period starting one year after the cessation of menstrual periods. Leading up to menopause, there is a variable period of perimenopause where there may be regular or irregular periods but always a fluctuation of sex hormones. In menopause, progesterone levels are almost nonexistent since it is produced primarily from the corpus luteum after ovulation, which is no longer occurring. Eventually, there is a gradual decrease of production in testosterone and estrogen from the ovary as well. The results of these hormonal changes can vary from individual to individual depending on the degree of hormone decline, overall physical condition, and efficiency of hormone receptors probably determined by genetics.

We have seen that progesterone can be slightly inhibitory for sexual desire and sex drive in women. Without the effects of progesterone in menopause, women do not experience this inhibition. They do, however, experience both physical and psychological symptoms related to the decline in estrogen. Remember that estrogen helps prime the brain for release of excitatory neurotransmitters, so it contributes to sexual responsiveness and arousal at the level of the CNS. Estrogen also supports the vaginal tissues and the blood flow to the genitals, so lack of sufficient estrogen will eventually result in vulvovaginal atrophy (VVA), decreased blood flow to the erectile tissue, and decreased overall sensation during sexual activity. When women continue to have regular sexual activity several times a week, it helps maintain the blood flow needed to the area, and they have a later onset of VVA as well as less sexual dysfunction.

Although the ovaries no longer produce the same levels of testosterone and estrogen, the adrenal gland continues to produce androgens. During menopause, more than 50% of androgens originate in the adrenal gland, which can then be converted to estrogen in adipose tissue, breast tissue, or other target tissues. Regardless, there is a significant decline in estrogen levels, which results in systemic side effects apart from the sexual symptoms. Some of these include hot flashes, memory loss, insomnia, loss in bone density, skin laxity, and decreased responsiveness of blood vessels. Those women who experience a surgically induced menopause (oophorectomy) will have increased severity of both sexual and systemic symptoms due to the suddenness of estrogen withdrawal. It is important to note that even women who have had a hysterectomy without oophorectomy may experience early menopause (premature ovarian failure) since up to 50% of the blood supply to the ovary is collateral circulation with the uterus. Once the uterine vessels are removed, there is a significant decline in crucial circulation to the ovaries.

Another interesting effect of menopause related to the decrease in estrogen is the concomitant decrease in SHBG. Recall that SHBG carries both estrogen and testosterone through the circulation, but it has a stronger binding affinity to testosterone. When estrogen declines in menopause, so does the amount of SHBG produced in the liver. This means that there is less SHBG to bind testosterone and estrogen. Consequently, although overall testosterone levels are lower than before menopause, the amount of free testosterone is often similar to premenopausal levels.

Many women find that their sexual pleasure and sexual relationships are significantly impacted by the sexual side effects of low estrogen after menopause. For this reason, as well as many other positive effects of estrogen replacement, many women choose to use hormone replacement to alleviate the sexual symptoms and regain their sexual vitality. When considering hormone replacement, it is important to note that synthetic estrogens given orally can significantly increase SHBG levels, which has negative consequences for free testosterone levels.

Transdermal estradiol, however, does not cause the same elevation of SHBG because it avoids a first pass effect through the liver. Menopausal women often find that replacement of estrogen can also help with their libido (for reasons discussed earlier). Testosterone replacement appears to enhance overall mood, energy level, and motivation but is not directly correlated with increases in libido as estrogen is. The decision to use estrogen after menopause in women with their uterus intact should include continuous micronized progesterone (different side effect profile from progestins) to protect the endometrium from the stimulating effects of estrogen.

Environmental Factors

An overview of female sexual biochemistry would not be complete without a discussion of environmental factors impacting women's physiology and sexual functioning, which is considerable. We know that overall hormone levels are affected by stress, proper amounts sleep, diet and nutritional deficiencies, and genetic susceptibilities. Truly, anything that can have an effect on hormone receptors has the potential to change or hinder a woman's sexual physiology and responsiveness. The main categories to keep in mind are use of pharmaceuticals and exposure to chemicals in the environment. Basically, things that women purposely put into their bodies as well as things that may enter a woman's body by incidental exposure can work against her sexual biochemistry to cause sexual dysfunction or disease.

One of the most common culprits of sexual difficulties is hormonal birth control. Although well tolerated by many women, synthetic hormones used for contraception, especially those with synthetic estrogens, can wreak havoc on some women's sex lives. Since synthetic estrogens increase SHBG levels considerably, this can leave many women with practically no free testosterone available to bind to androgen receptors. This can precipitate vestibulodynia in women with inefficient androgen receptors or excessive increases in SHBG. While this can normalize after discontinuation of hormonal birth control, for some women the increase in SHBG does not correct back to normal levels even after several years (Panzer, 2006).

The pill has also been found to affect women's levels of arousal as well as what qualities they find attractive in a prospective partner. Because women are not getting the same increases in estrogen at the midpoint of their menstrual cycles, they may not be as sensitive or sexually responsive as they would be off of hormonal contraception. This also poses a challenge for women having chosen a partner when on hormonal birth control who then find themselves less attracted to that same partner when they discontinue hormonal birth control in an effort to become pregnant.

More and more attention is also being paid to the hundreds of chemicals in the environment that women are exposed to on a daily basis that work as endocrine disruptors. Some examples of foreign substances acting on ERs (xenoestrogens) are plastic byproducts, fragrances, cleaning products, pesticides and insecticides, herbicides, and bleach byproducts. They are found in cosmetics, soaps, shampoos, perfumes, most household products, air fresheners, laundry detergents, and food and beverage containers; they even contaminate the water supply, the food supply, and dairy products. The additive effect of xenoestrogens exposure has consequences for hormone balance, onset of puberty, fertility, and cancer risk. Understanding the way that these xenoestrogens affect a woman's sexual biochemistry is paramount in making educated lifestyle decisions.

REFERENCES

Argiolas A, Melis MR, Murgia S, Schiöth HB (2000). ACTH- and α-MSH-induced grooming, stretching, yawning and penile erection in male rats: site of action in the brain and role of melanocortin receptors. *Brain Res Bull.* 51(5):425–434.

Atis G, Dalkilinc A, Altuntas Y, Atis A, Caskurlu T, Ergenekon E (2010). Sexual dysfunction in women with clinical hypothyroidism and subclinical hypothyroidism. *J Sex Med.* 7(7):2583–2590.

Auborn KJ, Fan S, Rosen EM, Goodwin L, Chandraskaren A, Williams DE, Chen D, Carter TH (2003). Indole-3-Carbinol is a negative regulator of estrogen. *J Nutr.* 133(7):2470S–2475S.

Bancroft J, Graham CA, Janssen E, Sanders SA (2009). The Dual Control Model: Current status and future directions. *J Sex Res.* 46(2–3):121–142.

Basson R (2000). The female sexual response: A different model. *J Sex Marital Ther.* 26(1):51–65.

Bolton JL, Thatcher GRJ (2008). Potential mechanisms of estrogen quinone carcinogenesis. *Chem Res Toxicol.* 21(1):93–101.

Bortolozzi A, Díaz-Mataix L, Scorza MC, Celada P, Artigas F (2005). The activation of 5-HT receptors in prefrontal cortex enhances dopaminergic activity. *J Neurochem.* 95:1597–1607.

Bulun, SE, Lin Z, Zhao H, et al. (2009). Regulation of aromatase expression in breast cancer tissue. *Ann N Y Acad Sci.* 1155:121–131.

Carmichael MS, Warburton VL, Dixen J, Davidson JM (1994). Relationships among cardiovascular, muscular, and oxytocin responses during human sexual activity. *Arch Sex Behav.* 23(1):59–79.

Dawling S, Roodi N, Mernaugh RL, Wang X, Parl FF (2001). Catechol-O-methyltransferase (COMT)-mediated metabolism of catechol estrogens: comparison of wild-type and variant COMT isoforms. *Cancer Res.* 61(18):6716–6722.

Doufas AG, Mastorakos G (2000). The hypothalamic-pituitary-thyroid axis and the female reproductive system. *Ann N Y Acad Sci.* 900:65–76.

Feijo Fde M, Bertoluci MC, Reis C (2011). Serotonin and hypothalamic control of hunger: a review. *Rev Assoc Médica Brasileira.* 57(1):74–77. [Article in Portuguese]

Friedrich EG, Kalra PS (1984). Serum levels of sex hormones in vulvar lichen sclerosus, and the effect of topical testosterone. *N Engl J Med.* 310:488–491.

Gangestad SW, Thornhill R, Garver-Apgar CE (2005). Adaptations to ovulation: implications for sexual and social behaviour. *Curr Dir Psych Sci.* 14(6):312–316.

Georgiadis JR, Kortekaas R, Kuipers R, Nieuwenburg A, Pruim J (2006). Regional cerebral blood flow changes associated with clitorally induced orgasm in healthy women. *Eur J Neurosci.* 24(11):3305–3316.

Glinoer D (1997). The regulation of thyroid function in pregnancy: pathways of endocrine adaptation from physiology to pathology. *Endocr Rev.* 18(3):404–433.

Goldey KL, van Anders SM (2011). Sexy thoughts: effects of sexual cognitions on testosterone, cortisol, and arousal in women. *Horm Behav.* 59(5):754–764.

Goldstein A, Goldstein I, Pukall C, eds. (2009). *Female sexual pain disorders: Evaluation and management,* 1st ed. Hoboken, NJ: Wiley-Blackwell.

Guay A, Davis SR (2002). Testosterone insufficiency in women: fact or fiction? *World J Urol.* 20(2):106–110.

Habayeb OM, Taylor AH, Finney M, Evans MD, Konje JC (2008). Plasma anandamide concentration and pregnancy outcome in women with threatened miscarriage. *JAMA.* 299(10):1135–1136.

Holloway AC, Anger DA, Crankshaw DJ, Wu M, Foster WG (2008). Atrazine-induced changes in aromatase activity in estrogen sensitive target tissues. *J Appl Toxicol.* 28(3):260–270.

Kaplan HS (1995). *Sexual desire disorders: dysfunctional regulation of sexual motivation.* New York: Routledge Press.

Klein KO, Baron J, Colli MJ, McDonnell DP, Cutler GB (1994). Estrogen levels in childhood determined by an ultra sensitive recombinant cell bioassay. *J Clin Invest.* 94(6):2475–2480.

Lasley BL, Crawford S, McConnell DS (2011). Adrenal androgens and the menopausal transition. *Obstet Gynocol Clin North Am.* 38(3):467–475.

Mast N, Linger M, Pikuleva IA (2013). Inhibition and stimulation of activity of purified recombinant CYP11A1 by therapeutic agents. *Mol Cell Endocrinol.* 371(1–2):100–106.

Masters W, Johnson V (1966). *Human sexual response.* Boston, MA: Little, Brown.

Medina F, Siddiqui A, Scimonelli T, Fenske C, Wilson CA, Celis ME (1998). The interrelationship between gonadal steroids and POMC peptides, β-endorphin and α-MSH, in the control of sexual behavior in the female rat. *Peptides.* 19(8):1309–1316.

Noble LS, Takayama K, Zeitoun KM, et al. (1997). Prostaglandin E2 stimulates aromatase expression in endometriosis-derived stromal cells. *J Clin Endocrinol Metab.* 82(2):600–606.

Ottesen B, Pedersen B, Nielsen J, Dalgaard D, Wagner G, Fahrenkrug J (1987). Vasoactive intestinal polypeptide (VIP) provokes vaginal lubrication in normal women. *Peptides.* 8(5):797–800.

Panjari M, Davis RS (2007). DHEA therapy for women: effect on sexual function and wellbeing. *Hum Reprod Update.* 13(3):239–248.

Panzer C, Wise S, Fantini G, et al. (2006). Impact of oral contraceptives on sex hormone-binding globulin and androgen levels: a retrospective study in women with sexual dysfunction. *J Sex Med.* 3(1):104–113.

Park K, Goldstein I, Andry C, Siroky MB, Krane RJ, Azadzoi KM (1997). Vasculogenic female sexual dysfunction: the hemodynamic basis for vaginal engorgement insufficiency and clitoral erectile insufficiency. *Int J Impot Res.* 9(1):27–37.

Perelman MA (2009). The sexual tipping point®: A mind/body model for sexual medicine. *J Sex Med.* 6(3):629–632.

Pezzani I, Reis FM, Di Leonardo C, et al. (2000). Influence of non-gonadotrophic hormones on gonadal function. *Mol Cell Endocrinol.* 161(1–2):37–42.

Pfaus JG (2009). Reviews: Pathways of sexual desire. *J Sex Med.* 6(6):1506–1533.

Rone MB, Fan J, Papadopoulos V (2009). Cholesterol transport in steroid biosynthesis: role of protein-protein interactions and implications in disease states. *Biochim Biophys Acta.* 1791(7):646–658.

Roney JR, Simmons ZL (2013). Hormonal predictors of sexual motivation in natural menstrual cycles. *Horm Behav.* 63(4):636–645.

Saso L (2002). Effects of drug abuse on sexual response. *Ann Ist Super Sanita.* 38(3):289–296.

Sayle AE, Savitz DA, Thorp JM, Hertz-Picciotto I, Wilcox AJ (2001). Sexual activity during late pregnancy and risk of preterm delivery. *Obstet Gynecol.* 97(2):283–289.

Soltysik K, Czekaj P (2013). Membrane estrogen receptors—is it an alternative way of estrogen action? *J Physiol Pharmacol.* 64(2):129–142.

Tanco SA, Watson NV, Gorzalka BB (1993). Lack of effects of 5-HT3 antagonists on normal and morphine-attenuated sexual behaviours in female and male rats. *Experientia.* 49(3):238–241.

Toran-Allerand CD, Guan X, Maclusky NJ, et al. (2002). Putative estrogen receptor that is regulated during revelopment and after ischemic brain injury. *J Neurosci.* 22(19):8391–8401.

van Anders SM, Hamilton LD, Schmidt N, Watson NV (2007). Associations between testosterone secretion and sexual activity in women. *Horm Behav.* 51(4):477–482.

Walsh SW, McCarthy MS (1981). Selective placental secretion of estrogens into fetal and maternal circulations. *Endocrinology*. 109(6):2152–2159.

Wolfgang R, Gessl A, Nowotny P, Vierhapper H (2003). Hyperprolactinaemia in hypothyroidism: clinical significance and impact of TSH normalization. *Clin Endocronol*. 58(2):185–191.

2

Midlife and Beyond: Female Hormonal Status/Menopause and Its Effects on Sexual Health

JANINE A. BLACKMAN

Hormone Changes through the Reproductive Stages as Women Age

Hormonal changes in women at midlife occur over many years, sometimes decades, before and after the day a woman experiences her final menstrual period (FMP). Writing from the perspective of a former epidemiologist involved in researching hormone replacement in women, a postmenopausal woman and grandmother, and a full-time integrative medicine physician who specializes in preventive and wellness care of women and men at midlife, including hormone therapy, it is very common to hear a woman exclaim, "No one ever told me this was going to happen to my body at the age of ___!" She is usually between 35 to 65 years old and is distraught over some combination of lack of sleep, fatigue, weight gain, vaginal dryness, loss of libido, loss of concentration and memory, loss of bladder control, and worse. She is often not complaining about hot flashes as her primary concern because she expected these to occur at some point in her life, and she expects them to go away in a reasonable amount of time. It is everything else that she was not expecting. And if hot flashes are not occurring, and if her period has not completely stopped yet, she generally does not believe she is peri- or almost postmenopausal, leaving her to conclude that her problems must be due to a new medical problem. While the focus of this book is on sexual health, no one will deny that all aspects of a person's health and life stress can also influence sexual health. It is similarly true that a decline in a person's sexual health at any age may also represent a decline in functioning of some aspect of his or her hormonal or total health, male and female alike.

Hormones represent specific molecules produced by endocrine organs: pituitary, adrenals, thyroid, testes, and ovaries. The focus in this chapter is primarily limited to estrogen, progesterone, testosterone, and thyroid as they relate to the changes in women's health through her lifespan. In overall health and in sexual health, all hormones are individually and wholly integral participants to the maintenance of physiological function and homeostasis. Hormone levels undergo diurnal variation and production changes in response to our environment, thought processes, stress levels, and food intake. Environmental toxins, medications, and pollutants also significantly affect hormone balance. With the aging process, hormones levels and/or hormone receptor sensitivity decrease. Problems with health maintenance due to declining levels of hormones can arise at any time through the lifespan. Further, the effects of decreased hormones that occur as a result of aging may or may not be confounded by concomitant disease states and environmental or other stress factors. In this chapter, the focus is on the hormone changes women experience through their lifespan related to age and menopause, with the purpose of relating these changes to effects on quality of life and sexual health.

The onset of the first menstrual cycle, referred to as menarche, occurs at a median age of 12–43 years in the United States (Chumlea, 2003). After menarche, women learn to understand their lives with the awareness that their hormones are changing on a monthly cycle with their menstrual periods. During pregnancy, their hormones change again to support the developing fetus and to support breastfeeding thereafter. However, hormones are also changing as part of the aging process, which includes the transition through menopause, beginning as early as a woman's 20s and continuing through her entire lifespan. As shown in Table 2.1, the reproductive aging process for women can be understood as 10 stages from the "early reproductive" stage beginning at the onset of her first period to a "late postmenopause" stage beginning approximately 10 years after the FMP. Theses stages were delineated by the Stages of Reproductive Aging Workshop, which consisted of scientists from five countries and multiple disciplines who evaluated data from cohort studies of women in midlife on change in menstrual, endocrine, and ovarian follicle count. Three primary criteria were used to define the stages of reproductive aging for women, namely, characteristics of the menstrual cycle, changes in follicular stimulating hormone (FSH), and antral follicle count (AFC), as shown in Table 2.1. According to this staging, the AFC begins to decline in a woman's 30s, about 20 years prior to her FMP. Also, the AFC does not drop to zero but commonly persists at a low level for at least 10 years after the FMP. Further studies are needed to determine whether an additional stage is warranted near the end of life, beyond what is currently designated as Late Postmenopause (Harlow, 2012).

Table 2.1. Age and Menopause-Related Hormone Changes in Women

Terminology (Harlow, 2012)	*Reproductive*				*Menopausal Transition*				*Postmenopause*	
					Perimenopause					
	Early *−5*	*Peak* *−4*	*Late* *−3b*	*Late* *−3a*	*Early* *−2*	*Late* *1*	*FMP* *0*	*Early* *+1a*	*Early* *+1b*	*Early* *+1c*
Estimated Age[a]	12–19	20–29	30–35	36–39	40–46	47–49	50	50–51	51–52	53–59
Menstrual Cycles (Harlow, 2012)	Variable to regular	Regular	Regular	Subtle changes in flow and length	Variable length in cycles	Intervals of amenorrhea of >= 60 days.	None			
FSH (Harlow, 2012)		Normal	Normal	Variable	↑ Variable	↑ >= 25 IU/L	↑ Variable >= 25 IU/L		Stabilizes >=25	
AFC (Harlow, 2012)		Normal	Low	Low	Low	Low	Very low			
Symptoms Attributed to Menopause (Harlow, 2012)						Vasomotor symptoms likely	Vasomotor symptoms most likely		Symptoms of urogenital atrophy likely	

<table>
<tr><th colspan="8">Hormone Changes</th></tr>
<tr><td>Thyroid
(Canaris, 2000)</td><td colspan="6">Prevalence of clinical and subclinical hypothyroidism increases throughout the lifespan. Virtually all studies show higher prevalence rates for hypothyroidism in women compared to men and with advancing age. One cross-sectional study of 25,862 people reported 40% of those taking prescribed thyroid medication were undertreated according to laboratory testing.</td><td></td></tr>
<tr><td>Testosterone
(Guay, 2002;
Labria, 1998;
Zumoff, 1995)</td><td>Testosterone levels are at their peak</td><td colspan="2">Testosterone begins to decline in a woman in her 20s</td><td colspan="2">Testosterone levels in a woman's 40s drop by 50% compared to that in her 20s</td><td>Ovaries continue to produce testosterone. However, testosterone declines as adrenal production declines with age</td><td></td></tr>
<tr><td>Progesterone
(Prior, 2011;
Santaro, 2008)</td><td></td><td>Normal</td><td>Gradual decline (see Table 6.3 for symptoms)</td><td colspan="3">very low to none</td><td></td></tr>
<tr><td>Estradiol
(Harlow, 2012)</td><td></td><td>Normal</td><td>Normal</td><td>Normal</td><td colspan="2">Erratic surges 20%–30% higher than previous peak levels. Followed by rapid declines, sometimes within the same cycle</td><td>Low to very low</td></tr>
</table>

Note: ↑ = increased; FMP = final menstrual period; AFC= antral follicle count; FSH = follicular stimulating hormone.

[a] In years. Estimated age is based on average age of menopause at 51 years, defined as 12 months after the FMP, the median age of onset of menstruation of 12–43 years, and the duration of each reproductive stage.

SEVERAL HORMONES PLAY A ROLE AS WOMEN AGE

The three primary sex hormones in women are estrogen, progesterone, and testosterone. Estrogen in the human body is produced in three different forms: estradiol, estriol, and estrone. Estradiol is the more potent form of estrogen in humans, male and female alike. Changes in estradiol, progesterone, and testosterone are summarized in Table 2.1. Although not a sex hormone, thyroid hormone is also included in Table 2.1 because prevalence of clinical and subclinical hypothyroidism (low thyroid) increases as women age, because many cases of clinical and subclinical hypothyroidism go unrecognized and untreated, and because thyroid dysfunction can greatly confound the symptoms and experience of women as they age and transition through menopause. (Canaris, 2000; Morganti, 2005; Peeters, 2008).

Testosterone

Interestingly, testosterone begins to decline as early as a woman's 20s, and the decline is independent of the transition through menopause. The symptoms of declining testosterone can be subtle, affecting a woman's sexual desire, satisfaction, and mood, commonly mistaken as signs of an underlying depression, prompting referrals for counseling and psychotherapy, when hormone measurements would be more appropriate and more fruitful (Guay, 2002).

In women, testosterone arises from three sources. Approximately 50% to 60% is derived from the peripheral conversion of the prohormones androstenedione (AD) and dehydroepiandrosterone (DHEA) and its sulphate form (DHEA-S). Approximately 25% is secreted directly by the ovary and 25% by the adrenal gland (Longscope, 1986). Testosterone circulates both in its free form (bioactive form) and bound to protein including albumin and sex steroid binding globulin (SHBG), the levels of which are an important determinant of free (bioactive) testosterone. For example, increases in estrogen from estrogen replacement therapy will lead to an increase in SHBG, which will cause a decrease in free testosterone, possibly leading to symptoms of low testosterone. Similarly, oral contraceptives of all doses combinations are known to decrease circulating levels of total and free testosterone and to increase SHBG. Due to the increase in SHBG, free testosterone levels decrease twice as much as total Ttestosterone (Zimmerman, 2014). Although the amount of testosterone produced by the ovary does not specifically decline once the ovaries stop producing eggs after menopause, SHBG does increase with age, so bioactive levels of testosterone decline with age. Other causes of testosterone deficiency include Addison's disease, corticosteroid therapy, chronic illness, premenopausal ovarian failure, or oophorectomy (Burger, 2002).

The most common symptom of low testosterone for women is decreased sexual desire, which is often characterized by a decrease in sexual thoughts and fantasies, as well as actions. (In contrast, in the author's experience, a decrease in libido is not the most common symptom in men with declining testosterone; rather afternoon fatigue, decreased memory, focus, and concentration top the list.) Muscle weakness is another frequent complaint, especially in athletic women. It appears that genital arousal and orgasmic response may also be affected and vaginal lubrication may decrease, even in women who are menstruating regularly, with presumably normal levels of estrogen. There is some evidence that testosterone, apart from estrogen, may have a direct function in genital arousal and orgasmic physiology (Guay, 2002). So a premenopausal woman in her 20s or 30s with normal estrogen levels may report decreased vaginal lubrication and decreased sexual satisfaction that is likely due to decreased testosterone. The academic medical community has labeled symptoms of reduced sexual satisfaction in women as hypoactive sexual desire disorder (HSDD; Leiblum, 2006). Estimates of the prevalence of HSDD range from 9% in naturally postmenopausal women to 26% in younger surgically induced postmenopausal women. For many women, HSDD is associated with emotional and psychological distress as well as significantly lower sexual and partner satisfaction. HSDD is also associated with significant decrements in general health status, including aspects of mental and physical health. One study concluded that HSDD is prevalent among women at all reproductive stages, with younger surgically postmenopausal women at greatest risk, and is associated with a less active sex life and decreased sexual and relationship satisfaction (Leiblum, 2006). This study on HSDD did not consider testosterone levels.

Testosterone deficiency in women is a topic that the medical community has been slow to address. Another study on sexual desire in women reported in 1999 that the percentage of women, categorized by age, education, and ethnic background, who complained of decreased sexual desire in the different categories ranged from 22% to 44%, with a mean of 32%. This would put the number of women in the United States with decreased libido in the tens of millions. How many of these women have decreased testosterone is unknown since measuring testosterone is not a routine practice in primary or gynecological care of women (Guay, 2002).

Another complaint of women with low testosterone is fatigue. In reality, women may develop symptoms of testosterone deficiency at any age, from their teens through late adulthood. Testosterone begins to decline in the 20s, and by the 40s a woman's testosterone level has dropped by 50 percent. Perimenopausal women who experience decreased sexual desire at the onset of their lifecycle changes may attribute it to menopause, when in fact other hormonal changes are also occurring. Unfortunately, there are no clear guidelines in conventional medicine today for doctors to evaluate for testosterone deficiency in women (Guay, 2002). In the author's experience, integrative medicine physicians who prescribe hormone therapy routinely include measurements of total and free testosterone in their laboratory assessment of women when they are investigating any sign or symptom that may be related, such as fatigue, decreased muscle strength or low lean body mass, vaginal dryness, night sweats, depression, decreased libido, bone loss, and more. There are no Food and Drug Administration (FDA) approved formulations of testosterone for women in the United States. Approaches to testosterone and other hormone therapies are discussed later in this chapter.

Progesterone

Progesterone is produced by the follicles in a woman's ovaries, according to a specific timing in her menstrual cycle. Women begin puberty with about 400,000 follicles, each with the potential to release an egg (oocyte) at ovulation for fertilization. For women in the reproductive stage of their life, the pituitary gland produces FSH midcycle every month, which initiates follicular growth. The midcycle rise and fall in FSH every month stimulates the most advanced follicle to proceed to ovulation. Following the release of an egg from the follicle (representing ovulation), a corpus luteum develops from the ovarian follicle within the ovary. While the oocyte traverses the fallopian tube into the uterus, the corpus luteum remains in the ovary and produces progesterone in response to luteinizing hormone (LH) from the pituitary gland. The purpose of the progesterone is to stimulate the uterine lining for implantation of a fertilized oocyte, now called a zygote. When pregnancy occurs, progesterone increases rapidly, but it begins to be produced by the newly developing placenta. When pregnancy does not occur, the corpus luteum involutes and progesterone production diminishes and eventually disappears, triggering release of the uterine lining (i.e., menstruation or menses).

As a woman enters her 30s, the quality and number of the remaining follicles in her ovaries begins to decrease. This is reflected by a decreased AFC in the late reproductive stage, as indicated in Table 2.1. Because progesterone is produced by the follicles, a reduction in the quality of follicles results in a decline in progesterone levels in the late reproductive stage, commonly in a woman's 30s. AFC and progesterone continue to decline into the early menopause transition (or perimenopause), commonly in a woman's 40s. By her late 40s, anovulatory (no ovulation) cycles are common, although periods may still be mostly regular. Due to the decline in AFC and progesterone, women experience gradual changes in the length of their cycles and commonly skip cycles, or experience more than 60 days between periods, one to three years before their FMP.

In the decade prior to the FMP, laboratory tests commonly show increased variability in the level of FSH. When no egg is released, the anterior pituitary produces larger amounts of FSH to stimulate a follicle for ovulation because it did not receive the feedback signal that an egg was released. Persistent measurements of FSH greater than 25 are commonly used to determine if a woman is still fertile or if she is postmenopausal. Although a common clinical definition of postmenopause requires that a woman must go a full 12 months without a period before she can be diagnosed as "postmenopausal," this clinical definition becomes problematic when a woman is still menstruating occasionally into her mid- or late 50s, or if she is exhibiting signs and symptoms of menopause at a very early age. Measuring FSH in these situations can help to determine a woman's menopausal status. In young premenopausal women, FSH generally varies from a low of 1.0 IU/L during menses up to 20 IU/L midcycle to stimulate ovulation. As shown in Table 2.1, measurements of FSH >=25 IU/L are expected in the late perimenopause stage, when ovulation does not occur and no feedback signal is sent to the pituitary to turn off the FSH signal. FSH generally continues to climb, up to values of 100 to 200 IU/L in early and late postmenopause, and generally stabilize approximately five years after the FMP, as indicated in Table 2.1.

Aside from its role in fertility, progesterone receptors exist throughout a woman's body, and the effects of progesterone are extensive. As shown in Table 2.2, progesterone is important for breast health, healthy bone remodeling, mood, sleep quality, blood sugar normalization, libido, and more (Lee, 2005). As progesterone levels decline in the late reproductive stage and early menopausal transition years, women may experience increased breast tenderness, breast swelling, or lumpy breasts one to two weeks prior to onset of her period each month; she may also notice increased difficulty falling asleep or midsleep wakening, new or increased problems with anxiety, new or increased problems with headaches or migraines, increased menstrual cramps, shorter menstrual cycles (<=25 days), heavier or longer flow, new or increased problems with premenstrual mood swings, and

Table 2.2. Physiological Effects of Progesterone in Women

Maintains secretory endometrium	Protects against breast fibrocyst formation
Helps use fat for energy	Natural diuretic
Elevates mood and calms anxiety	Prevents cyclical migraines
Promotes normal sleep patterns	Facilitates thyroid hormone function
Helps normalize blood sugar levels	Helps maintain and restore normal libido
Normalizes zinc and copper levels	Restores proper cell oxygen levels
Prevents endometrial cancer	Helps prevent breast cancer
Stimulates bone formation (osteoblast activity)	Improves vascular tone
Prevents autoimmune diseases	Increases sensitivity of estrogen receptors
Prevents coronary artery spasm and atherosclerotic plaque	

weight gain without changes in exercise or eating. New-onset night sweats, in particular in the premenstrual time of a woman's cycle, are also associated with a decline in progesterone (Prior, 2011). Night sweats that occur throughout the month are more likely associated with low testosterone, rather than low progesterone, as discussed in the previous section on testosterone. Of course a woman in her late reproductive stage (~35–39 years old) or in the menopausal transition (~ 40–50 years old) may be experiencing symptoms from both low progesterone and low testosterone.

Since progesterone is produced by the corpus luteum that develops within the ovary at the time of ovulation, anything that inhibits ovulation in young women will also eliminate progesterone production. Hence all oral contraceptives, by definition, will keep a woman's progesterone level very low. Similarly, if a young woman experiences extended periods of time without her period, possibly caused by anorexia, very low body weight and extreme exercise, or other significant physical or emotional stress, progesterone levels will be very low. Estrogen is generally also low under these circumstances, so symptoms of progesterone deficiency may be less apparent, as these two hormones have many complementary and antagonistic roles, which is discussed in the next section on estrogen.

Although a decrease in sexual health is not well described in the literature with respect to declining progesterone, this is most likely because the question has not been asked in any of the research studies. If a woman is experiencing three or more of the symptoms listed in Table 2.3 and she is still menstruating, she meets the criteria for a clinical diagnosis of early perimenopause (Prior, 2011). This phase can last over a decade for some women. With heavier, more painful, and more frequent menstrual cycles, and a few of the other symptoms listed in Table 2.2, it is not hard to imagine that a woman's sexual health is very likely

Table 2.3. Diagnosis of Early Perimenopause in a Woman with Regular Menstrual Cycles Can Be Made If She Is Experiencing at Least Three of the Following Nine Changes

1. New onset heavy and/or longer flow
2. Shorter menstrual cycles (<=25 days)
3. New sore, swollen, or lumpy breasts
4. New mid-sleep wakening
5. Increased cramps
6. Onset of night sweats, in particular just prior to onset of menses (premenstrual)
7. New or markedly increased migraine headaches
8. New or increased premenstrual mood swings and anxiety
9. Weight gain without changes in exercise or eating

impaired, diminished, or suppressed. Paraphrasing the many women in perimenopause seen by the author in clinic, "I'm either bleeding for half the month, or my breasts are too sore to touch, I can't fall asleep without two (or more) glasses of wine to calm my nerves every night, I wake up with the bed drenched in sweat, and I'm constantly bloated. So thoughts of sex are simply out of question! I can't even think about." Upon inquiry, most women want their sex life back, but until the perimenopausal symptoms are resolved, they simply do not feel desirable and do not feel well enough to enjoy it.

It is worth pointing out that all oral contraceptives contain a patented progesterone-like molecule that does not have identical actions in the female body as endogenously produced or human bioidentical progesterone. Similarly, medoxyprogesterone is a patented progesterone-like medication commonly prescribed to postmenopausal women under the tradename of Provera®, which does not have behave as a true, bioidentical progesterone. These patented progesterone-like prescription medications are referred to as "progestins," which should not be confused with human-identical progesterone. Unfortunately, the medical literature, pharmaceutical literature, and media often refer to progesterone in their title or headlines and conclusions, although the research study referenced was only looking at the effects of a particular patented progestin, not bioidentical progesterone. Similarly, conclusions about the negative effects of a patented progestin medication should not be assumed to apply to bioidentical progesterone. For example, the national Women's Health Initiative (WHI) trial in which women were randomized to receive either placebo or Premarin®, an estrogen compound sourced from pregnant horses. A subset of the women treated with Premarin® was also treated with Provera® if they still had their uterus. (It was standard practice to add Provera if a woman still had her uterus because the epidemic of endometrial cancer in the 1960s and 1970s had been attributed to Premarin®-only therapy, representing one of the largest epidemics of iatrogenic disease that had ever occurred in the United States; Jick, 1980; Ratner, 2001). Provera® (medoxyprogesterone) had been shown to antagonize the stimulating effect of estrogen on the endometrium and consequently protect woman from endometrial cancer caused by unopposed Premarin® therapy. Unfortunately, Provera®, unlike bioidentical progesterone, did not antagonize or balance the stimulating effects of Premarin® on breast tissue, and increased rates of breast cancer were seen in the WHI trial among the women treated with Provera® but not among the women treated only with Premarin®. It is now understood that Provera® led to the increased rates of breast cancer, and that it is unfortunate that bioidentical progesterone was not used in the WHI trial, as it known to protect women against breast cancer (Holtorf, 2009).

There is one FDA-approved oral bioidentical progesterone available in the United States under the tradename of Prometrium®. Progesterone, used in Prometrium®, is synthesized from a starting material from a plant source, typically yams or soy, and is molecularly identical to that of human ovarian origin.

Prometrium®, available as 100 and 200mg soft gelatin capsules, contain a suspension of progesterone in a fluid consisting primarily of peanut oil. The inactive ingredients in Prometrium® include peanut oil, gelatin, lecithin, titanium dioxide, FD&C Red No 40, and D&C Yellow No, 10 and 6 (Drugs.com, 2016). Custom-compounded progesterone capsules (and other forms of delivery, such as sublingual tablets for direct absorption in the blood stream to reduce metabolism by first-pass through the liver, and creams for absorption through the skin) can be made in any dose, and without peanut oil and dyes, as specified by the prescribing physician, for the specific needs the patient. Bioidentical progesterone is recognized as safe and effective for treating pre- and postmenopausal women with signs of progesterone deficiency (Prior, 2011).

Estrogen

Human estrogen is made in the ovaries, the corpus luteum, adrenal glands, and fat cells. Estrogen is not one specific molecule. There are three different types of estrogen molecules produced in the human female, namely estradiol, estrone, and estriol, with different actions and potencies (Gruber, 2002). Since the medical literature, media, and pharmaceutical industries often refer to estrogens as all being the same thing, it is worth noting that pharmaceutically created estrogens, and those obtained from other animal species, do not contain the same types of estrogen molecules produced naturally in humans. For example, Premarin®, the FDA-approved estrogen medication first marketed in the United States in 1942, is extracted from the urine of pregnant horses, and it contains numerous types of estrogen molecules, none of which are naturally found in human females (FDA, 2005).

Estradiol is the most active form of estrogen in the human body. It is primarily produced in the ovaries, and it directly affects a wide range of cellular functions, as estrogen receptors are ubiquitous in the body. Estradiol is the predominant form of estrogen during a woman's reproductive years. As shown in Table 2.1, estradiol levels become erratic and erratic surges of 20% to 30% above the usual level are common during the menopausal transition. After the FMP, estradiol levels decline to very low levels.

Estrone is a weaker estrogen than estradiol and is produced in one of two ways. The primary source is by conversion of testosterone derivatives (androstenedione) to estrone in fat cells. Estrone is also produced by conversion of estradiol to testosterone, which is then converted to estrone. Estrone becomes the predominant estrogen after menopause (when estradiol levels are low), and levels of estrone tend to rise with increasing body fat. Overweight older women tend to have high circulating levels of estrone.

Although less potent than estradiol, both estradiol and estrone activate estrogen receptors found in the reproductive tract, breast, pituitary, hypothalamus, liver, and bone, to varying degrees. And they both increase the synthesis of SHBG in the liver (reducing free testosterone) and increase synthesis of thyroid-binding globulin (reducing free thyroid hormone) and other serum proteins, as well as suppress the release of FSH and LH from the anterior pituitary. It is expected that the level of FSH will fall to premenopausal levels (below 25 IU/L) when a postmenopausal woman is treated with therapeutic levels of estradiol. She is still postmenopausal, even though her FSH will indicate otherwise. If she discontinues the estradiol, her FSH will return to its previous postmenopausal level.

Estriol is the weakest of the estrogens. It is not secreted by the ovary but is produced in small amounts by conversion from estrone in the liver. Large amounts of estriol are only seen in pregnant women, because large amounts of estriol are produced by the placenta. Estriol attaches to cell receptors affecting hair, nails, and skin. Levels of estriol in nonpregnant women are consistently low throughout a woman's lifespan, regardless of menopausal status (Gruber, 2002). Limited research on the functions of estriol in nonpregnant women show effects are limited mainly to skin and the vaginal wall, with minimal effect on the heart and bones in nonpregnant women (Schwartz, 2008).

Roles and effects of estradiol in women include bone building, vaginal lubrication, vaginal flora, sleep quality, cardiovascular protection, and brain protection (Files, 2011; Schwartz, 2008). With respect to reproduction, estradiol is responsible for increasing cellular proliferation in the endometrial lining of the uterus, in order to prepare it for implantation of an embryo. It is important to realize that estrogens and progesterone are antagonists. Their actions serve to balance and complement each other throughout the body. At no time do these hormones act entirely independently under normal circumstance in a healthy body. For example, as estrogen increases the endometrial lining, progesterone keeps it under control, which helps prevent endometrial hyperplasia in the uterus. (Levels of estradiol and progesterone fall if an embryo does not successfully implant in the uterus, causing the lining to slough off, resulting in a period. This process repeats on a monthly basis through the reproductive years.) Other examples are bone and breast health. Estrogen stimulates the cells of the bone leading to bone building (osteoblasts), while progesterone inhibits the cells that the break down bone (osteoclasts) during the remodeling process. With respect to breast health, estrogen stimulates breast cells, while progesterone protects against fibrocystic changes in the breast. In this way, the two hormones complement each other in the maintenance of healthy bone and healthy breast tissue. If estrogen levels are above normal, such as in the early menopause transition, and/or if progesterone levels are low relative to estrogen, women can exhibit signs and symptoms of estrogen dominance. Physiological effects of estrogen, and of estrogen dominance, are listed in Table 2.4.

Table 2.4. Physiological Effects of Estrogen in Women

Creates a proliferative endometrium	Increases risk of endometrial cancer without sufficient progesterone to balance the effects[a]
Dilates blood vessels (blood pressure may increase postmenopause	Stimulates breast cells (can cause fibrocystic breasts when progesterone is low[a])
Helps with sleep sustainability and improves sleep disorders	Salt and fluid retention (can be excessive without the diuretic effects of progesterone[a])
Maintains vaginal lubrication and healthy pH and flora	Prevents and reverses vaginal atrophy
Relieves hot flushes	Improves health (and flora) of urinary tract
Inhibits natural bone loss (inhibit osteoclasts)	Prevents and reverses osteoporosis
Improves memory	Prevents or delays Alzheimer's disease
Improves cholesterol profile	Prevents or delays cardiovascular disease

[a]Caused by too much estrogen or by a "relative estrogen dominance" if progesterone is inadequate.

By the time a woman is one or more years past her FMP, estradiol levels remain very low and vaginal symptoms often become very significant. Estrogen is the dominant regulator of vaginal physiology. The vagina is composed of an inner stratified squamous epithelium, a middle muscular layer, and an outer fibrous layer. Before menopause, in the presence of endogenous estrogen, the vagina is characterized by a thickened, rugated vaginal surface, increased vaginal blood flow, and vaginal lubrication. Before menopause, estrogen stimulates the exfoliation of cells that line the vaginal wall. As these cells exfoliate and die, they release large amounts of glycogen. This glycogen is hydrolyzed to glucose, and lactobacillus converts the glucose to lactic acid, which helps maintain vaginal pH between 3.5 to 4.5. After menopause, with declining levels of estradiol, exfoliation is less active and the pH increases to 5.0 to 7.5. At these higher pH levels, lactobacillus survival is hindered, and vaginal colonization by other bacterial species predisposes women to symptomatic vaginal infection or urinary tract infection.

A gynecological examination of a woman with vaginal atrophy reveals a vagina that is dry with pale, frail tissue and lacking the normal mucosal ridges and folds. The expected elasticity and pliability associated with a well-estrogenized vagina is absent. There is minimal lubrication due to decreased blood flow, and the tissues are easily traumatized during examination. There may also be petechiae or small hemorrhages on the vaginal lining. The vaginal introitus may be narrowed; the epithelial surface is typically very friable and may be ulcerated. The interior walls of the vagina can literally be stuck to each other, and bleeding from the walls can occur as the walls are separated during the examination. Shrinkage of the clitoral tissue may be evident. Prolapse of pelvic organs—uterus, bladder, and rectum—into the vagina is also not uncommon (Lindahl, 2014).

It is estimated that up to 45% of women are affected by vaginal atrophy significant enough to seek medical attention (Lindahl, 2014). Estimates of some degree of urinary incontinence are as high as 73% of postmenopausal women (Stenberg, 1996) Symptoms reported by women with vaginal atrophy include vaginal dryness, vulvovaginal irritation and itching, pain and bleeding during intercourse, burning and discharge, as well as urinary problems. These symptoms are usually progressive and unlikely to resolve spontaneously. Left untreated, vaginal atrophy can result in years of vulvovaginal discomfort, with a significant negative impact on sexual health and quality of intimacy and quality of life. Additional medical complications of low estrogen and vaginal atrophy can include frequent urinary tract infections, vaginal infections, and prolapse of pelvic organs—uterus, bladder, and rectum—into the vagina (Lindahl, 2014; North American Menopause Society [NAMS], 2007).

As indicated in Table 2.1, vasomotor symptoms, commonly known as hot flashes, are common in the early years of postmenopause. These symptoms usually resolve in late menopause, when symptoms of vaginal atrophy and urinary problems, as discussed previously, become predominate and persist for a woman's remaining lifespan (Harlow, 2012). Other problems that can become persistent in the older postmenopausal woman that are associated with low hormone levels include insomnia, reduced memory and focus, low libido, weight gain and bloating, aging skin, increasing blood sugar, increasing cholesterol, bone loss, muscle weakness and decrease in skeletal muscle mass, and increasing blood pressure. These are the concerns postmenopausal women bring to their primary care physician, gynecologist, and/or integrative medical doctor on a daily basis. No studies exist that explore the combined impact of these hormonally related problems on quality of life in women in late midlife, along with the negative side effects of medications commonly prescribed for these problems. When the additional demands of midlife are also considered, such as higher demands at work, college-aged children and/or grandchildren, elderly parents needing care, and divorce and the fear of dating again, the stress on a hormonally depleted woman with insomnia and other energy-depleting issues easily becomes overwhelming. Fortunately, there are safe and effective hormone therapies, as discussed later in the chapter.

THYROID

Hypothyroidism is a common condition with inadequate amounts of active thyroid hormone, known as triiodothyronine (referred to as T3—it contains three molecules of iodine) at the cellular level. The thyroid gland largely produces a hormone called thyroxine (referred to as T4, containing four molecules of iodine), which is an inactive form of thyroid hormone that is activated to T3 throughout the body at the cellular level where and when it is needed. The production of T4 from the thyroid gland is triggered by thyroid stimulating hormone (TSH) produced by the pituitary gland. Every metabolic function throughout the human body is affected by the presence, or lack thereof, of active thyroid hormone. The range of symptoms of inadequate T3 in adults can include fatigue, weakness, weight gain, headaches, decreased libido, depression, reduced mental focus and cognitive dysfunction, reduced sleep quality, diminished exercise tolerance, dry skin, and hair loss, all of which can also be caused by declines in testosterone, progesterone, and/or estrogen, as discussed earlier. Additional signs and symptoms generally more specific to hypothyroidism than changes in the sex hormones include cold intolerance, pallor, loss of temporal eyebrow hair, constipation, and cold extremities (Brent, 2012; Nussey, 2001).

The production of T4, and the conversion of T4 to active T3, depends on several optimal conditions, including specific micronutrients, such as iodine, zinc, and selenium, and optimal bowel bacteria. Approximately 20% of the active T3 in the human body is produced by desirable bacteria in our large intestines. These desirable bacteria can be killed off by antibiotics, infections, and other medications. There is a two- to three-fold increase in the incidence of thyroid dysfunction with age, including overt and subclinical hypothyroidism (elevated TSH with normal T4 and T3 levels; Canaris, 2000). It has been proposed that subclinical hypothyroidism be replaced with the term "mild thyroid failure" (MTF) and should be treated despite T4 and T3 levels in the laboratory's reference range of normal (McDermott, 2001). It is worth knowing that laboratory reference ranges for T4 and T3 represent where 95% of people fell who received that test in the past year in the geographic range of the laboratory, without regard for illness or treatment, and do not represent anything to do with what would be an optimal range for a healthy individual. The same is true for virtually all laboratory test reference ranges.

One conclusion is that several types of low thyroid function exist, the incidence of declining thyroid function increases with age, and it is often underdiagnosed and undertreated in all ages, contributing to significant comorbidities and decreased quality of life and possibly contributing to a range of other changes women experience through midlife and beyond. Numerous studies have demonstrated increased cholesterol levels in patients with MTF and show that thyroid replacement results in a significant reduction in cholesterol levels. In addition to the increase in total and low-density cholesterol seen with MTF, endothelial function with impaired vasodilation have also been demonstrated, further increasing the risk of cardiovascular events. In one 20-year longitudinal study of over 2,000 men and women between the ages of 17 and 89, MTF was associated with a 2.2-fold increased risk of coronary artery disease and a 1.5-fold increased risk of cardiovascular mortality after adjustment for other known cardiovascular risk factors (Walsh, 2005).

Approaches and Controversies in Hormone Therapy for Women at Midlife and Beyond

The following are case stories of Sue and Joan—identical twin sisters with identical initial health concerns who sought conventional and integrative medical approaches, respectively, resulting in very different outcomes over their lifetime.

At 35 years old, Sue and Joan, identical twins, were suffering from the same symptoms of mild depression, new onset anxiety, irritability, mood swings, fatigue, weight gain, low libido, and sleeping difficulties. Sue visited her primary care doctor who prescribed an antidepressant to address her primary concerns, even though the class of medication has been shown to decrease libido and lead to weight gain. In contrast, Joan sought the advice of an integrative medicine physician who specialized in hormone therapy. After an in-depth consultation and lab work to evaluate her sex hormones and thyroid hormone, it was revealed that she had low testosterone. Joan received custom-compounded bioidentical testosterone that was inserted under her skin for continuous subcutaneous delivery. In six weeks, when Joan returned for follow-up, she reported the depression, anxiety, mood swings, fatigue, and sleep issues were all gone. And her libido had increased wonderfully. She was beaming and reported her husband was very happy as well. Joan had also started working out and reported feeling like her old self again.

At 45, Sue returned to her primary care doctor, still depressed and more overweight. Her labs showed her cholesterol had also increased. Sue was prescribed a statin medication for her cholesterol. She subsequently reported increased muscle pain, making it impossible to participate in any exercise. The statin was continued. In contrast, Joan reported to her integrative physician that at the age of 45 she no longer resembles her identical twin. Joan is fit, exercising, and on no extra prescription medications. Her cholesterol continues to be normal. Joan continued with the bioidentical testosterone therapy.

At 50, Sue complained of hot flashes and night sweats. She was put on Duavee®, an oral combination or Premarin® and a synthetic drug called bazedoxifene, which is an estrogen antagonist in the uterus. She was not put on progesterone to protect her breast from stimulation by the estrogenic effects of Premarin®. It had been known from the WHI trial and other studies that Premarin® increases risk of blood clots and cardiovascular disease (Grady, 1997; Writing Group for the Women's Health Initiative Investigators [WGWHII], 2002). She was told to discontinue the Duavee® four to six weeks prior to any future surgery due to increased risk of venous-thrombo-emboli associated with form of estrogen in this medication. On the Duavee®, Sue reported that her hot flashes disappeared; however, she still had no sex drive or energy, was overweight, and was unable to lose weight. In contrast, at 50, Joan continued to feel well on the testosterone replacement, and she had no new complaints.

At 55, Sue told her primary care doctor that she quit the Duvaee® because, despite it helping with the disappearance of the hot flashes, she was feeling worse over the recent years in every other respect about her health. She was also worried that she was getting Alzheimer's because she could not remember anything. Sue's blood pressure had also been elevated in several recent visits, so her primary care doctor prescribed an antihypertensive medication. Blood tests showed that Sue has developed type II diabetes, so she was placed on another medication for that. A DEXA scan showed Sue had significant bone loss, meeting the diagnostic criteria of osteopenia, so another medication was prescribed to improve the results of her bone density scan. The dose of her antidepressant was also increased. Nothing else was suggested to address Sue's new concerns about her worsening memory. By the age of 55, Joan began to experience hot flashes, infrequent periods, and decreased vaginal lubrication, five years later than her sister began experiencing the same symptoms. Her integrative physician evaluated her labs and explained her estradiol levels and FSH were now consistent with postmenopause levels—which was expected—so custom-compounded estradiol was added with her subcutaneous testosterone therapy and oral, custom-compounded progesterone was added to balance with her estrogen and provide all the health and safety benefits of balanced therapy. Within four weeks, Joan's hot flashes were gone, her vaginal lubrication was back to normal, and she was feeling great again. If Joan had reported symptoms of progesterone deficiency earlier (as listed in Table 2.3), oral progesterone would have been added at that time, as it is common to experience symptoms of estrogen dominance as estrogen surges occur in the early menopausal transition (see Table 2.1). In fact, Joan sailed through perimenopause without any concerns, only relying on the subcutaneous testosterone therapy until the age of 55, without any other medication. Moreover, Joan's bone density scan was normal, as were her cholesterol levels and blood pressure. She showed no indication of diabetes, and her body weight was normal. Joan continued to report she cannot believe the differences between she and her identical twin, and she wished her sister would have not have listened to her primary care doctor who told Sue that custom-compounded hormone therapy is not safe.

By 60, Sue was on a long list of medications for depression, hypertension, high cholesterol, diabetes, and osteopenia, and her quality of life was very poor, with persistent fatigue, body aches, poor memory and no sex life. Sue was never evaluated by a doctor who tested her sex hormone status and addressed her original signs and symptoms at the root cause. Joan, on the other hand, felt very fortunate to have had a doctor who had the expertise to test hormones properly and knew how to safely dose and deliver the bioidentical hormones in an effective and simple way. At 60, Joan felt and looked much younger than her twin sister. She enjoyed plenty of energy, mental clarity, sharp focus, and an active, satisfying sex life, all without the need for any patented pharmaceutical drugs.

The story of the identical twin sisters illustrates two radically different medical approaches available to women today who are experiencing problems related to hormonal changes and sexual health issues at midlife. Unfortunately, finding an integrative doctor with expertise in hormone therapy does not happen through the usual specialty referral system in the American medical system today. Instead, it requires a woman to personally research her options outside of those endorsed by her primary care doctor and her gynecologist. And, if she chooses to obtain a consultation with a conventional endocrinologist, she will also not receive any advice different from what was offered by her previous doctors. This leaves women (and men) with the quandary, "If bioidentical hormones are so safe and helpful for maintaining our health and quality of life, why are they not recommended by my usual doctors, who I trust for everything else regarding my healthcare?" There is no truly logical or simple answer to this very legitimate question, and a thorough discussion is beyond the scope of this book. The simple answer is that the findings of the WHI trial published in 2002 dramatically changed the prescribing practices of physicians in the United States, which drove women to seek alternative therapies thought to be safer (Hersh, 2004). This is true, but women have been seeking and receiving custom-compounded bioidentical hormone therapy since the 1940s (Donovitz, 2015), although its popularity certainly increased after 2002 when the public was made aware of increased risks of breast cancer and stroke with the conventional (nonbioidentical) prescriptions of equine estrogen (Premarin®) and medoxyprogesterone (Provera®). The reader is referred to a few peer-reviewed articles that discuss the controversy and confusion surrounding age-related hormone loss and the use of estrogen, progesterone, testosterone, and thyroid hormones and one book written by a gynecologist who is a leader in training physicians in the United States in subcutaneous bioidentical hormone therapy (Donovitz, 2015; Files, 2011; Holtorf, 2009; Schwartz, 2008).

UNDERSTANDING HORMONE TERMINOLOGY AND DIFFERENT FORMS OF THERAPY

Understanding the terminology and commonly misinterpreted terms surrounding the subject of hormone replacement therapy will be helpful to anyone (patient, doctor, researcher, journalist, FDA employee, etc.) who is trying to interpret the literature and make a decision for public policy or for her personal health. Additionally, the form or delivery method of the hormones—oral, transdermal, sublingual, injection, subcutaneous—also makes a difference in how the human body receives and responds to the hormone and thus affects the benefit and risk profile accordingly.

Bioidentical Hormones

The term "bioidentical" is not a marketing term. The Endocrine Society defines bioidentical hormones as "compounds that have the exact same chemical and molecular structure as hormones that are produced in the human body" (Files, 2011). This definition does not address the manufacturing source, or delivery method, or other compounds that may be used as fillers in the final product. The bioidentically manufactured molecules of estradiol, testosterone, and progesterone are, in fact, FDA-approved ingredients; however, this approval only applies to specific formulations: progesterone in the form of Prometrium® gel capsules (in peanut oil) and in vaginal cream formulations approved only for infertility; estradiol in pill form for oral consumption, a patch for transdermal delivery, a cream for treating vaginal atrophy; and testosterone in a lozenge for sublingual delivery and in a gel and patch for transdermal delivery. The FDA-approved bioidentical products for testosterone are only approved for men and are not available in doses appropriate for women. When doctors order a custom-compounded formulation of estradiol, progesterone, and/or testosterone, a nationally accredited compounding pharmacy uses the same FDA-approved bioidentical hormones to produce the dose and product as specified by the doctor—oral capsule, sublingual tablet, cream, vaginal suppository, or pellet (to be inserted subcutaneously by the doctor). The end product of a custom-compounded prescription—with the exact same active ingredients of products mass produced by a pharmaceutical company—no longer meets the criteria of an FDA-approved product. As such, most insurance companies consider compounded hormones "investigational" and choose to not reimburse for them (Blue Cross Idaho, 2015).

The term "bioidentical" also does not imply natural. Source material for bioidentical hormones are obtained from plants and chemically modified to create molecular structures identical to the specific human hormone. Premarin®, on the other hand, is natural—but it is only natural for a horse. And progestins used in oral contraceptives, in Depo- Provera®, and in Provera® are not natural for any living creature, and they each differ widely in their chemical structures, metabolism, and potencies (Stanczyk, 2003).

Custom Compounding

Originally all prescriptions in the United States were custom compounded. In the 1950s and 1960s, compounding declined with the advent of mass drug manufacturing. The FDA approval process for medications was intended only for quality control on mass-produced drugs made by pharmaceutical manufacturers. Pharmaceutical marketing campaigns have been effective at making everyone believe in the superiority of pharmaceutical drugs and forgetting that pharmaceutical companies are not the doctor who is listening to the individual patient and designing the best treatment on an individual basis. Unfortunately, training on custom-compounding was phased out of medical and pharmaceutical schools several decades ago. And human nature being what it is, if something is not taught in medical school or in subspecialty residency training, then the common misperception is it must not be a legitimate form of healthcare. One of the underlying reasons why the specialty of integrative medicine was started in the first place is to bring an open mind and consider the merits of anything that brings health and prevents or reverses illness, so as to not be limited by what the pharmaceutical industry dictates as the standard of care. Today, if self-motivated pharmacists or physicians want to learn about custom-compounding, they often seek apprentice-style training to learn from older practitioners willing to share their skills. During a fellowship training program in integrative medicine at the University of Maryland School of Medicine in 2002, the author had the good fortune of being mentored in custom-compounding by two family medicine physicians who had over 20 years' experience prescribing custom hormones and by a compounding pharmacist with over 30 years' experience.

Unfortunately, there is ample negative press about custom-compounded hormones (Bhavnani, 2012; NAMS, 2015; Pinkerton, 2012). But while authors and organizations who represent the larger pharmaceutical industry and other organizations supported by the pharmaceutical industry criticize the fine points of the variability in custom-compounded hormones, they simultaneously downplay all the well-documented problems causally linked to conventional, nonbioidentical hormones and other FDA-approved pharmaceutical drugs that all have significant negative side effects. One only needs to listen to a single TV commercial for an FDA-approved drug and realize they all have significant side effects and risks. In addition, serious epidemics of pharmaceutically induced cancers by an FDA-approved hormone is no longer a matter of debate (Hersh, 2004; Jick, 1980). In the end, it comes down to differences in core beliefs and philosophies about modern medicine, healthcare, and wellness in the United States. If readers have grown with or are trained with the belief that modern medicine (in collaboration with the pharmaceutical industries) will develop an FDA-approved drug for every disease, and any significant negative side effects can be addressed with additional FDA-approved drugs, then they will fall on the side against non-FDA-approved treatments of any kind. On the other hand, if they come to understand that treatment of a disease does not equate to wellness, and that the objective of custom-compounded bioidentical hormone therapy is to sustain physiological levels of previously naturally produced hormones in order to optimize wellness and prevent disease before it starts, then they are more likely to fall on the opposite side of the argument. The fact that compounded hormones using the same active ingredients of FDA-approved drugs produced by a pharmaceutical company do not maintain FDA-approval status is purely due to the legal jurisdiction of how and why the FDA was created in 1906 in the first place. The FDA was founded to ensure the quality and safety of mass-produced products. It was not founded to control or limit custom-compounding prescribed by a licensed doctor for an individual patient. It is a personal choice whether or not one wants to limit one's choices in healthcare, whether one is a prescribing physician or a patient/consumer.

Different Meanings of Estrogen Replacement Therapy, Hormone Replacement Therapy, and Bioidentical Hormone Therapy

In general, the terms estrogen replacement therapy (ERT) or estrogen therapy (ET) in the medical literature and media refer to the use of equine estrogen in the form of Premarin® alone, without a progestin, such as Provera®, and without bioidentical progesterone Prometrium®. Most commonly, the equine estrogen was prescribed in a pill to be taken orally. Fewer studies use equine estrogen delivered as a transdermal patch. Importantly, the terminology of "estrogen" therapy almost always refers to equine estrogen. It does not refer to bioidentical estradiol replacement in any form. Analogously, the term "hormone replacement therapy" (HRT) generally refers to the use of Premarin® combined with either Provera® or Prometrium®. This distinction is important, as virtually all the conclusions of negative effects reported in the WHI trial have been inappropriately applied to all hormone therapies, including bioidenticals. The harmful effects found in the WHI trial apply exclusively to *oral* Premarin® and Provera®. The emphasis on *oral* Premarin® is made because increased risk of blood clots and stroke are associated with all forms of *oral* estrogen prescriptions (including oral contraceptives), and not with delivery methods of estrogen that do not require a first pass through the liver before they enter the blood stream and reach estrogen receptors in peripheral target tissues. The liver upregulates clotting factors when it metabolizes orally administered estrogen, which then leads to increased risk of clotting (Committee on Gynecologic Practice, 2013).This response by the liver is dose-related and has been a primary reason why the dose of estrogen in oral contraceptives has been gradually decreased since they were originally developed in 1960. This is also a primary reason why most doctors trained in compounded bioidentical hormones never prescribe estradiol to be given orally. Transdermal, transvaginal, and subcutaneous delivery methods are preferred, so increased risk of blood clots and stroke are not a concern, and, in general, when prescribed at normal physiological levels, women do not need to discontinue non-oral bioidentical estradiol therapies prior to surgery.

Therapies for Vaginal Atrophy

CASE STUDY OF HORMONAL THERAPY FOR VAGINAL ATROPHY

Ann is a 63-year-old female who had never used any form of HRT. She has two grown children, has been divorced for several years, and is not currently sexually active. She works full-time in the computer industry. Ann self-referred herself to the author's clinic with two primary complaints: fatigue and a vagina "that feels like the Sahara Desert!" She went on to say that "everything down there is dry and stuck together." Other than being mildly overweight and having moderate irritable bowel symptoms and mild to moderate joint aches, Ann reported no other health concerns and was not on any medications. Blood testing showed very low levels of estradiol, progesterone, testosterone, thyroid, and Vitamin D. Blood tests also showed elevated C-reactive protein, a marker of inflammation, and elevated HbA1c, indicative of insulin resistance and blood sugar in the prediabetes range. Both correlate with increased risk of cardiac disease. Gynecological examination showed all the signs consistent with vaginal atrophy as discussed in the section on estrogen.

To address Ann's immediate problem pain due to vaginal atrophy, Ann was prescribed custom-compounded estradiol cream and suppositories. She was instructed to apply a small amount of cream to the external labia twice daily until such time the labia separated and she could comfortably insert the estradiol suppository into the full depth of the vaginal canal. Within two weeks she was able to insert the suppository, which she was instructed to do every night for two weeks, then decrease to two to three times weekly, as needed to maintain normal moisture levels. Within four weeks, Ann reported that her vagina felt normal again and she was no longer experiencing daily discomfort.

To address her complaint of fatigue, and to address the elevated C-reactive protein and HbA1c, Ann also agreed to begin systemic hormone therapy. She was prescribed estradiol and testosterone in the form of compounded bioidentical pellets that were inserted subcutaneously and were expected to last three to five months before needing a new set to be reinserted. She was also prescribed oral progesterone and thyroid to be taken daily. Vitamin D3 was also recommended to bring her levels to an optimal range. Within three months on the new hormone regimen, Ann's fatigue was gone, she was enjoying yoga on a daily basis, and she had lost weight; her joint pain and irritable bowel symptoms were also gone. She was able to discontinue the estradiol suppositories because the estradiol pellet therapy was maintaining adequate estrogen levels to prevent the return of vaginal atrophy. At five months, when she returned for new estradiol and testosterone subcutaneous pellet insertion, her labs showed the C-reactive protein and HbA1c were in an ideal range, eliminating her previously increased risk for cardiac disease. Ann is leading a very active, pain-free life and is looking forward to a very active retirement.

RADIOFREQUENCY VULVO-VAGINAL REJUVENATION FOR VAGINAL ATROPHY AND URINARY LEAKAGE

Prior to 2015, if a woman sought therapy for vaginal dryness, vaginal atrophy, decreased sexual pleasure, or decreased orgasms, there were few to no treatments available that did not involve some form of hormone therapy—conventional or bioidentical. And if she had any signs of bladder or rectal prolapse, urinary leakage, or significant laxity or injury from childbirth, some type of surgery or ablative procedure with certain risks was a woman's only option. Some type of additional hormone therapy would also be suggested to promote healing from the procedure and/or prevent recurrence. Unfortunately, for women with a history of a hormone-dependent cancer, or with high family risk of such cancers, the thought of hormone therapy is frightening, even though there are some relatively safe hormone therapies available even in these situations.

In 2015 a new FDA-approved technology for rejuvenating the entire vaginal canal and vulvar region became available, and it does not involve any surgery or hormone therapy. The treatment is painless, involves no ablative injury to the tissue, requires no recovery time, and involves only three 30-minute treatment sessions performed a month apart to achieve rejuvenation of the entire vulvo-vaginal area. Normal blood flow is restored, nerves regenerate, existing collagen is restructured, and new collagen production is stimulated. With restoration of blood flow, natural lubrication returns to the vaginal tissue. The treatment uses radiofrequency heat delivered in a slender, temperature-controlled probe that elicits a healing response deep within the entire vaginal canal and external labial and clitoral tissues. The heat is comfortable, and no anesthesia is necessary. The technology includes measurement of the temperature at the skin surface and is computer-controlled to prevent the possibility of overheating. The clinical trial showed that 100% of women reported improvement if they had rated a need for improvement in any of the following areas: vaginal laxity, satisfaction from vaginal intercourse, ability to climax, vaginal moisture during sexual activity, ability to control urine when coughing, ability to control the stream of urine (start and stop), and feeling of urinary urgency. The timing of these improvements occurred anywhere between immediately after the initial treatment to within 30 days after the third treatment session. Currently, Thermi is the only manufacturer of the FDA-approved radiofrequency treatment for vulvo-vaginal rejuvenation. The treatment is marketed under the name of ThermiVa®. The author has performed ThermiVa® on women aged 28 to 68 years old. The younger women were eight or more weeks postpartum, and the procedure was performed to accelerate healing of tears or incisions from a vaginal delivery and to resolve vaginal laxity and improve satisfaction during intercourse. For the women in their later 60s, ThermiVa® was performed to resolve vaginal atrophy, urinary leakage, and mild bladder prolapse. Women between 40 to 60 who were already on bioidentical hormone therapy with no problems of vaginal dryness also underwent ThermiVA® to improve sexual pleasure and intensity of their orgasms, which they felt had declined over the years. More information is available at ThermiVa.org.

Women at Midlife: Understanding Our Health, New Opportunities, and Making Choices for Overall Well-Being and Sexual Health

Maintaining our health as we age is often a forgotten priority for many women until we "wake-up" at midlife and realize we have lost our vitality and we really do not know how to get it back. We were focused on work, raising babies, teenagers, husbands, our home, and elderly parents, leaving little time or energy for hobbies, making love, and self-care. And we were completely unaware that our testosterone had been declining since our 20s. Who of us even knew that women needed testosterone in the first place? And the birth control or antidepressant we took for 15 years also knocked out our libido. No wonder sex lost its appeal. But now, at midlife, sex has become painful without a tub of K-Y® Jelly nearby and is more something to endure rather than enjoy. How did this happen? And where did these extra 30 pounds come from?

Women have been told to expect midlife to be a time to celebrate their empty nest and begin new careers, start a business, write that book, and say what they think—to finally do what they always wanted to do. But for women who waited to complete their family until their late 30s or 40s, there is no empty nest on the near horizon at midlife. Children are returning home after college and taking more time to get on their own two feet. The accumulated stress of modern life and depleted hormones can make the horizon feel like a burden rather than an exciting opportunity for a personal overhaul and a second chance at life. Loss of a sexual life becomes an accepted consequence of middle age.

The author has personally helped hundreds of women at midlife who were feeling significantly "menopausal and burned-out" regain their well-being, sexual health, and passion in life through an integrative medical approach of customized bioidentical hormones, nutritional recommendations, and lifestyle changes. Over the years, there have been a small subset of these same women who discontinued their hormone therapy because another doctor told them that "older" women should not be on hormones. These very same women returned to restart bioidentical hormone therapy a few months to a few years later because symptoms of fatigue, insomnia, weight gain, loss of memory and concentration, loss of libido, vaginal dryness, bone loss, and other symptoms all reoccurred and left them unable to live the life they wanted. As one woman in her late 50s said when she returned for bioidentical hormones, "I really thought I did not need the hormones any longer because I was finally eating right, not drinking too much, exercising frequently, sleeping well, maintaining my ideal weight, and I felt great! But within a few months off the hormones, I stopped caring about what I ate, I stopped exercising, I stopped sleeping well, I gained weight, I lost interest in sex, and my life was out of control again!" She concluded, "The hormones feel like an essential key to keeping my life on track."

It is worth noting that in the late 1970s the stated goal of HRT for women was "to help women feel better as they aged and not to harm them in the process" (Greenblatt, 1978). At that time in US history, oral equine estrogen (Premarin®)—combined with a synthetic progestin (Provera®) if a woman still had her uterus—was the mainstay of HRT (Hersh, 2004). By 1997, sales of Premarin® reached $1 billion, and they continue to exceed $1 billion today, although the family of Premarin® medications has expanded to Premarin Vaginal® and Duavee® (Benson, 2015). Unfortunately, when serious harmful effects of these particular types of FDA-approved hormones were announced in 2002 based on the WHI trial (WGWHII, 2002), almost everyone (researchers, doctors, media, and lay public) inappropriately applied these results to all forms of estrogen therapy and all forms of progestins and progesterone. As explained in this chapter, nothing could be further from the truth, nor further from good science. Abundant evidence exists on the safety of bioidentical hormones and the benefits for total well-being (Donovitz, 2015; Holtorf, 2009; Schwartz, 2008). Unfortunately, many women have suffered and will continue to suffer because of this misinformation (Sarrel, 2013; Sifferlin, 2013). The training of doctors who want to understand the science and value of bioidentical hormones for our health and well-being is occurring, however, and began with doctors who independently started practicing and researching bioidentical hormones as early as the 1940s (Donovitz, 2015).

The training programs for bioidentical hormones in the United States are run by doctors (not pharmaceutical institutions) for doctors and are growing throughout the country. Largely, it is doctors who identify themselves as "integrative" who are stepping beyond the boundaries of their respective board specialties to obtain evidence-based training in compounded bioidentical hormones and other areas of wellness, disease prevention, and healthy aging. Historically, the specialty of endocrinology in the United States was considered to hold the domain on the study and treatment of pathological (i.e., disease) conditions related to hormones that occur in the human body. Currently, there is no other board-certified specialty in medicine that is designated to study and analyze the effects of hormones on wellness and disease prevention in women or men. The relatively young specialty of integrative medicine in the United States is most likely to continue the study of hormones on all aspects of health and well-being for both women and men, including sexual health. One of the underlying tenets of integrative medicine is to utilize all evidence-based approaches, inside and outside the boundaries of modern medicine, to improve and sustain health at the root issue. Without being tethered to pharmaceutical and surgical approaches, the field of integrative medicine is free to examine the merits of modalities generally considered "alternative" or "complementary," such as herbs, nutrition, acupuncture, meditation, massage, prayer, and so on.

The focus in this chapter has been on the hormonal changes in women and how they influence a woman's mental, physical, and sexual health through her entire adult lifespan. Gradual and sudden changes in hormone levels occur over the five decades from a woman's 20s to her 60s, as shown in Table 2.1. And many prescribed medications, environmental stressors, and life events can cause additional disruptions in a woman's hormonal symphony. The history of hormone treatments available to women in the United States is filled with tremendous harms, tremendous benefits, and unnecessary controversy. The growth in training of integrative doctors in custom-compounded bioidentical hormones throughout the United States offers a very credible opportunity for women to take control of their health. Armed with this awareness, women of all ages can make more informed decisions about the professional opinions they seek and the choices they make in order to optimize their sexual health, and their overall well-being and productivity, for their lifetime.

REFERENCES

Benson M (2015). Market Realist. http://marketrealist.com/2015/03/products-affect-pfizers-revenue

Bhavnani BR, Stanczyk FZ (2012). Misconception and concerns about bioidentical hormones used for custom-compounded hormone therapy. *J Clin Endorcrinol Met.* 97(3):756–759.

Blue Cross of Idaho (2015). Implantable hormone pellets. January. https://www.bcidaho.com/providers/medical_policies/pd/mp_501109.asp.

Brent GA (2012). Mechanisms of thyroid hormone action. *J Clin Invest.* 122(9):3035–3043.

Burger H (2002). Androgen production in women. *Fertil Steril.* Suppl 4:S3–S5.

Canaris G, Manowitz N, Mayor G, Ridgeway C (2000). The Colorado Thyroid Disease Prevention Study. *Arch Intern Med.* Feb;160:526–534.

Chumlea W, Schubert C, Roche A, et al. (2003). Age at menarche and racial comparisons in US girls. *Pediatrics.* 111(1):110–113.

Committee on Gynecologic Practice (2013). American College of Obestetrics and Gynecology website. http://www.acog.org/Resources-And-Publications/Committee-Opinions/Committee-on-Gynecologic-Practice/Postmenopausal-Estrogen-Therapy.

Donovitz G (2015). *Age healthier, live happier.* Orlando, FL: Celebrity Press

Drugs.com (2016). Prometrium. http://www.drugs.com/pro/prometrium.html.

Files JA, Ko MG, Pruthi S (2011). Bioidentical hormone therapy. *Mayo Clinic Proceed.* 86(7):673–680.

Food and Drug Administration (2005). Drug safety. http://www.fda.gov/Drugs/DrugSafety/InformationbyDrugClass/ucm168838.htm.

Grady D, Hulley SB, Furberg C (1997). Venous thromboembolic events associated with Hormone Replacement Therapy. *JAMA.* 278(6):477.

Greenblatt R, Stoddard L (1978). The estrogen controversy. *J Am Geriatr Soc.* 26(1):1–8.

Gruber C (2002). Production and actions of estrogens. *N Eng J Med.* 346(5):340–352.

Guay A, Davis S (2002). Testosterone insufficiency in women: fact or fiction? *World J Urol.* 20:106–110.

Harlow S, Gass M, Hall J, et al. (2012). Executive summary of the Stages of Reproductive Aging Workshop + 10: addressing the unfinished agenda of staging reproductive aging. *Menopause.* Nov;19(4):1–9.

Hersh AL, Stefanick ML, Stafford RS (2004). National use of postmenopausal hormone therapy. *JAMA.* 291(1):47–52.

Holtorf K (2009). The bioidentical hormone debate: are bioidentical hormones (estradiol, estriol, and progesterone) safer or more efficacious than commonly used synthetic versions in hormone replacement therapy? *Postgrad Med.* 121(1):1–13.

Jick H, Walker AM, Rothman KJ (1980). The epidemic of endometrial cancer: a commentary. *Am J Pub Health.* 70:264–267.

Labria F, Belanger A, Luu-The V, et al. (1998). DHEA and the intracrine formation of androgens and estrogens in peripheral target tissues: its role during aging. *Steroids.* May-Jun;63(5–6):322–328.

Lee J, Zava D, Hopkins V (2005). *What your doctor may not tell you about breast cancer.* New York: Grand Central Publishing.

Leiblum S, Koochaki P, Rodenberg C, Barto I, Rosen R (2006). Hypoactive sexual desire disorder in postmenopausal women: US results from the Women's International Study of Health and Sexuality (WISHeS). *Menopause.* 13(1):46–56.

Lindahl SH (2014). Reviewing the options for local estrogen treatment of vaginal atrophy. *Int J Womens Health.* 6:307–312.

Longscope C (1986). Adrenal and gonadal androgen secretion in normal females. *Clin Endocronol Metab.* 15:213–228.

McDermott MT, Rigeway EC (2001). Subclinical hypothyroidism is mild thyroid failure and should be treated. *J Clin Endocrinol Metab.* 86(10):4585–4590.

Morganti S, Ceda G, Saccani M, et al. (2005). Thyroid disease in the elderly: sex-related difference in clinical expression. *J Endocrinol Invest.* 28(11 Suppl Proc):101–104.

North American Menopause Society (2007). The role of local vaginal estrogen for treatment of vaginal atrophy in postmenopausal women: 2007 position statement of the North American Menopause Society. *Menopause.* 14(3):357–369.

North American Menopause Society (2015). Use of compounded hormone therapy in the United States: report of the North American Menopause Society Survey. *Menopause.* 22(12):1–9.

Nussey S, Whitehead S (2001). NCBI Bookshelf. http://www.ncbi.nlm.nih.gov/books/NBK28/?report=printable.

Peeters R (2008). Thyroid hormones and aging. *Hormones*. 27(1):28–35.

Pinkerton JV (2012). The truth about bioidentical hormone therapy. *Female Patient*. 37:16–20.

Prior J (2011). Progesterone for symptomatic perimenopause treatment—progesterone politics, physiology and potential for perimenopause. *ObGYN*. 3(2):109–120.

Ratner S, Ofri D (2001). Menopause and hormone-replacement therapy: Part 2. Hormone-replacment therapy regimens. *West J Med*. 175:32–34.

Santaro N, Crawford S, Lasley W, et al. (2008). Factors related to declining luteal function in women during the menopause transition. *J Clin Endocrinol Metab*. 93(5):1711–1721.

Sarrel PM, Njike VY, Vinante V, Katz DL (2013). The mortality toll of estrogen avoidance: an analysis of excess deaths among hysterectomized women aged 50 to 59 years. *Am J Public Health*. 103(9):1583–1588.

Schwartz E, Holtorf K (2008). Hormones in wellness and disease prevention: common pratices, current state of the evidence, and questions for the future. *Prim Care Clin*. 35:669–705.

Sifferlin A (2013). Hormone-replacement therapy: could estrogen have saved 50,000 lives? *Time*. http://healthland.time.com/2013/07/20/hormone- replacement-therapy-could-estrogen-have-saved-50000-lives

Stanczyk FZ (2003). All progestins are not created equal. *Steroids*. 68:879–890.

Stenberg A, Heimer G, Ulmsten U, Cnattingius S (1996). Prevalence of genitourinary and other climacteric symptoms in postmenopausal women. *Maturitas*. 24(1–2):31–36.

Walsh JP, Bremner AP, Bulsara MK, et al. (2005). Subclinical thyroid dysfunction as a risk factor for cardiovascular disease. *Arch Intern Med*. 165:2467–2472.

Writing Group for the Women's Health Initiative Investigators (2002). Risks and benefits of estrogen plus progestin in healthy postmenopausal women. *JAMA*. 288(3):321–333.

Zimmerman Y, Eijkemans M, Coelingh Bennink H, Blankenstein M, Fauser B (2014). The effect of combined oral contraception on testosterone levels in healthy women: a systematic review and meta-analysis. *Hum Reprod Update*. 20(1):76–104.

Zumoff B, Strain G, Miller L, Rosner W (1995). Twenty-four-hour mean plasma testosterone concentration declines with age in normal premenopausal women. *J Clin Endocronol Metab*. 80(4):1429–1430.

3

Integrative Treatment of Female Sexual Dysfunction

JACQUELYN M. PAYKEL

"Sexual expression is a normal and healthy part of human behaviour. Positive sexual experiences are related to health and well-being throughout life; hence, there is a need to think about sexual health as not simply the absence of sexual disorders, but as a key factor affecting the quality of life."

—C. H. Mercer (2013)

Jennifer is a 48-year-old professional florist and breast cancer survivor. Since diagnosis of estrogen-receptor-positive breast cancer two years ago, Jennifer has undergone a mastectomy, chemotherapy, and initiation of anastrozole. She presented with severe dyspareunia, anorgasmia, and decreased libido. Her most distressing concern was that of decreased sex drive. She was in a stable heterosexual relationship for the past 12 years. Until two years prior she had a "very active" monogamous relationship with sexual activity "most days of the week." Now she was a reluctant participant in intimacy one to two times per month. She expressed personal distress but also relationship disruption due to her lack of interest. During her initial clinic visit she expressed that that her number one goal was "to feel close and intimate" with her partner again. Given her cancer history, she was very interested in an integrative approach to treat her sexual dysfunction.

Female Sexual Dysfunction

INTRODUCTION

Sexual satisfaction in women is associated with general well-being (Kingsberg, 2011). Meanwhile, female sexual dysfunction (FSD) is strongly associated with low feelings of physical and emotional satisfaction, decreased happiness, reduced quality of life, and impaired interpersonal relationships (Warnock, 2002).

FSD is a common health concern. Approximately 40% of women living in the United States have experienced sexual difficulties at least once in their life, the prevalence of which is highest during midlife. A sexual complaint becomes a disorder when it causes personal distress (American Psychiatric Association [APA], 2013). Among women reporting symptoms of FSD, 12% have reported personal distress (Laumann, 1999; Shiffren, 2008).

CHANGE IN CLASSIFICATION OF FSD

The *Diagnostic and Statistical Manual of Mental Disorders*, 5th edition (DSM-V), published by the APA in December of 2013, offers revised diagnostic criteria for FSD. The new criteria for FSD state the experience of the disorder must be present at least 75% of the time and must have a minimum duration of six months. In addition, the condition must cause "significant personal distress" rather than "marked distress and interpersonal difficulty" as was required by the fourth edition DSM-IV.

The DSM-V also simplifies the classification of FSD. Female hypoactive sexual desire disorder (HSDD) and female arousal disorder (FAD) were combined into a single classification: Sexual interest or arousal disorder. Likewise, the formerly separate dyspareunia and vaginismus are now called "genitopelvic pain or penetration disorder." Female orgasmic disorder remains unchanged (APA, 2013, Classification of Female Sexual Dysfunction).

Female sexual interest or arousal disorder is based on at least three diagnostic criteria of a possible six: (1) sexual activity is diminished due to absent or reduced interest; (2) patients' sexual or erotic thoughts and fantasies have diminished; (3) patients are less interested in sexual activity, whether initiated by themselves or a partner; (4) sexual excitement or pleasure is very reduced in most sexual encounters; (5) internal or external sexual or erotic cues fail to stir sexual interest or arousal; (6) genital or nongenital sensations during sex are greatly reduced.

The diagnosis of female orgasmic disorder is made if—in all or almost all sexual activity—there is marked delay, infrequency, or absence of orgasm and greatly reduced intensity of sensations caused by orgasm.

The diagnosis of genitopelvic pain or penetration disorder occurs when certain persistent or recurrent difficulties are present: with vaginal penetration during intercourse in general; intercourse-related vulvovaginal or pelvic pain; notable fear or anxiety about vulvovaginal pain when anticipating or experiencing penetration; and, during attempted vaginal penetration, marked tensing of pelvic floor muscles.

Why the change? The DSM-IV, originally published in 1994, utilized the Masters and Johnson linear sexual response model when creating the classifications of FSD: desire—arousal—orgasm—resolution (Masters, 1966). Since then, Rosemary Basson has published an alternative model of a woman's sexual response. Taking this new biopsychosocial model into consideration, various researchers on the DSM-V committee suggested that the separation of HSDD and FAD may have been artificial because, as Basson points out, desire may *not* precede arousal. Many women in long-term relationships enter into a sexual interface for reasons other than desire or pursuit of sexual satisfaction but can quickly evolve into that state of mind once aroused.

"Normal" Female Sexual Response

NORMALCY AND MODELS OF FEMALE SEXUAL RESPONSE

What is normal sexual function? The concept of "normal" sexual function (in both women and men) is evolving and can only be approximated by current epidemiological data and present-day cultural norms (e.g., Mercer, 2013; 2014). On the individual level, "normal" female sexual function depends upon age, gender identity, genetics, upbringing, health status, the novelty of a relationship, current life circumstances, and prior sexual experiences—both recent and remote. For this reason, each woman will have her own definition of "normal sexual function." For provider and patient both, it is beneficial to explore the origin of an individual's sexual norms in effort to understand sexual dysfunction and/or distress.

MASTERS AND JOHNSON MODEL OF SEXUAL RESPONSE

William H. Masters and Virginia E. Johnson pioneered laboratory research of the nature of human sexual *response* and the diagnosis and treatment of sexual disorders from 1957 through the 1990s. In 1966, they published the familiar linear female sexual response model (Masters, 1966) (Figure 3.1).

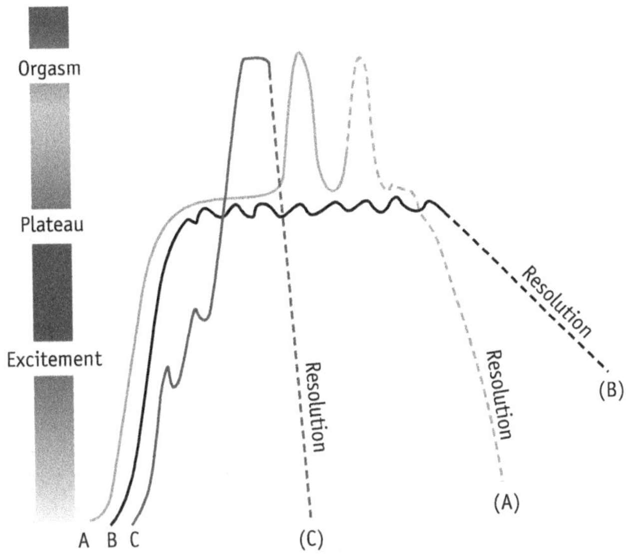

FIGURE 3.1. Masters and Johnson model of female sexual response cycle.
Source: Masters WH, Johnson VE. (1966). *The human sexual response.* Boston: Little, Brown.

This graph resulted from the data accumulated by observing and recording over 10,000 completed sexual response cycles between a heterosexual dyad. The female participants were of a small subset of the population: They were willing to be observed in a laboratory setting, had multiple sexual partners, and were orgasmic with vaginal intercourse. Therefore, this model is not readily applicable to most women. Women can often relate to this model during the beginning of a new relationship or after a prolonged absence from their sexual partner, but, in general, it does not apply to women in monogamous, long-term relationships.

BASSON'S NONLINEAR MODEL OF FEMALE SEXUAL RESPONSE

(Figure 3.2) Basson's non-linear model of female sexual response, originally published in 2001, introduced two new concepts regarding female sexual response: (a) It is normal for women to experience "responsive desire," which is experienced *after* arousal, thereby entering into a sexual interface for reasons other than desire (i.e., from a position of sexual neutrality and/or willingness to participate) and (b) women's sexual function is nonlinear and complex, incorporating the

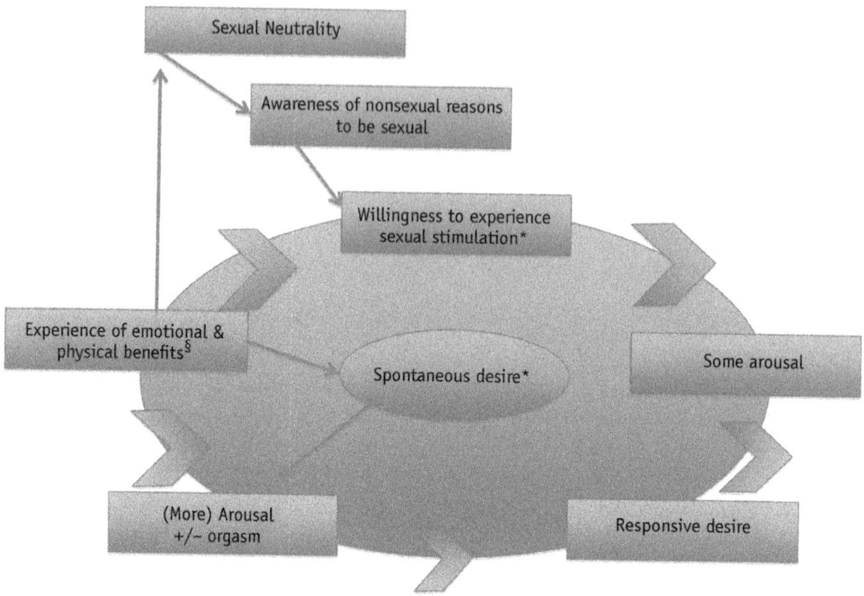

FIGURE 3.2. Basson's nonlinear model of female sexual response.
Nonsexual reasons to be sexual: Emotional and physical benefits (§)
* Willingness depends upon a number of items including trust, level of intimacy, ability to be vulnerable, respect, communication, affection, pleasure from sensual touch, and physical comfort.
§ Emotional and physical benefits of sex: desire to feel close, bonding, commitment, love, affection, acceptance, toleration, physical well-being, and well-being of partner.
**Components of spontaneous desire: sexual thoughts, sexual dreams, fantasies, need to self-stimulate.
Modified from Basson R. (2000). The female sexual response: A different model. *J Sex Marital Ther.* 26(1):5165.

importance of "trust, intimacy, the ability to be vulnerable, respect, communication, affection and pleasure from sensual touching" (Basson, 2000).

Basson mentions in her original paper a 1996 study by Cawood in which 140 women between the ages of 40 and 60 were randomly selected from a community to be interviewed about their sexual well-being and have blood drawn to measure steroid hormones once per week for five consecutive weeks. Of 50 premenopausal women in the cohort, only two reported sexual thoughts more than once per week, and 23 reported sexual thoughts less than once per month or never (Cawood, 1996).

Many women believe there is something wrong with their sexual abilities when spontaneous desire does not precede arousal. However, the reality is that most women lose their spontaneous sexual drive within months of entering into a new relationship. Therefore, a baseline of "responsive desire" should also be considered normal. Learning about this model can be tremendously reassuring to a woman and her sexual partner.

NEUROTRANSMITTERS AND HORMONES INVOLVED IN THE SEXUAL RESPONSE

Normal sexual function is dependent upon the effects of sex hormones and neurotransmitters on the central and peripheral nervous system (Table 3.1). It is helpful to keep this information in mind when considering the etiology and treatment of a given female patient's sexual dysfunction.

Table 3.1. Sex Hormones and Neurotransmitters Involved in Sexual Function

Sex Hormone/ Neurotransmitter	*Function Affected*	*Type of Effect*
Dopamine	Desire, arousal	Positive
Estrogen	Arousal, desire	Positive
Nitric oxide	Vasocongestion of clitoris	Positive
Norepinephrine	Arousal	Positive
Oxytocin	Receptivity, orgasm	Positive
Progesterone	Receptivity	Positive
Prolactin	Arousal	Negative
Serotonin	Arousal, desire	Positive and negative
Testosterone	Desire, initiation of sexual activity	Positive
Vasoactive Intestinal Peptide	Vasocongestion of clitoral tissue	Positive

Modified from Frank JE, Mistretta P, Will J (2008). Diagnosis and treatment of sexual dysfunction. *Am Fam Physician.* 77(5):635–42.

More on Female Sexual Dysfunction

PREVALENCE

The PRESIDE (Prevalence of Female Sexual Problems Associated with Distress and Determinants of Treatment Seeking) study published in 2008 reported results of 31,851 completed surveys of women aged 18 years and older from 50,002 households in the United States. The prevalence of any sexual problem was 44.2%, whereas 12% of women reported a *distressing* sexual problem. Middle-aged (35–64 years) women were most likely to report a distressing sexual dysfunction with a prevalence of 14.8%. The most prevalent distressing sexual dysfunction in women across all ages was decreased sexual desire (10%; Shifren, 2008). The PRESIDE results are similar to those reported by Leiblum et al. in the Women's International of Health and Sexuality study (Leiblum, 2006).

PATHOPHYSIOLOGY

The pathophysiology of FSD is usually multifactorial and is best understood when approached from a biopsychosocial perspective (Table 3.2). Biological factors can include underlying normal aging, medical conditions, and medications (Table 3.3). Psychological factors include psychiatric conditions, childhood and/or adolescent maltreatment, current stressors, and body image. Social factors include relationship status and satisfaction and one's upbringing (Leiblum, 2006; Shifren, 2008; West, 2008).

Subcategories of FSD

FEMALE SEXUAL INTEREST OR AROUSAL DISORDER: HSDD AND FAD

Hypoactive Sexual Desire Disorder

The prevalence of HSDD is 6.6% to 12.5% in US women; results vary among epidemiologic studies due to methods, age of participants, and definition of the disorder. HSDD is twice as common in the postmenopausal as compared to the premenopausal period (West, 2008); the highest prevalence was found in women who underwent surgical menopause (Leiblum, 2006; West, 2008). It was found that HSDD impairs health-related quality of life (Biddle, 2009). Distressing HSDD has been associated with aging, postmenopausal status, currently being involved

Table 3.2. Common Conditions Associated with Female Sexual Dysfunction

Cardiovascular/Hematologic: Anemia, hypertension, metabolic syndrome, post-myocardial infarction, vascular insufficiency
Chronic Pain: Chronic pelvic pain, endometriosis, fibromyalgia, hip pain, low back pain, pelvic floor dysfunction
Endocrinologic/Hormonal: Diabetes mellitus, hyperprolactinemia, hypoandrogenism, lactation, peri-menopause, polycystic ovarian syndrome, postpartum, post-menopause, thyroid dysfunction
Gastrointestinal: Constipation, irritable bowel syndrome
Neurologic: Aging, disorders of peripheral and central nervous system, spinal cord injury
Post surgical: colorectal surgeries, gynecologic surgeries (for cancer and non-cancerous pathologies), surgical menopause
Psychological: Anxiety, chronic stress, decreased self-esteem, depression, fear of getting pregnant, history of childhood physical and/or sexual abuse, history of sexual assault at any age, personality disorders, poor body image, PTSD, substance dependency.
Social: Day-to-day stressors, financial stressors, long-term relationships, partner's sexual dysfunction, relationship difficulties, single status, social morays from upbringing
Urogenital: Cervical dysplasia, cervicitis, fecal incontinence, fibroids, history of traumatic childbirth, interstitial cystitis, lichen sclerosus, pelvic floor dysfunction, pelvic organ prolapse, recurrent urinary tract infections, sexually transmitted infections, urinary incontinence, vaginal dryness, vaginismus, vaginitis, vestibulitis, vulvo-vaginal atrophy
Other Biologic: Arthritis, asthma, chronic fatigue, obstructive sleep apnea, sleep dysfunction

Compiled from: Attaallah (2014); Ayers (2006); Bazzichi (2012); Benetti-Pinto (2015); Blazquez (2015); Frank (2008); Hamilton (2013); Kalmbach (2015); Kingsberg (2015); Leiblum (2006); Liu (2015); Ma (2015); Otunctemur (2015); Pauls (2005); Shifren (2008); Tran (2015); West (2008).

in a long-term relationship or no relationship at all, and worsening self-assessed health (Leiblum, 2006; Shifren, 2008).

Similar to FSD, the pathophysiology of HSDD is usually complex, multifactorial, and variable from one woman to the next. HSDD diagnosis and treatment should be approached from a biopsychosocial perspective. (See Table 3.3.) It is particularly pertinent to address perceived "normal" sexual function and social context with women suffering from HSDD. "Cultural, social, and religious values and mores can negatively influence women's sexual desire, especially in women raised

Table 3.3. Medications Associated with Female Sexual Dysfunction

Cardiovascular	ACE inhibitors, amiodarone, beta-blockers, calcium channel blockers, clonidine, digoxin, diuretics, lipid-lowering drugs
Drugs of abuse	Alcohol, amphetamines, cocaine, heroin, marijuana, tobacco
Hormonal mediators	Anti-androgens, anti-estrogens, estrogens, GnRH agonists, oral contraceptive pills, flexible combined contraceptive vaginal ring
Neuroleptics	Carbamazepine, phenytoin, primidone
Pain relievers	NSAIDs, opioids
Psychotropics	Antipsychotics, anxiolytics, SNRIs, SSRIs, tricyclic antidepressants
Others	Antihistamines, chemotherapeutic agents, metformin

Modified from Kingsberg SA, Woodard T (2015). Female sexual dysfunction: focus on low desire. *Obstet Gynecol.*125:477–486.

in highly restrictive cultures or religions" (Kingsberg, 2015a, b). As Christiane Northrup, MD, writes in *Women's Bodies, Women's Wisdom*, "Most Western religions seldom perceive female sexuality and motherhood as component parts of the same whole. (They) have a very long history of separating motherhood and sexuality, resulting in a virgin/whore split in our psyches that has caused much distress for many women" (Northrup, 2006).

Arousal Disorder with Focus on Objective Form (Vaginal Dryness)

The DSM-IV defined FAD as "Persistently or recurrently deficient sexual fantasies and desire for sexual activity that causes marked distress or interpersonal difficulty" (APA, 2000). FAD has been categorized as subjective, objective, or both. Subjective FAD is described as the lack of feeling aroused by any type of sexual stimulation; however, physiologic signs of arousal are present such as genital vasocongestion and vaginal lubrication. Objective FAD is described as a lack of a physiologic response to sexual stimulation albeit feeling aroused. One objective sign of arousal is a lack of vaginal lubrication in response to sexual stimuli.

Vaginal dryness is the most common urogenital symptom associated with menopause, followed by irritation/itching, discharge, and dysuria. These symptoms are usually the result of vaginal atrophy, which is caused by reduced transudation through the vaginal epithelium and reduced cervical gland secretions resulting from postmenopausal sex hormone depletion. Vaginal atrophy can occur early in the menopausal transition but generally occurs four to five years after menopause and progressively increases in prevalence and severity in the subsequent years (Kokot-Kierepa, 2012).

Treatments for vaginal dryness include vaginal lubricants (used with each episode of vaginal penetration), vaginal moisturizers (used prophylactically every three days), pelvic floor toning, and hormonal therapy (oral and/or topical). The most recent Food and Drug Administration (FDA)-approved addition has been ospemifene, a selective estrogen receptor modulator, which acts similarly to estrogen on the vaginal epithelium, increasing vaginal wall thickness, thereby reducing pain associated with dyspareunia.

Female Orgasmic Disorder

Female orgasmic disorder is the second most commonly reported FSD. Orgasm is a sensation of intense pleasure creating an altered consciousness state accompanied by pelvic striated circumvaginal musculature and uterine/anal contractions and myotonia that resolves sexually induced vasocongestion and induces well-being/contentment (Meston, 2004a, b). Prevalence of female orgasmic disorder ranges from 4% to 42% depending on the study (Frank, 2008). Orgasmic disorders can be classified as primary (never experienced), secondary (currently unable to experience), situational, or diminishing in frequency/intensity.

Many women are under the impression they have an orgasmic disorder if they are unable to achieve a vaginal orgasm by stimulation of the controversial G-spot (located 2 to 3 cm proximal to the anterior vaginal introitus) during intercourse. Reassurance can be given that, dependent upon the study cited, only 15% to 30% of women are able to consistently achieve a vaginal orgasm. Most women achieve orgasm through clitoral stimulation (Buisson, 2013).

Current recommended treatments for female orgasmic disorder include cognitive-behavioral therapy (CBT), sexual therapy, resolution of underlying sexual pain disorders, cessation or reduction of disruptive medications, and attention to underlying vascular or neurologic conditions that may be the cause of impaired orgasm. Off-label use of sildenafil citrate and androgen therapy have also been found to be helpful (Akbarzadeh, 2014; Fernandes, 2014).

Genitopelvic Pain/Insertion Disorder

Dyspareunia

Dyspareunia is defined as persistent or recurrent pain with attempted or completed vaginal entry and/or penile-vaginal intercourse. Estimated prevalence ranges from 3% to 46% (Frank, 2008). Etiology can be due to a single factor but most often is multifactorial in nature. The most common causes of dyspareunia include external genital hypersensitivity, vulvovaginal atrophy, pelvic floor hypertonicity, pelvic organ prolapse, anatomic positioning of pelvic organs, infection, systemic pain disorders, vasculogenic changes, and sex hormone deficiencies.

Treatments for dyspareunia include insuring adequate vaginal lubrication (with vaginal lubricants/moisturizers, natural oils, adequate foreplay, hormonal therapy), treatment of infections, pelvic floor physical therapy, muscular relaxation, various antidepressant/antipain medications, and mindfulness. This topic is covered more extensively elsewhere in this text.

Vaginismus

Vaginismus is defined as recurrent or persistent involuntary spasms of the musculature of the outer one-third of the vagina that interferes with vaginal penetration and causes personal distress. Estimated prevalence ranges from less than 1% to up to 30% (Frank, 2008). Etiology includes ongoing dyspareunia due to inadequate lubrication, vulvovaginal atrophy, infections, endometriosis, vestibulitis, cystitis, history of sexual trauma, and history of sexual abuse. Treatment generally consists of treating the underlying etiology, pelvic floor physical therapy with or without utilization of sequential vaginal dilators, and sexual therapy.

Evaluation of FSD

SCREENING

Many screening tools are available to assess for FSD. The Female Sexual Function Index 6 (FSFI-6) can be used for rapid screening. This tool is self-administered, has been validated in a number of languages, and is best used in the primary care setting to determine if additional conversation regarding sexual dysfunction may be of benefit for the patient (Isidori, 2010).

The full-length FSFI 19 measures sexual functioning in women. It was developed as a research tool with the purpose of assessing specific domains of sexual functioning. It is not a diagnostic tool and should be used in conjunction with a complete sexual history (Rosen, 2000).

The Female Sexual Distress Scale (FSDS) is a validated clinical tool to assess distress associated with impaired sexual function (Derogatis, 2002). The FSDS–Revised has been validated to identify distress in women with HSDD (Derogatis, 2007; 2008).

HISTORY

Prior to gathering a patient's history, normalization of sexual concerns can be very reassuring, especially for women who feel uncomfortable discussing the topic. A detailed medical, surgical, gynecological, social (including tobacco, alcohol, and recreation drug use), and sexual history should be acquired. In addition to the patient's specific concerns, important aspects of the sexual history include sexual orientation, availability of current partner(s), past sexual experiences, beliefs and values about sexuality learned from family of origin, expressed impact of sexual dysfunction on personal and relational quality of life, and history of sexual trauma.

PHYSICAL EXAM

Clinical evaluation is dictated by the history. In general, a thorough physical exam including thyroid, breast, and genital examination is recommended. The genital exam is particularly helpful when sexual pain disorders are part of the women's concerns. Assessment of external genitalia includes the general appearance, color, presence of skin conditions, sensation, muscular tone, tenderness, signs of atrophy, integrity of the perineum, and presence of hemorrhoids. The vulvar vestibule should be assessed for erythema and tenderness. A speculum examination is important to view the status of the vaginal discharge, mucosa, and cervix and to obtain swabs to assess for vaginal pH and infectious etiology. In addition, single-bladed speculum examination is important to assess for prolapse. A thorough bimanual examination assesses tenderness, tonicity, pelvic organ location, rectovaginal exam, surrounding pelvic structures, and involvement of the anterior abdominal wall.

LABORATORY EVALUATION

The author has utilized a standardized laboratory panel for most patients presenting with sexual dysfunction. In postmenopausal patients the laboratory evaluation includes FSH, LH, CBC, AST, ALT, CRP, TSH, free T3, free T4, fasting lipids, HgbA1C, estradiol, free testosterone, total testosterone, DHEA-S, and prolactin. Women currently pre- and perimenopausal have all of the aforementioned labs drawn on the 21st day of the menstrual cycle except for FSH and LH, which are drawn on the third day.

IMAGING STUDIES

Ultrasound evaluation of pelvic organs is advised if pain, postcoital bleeding, abnormal menstrual bleeding, postmenopausal bleeding, excessive vaginal discharge, or hyperandrogenism are part of the history. Some providers will obtain an annual pelvic ultrasound for uterine assessment if initiating sex hormone therapy to monitor the endometrial lining for overgrowth. Annual breast imaging with mammography is recommended for all women considering or currently utilizing systemic hormonal therapy.

Treatment of FSD

LARGE PLACEBO EFFECT CONFOUNDING WOMEN'S SEXUALITY STUDIES

Following are the most common treatment modalities recommended for FSD. This area of science is relative new, and, for many treatments, robust research is lacking. In addition, the placebo effect of 30% (and at times greater than 40%) for various therapies clouds the true efficacy of the modalities or substances in question (NAMS, 2015). "This suggests that just paying greater attention to sexuality—as women do when enrolling in a clinical study, or when couples jointly seek counseling—can have positive effects even when no active treatment is used" (Bradford, 2007).

EDUCATION

Readily available, nonoffensive reading materials in the office space encourage women to ask questions during their medical office visit or to take them home and review with their partners. Opening the dialog is the initial step toward meaningful discussion of sexual health between patients and their healthcare provider.

Once in the private exam room, initial conversations about a patient's sexual function should include descriptions of what "normal" sexual function constitutes, setting realistic goals and dispelling myths. Introduction of the Basson's biopsychosocial model and discussion of "normal" variability between women and during different phases of the same woman's sexual life can be very reassuring.

LIFESTYLE MODIFICATION

Nutrition

Although no specific dietary regimen has been found the most beneficial, a well-balanced diet, which promotes overall health and weight normalization, may optimize all aspects of health including sexual function. Certain foods such as figs, bananas, avocadoes, and chocolate have been promoted as aphrodisiacs. There is no evidence to support these claims. However, all of these foods may, in a limited quantity, contribute to overall health. Alternatively, red wine consumption of one to two glasses daily has been associated with higher FSFI scores (i.e., better sexual function) concerning sexual desire, lubrication, and overall sexual function (Mondaini, 2009).

Regular Exercise

Exercise is an important part of overall well-being and can therefore contribute to sexual functioning. Intense exercise just prior to sexual activity may improve vaginal lubrication but has not been associated with increased sexual drive (Lorenz, 2012). Pelvic floor exercises, such as Kegel exercises, do improve vaginal sensations. The strength of the pelvic muscles has been positively correlated with a woman's orgasmic response (Graber, 1979).

Stress Reduction

Animal models have demonstrated a reciprocal relationship between the hypothalamic-pituitary-adrenal (HPA) axis and the hypothalamic-pituitary-gonadal axis. Estrogen and testosterone modulate the response of the HPA axis, whereas the activation of recurrent or chronic activation of the stress axis inhibits estrogen and testosterone secretion, thereby altering sexual response and emotional well-being (Toufexis, 2014).

Other

Sleep improvement, decreasing boredom, and tobacco cessation are also associated with improvement of sexual drive (Kingsberg, 2011).

PSYCHOLOGICAL THERAPY

Psychological interventions for FSD include traditional sexual therapy (TST) and CBT. Although there remains a dearth of clinical trials for psychological interventions to treat FSD, the available literature demonstrates an improvement of sexual function when a woman participates in TST, CBT, or both. Psychological therapies are most effective when low libido is acquired, situational, or both.

Traditional Sex Therapy

TST is a behavioral therapy that is meant to improve an individual or couple's sexual experiences while reducing self-consciousness and anxiety about erotic activity (Kingsberg, 2015a, b). TST has been found to be effective for treatment of female orgasmic disorder, sexual aversion disorder, and vaginismus. There is not adequate evidence to support sexual therapy for HSDD, unless the loss of desire is mainly due to relationship, emotional, or psychological factors. Sexual therapists have numerous techniques to employ with sexually distressed individuals or couples including sensate focus (Table 3.4), modifying perceptions of sexual performance toward one of curiosity instead of goal orientation, introduction of erotica, other bibliotherapy, and instructions in directed masturbation.

Table 3.4. Steps of Sensate Focus

Step 1	Each partner spends about 30 minutes caressing the other's naked body, head to toe, and front and back, excluding the genitals and breasts. The emphasis is on the "giver" touching in a way that he or she enjoys. The focus is on the sensation of touch and the recipient's reaction to it.
Step 2	The couple repeats the exercise from Stage 1, except the "receiver" gives direction on what kind of touching he or she wants and what is most enjoyable.
Step 3	The couple expands the Step 2 exercise to include touching of the genitals and breasts. The focus remains upon sensations and expressing wants and pleasures without the goal of orgasm.
Step 4	The couple now may engage in mutual touching and stimulation to the point of orgasm, including intercourse if desired.

Modified from North American Menopause Society website: http://www.menopause.org/for-women/sexual-health-menopause-online/effective-treatments-for-sexual-problems/sex-therapy-and-counseling

Directed Masturbation

LoPiccolo and Lobitz introduced directed masturbation for treating female primary orgasmic dysfunction in 1971 (LoPiccolo, 1971). It is a nine-step program consisting of education (individual, couple, or group therapy) followed by self-exploration and body awareness, masturbation, and sensate focus exercises completed in the privacy of their own homes. Numerous studies have demonstrated high success rates of directed mastrubation in women with primary anorgasmia with 20% to 60% able to experience orgasm with their partners and up to 80% being able to experience orgasm with masturbation (Heiman, 2007).CBT

CBT focuses on the "here and now" and works toward modifying beliefs or thought patterns that may interfere with intimacy and sexual pleasure, which may include developing positive personal attributes, assertion training, and improvement of women's body-image dissatisfaction (Butters, 1987; Kingsberg, 2015a, b).

SUPPLEMENTS

Iron for FSD

Iron deficiency is a common micronutrient deficiency throughout the world. While results are mixed, research suggests that there is a relationship between low iron status and mood. (Lomagno, 2014; Shariatpanaahi, 2007) Iron deficiency with anemia (IDA) contributes to anxiety, which is a major component of FSD. One 2014 pilot study including 207 reproductive-age women treated IDA to correction. The Beck Anxiety Inventory and FSFI scores were statistically improved in the treatment cohort, whereas quality of life scores were not. Additional studies are needed (Gulmez, 2014).

S-Adenosyl Methionine as an Alternative Therapy for Depression

S-adenosyl methionine in some studies has been effective in treating mild to moderate depression without the unwanted sexual side effects of selective serotonin reuptake inhibitors (SSRIs) and serotonin norepinephrine reuptake inhibitors (SNRIs). Additional studies are needed regarding its efficacy in treating depression.

Zinc for FSD

Serum zinc levels are positively correlated to testosterone production in men (Khedun 1995; Prasad 1996). Women's sexual function is also reliant upon testosterone levels. Although no research has yet been conducted on zinc supplementation and FSD, zinc-deficient diets contribute to depressive symptoms in women (Lomagno, 2014; Maserejian, 2012). Zinc is effective as an adjunct to antidepressant treatment for depression in premenopausal women (Nowak, 2003; Siwek, 2009). (It has not yet been studied as a monotherapy for treatment of depression or sexual dysfunction.)

BOTANICALS

Ashwagandha for FSD

A small randomized, double-blind placebo controlled pilot study of high concentrated ashwagandha root extract (HCARE) 200mg twice daily for eight weeks in premenopausal women with FSD were randomized into two groups: 25 women received counseling + HCARE, and 25 counseling + placebo. The treatment arm demonstrated statistical improvement in FSFI total score, arousal, vaginal lubrication, orgasm, sexual satisfaction and reduction in FSDS but no change in desire (Dongre, 2015).

Ginkgo Biloba for Female Sexual Arousal Disorder

Ninety-nine women with sexual arousal disorder (33 on antidepressants) were recruited through a radio announcement for a double-blind, placebo-controlled eight-week trial. They were randomly assigned to ginkgo biloba extract (GBE) 300mg/day, sex therapy (with focus on instruction of genital sensations), GBE + sex therapy, or placebo. GBE + sex therapy improved desire and contentment; therapy alone enhanced orgasm (Meston, 2008).

Isoflavones for Vaginal Dryness

- Red clover (*Trifolium pratense*): Multiple studies have demonstrated the benefits of red clover on menopausal symptoms, including vasomotor symptoms and vaginal dryness. In one randomized double-blind placebo-controlled trial, 80mg/day for 90 days had a positive effect on vaginal cytology. The authors did not mention the subjective effect on vaginal dryness (Hidalgo, 2005). In a more recent systematic review, both vaginal dryness and cytology maturation showed a significant improvement with an 80mg dose of red clover daily (Ghazanfarpour, 2015).

- Soy-derived isoflavones: In one randomized, double blind, placebo-controlled, washout trial, 114 mg of soy-derived isoflavones taken daily did not improve vaginal dryness or alter vaginal cytology after three months of therapy in 64 postmenopausal breast cancer survivors. In addition, no changes were noted in the endometrium (Nikander, 2005).
- Soy-enriched diet: 36 Thai peri- and postmenopausal women participated in a randomized cross-over study for 12 week period. Consuming a soy-enriched diet containing at least 50mg of soy isoflavones did not relieve urogenital symptoms, restore vaginal epithelium, or "improve the vaginal health" in perimenopausal and postmenopausal Thai women (Manonai, 2006).

*Maca Root (*Lepidium meyenii*) for Antidepressant-Induced FSD*

- In one randomized, double-blind dose-finding pilot study, 10 women in remission from depression but suffering from SSRI-induced sexual dysfunction received either 3 grams or 1.5 grams of maca root daily. Women receiving the higher dose demonstrated improved sexual function scores, including enhanced libido (Dording, 2008).
- A subsequent study by the same author of 45 patients with SSRI/SNRI-induced sexual dysfunction treated with 3 grams of maca root daily for 12 weeks showed the women experienced significant improvement in medication-induced sexual dysfunction (Dording, 2015).
- A systematic review identified limited evidence of maca in improving sexual function, stating that the total number of trials, sample size, and methodological quality of available studies was too limited to draw firm conclusions (Shin, 2010).

*Rhodiola (*Rhodiola Rosea*) for Decreased Libido*

Rhodiola, an adrenal adaptogen, is usually used in combination with other herbs in traditional Chinese medicine to improve female libido. Definitive studies are lacking.

Russian Olive (Elaeagnus Angustifolia) for Female Orgasmic Disorder

E. angustifolia flower extract (4.5 grams divided into two doses for 35 consecutive days), when compared to sildenafil and placebo, improved female orgasmic disorder better than placebo but not as well as sildenafil (Akbarzadeh, 2014).

*Saffron (*Crocus Sativus L.*) for Antidepressant-Induced FSD*

Saffron, a spice derived from the flower of *Crocus sativus*, has been studied as a treatment for depression.

- A meta-analysis including six small studies suggests that saffron has a similar effect on mild to moderate depression compared to standard antidepressant therapy. The antidepressant effects of saffron are potentially due to its neuroendocrine, serotonergic, antioxidant, and anti-inflammatory effects (Lopresti, 2014).
- One randomized, double-blind, placebo-controlled trial recruited 38 women with major depressive disorder who were adequately treated with fluoxetine 40mg/day for a minimum of six weeks and experienced subjective, drug-induced sexual dysfunction. Participants were randomly assigned to saffron 30mg per day or placebo for four weeks (while continuing fluoxetine therapy). The FSFI was administered at baseline and at two and four weeks. The saffron group had significant improvement in arousal, vaginal lubrication, and pain but no improvement in desire, satisfaction, or orgasm. No side effects were noted when comparing the treatment to control group (Kashani, 2013).

Tribulus Terrestres *for FSD*

Tribulus terrestres, a spice of Iranian origin, has been known as an aphrodisiac for centuries. A 2014 double-blind, placebo-controlled trial involving 60 Iranian women with HSDD were randomized to take either placebo or 7.5 mg twice daily of tribulus for four weeks. The FSFI was administered prior to initiation of medications and after four weeks of treatment. The treatment group experienced significant improvement in desire, arousal, lubrication, sexual satisfaction, and pain (Akhtari, 2014).

Yohimbine for Female Sexual Arousal Disorder

Yohimbine is an indole alkaloid derived from the bark of the *Pausinystalia yohimbe* tree in Africa traditionally used in male erectile dysfunction in men due to its vasodilatory effects. Evidence is lacking for use of yohimbine in FSD.

MIND-BODY MEDICINE

Hypnotherapy for FSD

Although hypnotherapy is commonly used in the treatment of sexual dysfunction, there is little evidence to support its efficacy (Almas, 2010). A study published by Sotile in 1977 evaluated four studies of hypnotherapy in treating psychogenic FSD. Three of the studies reported successful outcomes (Sotile, 1977). Since then, very little additional research has been published. Additional information is needed to make recommendations regarding hypnosis for FSD.

Mindfulness for FSD

- Ninety-five women participated in four 90-minute group sessions that included mindfulness meditation, cognitive therapy, and education. Treatment improved sexual desire, arousal, lubrication, satisfaction, and overall sexual function. Distress significantly decreased (Brotto, 2014).
- In another study, 30 women were randomized to participate in either a 12-week meditation laboratory or control meetings without meditation instruction. Meditation improved self-awareness of responses to sexual stimuli, attention, self-judgment, and symptoms of anxiety and depression, concluding that mindfulness-based improvements in interoceptive awareness highlight the potential of mindfulness training as a treatment of FSD (Silverstein, 2011).

Yoga for Female Sexual Arousal Disorder

In one study a total of 41 women age 30 to 60 with metabolic syndrome were randomized to either 12-week yoga intervention (n = 20) versus being wait listed (n = 21). The treatment group participants experienced a significant improvement in arousal and lubrication compared with the control group (Kim, 2013).

MANUAL MEDICINE

Pelvic Floor Physical Therapy

Constipation, urinary incontinence, and pelvic floor dysfunction are effectively treated with pelvic floor physical therapy. Therapeutic approaches include trigger point massage, myofascial release, joint mobilization, strain-counterstrain, biofeedback, education about lifestyle modifications, and home-based exercises. Treatment depends upon presenting symptoms and usually includes strategies to optimize lumbopelvic and spinal function to improve bladder, bowel, and sexual function (Faubion, 2012).

Urinary incontinence is known to negatively impact sexual function. In one secondary analysis of a randomized trial it was demonstrated that the 18 women with sexual dysfunction and urinary incontinence who participated in three physical therapy sessions that included pelvic floor muscle exercises for the treatment of female urinary incontinence had statistically improved sexual function; predictors of improvement in sexual function included higher adherence to pelvic floor muscle exercises, improvement in the strength of pelvic floor muscles, and a decrease in urine leakage (Sacomori, 2015).

Physiotherapy

Ninety Iranian patients between the ages of 25 and 55 with mild pelvic organ prolapse and sexual dysfunction in a urogynecology clinic were randomized to surgery (rectocele repair and perineorrhaphy) versus twice-weekly physiotherapy (electrical stimulation and Kegel exercises) for eight weeks. Eight weeks after the intervention both groups experienced statistical improvement in libido and arousal; only the physiotherapy group experienced improvement in orgasm and dyspareunia (Eftekhar, 2014).

MEDICATIONS: HORMONE THERAPY

Vaginal Estrogen

Vaginal estrogen restores the pH, cytology, and vascularity of the vagina resulting in resolution of vaginal dryness and associated dyspareunia. Various vaginal preparations include conjugated equine estrogen and estradiol vaginal creams, a sustained-release estradiol ring, and low-dose estradiol tablets.

Systemic Estrogen With or Without Progestogens

A 2013 Cochrane review indicates that estrogen alone or with progestogens does have a small to moderate effect on sexual dysfunction, particularly in dyspareunia when used in women with menopausal symptoms or who are in early menopause (within five years of amenorrhea; Nastri, 2013). Systemic estrogen levels below 50 pg/mL alter vaginal lubrication and sensitivity (Basson, 2002).

Estrogen + Testosterone

Topical application of estrogen + testosterone cream in conjunction with cessation of combined hormonal contraceptives (CHC) is effective for treatment of CHC-provoked vestibulodynia (Burrows, 2013).

Testosterone

In general, serum testosterone levels in women do not correlate with sexual function/dysfunction. However, in hypoandrogenic women (total testosterone < 20 ng/dL), supplemental testosterone has been shown to improve sexual function.

- Postmenopausal: Numerous randomized studies have verified the benefit of either topical or oral testosterone supplementation for surgically postmenopausal women who suffer from HSDD (Poels, 2013; van Rooij, 2013). A 2005 meta-analysis of 35 trials lasting 1.5 to 24 months and including 4,768 women demonstrated that supplemental testosterone improved sexual function scores and increased the number of sexually satisfying encounters. Significant side effects included reduction in high-density lipoprotein, facial hair growth, and acne (Palacios, 2011).
- Vaginal application of testosterone has been shown to increase sexual desire, lubrication, sexual satisfaction, and orgasm and reduce dyspareunia (Fernandes, 2014).
- Premenopausal testosterone supplementation is controversial.

Dehydroepiandrosterone

Dehydroepiandrosterone (DHEA) and it sulfate form (DHEAS) are the most abundant circulating sex hormones in women. DHEA is produced by the ovary and adrenal glands and declines with age. It is widely speculated that the decline of such androgens may contribute to decreased libido, loss of well-being, and deterioration in cognition.

- The available literature on DHEA for the treatment of low libido in women is limited by studies of inadequate sample size and short treatment duration. The potential value of DHEA requires additional exploration in adequately powered studies (Panjari, 2007).

- However, topical DHEA for correction of vaginal dryness and atrophy is well documented. One recent study of 521 postmenopausal women received daily intravaginal administration of 0.5% (6.5 mg) DHEA in an open label, phase III study for 12 weeks to 52 weeks. Statistical improvement in dyspareunia, vaginal dryness, and irritation/itching were reported compared to baseline ($p < 0.0001$). Beneficial effects were also observed on gynecologic examination of vaginal secretions, color, epithelial surface thickness, and epithelial integrity. ($p < 0.0001$) without increase in serum levels of steroids or development of endometrial hyperplasia (Archer, 2015; Labrie, 2015). DHEA loses efficacy if dosed twice weekly (Bouchard, 2015). Women may find it unpalatable to adhere to daily dosing with a cream preparation (Panjari, 2011). Using a trusted brand of an over-the-counter DHEA supplement of a 5 mg tablet moistened and placed in the vagina may also be considered. The dose can be titrated by the patient given her comfort and sexual needs.

Ospemifene

Ospemifene, a selective estrogen receptor modulator, is FDA approved to treat dyspareunia secondary to genital atrophy in postmenopausal women. A dose of 60mg daily has been found to be more effective than placebo, to be generally safe, and to have minimal side effects (Cui, 2014). Common side effects include vasomotor symptoms, muscle spasms, and vaginal discharge. Rare but serious side effects include increased risk of endometrial cancer, stroke, and blood clot formation.

OTHER MEDICATIONS

Bupropion

A 2013 Cochrane review included randomized controlled trials that compared management strategies for antidepressant-induced sexual dysfunction versus placebo or any alternative strategy. Three studies indicate that 150mg twice daily of bupropion demonstrates a benefit for SSRI/SNRI-induced sexual dysfunction in women. Two studies of bupropion 150mg daily demonstrated no statistical significance (Taylor, 2013).

Diazepam

Multiple small studies have demonstrated efficacy of intravaginal diazepam as an adjunctive therapy with pelvic floor physical therapy for treating high-tone pelvic floor dysfunction and urogenital pain. Additional studies are needed to identify the best form of delivery, dose, and optimal duration of treatment (Carrico, 2011; Rogalski, 2010).

Flibanserin

Flibanserin is a new FDA-approved drug intended to treat sexual interest/arousal disorder in premenopausal women. Flibanserin is a 5-HT1A agonist and 5-HT2A, 2B, and 2C antagonist. One randomized placebo-controlled trial lasting 24 weeks demonstrated improvement in sexual desire, number of satisfying events, and reduced sexual distress of 468 in the treatment arm when compared to 481 women in the placebo arm. Flibanserin was dosed at 100mg nightly. The most frequent adverse events were dizziness, nausea, somnolence, and headache. Although flibanserin has been FDA approved for premenopausal women with HSDD, it has also been found to be effective and safe for postmenopausal women (Simon, 2014).

Lidocaine

In one small trial (n = 61), lidocaine ointment 5% was applied nightly to the vestibule for seven week with statistical reduction in pain during intercourse and increased number of women "able to have intercourse" (Zolnoun, 2003).

Methylphenidate

A small cross-over study (N = 30) compared 40mg methylphenidate (a dopamine psychostimulant) to 75 mg of 3,4-methylenedioxymethamphetamin (MDMA, also known as ecstasy, a serotonergic drug). Methylphenidate, unlike MDMA, resulted in increased sexual drive and arousal (Schmid, 2015).

Phosphodiesterase Inhibitors

Sildenafil did not improve sexual arousal disorder more effectively than placebo in a randomized trial of 577 estrogenized and 204 estrogen-deficient women

(Basson, 2002). Sildenafil does improve female orgasmic disorder. In one trial 125 women ranging in age from 18 to 40 were randomized to E. angustifolia 4.5 grams in divided doses twice daily for 35 days, sildenafil citrate 50mg (one for four weeks one hour before sexual relations), or control group. Women in both treatment groups realized improvement. Sildenafil citrate was more effective in reducing female orgasmic disorder (a reduction from 40.5% to 16.7%), whereas the E. angustifolia group reduced from 41.5% to 29.3% and the control group reduced from 57.1% to 50% ($p = 0.004$). Sildenafil has been shown to improve clitoral blood flow (Cavalcanti, 2008).

SURGICAL INTERVENTION

Surgical intervention may be helpful in ameliorating physiologic sexual pain disorders such as refractory vestibulodynia, recurrent fissures of the posterior fourchette, and reduction labioplasty. Surgical intervention for reduction of dyspareunia may be helpful. However, caution should be exercised when counseling women requesting genital cosmetic surgery (i.e., vaginal reconstruction) for sexual enhancement. There is no robust evidence supporting the effectiveness of female genital cosmetic surgery in the improvement of female sexual function (Barbara, 2015).

Sequelae of Cancer Treatment

It is very common for women with a history of cancer, especially cancers of the female organs, to experience sexual dysfunction. Many times women will feel as though their female body has betrayed them or is no longer whole. To optimally support women who have experienced cancer or are in remission, a biopsychosocial care plan is recommended.

For example, most breast cancer survivors who take anti-estrogen therapy will experience vaginal dryness and genital atrophy resulting in severe dyspareunia. If topical lidocaine, vaginal lubricants, moisturizers, vulvar hygiene principles, daily sitz baths, and water barrier application are not sufficient to restore tissue integrity, small doses of topical hormone therapy may be considered. Working closely with the patient and her oncologist to suggest topical application of testosterone, estradiol, or compounded estriol may be helpful.

Because of anticipated dyspareunia, pelvic floor dysfunction is common in female cancer survivors. Pelvic floor re-education and physical therapy may be very helpful. If depression, anxiety, or relationship disruption is a concern, referral to a behavioral health professional experienced in treating sexual dysfunction is recommended. Mindfulness-based cognitive-behavioral intervention has been successful in improving sexual desire and arousal problems in gynecologic cancer survivors (Brotto, 2012).

FOLLOW-UP

Jennifer, the professional florist introduced at the beginning of this chapter, is doing well. When she initially came to our sexual health clinic, she was very concerned about decreased libido, dyspareunia, and anorgasmia. Her FSFI score was 12 and FSDS-R score was 32. Her physical exam at presentation was significant for external genital atrophy, partial clitoral phimosis, labial agglutination, severe provoked vestibulodynia, vaginal atrophy, and severe pelvic floor hypertonicity. Working closely with Jennifer and her oncology team, we developed and implemented the following plan:

- Daily sitz baths in warm, plain water for 15 minutes
- Implementation of vulvar hygiene principles
- Couples sexual therapy, including mindfulness instruction
- 5% lidocaine ointment applied to the vaginal introitus nightly to address vestibulodynia
- Intravaginal diazepam 2 mg nightly for one month then as needed for pelvic floor hypertonicity
- Pelvic floor physical therapy and re-education
- Coconut oil for vaginal lubrication with vaginal intercourse
- Compounded estriol cream, a pea-sized amount applied to labia and clitoris to treat genital and clitoral atrophy
- Follow up every three to six months to monitor improvement of symptoms and sex hormone serum levels and to reduce or eliminate unneeded therapies

REFERENCES

Akbarzadeh M, Zeinalzadeh S, Zolghadri J, et al. (2014). Comparison of Elaeagnus angustifolia extract and sildenafil citrate on female orgasmic disorders: a randomized clinical trial. *J Reprod Infertil.* Oct;15(4):190–198.

Akhtari E, Raisi F, Keshavarz M, et al. (2014). Tribulus terrestris for treatment of sexual dysfunction in women: randomized double-blind placebo-controlled study. *Daru.* 22(1):40.

Almas E, Landmark B (2010). Non-pharmacological treatment of sexual problems—a review of research literature 1970–2008. *Sexologies.*19:202–211.

American Psychiatric Association (2000). *Diagnostic and statistical manual of mental disorders* (4th ed.). Washington, DC: American Psychiatric Association.

American Psychiatric Association (2013). *Diagnostic and statistical manual of mental disorders* (5th ed.). Washington, DC: American Psychiatric Association.

Archer DF, Labrie F, Bouchard C, et al. (2015). Treatment of pain at sexual activity (dyspareunia) with intravaginal dehydroepiandrosterone (prasterone). *Menopause.* Sep;22(9):950–963.

Attaallah W, Ertekin C, Tinay I, Yegen C (2014). High rate of sexual dysfunction following surgery for rectal cancer. *Ann Coloproctol.* Oct;30(5):210–215.

Ayers S, Eagle A, Waring H (2006). The effects of childbirth-related posttraumatic stress disorder on women and their relationships: a qualitative study. *Psychol Health Med.* Nov;11(4):389–398.

Barbara G, Facchin F, Meschia M, Vercellini P (2015). The first cut is the deepest: a psychological, sexological and gynecological perspective on female genital cosmetic surgery. *Acta Obstet Gynecol Scand.* Sep;94(9):915–920.

Basson R (2000). The female sexual response: a different model. *J Sex Marital Ther.* Jan–Mar;26(1):51–65.

Basson R, McInnes R, Smith MD, et al. (2002). Efficacy and safety of sildenafil citrate in women with sexual dysfunction associated with female sexual arousal disorder. *J Womens Health Gend Based Med.* May;11(4):367–377.

Bazzichi L, Giacomelli C, Rossi A, et al. (2012). Fibromyalgia and sexual problems. *Reumatismo.* Sep;64(4):261–267.

Benetti-Pinto CL, Ferreira SR, Antunes A, Yela DA (2015). The influence of body weight on sexual function and quality of life in women with polycystic ovary syndrome. *Arch Gynecol Obstet.* 291(2):451–455.

Biddle AK, West SL, D'Aloisio AA, et al. (2009). Hypoactive sexual desire disorder in postmenopausal women: quality of life and health burden. *Value Health.* 12:763–772.

Blazquez A, Ruiz E, Aliste L, et al. (2015). The effect of fatigue and fibromyalgia on sexual dysfunction in women with chronic fatigue syndrome. *J Sex Marital Ther.* 41(1):1–10.

Bouchard C, Labrie F, Archer DF, et al. (2015). Decreased efficacy of twice-weekly intravaginal dehydroepiandrosterone on vulvovaginal atrophy. *Climacteric.* 18(4)590–607.

Bradford A, Meston C (2007). Correlates of placebo response in the treatment of sexual dysfunction in women: a preliminary report. *J Sex Med.* Sep;4(5):1345–1351.

Brotto LA, Erskine Y, Carey M, et al. (2012). A brief mindfulness-based cognitive behavioral intervention improves sexual functioning versus wait-list control in women treated for gynecologic cancer. *Gynecol Oncol.* 125:320–325.

Brotto LA, Basson R (2014). Group mindfulness-based therapy significantly improves sexual desire in women. *Behav Res Ther.* Jun;57:43–54.

Buisson O, Jannini EA (2013). Pilot echographic study of the differences in clitoral involvement following clitoral or vaginal sexual stimulation. *J Sex Med.* Nov;10(11):2734–2740.

Burrows LJ, Goldstein AT (2013). The treatment of vestibulodynia with topical estradiol and testosterone. *J Sex Med.* 1:30–33.

Butters JW, Cash TF (1987). Cognitive-behavioral treatment of women's body-image dissatisfaction. *J Consult Clin Psychol.* Dec;55(6):889–897.

Carrico DJ, Peters KM (2011). Vaginal diazepam use with urogenital pain/pelvic floor dysfunction. *Urol Nurs.* 31(5):279–284.

Cavalcanti AL, Bagnoli VR, Fonseca AM, et al. (2008). Effect of sildenafil on clitoral blood flow and sexual response in postmenopausal women with orgasmic dysfunction. *Int J Gynaecol Obstet.* Aug;102(2):115–119.

Cawood EH, Bancroft J (1996). Steroid hormones, the menopause, sexuality and well-being of women. *Psychol Med.* Sep;26(5):925–936.

Cui Y, Zong H, Yan H, Zhang Y (2014). The efficacy and safety of ospemifene in treating dyspareunia associated with postmenopausal vulvar and vaginal atrophy: a systematic review and meta-analysis. *J Sex Med.* 11(2):487–497.

Derogatis LR, Rosen R, Leiblum S, et al. (2002). The Female Sexual Distress Scale (FSDS): initial validation of a standardized scale for assessment of sexually related personal distress in women. *J Sex Marital Ther.* Jul-Sep;28(4):317–330.

Derogatis LR, Burnett AL (2007). Key methodological issues in sexual medicine research. *J Sex Med.* May;4(3):527–537. Review.

Derogatis L, Clayton A, Lewis-D'Agostino D, et al. (2008). Validation of the Female Sexual Distress Scale–Revised for assessing distress in women with hypoactive sexual desire disorder. *J Sex Med.* 5(2):357–364.

Dongre S, Langade D, Bhattacharyya S (2015). Efficacy and safety of ashwagandha (Withania somnifera) root extract in improving sexual function in women: a pilot study. *Biomed Res Int.* 2015:1–9.

Dording CM, Fisher L, Papakostas G, et al. (2008). A double-blind, randomized, pilot dose-finding study of maca root (L. meyenii) for the management of SSRI-induced sexual dysfunction. *CNS Neurosci Ther.* 14(3):182–191.

Dording CM, Schettler PJ, Dalton ED, et al. (2015). A double-blind placebo-controlled trial of maca root as treatment for antidepressant-induced sexual dysfunction in women. *Evid Based Complement Altern Med.* 2015:949036.

Eftekhar T, Sohrabi M, Haghollahi F, et al. (2014). Comparison effect of physiotherapy with surgery on sexual function in patients with pelvic floor disorder: a randomized clinical trial. *Iran J Reprod Med.* 12(1):7–14.

Faubion SS, Shuster LT, Bharucha AE (2012). Recognition and management of nonrelaxing pelvic floor dysfunction. *Mayo Clin Proc.* Feb;87(2):187–193.

Fernandes T, Costa-Paiva LH, Pinto-Neto AM (2014). Efficacy of vaginally applied estrogen, testosterone, polyacrylic acid on sexual function in postmenopausal women: a randomized controlled trial. *J Sex Med.* May;11(5):1262–1270.

Frank JE, Mistretta P, Will J (2008). Diagnosis and treatment of female sexual dysfunction. *Am Fam Physician.* Mar 1;77(5):635–642.

Ghazanfarpour M, Sadeghi R, Roudsari RL, et al. (2015). Red clover for treatment of hot flashes and menopausal symptoms: a systematic review and meta-analysis. *J Obstet Gynaecol.* Oct 15:1–11.

Graber B, Kline-Graber G (1979). Female orgasm: role of the pubococcygeus. *J Clin Psych.* 30:34–39.

Gulmez H, Akin Y, Savas M, et al. (2014). Impact of iron supplementation on sexual dysfunction of women with iron deficiency anemia in a short term: a preliminary study. *J Sex Med.* Apr;11(4):1042–1046.

Hamilton LD, Meston CM (2013). Chronic stress and sexual function in women. *J Sex Med.* October;10(10):2443–2454.

Heiman JR (2007). Orgasmic disorders in women. In S.R. Leiblum (Ed.), *Principles and practices of sex therapy* (4th ed., pp. 84–123). New York: Guilford.

Hidalgo LA, Chedraui PA, Morocho N, Ross S, San Miguel G (2005). The effect of red clover isoflavones on menopausal symptoms, lipids and vaginal cytology in menopausal women: a randomized, double-blind, placebo-controlled study. *Gynecol Endocrinol.* Nov;21(5):257–264.

Isidori AM, Pozza C, Esposito K, et al. (2010). Development and validation of a 6-item version of the Female Sexual Function Index (FSFI) as a diagnostic tool for female sexual dysfunction. *J Sex Med.* Mar;7(3):1139–1146.

Kalmbach DA, Arnedt JT, Pillai V, Ciesla JA (2015). The impact of sleep on female sexual response and behavior: a pilot study. *J Sex Med.* 12(5):1221–1232.

Kashani L, Raisi F, Saroukhani S, et al. (2013). Saffron for treatment of fluoxetine-induced sexual dysfunction in women: randomized double-blind placebo-controlled study. *Hum Psychopharmacol.* Jan;28(1):54–60.

Khedun SM, Naicker T, Maharaj B (1995). Zinc, hydrochlorothiazide and sexual dysfunction. *Cent Afr J Med.* Oct;41(10):312–315.

Kim HN, Ryu J, Kim KS, Song SW (2013). Effects of yoga on sexual function in women with metabolic syndrome: a randomized controlled trial. *J Sex Med.* Nov;10(11):2741–2751.

Kingsberg SA, Althof SE (2011). Satisfying sexual events as outcome measures in clinical trial of female sexual dysfunction. *J Sex Med.* Dec;8(12):3262–3270.

Kingsberg SA, Woodard T (2015a). Female sexual dysfunction: focus on low desire. *Obstet Gynecol.* Feb;125(2):477–486.

Kingsberg SA, Clayton AH, Pfaus JG (2015b). The female sexual response: current models, neurobiological underpinnings and agents currently approved or under investigation for the treatment of Hypoactive Sexual Desire Disorder. *CNS Drugs.* Nov;29(11):915–933.

Kokot-Kierepa M, Bartuzi A, Kulik-Rechberger B, Rechberger T (2012). Local estrogen therapy—clinic implications—2012 update. *Ginekol Pol.* Oct;83(10):772–777.

Labrie F, Archer DF, Bouchard C, et al. (2015). Prasterone has parallel beneficial effects on the main symptoms of vulvovaginal atrophy: 52-week open-label study. *Maturitas.* May;81(1)46–56.

Laumann EO, Paik A, Rosen RC (1999). Sexual dysfunction in the United States: prevalence and predictors. *JAMA.* 281:537–544.

Leiblum SR, Koochaki PE, Rodenberg CA, et al. (2006). Hypoactive sexual desire disorder in postmenopausal women: US results from the Women's International of Health and Sexuality (WISHeS). *Menopause.* 13:46–56.

Liu L, Kang R, Zhao S, et al. (2015). Sexual dysfunction in patients with obstructive sleep apnea: a systematic review and meta-analysis. *J Sex Med.* 12(10):1992–2003.

Lomagno KA, Hu F, Riddell LJ, et al. (2014). Increasing iron and zinc in pre-menopausal women and its effects on mood and cognition: a systematic review. *Nutrients.* Nov; 6(11):5117–5141.

LoPiccolo J, Lobitz C (1971). The role of masturbation in the treatment of orgasmic dysfunction. Presented at meeting of Western Psychological Association. Portland. Ore.

Lopresti AL, Drummond PD (2014). Saffron (Crocus sativus) for depression: a systematic review of clinical studies and examination of underlying antidepressant mechanisms of action. *Hum Psychopharm: Clin Exp.* Nov;29(6):517–527.

Lorenz TA, Meston CM (2012). Acute exercise improves physical sexual arousal in women taking antidepressants. *Ann Behav Med.* Jun; 43(3):352–361.

Ma J, Kan Y, Zhang A, et al. (2015). Female sexual dysfunction in women with non-malignant cervical diseases: a study from an urban Chinese sample. *PLoS One.* Oct 16;10(10):e0141004.

Manonai J, Songchitsomboon S, Chanda K, et al. (2006). The effect of a soy-rich diet on urogenital atrophy: a randomized, cross-over trial. *Maturitas.* May 20;54(2):135–140.

Maserejian NN, Hall SA, McKinlay JB (2012). Low dietary or supplemental zinc is associated with depression symptoms among women, but not men, in a population-based epidemiological survey. *J Affect Disord.* Feb; 136(3):781–788.

Masters EH, Johnson VE (1966). *Human sexual response.* Boston: Little, Brown;

Mercer CH (2014). Sexual behaviour. *Medicine (Abingdon).* Jun; 42(6):291–293

Mercer CH, Tanton C, Prah P, Erens B, Sonnenberg P et al. (2013). Changes in sexual attitudes and lifestyles in Britain through the life course and over time: findings from the National Surveys of Sexual Attitudes and Lifestyles (Natsal). *Lancet.* Nov 30;382(9907):1781–1794. .

Meston CM, Hull E, Levin RJ, Sipski M (2004a). Disorders of orgasm in women. *J Sex Med.* Jul;1(1):66–68. Review.

Meston CM, Levin RJ, Sipski ML, Hull EM, Heiman JR (2004b). Women's orgasm. *Annu Rev Sex Res.* 15:173–257. Review.

Meston CM, Rellini AH, Telch MJ (2008). Short- and long-term effects of Ginkgo biloba extract on sexual dysfunction in women. *Arch Sex Behav.* Aug;37(4):530–547

Mondaini N, Cai T, Gontero P, et al. (2009). Regular moderate intake of red wine is linked to a better women's sexual health. *J Sex Med.* Oct;6(10):2772–2777.

Nastri CO, Lara LA, Ferriani RA, et al. (2013). Hormone therapy for sexual function in perimenopausal and postmenopausal women. *Cochrane Database Sys Rev.* 6: 20130605.

Nikander E, Rutanen EM, Nieminen P (2005). Lack of effect of isoflavonoids on the vagina and endometrium in postmenopausal women. *Fertil Steril.* Jan;83(1):137–142.

North American Menopause Society (NAMS) (2015). Nonhormonal management of menopause-associated vasomotor symptoms: 2015 position statement of The North American Menopause Society. *Menopause.* Nov;22(11):1155–1172.

Northrup C (2006). *Women's bodies, women's wisdom: creating physical and emotional health and healing.* New York: Bantam Dell.

Nowak G, Siwek M, Dudek D, et al. (2003). Effect of zinc supplementation on antidepressant therapy in unipolar depression: a preliminary placebo-controlled study. *Pol J Pharmacol.* 55:1143–1147.

Otunctemur A, Dursun M, Ozbek E, et al. (2015). Effect of metabolic syndrome on sexual function in pre- and postmenopausal women. *J Sex Marital Ther.* 41(4):440–449.

Palacios S (2011). Hypoactive sexual desire disorder and current pharmacotherapeutic options in women: Pharmacotherapy of HSDD. *Womens Health; CME.* Jan;7(1):95–107. Available at https://www.medscape.org/viewarticle/734772_7.

Panjari M, Davis SR (2007). DHEA therapy for women: effect on sexual function and wellbeing. *Hum Reprod Update.* 13(3):239–248.

Panjari M, Davis SR (2011). Vaginal DHEA to treat menopause related atrophy: a review of the evidence. *Maturitas.* Sep;70(1):22–25.

Pauls RN, Kleeman SD, Karram MM (2005). Female sexual dysfunction: principles of diagnosis and treatment. *Obstet Gynecol Survey.* 60(3):196–205.

Poels S, Bloemers J, van Rooij K, et al. (2013). Toward personalized sexual medicine (part 2): testosterone combined with a PDE5 inhibitor increased sexual satisfaction in women with HSDD and FSAD, and a low sensitive system for sexual cues. *J Sex Med.*10:810–823.

Prasad AS, Mantzoros CS, Beck FW, et al. (1996). Zinc status and serum testosterone levels of healthy adults. *Nutrition.* May;12(5):344–348.

Rogalski MJ, Kellogg-Spadt S, Hoffmann AR, et al. (2010). Retrospective chart review of vaginal diazepam suppository use in high-tone pelvic floor dysfunction. *Int Urogynecol J.* Jul;21(7):895–899.

Rosen R, Brown C, Heiman J, et al. (2000). The Female Sexual Function Index (FSFI): a multidimensional self-report instrument for the assessment of female sexual function. *J of Sex Marital Ther.* 26:1 91–208.

Sacomori C, Cardoso FL (2015). Predictors of improvement in sexual function of women with urinary incontinence after treatment with pelvic floor exercises: a secondary analysis. *J Sex Med.* Mar;12(3):746–755.

Schmid Y, Hysek CM, Preller KH, et al. (2015). Effects of methylphenidate and MDMA on appraisal of erotic stimuli and intimate relationships. *Eur Neuropsychopharmacol.* Jan;25(1)17–25.

Shariatpanaahi MV, Shariatpanaahi ZV, Moshtaaghi M, et al. (2007). The relationship between depression and serum ferritin level. *Eur J Clin Nutr.* 61:532–535.

Shifren JL, Monz BU, Russo PA, et al. (2008). Sexual problems and distress in United States women: prevalence and correlates. *Obstet Gynecol.* 112(5):970–978.

Silverstein RG, Brown AC, Roth HD, Britton WB (2011). Effects of mindfulness training on body awareness to sexual stimuli: implications for female sexual dysfunction. *Psychosom Med.* Nov-Dec;73(9):817–825.

Simon JA, Kingsberg SA, Shumel B, et al. (2014). Efficacy and safety of flibanserin in postmenopausal women with hypoactive sexual desire disorder: results of the SNOWDROP trial. *Menopause.* Jun;21(6):633–640.

Siwek M, Dudek D, Paul IA, et al. (2009). Zinc supplementation augments efficacy of imipramine in treatment resistant patients: a double blind, placebo-controlled study. *J Affect Disord.* 118:187–195.

Sotile WM, Kilmann PR (1977). Treatments of psychogenic female sexual dysfunctions. *Psychol Bull.* Jul;84(4):619–633.

Taylor MG, Rudkin L, Bullemor-Day P, et al. (2013). Strategies for managing sexual dysfunction induced by antidepressant medication. *Cochrane Database Syst Rev.* May 31;5:CD003382.

Toufexis D, Rivarola MA, Lara H, Viau V (2014). Stress and the reproductive axis. *J Neuroendocrinol.* Sep;26(9):573–586.

Tran JK, Dunckel G, Teng EJ.(2015). Sexual dysfunction in veterans with post-traumatic stress disorder. *J Sex Med.* Apr;12(4):847–855.

van Rooij K, Poels S, Bloemers J, et al. (2013). Toward personalized sexual medicine (part 3): testosterone combined with a serotonin 1A receptor agonist increases sexual satisfaction in women with HSDD and FSAD, and dysfunctional activation of sexual inhibitory mechanisms. *J Sex Med.*10:824–837.

Warnock JJ (2002). Female hypoactive sexual desire disorder: epidemiology, diagnosis, and treatment. *CNS Drugs.* 16(11):745–753.

West SL, D'Aloisio AA, Agans RP, et al. (2008). Prevalence of low sexual desire and hypoactive sexual desire disorder in a nationally representative sample of US women. *Arch Intern Med.* 168:1441–1449.

Zolnoun DA, Hartmann KE, Steege JF (2003). Overnight 5% lidocaine ointment for treatment of vulvar vestibulitis. *Obstet Gynecol.* Jul;102(1):84–7.

4

Male Sexual Function and Dysfunction

JEFFREY A. ALBAUGH

Sexual dysfunction is a common problem, impacting about one-third of men ages 18 to 59 (Laumann, 2000) and is even more common in older adult men (Laumann, 2008). There are several types of male sexual dysfunction. The main types of sexual dysfunction problems reported by men are hypoactive sexual desire (lack of libido), erectile dysfunction (ED), premature ejaculation, delayed ejaculation, and Peyronie's disease.

Sexual desire may be defined differently by different individual men, but for clinical purposes it has been defined as "the sum of forces that lean us toward and push us away from sexual behavior" (Levine, 2003, p. 280). It is a complex concept and has been delineated into three main components: drive, expectations/wishes, and motivation (Levine, 2003). Drive is the biological component of desire mediated by neurological and endocrine physiology and relates to the chemical mediators in the nervous system such as dopamine and hormones (such as testosterone) that biologically motivates a man toward sex. Drive may diminish with aging related to changes in key hormones such as testosterone. Testosterone may be greatly impacted by common medical conditions that negatively affect hormones that occur more as men age. Some of these conditions that negatively impact testosterone are diabetes, obesity, and metabolic syndrome. Some medications may also impact drive by effecting the nervous system and/or hormones. Although the primary role of testosterone is to increase libido, it also plays a smaller role in erectile function and orgasm/climax. Testosterone deficiency can therefore impact libido, erections, and ability to climax. Expectations and wishes comprise the cultural component of desire that reflects a person's beliefs and values about sex and shape the nature of his or her sexual experiences. Motivation is the final component of desire, and this encompasses the psychological dimensions of desire. Motivation reflects the willingness to have sex related to psychological factors such as a person's mental state or mood, interpersonal context (such as communication, disagreement, and affection), and social context (such as the relationship duration or the nature of the relationship itself with a sex partner). Each of these three aspects of desire must be examined when a man has a desire disorder to determine if the problem is more physiologically based in terms of drive, socially/culturally derived in terms of expectations/wishes, or psychologically based in terms of motivation. Once the underlying problems have been identified, treatment can be aimed at resolving any underlying issues and moving forward toward a healthy sex life. Treatment of desire disorder may include sex therapy, sex education, focused masturbation, bibliotherapy (reading instructive or erotic books), relationship therapy, and/or testosterone replacement therapy.

ED is the most common sexual dysfunction reported by men 40 and older, and, due to stigma surrounding sexual dysfunction, it is often underreported and underdiagnosed by clinicians. In the landmark population survey titled the Massachusetts Male Aging Study, over half of the men between the ages of 40 and 70 reported issues with ED (Johannes, 2000). ED is defined as the inability to attain and/or sustain an erection sufficient for satisfactory sexual performance (National Institutes of Health, 1993). The sad truth is that although ED is a treatable condition, many men suffer in silence and the problem is not always identified and therefore is left undertreated.

Erectile Function

There are psychological, hormonal, neurologic, and vascular factors necessary for a normal erection to occur in a man. Each of these factors can impact erectile function. A normal erection begins with sexual thoughts and/or tactile stimulation. The brain plays a powerful role in integrating sexual information and mediating an erectile response. Many things can impact erectile function including religion, social and cultural norms, self-image, increased stress, past sexual experiences, depression, and /or anxiety (Carson, 2009). The complex bodily processes leading to erections are not completely understood, but some aspects are clear. An erection involves a neurovascular event in the penis triggered by the central nervous system (Carson, 2009). Some of the areas of the brain involved in erections are the paraventricular nucleus, hypothalamus, medial amygdala, medial preoptic area, and hippocampus (Gratzke, 2010). Parasympathetic input from the sacral spinal cord contributes to smooth muscle relaxation and the vasodilation within the penis. Chemical mediators cause the smooth muscle relaxation and perfusion of the erectile cylinders. Smooth muscle relaxation is mediated by nitric oxide and its second messenger, cyclic guanosine monophosphate (Gratzke, 2010). There are a pair of corpora cavernosa and the corpora spongiosum that make up the three erectile cylinders within the penis that become engorged. The penile artery is the source of increased blood within the corpora cavernosa. The sinusoidal venules drain the corpora cavernosa and form a plexus that eventually drains into the dorsal vein of the penis by way of the emissary veins (Carson, 2009). Once the three cylinders within the penis are engorged, they temporarily compress venous outflow by pressing against the surrounding noncompliant Buck's fascia, which allows erection maintenance. After sexual stimulation ceases or ejaculation has occurred, sympathetic innervation arises from T10-L2 and causes the penis to return to the flaccid state. Testosterone is the primary male androgen and is regulated via the hypothalamus-pituitary-testis axis. Testosterone is involved in the maturation and the maintenance of penile tissue composition (Carson, 2009) and therefore plays a role in erectile function. Testosterone plays a major role with sex drive but a smaller role in erectile function and ability to achieve climax.

ETIOLOGY OF ED

There are many causes for ED, and often the cause can be related to multiple factors including psychological, endocrine, neurological, iatrogenic (from medications, surgery, or radiation), or vasculogenic problems. Depression, stress, and anxiety can adversely affect erectile function, and these underlying issues need to be addressed. Although there is often a psychological component to ED, findings show that the majority of men with ED (approximately 70%) have a physiologic component as well (Dean, 2008). Endocrine problems may lead to ED. As mentioned earlier, diabetes can contribute to ED and so can testosterone deficiency. The thyroid has an impact on hormones, and hypothyroidism may also play a role in diminished erectile function (Carson, 2009; Veronelli, 2006). Iatrogenic causes of ED can be from surgery and/or radiation to the groin or prostate.

Blood flow and nervous conduction are paramount to erectile function. Therefore, one of the main causes of ED is a problem with engorgement of the penis with blood and/or venous leakage of blood from the penis back into the body. ED and cardiovascular disease share the same common risk factors associated with endothelial dysfunction and vascular changes as a result of hyperlipidemia, high blood pressure, coronary artery disease, diabetes, and arteriosclerosis. Some lifestyle factors such as obesity, sedentary lifestyle, high-fat diet, and/or smoking may negatively impact erectile function. Uncontrolled diabetes can impact both blood flow and nervous conduction necessary for erectile function. Pelvic trauma to the groin area or prolonged sitting during bike riding may also damage the peripheral nerves or blood vessels and contributes to ED. Any disease that impacts the nervous system (such as spinal disc disease, Parkinson's disease, stroke, or multiple sclerosis) may impact erectile function. Men typically get several erections each night during deep sleep to maintain erectile structure and function; therefore, sleep apnea, insomnia, and other sleep disorders can negatively impact erectile function.

Some of the commonly prescribed medications negatively impact sexual and/or erectile function. The main groups of medications that are problematic are antihypertensives (beta blockers and diuretics are the worst) and antidepressants (the selective serotonin reuptake inhibitors [SSRIs] and tricyclics are the worst). Anticholinergic agents, antiarrhythmics, anticonvulsants, antihistamines, narcotics, analgesics, antianxiety medications, and antipsychotics are other groups of medications that may have a negative effect on erectile function. Nicotine, alcohol, and recreational drugs can also be problematic to erectile function. A full history and review of all medications, vitamins, and herbs may reveal underlying causes for ED. It is important for the clinician to collaborate with other prescribing providers, such as primary care and psychiatry, to make changes or modifications to any of the patient's prescribed medications.

Treatment of Sexual Dysfunction

All treatments have pros and cons, and treatments for sexual dysfunction are no different. Once the sexual issues have been clearly identified, the clinician needs to instruct the patient on the good and bad aspects of each treatment to help empower him to make informed decisions. The process involves really understanding each patient and (if applicable) his partner and the sexual goals they have in mind. Oftentimes patients come in believing there is a miracle cure for the sexual problem, and it is critical they understand that all treatments have positives and negatives and what works for one man can be totally wrong for another man. Treatment of ED may include addressing underlying medical issues such as hypogonadism, high blood pressure or diabetes, counseling and education of patient and partner, lifestyle modification, oral agents (i.e. sildenafil, tadalafil, varenafil, and avanafil), vacuum devices and/or constriction loops or rings, intraurethral suppositories, intracavernosal injections, and penile implant surgery.

LIFESTYLE CHANGES

Healthy lifestyle choices are important because they are associated with less metabolic syndrome, less cardiovascular disease, improved quality of life, and less ED (Esposito, 2006). The only treatment strategy for ED that can reverse the process of ED is lifestyle management in terms of healthy diet, weight control, and regular cardiovascular exercise (Esposito, 2009). A primary focus when treating men for ED should be prevention to decrease further progression of ED. Findings from cross-sectional and longitudinal studies have linked the development of ED to hypertension, diabetes mellitus, renal disease, hyperlipidemia, metabolic syndrome, depression, and lower urinary tract symptoms (Nehra, 2012). ED may be an early predictive sign of cardiovascular disease preceding a cardiovascular event (such as a heart attack or stroke) by an average of two to five years (Nehra, 2012). Adopting healthy lifestyle choices may ultimately improve sexual health (Gupta, 2011). Lifestyle changes recommended may include smoking cessation, regular exercise, weight loss, limiting alcohol, and following a cardiovascular low-fat, low-cholesterol, plant-based (Mediterranean) diet.

Although it is now thought that the majority of ED is physiologically based, there is often a psychological component as well. It is important to encourage clients to seek counseling for themselves and their partner. Counseling may include sex therapy, couple's therapy, and/or individual therapy. Sex therapy can help individuals and couples understand their feelings and illuminate issues about sex. It can help reduce performance anxiety and help couples to be more proficient in achieving their sexual goals. Counseling can also help couples work out relationship issues that impact sex.

Vacuum constriction devices are the least invasive ED treatment option. Vacuum devices consist of a pump, plastic cylinder, water-soluble lubricant, and tension/constriction band to hold the erection. Medical grade devices have a pressure release valve on the pump or cylinder to quickly release vacuum pressure. They are available in either a manual hand pump (Figure 4.1) or a battery-operated pump system (Figure 4.2).

The vacuum device produces an erection by creating negative pressure to pull blood into the penis. The erection is maintained outside the cylinder by applying the tension ring off the edge of the device to the base of the penis. No tension loop or band should be on the penis for more than 30 minutes maximum. Pros of the vacuum device are that, with proper training, practice, and instructions, they can be effective in the majority of men and they are completely noninvasive. Cons of the vacuum device are that they can be cumbersome and awkward, creating a less natural feeling and looking erection that stops at the point of the tension ring, creating wobbling at the base of the penis. Some men do not need the vacuum device to help them get hard enough for sex but only a tension device to hold his erections during sex. These tension loops or rings are helpful for men who have difficulty maintaining an erection. They are bands of various designs and materials that are placed around the base of the penis to hold the erection during sexual activity. One example of a device is the Venoseal® from TIMM Medical. It is an adjustable loop that is applied at the base of the penis to decrease venous flow return into the body. Again, no constriction device should be worn around the penis for longer than 30 minutes.

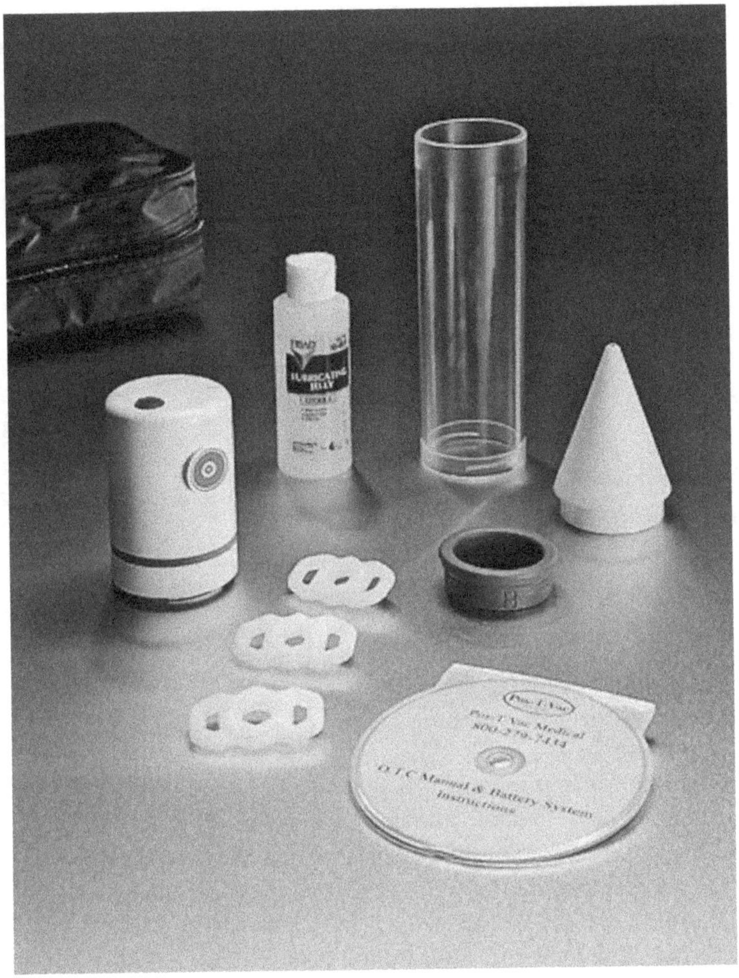

FIGURE 4.1. Pos-T-Vac ErecTech manual system.
Reprinted with Permission of American Medical Systems.

The most popular treatment for ED is oral therapy (sildenafil, vardenafil, tadalafil, or avanafil), due to the discreet convenience of taking a pill orally prior to sex, as compared to some of the other treatment options (Hatzichristou, 2000). Phosphodiesterase type 5 inhibitors (PDE5 inhibitors) inhibit PDE5, which plays an important role in triggering the cascade of chemical mediators leading to smooth muscle relaxation and vasodilatation within the penis (Porst, 2013). There are several Food and Drug Administration (FDA)-approved agents including sildenafil, vardenafil, avanafil, and tadalafil. Pros are that they are simple, easy, and

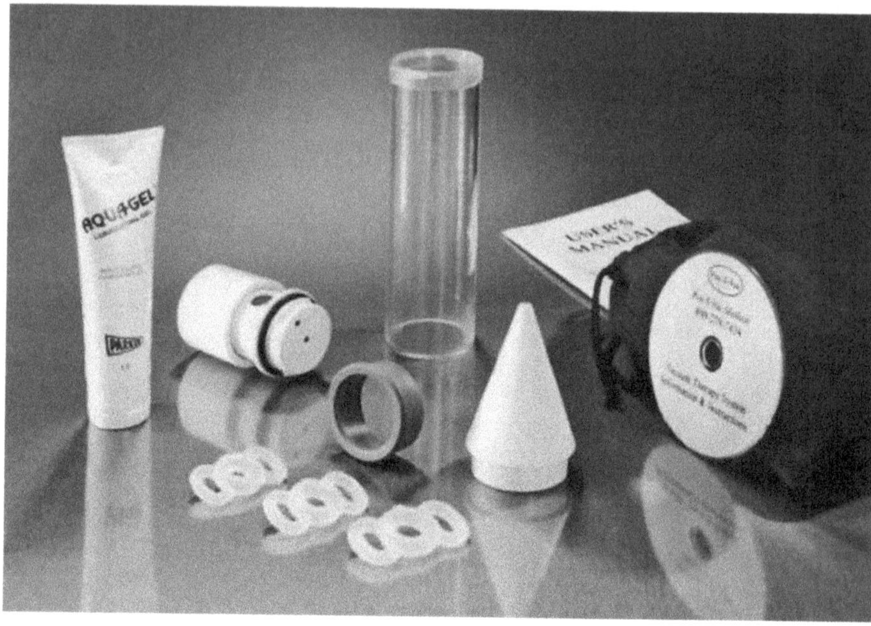

FIGURE 4.2. Pos-T-Vac ErecTech battery operated.
Reprinted with Permission of American Medical Systems.

discreet to use. PDE5 inhibitors enhance a patient's own ability to get harder erections in the presence of sexual stimulation. Cons are that they do not work in all men with ED, and, as ED continues to progress, they may become ineffective. Other cons are that they can have side effects, cannot be taken by some men (e.g., men on nitrates), and can be expensive.

Injection therapy involves self-injecting a vasoactive agent directly into the corpora cavernosa of the penis, leading to smooth muscle relaxation and vasodilation within the penis. It requires one-on-one training by a qualified professional and careful titration to determine an appropriate medication and dosage. There are FDA-approved alprostadil medications (such as Edex and Caverject) and off-label non-FDA-approved compounded injectable medications (such as bimix, trimix, and quadmix). Pros of injection therapy are that it creates a fairly natural feeling and looking erection that often lasts about an hour average and that it is effective in most men. Cons are that it cannot be used in all men (e.g., men with sickle cell anemia or multiple myeloma), it has side effects (most commonly pain, bleeding, or bruising but less often can lead to long-term deformity of the penis or permanent damage to the penis related to priapism), and it can be expensive.

Intraurethral alprostadil (MUSE®) works similar to the injectable alprostadil by causing smooth muscle relaxation and vasodilation, but it is delivered intraurethrally through the meatus. Pros to intraurethral alprostadil are that it is easy to use, and, when it works, it creates a fairly natural feeling and looking erection. Cons are that is does not work in the majority of men with significant ED, it cannot be used in all men with ED (e.g., men with sickle cell anemia or multiple myeloma), and it has some side effects (a significant amount of men will experience pain or burning and much fewer may experience dizziness, lightheadedness, and/or a drop in blood pressure). This product should be tried in the clinic the first time to determine safety and efficacy.

The penile implant prosthesis has been around longer than any other ED treatment. The implant surgery still provides a reliable treatment for ED, but the decision for surgery usually comes after unsuccessful treatment with the less invasive medical treatments. Implants come in various designs and may be either inflatable or malleable. The inflatable device is more often preferred since it can be inflated for sex and deflated afterwards, mimicking the erect and flaccid states of the penis. The inflatable device consists of two cylinders that are placed within each corpora cavernosum, a reservoir filled with saline, and an inflate-deflate pump in the scrotum (Figure 4.3).

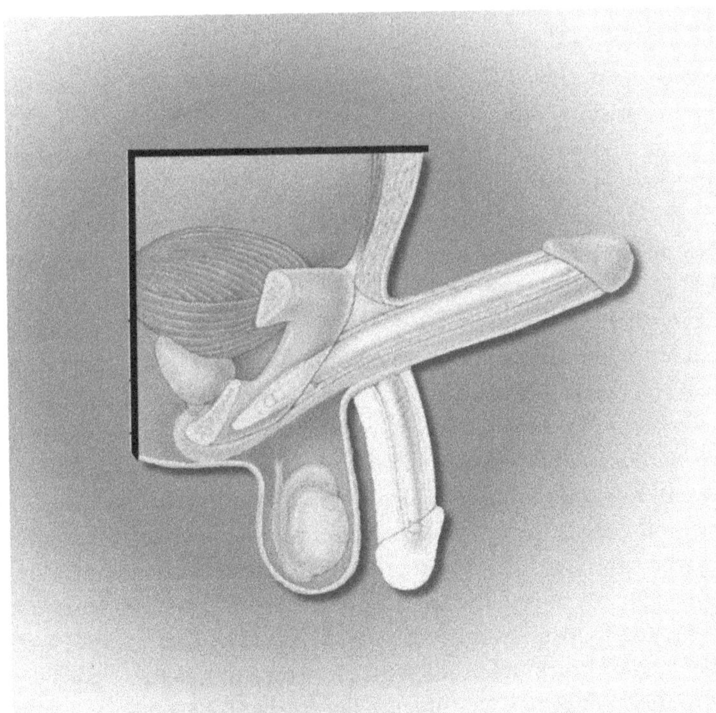

FIGURE 4.3. Titan penile implant.
Reprinted with permission of Coloplast.

To activate the device, the patient repeatedly squeezes the pump within the scrotum, which forces fluid from the reservoir into the cylinders. The patient can deflate the pump by pressing the release button on the pump in the scrotum, allowing fluid to flow back out of the cylinder and into the reservoir. The malleable penile prosthesis can be positioned upright for sexual relations and then positioned back down out of the way afterwards. The pros to the implant are that it typically works to provide an erection hard enough for sex, it is concealed within the body, and patient and partner satisfaction are generally fairly high (McLaren, 1992). The cons with the implant are that the erection produced by an implant is not exactly "natural" and is different particularly in length. In addition, the penile implant will permanently change the structure of the penis, so it is a considered a final option when other treatments have failed. Finally, the penile implant is a surgery and holds the same risks as any surgery. Complications may include device failure, perforation, infection, persistent pain, and aneurysmal dilation (inflatable devices).

TREATMENT OF LOW LIBIDO

Treatment of men with low libido accompanied by a low testosterone level has been historically done with androgen replacement therapy. This treatment is generally reserved for men with documented low serum total and/or free testosterone with symptoms such as low libido, ED, decreased energy level, decreased mood, and fatty weight gain. Unfortunately, the use of androgen replacement therapy may suppress natural androgen productions within the body. Testosterone replacement is not without negative effects and should be used judiciously in men with a repeated documented low testosterone level and symptoms only (Bhasin, 2010). Testosterone therapy comes in several available prescription formulas including topical/transdermal, intramuscular, pellets, and the buccal patch. The pros to testosterone therapy are that testosterone elicits an impact on normal body function all over a man's body and normalizing testosterone can not only impact libido, erections, and ability to climax but also improve blood counts, promote bone health, increase energy, heighten mood, and improve cognition. Cons are that testosterone is a steroid and can be dangerous. Replacing or increasing testosterone must be done very carefully under tight supervision to avoid long-term side effects such as infertility, cardiovascular events (such as heart attacks or strokes), worsening obstructive sleep apnea, and/or promotion of prostate or breast cancer growth. Testosterone replacement or manipulation with supplements is contraindicated in some men and should only be done under careful supervision.

PREMATURE EJACULATION

Orgasm is the feeling or sensation associated with climax. Ejaculation involves the release of fluid associated with the feeling of climax. Premature ejaculation is defined by the International Society of Sexual Medicine (ISSM) as a male sexual problem characterized by a shortened intravaginal ejaculation latency time (IELT), inability to control/delay ejaculation, and negative consequences such as distress, frustration, or bother from the condition (Althof, 2014). Delayed ejaculation is the persistent or recurrent delay in or absence in attaining orgasm with sufficient sexual stimulation (McMahon, 2004).

Premature ejaculation is the most common complaint of younger men, and this is likely due to the fact that, in general, women take an average of four times longer to reach climax than men (Laumann, 2000). Premature ejaculation is typically divided into lifelong premature ejaculation and acquired premature ejaculation (premature ejaculation symptoms begin after a period of normal ejaculatory function). Lifelong premature ejaculation occurs within a minute of intravaginal penetration, while acquired premature ejaculation typically involves a reduced ejaculatory latency period that is often three minutes or less (Althof, 2014). After comprehensive review of all the research concerning prevalence of premature ejaculation, both lifelong and acquired, the ISSM determined that approximately 5% of men in the general population have an IELT of two minutes or less (Althof, 2014). The etiology of premature ejaculation is not completely understood, and there is no definitive research to confirm etiologic factors. In the past, it was thought that premature ejaculation was caused by psychological issues such as anxiety or interpersonal/relationship issues such as rushed early sexual experiences. More recent etiological factors identified for premature ejaculation are hypersensitivity of the glans penis, robust cortical representation of the pudendal nerve, disturbances in central nervous system dopamine and/or serotonin neurotransmission, detoxification from prescribed medications, recreational drugs, chronic pelvic pain, erectile or other sexual issues, and/or thyroid problems (Althof, 2014).

Premature ejaculation is treated in a variety of ways. Assessment, including a complete history, will guide treatment since a complete understanding of both premature ejaculation and comorbid conditions is crucial to treatment. Questions need to be aimed at understanding the nature of the premature ejaculation and how, when, and where the issues are occurring. There are also two questionnaires that are commonly used to assess premature ejaculation: the Premature Ejaculation Profile (Patrick, 2009) and the Index of Premature Ejaculation (Althof, 2006). Addressing underlying issues, such as ED, prostatitis, and/or psychological factors, can be an essential component to treating premature ejaculation.

Behavioral treatment typically involves the stop/start technique with stopping penile stimulation until things settle down and then re-engaging in stimulation on the penis and the squeeze technique (squeezing the tip of the penis to delay ejaculation). Both these techniques involve a man training himself to delay ejaculation through masturbation work and work with his partner(s) to learn to delay ejaculation. Pharmacological techniques are typically aimed at either desensitizing the penis by using a numbing spray or gel or working through the central nervous system to impact the dopamine-serotonin balance. Desensitizing agents contain numbing medications and include Promescent spray (10mg of lidocaine per spray with typical dosing at between 1 to 10 sprays), Stud-100 spray (7.7 mg of lidocaine base per metered spray with three to eight sprays being the typical dose), numbing gels used under a condom, or EMLA cream (approved and used in Europe and containing 2.5% each of lidocaine and prilocaine applied 20 minutes prior to sex used under a condom). Central acting agents, such as SSRIs (sertraline, fluoxetine, paroxetine, citalopram) and tricyclic antidepressants (clomipramine) are prescribed off-label and have some research to support efficacy. These medications are typically most effective when taken daily, but some studies support on-demand use of dapoxetine, paroxetine, clomipramine, sertraline, or fluoxetine. The medications are not without side effects, and so the patient must determine the risk-benefit ratio of taking these medications for premature ejaculation.

Men less commonly complain of the opposite problem of delayed orgasm, and therefore knowledge about prevalence, etiology, and treatments is limited. It is typically characterized by persistent delayed or absent orgasm in the presence of sexual stimulation (McMahon, 2012). The etiology of diminished orgasm is not well understood but may be related to congenital disorders, neuropathy, medication, or psychological problems. Treatment includes psychological and educative strategies including sex education, reduction of sexual anxiety, masturbatory retraining, and exploring ejaculatory response on his own or in the presence of his partner and discovering different types of genital stimulation that lead to orgasm to incorporate that into loveplay with his partner. The main focus of psychological behavioral strategies is masturbatory exploration and training, including (but not limited to) exploring vibratory stimulation. There are no approved medications for delayed orgasm, with very limited anecdotal and uncontrolled studies to support the few medications that may be helpful. Buproprion has been used with mixed results in antidepressant-induced delayed orgasm (Clayton, 2006). Other agents such as cyproheptadine, yohimbine, apomorphine, and oxytocin have also been used with some success, but research is needed to confirm efficacy and safety (McMahon, 2102). There are no clear efficacious treatments for delayed climax, and this problem can be very distressing to some men.

PEYRONIE'S DISEASE

Peyronie's disease is a disorder characterized by penile deformity from fibrotic formation in the tunica albuginea of the corpora cavernosa within the penis, resulting in penile shortening, narrowing, curvature and/or hinging. It can also be accompanied by pain and occurs in about 3% to 10% of men (Sherer, 2014). It is considered to be a wound-healing disorder that occurs following penile trauma. If a man is able to engage in coitus despite the curve and is not restricted, Peyronie's disease may be left untreated since treatment may lead to further scarring, and a small amount of Peyronie's deformity will resolve on its own within the first six months. Treatment varies and may include many medical or surgical options. The more common options used by clinicians include off-label use of pentoxifylline, intralesional injections of collagenase, surgical management with tunica albuginea plication (with or without grafting), and prosthetic penile implant for men with both ED and Peyronie's disease. Although a large variety of oral agents have been tried, there is little to no research to support the use of most agents (Sherer, 2014). Pentoxifylline 400mg TID has been used off-label and showed a 36.9% overall positive response compared to 4.5% in the placebo group as well as improvement in penile curvature and decreased plaque volume as compared to patients receiving placebo (Safarinejad, 2009). Ultimately the guidelines for treatment of Peyronie's disease published in the *Journal of Sexual Medicine* suggest that there is no benefit with respect to deformity reduction with any oral therapies (such as colchicines, Vitamin E, carnitine, tomoxifen, or potassium aminobenzoate) based on grade B evidence (Honig, 2014). Intralesional injections with collagenase has been FDA approved in the United States with a statistically significant improvement in both mean penile curvature (34% vs. 17%, $p < 0.0001$) and change in symptom bother domain of the Peyronie's Disease Questionnaire (2.8 vs. 1.8, $p < 0.003$) in the treatment group versus placebo group (Gelbard, 2013). This treatment requires several injections into the palpable Peyronie's plaque. Multiple placation techniques have been used to surgically correct Peyronie's deformity. Surgical reconstruction aims at correcting the curvature and may include tunica placation, plaque incision, plaque excision, and/or grafting and is typically reserved for men with stable disease, compromised ability for coitus, extensive plaque calcification, and failed conservative treatment (Levine, 2013). Other less utilized and investigated treatment options are discussed in a later chapter on treating male sexual dysfunction.

SEXUAL DYSFUNCTION ONGOING FOLLOW-UP AND CARE

Treating sexual dysfunction involves an iterative process of follow-up visits and reassessment at regular intervals. Reassessment of progress or success is essential to ongoing success. Most treatments are associated with high rates of discontinuation if the patient is not followed systemically. The visit may include administration of a standardized scale such as the International Index of Erectile Function and the erection hardness scale to evaluate progress and success. Follow-up may involve ongoing patient communication and education, medication titration and adjustment, and medical and psychological reassessment. Treating sexual issues requires ongoing support and individualized care to help the patient continue to make informed, empowered decisions about the best treatment for him and his partner.

REFERENCES

Althof S, Rosen R, Symonds T, Mundayat R, May K, Abraham L (2006). Development and validation of a new questionnaire to assess sexual satisfaction, control, and distress associated with premature ejaculation. *J Sex Med.* 3(3):465–475.

Althof SE, McMahon CG, Waldinger MD, et al. (2014). An update of the International Society of Sexual Medicine's guidelines for the diagnosis and treatment of premature ejaculation (PE). *J Sex Med.* 11(6):1392–1422.

Bhasin S, Cunningham RG, Hayes JF, et al. (2010). Testosterone therapy in men with androgen deficiency syndromes: an endocrine society clinical practice guideline. *J Clin Endocrin Metab.* 95(6).

Carson CC, Kirby RS, Goldstein I, Wyllie MG (2009). *Textbook of erectile dysfunction*, 2nd ed. New York: Informa Healthcare.

Clayton AH, Montejo AL (2006). Major depressive disorder, antidepressants, and sexual dysfunction. *J Clin Psych.* 67(Suppl 6):33–37.

Dean J, Rubio-Aurioles E, McCabe M, et al. (2008). Integrating partners into erectile dysfunction treatment: improving the sexual experience for the couple. *Int J Clin Pract.* 62(1):127–133.

Esposito K, Ciotola M, Giugliano F, et al. (2006). Mediterranean diet improves erectile function in subjects with the metabolic syndrome. *Int J Impotence Res.* 18(4):405–410.

Esposito K, Ciotola M, Giugliano F, et al. (2009). Effects of intensive lifestyle changes on erectile dysfunction in men. *J Sex Med.* 6(1):243–250.

Gelbard M, Goldstein I, Hellstrom WJ, et al. (2013). Clinical efficacy, safety and tolerability of collagenase clostridium histolyticum for the treatment of peyronie disease in 2 large double-blind, randomized, placebo controlled phase 3 studies. *J Urology.* 190(1):199–207.

Gratzke C, Angulo J, Chitaley K, et al. (2010). Anatomy, physiology, and pathophysiology of erectile dysfunction. *J Sex Med.* 7(1 Pt 2):445–475.

Gupta BP, Murad MH, Clifton MM, Prokop L, Nehra A, Kopecky SL (2011). The effect of lifestyle modification and cardiovascular risk factor reduction on erectile dysfunction: a systematic review and meta-analysis. *Arch Int Med.* 171(20): 1797–1803.

Hatzichristou DG, Apostolidis A, Tzortzis V, Ioannides E, Yannakoyorgos K, Kalinderis A (2000). Sildenafil versus intracavernous injection therapy: efficacy and preference in patients on intracavernous injection for more than 1 year. *J Urology*.164(4):1197–1200.

Honig SC (2014). Intralesional collagenase in the treatment of Peyronie's disease. *Ther Adv Urol*. 6(2):47–53.

Johannes CB, Araujo AB, Feldman HA, Derby CA, Kleinman KP, McKinlay JB (2000). Incidence of erectile dysfunction in men 40 to 69 years old: longitudinal results from the Massachusetts male aging study. *J Urology*. 163(2):460–463.

Laumann EO, Gagnon JH, Michael RT, Michaels S (2000). *The social organization of sexuality: Sexual practices in the United States*. Chicago: University of Chicago Press.

Laumann EO, Waite LJ (2008). Sexual dysfunction among older adults: prevalence and risk factors from a nationally representative U.S. probability sample of men and women 57–85 years of age. *J Sex Med*. 5(10):2300–2311.

Levine LA, Burnett AL (2013). Standard operating procedures for Peyronie's disease. *J Sex Med*. 10(1):230–244.

Levine SB (2003). The nature of sexual desire: a clinician's perspective. *Arch Sex Behav*. 32(3):279–285.

McLaren RH, Barrett DM (1992). Patient and partner satisfaction with the AMS 700 penile prosthesis. *J Urology*. 147(1):62–65.

McMahon CG, Abdo C, Incrocci L, et al. (2004). Disorders of orgasm and ejaculation in men. *J Sex Med*. 1(1):58–65.

McMahon CG, Jannini E, Waldinger M, Rowland D (2012). Standard operating procedures in the disorders of orgasm and ejaculation. *J Sex Med*. 10:204–209.

Nehra A, Jackson G, Miner M, et al. (2012). The Princeton III Consensus recommendations for the management of erectile dysfunction and cardiovascular disease. *Mayo Clinic Proc*. 87(8):766–778.

National Institutes of Health, Consensus Development Panel on Impotence (1993). NIH consensus conference. Impotence. *JAMA*. 270(1):83–90.

Patrick DL, Giuliano F, Ho KF, Gagnon DD, McNulty P, Rothman M (2009). The Premature Ejaculation Profile: validation of self-reported outcome measures for research and practice. *British J Urology Int*. 103(3):358–364.

Porst H, Burnett A, Brock G, et al. (2013). SOP conservative (medical and mechanical) treatment of erectile dysfunction. *J Sex Med*. 10(1):130–171.

Safarinejad MR, Asgari MA, Hosseini SY, Dadkhah F (2009). A double-blind placebo-controlled study of the efficacy and safety of pentoxifylline in early chronic Peyronie's disease. *British J Urology Int*. 106(2):240–248.

Sherer BA, Warrior K, Levine LA (2014). 2013–2014 updates in Peyronie's disease management. *Curr Urology Rep*. 15(12):459.

Veronelli A, Masu A, Ranieri R, Rognoni C, Laneri M, Pontiroli AE (2006). Prevalence of erectile dysfunction in thyroid disorders: comparison with control subjects and with obese and diabetic patients. *Int J Impotence Res*. 18(1):111–114.

5

Midlife and Beyond: Male Hormonal Status/Andropause and Its Effects on Sexual Health

GEOVANNI ESPINOSA AND RALPH ESPOSITO

Andropause is often known by many other names including male menopause, late onset hypogonadism, androgen deficiency, male climacteric, androgen decline in aging male, and colloquially as grumpy old man syndrome.

The prevalence of androgen deficiency among men comes from two large studies: the Massachusetts Male Aging Study and the Boston Area Community Health survey estimated 6% to 12% of men between 30 to 70 years old experience androgen deficiency (Araujo, 2011). The latest guidelines from the Endocrine Society suggests androgen deficiency, as we refer to in this chapter as andropause, must exhibit signs and symptoms of androgen deficiency (Table 5.1) and total testosterone (T) levels below 300 ng/dl (9.8–10.4 nmol/liter) and a free T below 5 to 9 ng/dl (0.17–0.31nmol/liter; Bhasin, 2010). Although the rates of andropause range from 6% to 12% of the general male population, the likelihood of genuine andropause among men seeking a physician for related symptoms is 15% (Corona, 2013), suggesting there are asymptomatic men with low T, and those with symptoms and seeking medical help should be taken with more consideration.

Of all signs and symptoms, the three most significantly connected to low T and low free T are sexual symptoms: erectile dysfunction (ED), low libido (decreased sexual thoughts), and decreased morning erections (Wu, 2010). Therefore andropause encompasses the sexual health of the aging male.

Although on occasion physicians will insist on the diagnosis of andropause, late onset hypogonadism, testicular hypofunction, or whichever diagnostic terminology seems most fitting, men are often misdiagnosed due to the fact that most men with low T levels are asymptomatic and even men with normal T levels may

Table 5.1. Symptoms and Signs of Androgen Deficiency/Andropause in Men

A. Specific Symptoms and Signs	Incomplete of delayed sexual development, eunuchoidism
	Low libido and activity
	Decreased spontaneous erections
	Gynecomastia, breast discomfort
	Loss of body (axillary and pubic) hair, reduced shaving
	Very small (<5ml) or shrinking testes
	Low/zero sperm count
	Height loss, low trauma fracture, and low bone mineral density
	Hot flushes, sweats
B. Less specific symptoms and signs	Decreased energy, motivation, initiative, and self-confidence
	Feeling depressed, dysthymia
	Poor concentration and memory, poor mental clarity
	Sleep disturbance, increase sleepiness
	Mild anemia (normochromic, normocytic, in the female range)
	Reduced muscle bulk and strength
	Increased body fat, BMI
	Diminished physical or work performance

exhibit andropause symptoms. This has been exhibited in several studies, and it is essential to remember physicians need not treat whole populations but instead each individual. There are in fact genetic differences that may impact transcriptional activity of the androgen receptor, thus explaining not only the variance in symptoms with total and free T levels but also the possible variance and androgen insensitivity seen in men treated with exogenous T (Langdahl, 2003; Zitzmann, 2003, 2009). Low T in men does not necessarily categorize a male into andropause; neither does a male exhibiting decreased sexual function, therefore having both is necessary to consider.

The role of T in male sexual health is not completely understood especially with relation to common andropause symptoms including ED, low sexual desire (libido), and decreased morning erections (nocturnal penile tumescence).

When developing a treatment for andropause, multiple routes are possible; however, the underlying cause of the low T and associated symptoms must be revealed prior to developing a proper treatment. It is imperative to complete a hormonal panel including T, free T, sex hormone binding globulin (SHBG), estradiol (E2), prolactin, thyroid panel (TSH, fT3, T4, rT3), follicle stimulating hormone (FSH), luteinizing hormone (LH), prolactin (PRL), salivary cortisol, androstenedione, cholesterol (HDL, LDL, total, particle density size), and ferritin with high consideration to glycosylated hemoglobin (Hb A1c).

It has been shown in several studies in an age-wise fashion a decrease in total T, free T, dehydroepiandrosterone (DHEA), estrogen, cortisol and increases in PRL, FSH, LH, and DHT (Denti, 2000; Feldman, 2002; Harman, 2001). These hormones have an intricate relationship, as optimal function among each is essential for peak T levels, with thyroid hormone playing an essential role, as hyperthyroidism and hypothyroidism may increase SHBG thus decreasing free T (Dumoulin, 1995; Sarne, 1988; Selva, 2009). The role of PRL in men is fairly unclear, yet it is known to negatively feed back to the hypothalamus, reducing gonadotrophic releasing hormone (GnRH) in males thus reducing LH, consequently reducing Leydig cell activity and total T. Interestingly TRH, stress adaptive responses (which we discuss in more depth later), serotonin, and estrogens may increase PRL levels with cortisol being negatively correlated with PRL levels (Ben-Jonathan, 2008; Sobrinho 2003). In addition, elevated estradiol may increase SHBG, further reducing free T (Hammond, 2008; Kiecolt-Glaser, 2011).

Role of Testosterone in Sexual Function

Spontaneous erections (including morning erections), libido, sexual thoughts, nocturnal penile tumescence, and orgasmic function are all very closely correlated with T levels. A large meta-analysis reviewing the effects of testosterone supplementation found men with low T had decreased spontaneous erections and sexual-related erections, decreased orgasmic function, and decreased sexual desire/libido (Corona, 2014). Although Corona et al. have not been able to confirm that T supplementation improves erectile function in men, it is evident that T levels are positively associated with measures of erectile function, rigidity, and amplitude along with sexual function, sexual desire, and activity and in fact is necessary for erections and nitric oxide (NO) synthesis (Andersson, 2011; Cunningham, 2014; Rajfer, 2000; Wu, 2010). Although, it must be mentioned, some men with ED may not benefit from testosterone supplementation, and this suggests ED is not solely androgen or T dependent (Wu, 2010).

Therefore the goal to optimize male T levels along with other hormone parameters is justified and effective in enhancing male sexual health in andropausal men. Although "andropause" has been used to characterize men 40 years and older, the same features may present in men in their 20s and 30s.

Medical Nutraceuticals

DHEA

Dose: 50 to 100mg/day before bed

This hormone is a precursor to T and certainly does decrease in the aging male. Pregnenolone, derived from cholesterol, is the mother hormone to DHEA and its sulfated form, which are predominantly produced in the adrenal glands. DHEA is catalyzed by 3-Beta-hydroxysterioud dehydrogenase (3-βHSD) to androstenedione then to T via 17-Beta-hyrdoxysterioud dehydrogenase (17-βHSD). DHEA may also be catalyzed to androstenediol via 17-βHSD then to T via 3-βHSD.

DHEA in the aging male is negatively correlated to erectile function, sexual desire, orgasmic function, and intercourse satisfaction (Feldman, 1994; Reiter, 1999); as DHEA declines with age, these aspects of sexual function weaken. Other studies show increased T levels with DHEA supplementation in middle-aged and young men (Libè, 2004; Liu, 2013; Muller, 2006; Ostojic, 2010; Weiss, 2009). One particular study showed 50mg of DHEA significantly improved T levels in non-hypogonadic middle-aged and young men (21.4 +/− 0.3 years) performing high-intensity interval training (Liu, 2013) and another showing improvement in young (21–24 years old) male athletes consuming 100mg of DHEA (Ostojic, 2010). Therefore, these studies show the response to DHEA supplementation is significant enough to produce increased free T even in men without low T. This is particularly impressive considering T level improved in young men with T levels that are typically optimal. Studies indicating improved T levels in midlife and elder men on DHEA supplementation, who experienced no adverse effects, support using DHEA as a safe and effective tool to improve andropause symptoms and low T (Libè, 2004, Muller, 2006; Weiss, 2009).

ZINC

Dose: 15 to 30mg/day (preferably as zinc picolinate)

This mineral has an integral role in the function of androgen receptors and testicular steroidogenesis. It is suspected that zinc (Zn) deficiency leads to decreased utilization of cholesterol and steroid precursors in the synthesis of androgens (Lei, 1976). In addition to decreased androgen production, Zn deficiency may decrease activity of androgen receptors, resulting in decreased T activity even with normal physiologic levels (Chung, 1986; Lei, 1976; Om, 1996). Although testing for Zn levels is best done through red blood cell analysis, this is not a common practice and not offered by many labs. The Zinc Tally Test is a possible alternative; however, given the low risk of Zn supplementation, supplementing with Zn is likely safe and effective at the recommended dose (Age-Related Eye Disease Study Research Group, 2001; Samman, 1987). Clinically, Zn deficiency may be suspected with leukonychia and Beau's lines (Kelly, 1998).

VITAMIN D

Dose: 2,500 IU to 5,000 IU cholecalciferol daily with a fatty meal adjusted appropriately to 25(OH)D lab results.

This nutrient, which is considered more a hormone than a vitamin, has an intricate role and association with T and androgens. The link between low vitamin D (hypovitaminosis D) and low androgens has been elicited in some studies albeit with some conflicting evidence. However, 25(OH)D levels less than 50 nmol/L were significantly associated with secondary and compensated hypogonadism and correlated nonsignificantly to primary hypogonadism (as defined as T < 10.5 nmol/L and LH cutoff at 9.4 U/L; Lee 2012). Additionally, men with vitamin D deficiency were positively associated with T and calculated free androgen index, indicating men with higher 25(OH)D levels had significantly higher T levels (Pilz 2011; Wehr, 2010). The mechanism by which vitamin D may have an influence on male androgens may be explained by the presence of vitamin D receptors (VDR) and vitamin D metabolizing cytochrome enzymes in Leydig cells, which may potentially impact steroid production (Blomberg Jensen, 2010; Nangia, 2007). Interestingly, VDR knockout mice developed hypergonadtropic hypogonadism (Kinuta, 2000); combined with the discovery of VDR and 1-alpha-hydroxylase in the anterior pituitary and hypothalamus and 25-hydroxylase in Leydig cells, these findings together support the intricate involvement of vitamin D in male steroidogenesis (Eyles, 2005; Pérez-Fernandez 1996).

With the prevalence of vitamin D deficiency in American men it would be prudent to optimize male 25(OH)D levels to above 20 ng/mL (50 nmol/L) with optimal levels reaching above 50 ng/mL.

L-CITRULLINE

Dose: 1.5 g divided daily

This nonessential amino acid is a key precursor to L-arginine. L-Arginine is a precursor in the synthesis of NO catalyzed by endothelial NO synthase (NOS), which works to relax endothelial tissue and enhance blood flow. Arginine has been shown to improve scores on the International Index of Erectile Function

(IIEF), Erection Hardness Score, and Erectile Dysfunction Inventory of Treatment Satisfaction within one to two hours of intercourse (Lebret, 2002; Neuzillet 2013). Although arginine is commonly used as an over the counter treatment for ED, it has a very short half-life. However, citrulline provides a prolonged increased in NO and need not be taken prior to sexual activity, thus resulting in enhanced endothelial function, superior to arginine (Cormio, 2011; Cynober, 2010; Moinard, 2007; Romero, 2006; Shiota, 2013; Wijnands, 2012).

Medicinal Herbs

DOPA BEAN

Dose: 5 g of *Mucuna pruriens* daily with meals

Mucuna pruriens is an Ayurvedic herb traditionally known as an aphrodisiac, with modern research supporting its ability to improve sperm quality (Shukla, 2010). One of the main constituents of this herb is L-3,4-Dihydroxyphenylalanine (L-dopa), which is a precursor to dopamine (DA). In the central nervous system, DA has the ability to inhibit PRL thus preventing prolactin's inhibitory effect on the anterior pituitary's release of gonadotropins (FSH, LH). Essentially, the less PRL and the more LH, the more T production (Islam, 1998).

Two studies have shown *M. pruriens* to significantly improve T levels and LH levels in normo-, oligo-, and astheno-zoospermic males at doses of 5 g/day (Gupta, 2011; Shukla, 2009). The mechanism of action is suspected to work on the hypothalamic-pituitary-gonadal axis, as dopamine, epinephrine, and norepinephrine levels all were shown to be elevated in blood and seminal plasma (Shukla, 2009).

MACA ROOT

Dose: 1.5 to 3.5 g divided daily

The use of maca root, *Lepidium meyenii*, to enhance male libido has been touted in traditional cultures for decades, even centuries, yet medical research has only recently begun to unveil its validity. Maca has been shown to exert its libido-enhancing benefits independent of any hormonal influences, with studies showing maca improving sex drive in men as young as 21 years up to middle-aged men (Gonzales, 2002; Stone, 2009). A trial with a very small male sample size ($n = 3$) showed improved libido in men with selective serotonin reuptake inhibitor–induced sexual dysfunction; these findings are consistent with other studies of maca's benefits (Dording, 2008). A study utilizing the International

Index of Erectile Function–5 and the Satisfaction Profile (SAT-P, an indicator of quality of life) questionnaire showed maca to significantly improve ED and SAT-P (Gonzales, 2002). Although this study did not test maca's possible libido enhancement, it did indicate its ability to improve sexual function, which often is correlated with increased libido. It may be proposed that increasing libido may improve ED. Animal studies further support the use of maca for the ED and low libido often associated with andropause (Cicero, 2001; Zheng 2000). The mechanism of action of maca is not known, yet it is known that maca does not impact male hormonal levels.

NETTLE ROOT

Dose: 300 to 600mg three times daily

Urtica dioica radix does not directly increase steroidogenesis, yet it works on SHBG and aromatase. Nettle root increases free T levels by inhibiting the binding of SHBG to free T, thus increasing biologically active T. Additionally nettle root works as an aromatase inhibitor, decreasing conversion of T to estrogen and effectively increasing total T levels. Nettle root may act to lower estrogen levels, which tend to be elevated in andropausal men and overweight men (Gansser, 1995a, 1995b; Koch, 1995; Schmidt, 1983; Schöttner, 1997). The constituents secoisolariciresinol and 3,4-divanillyltetrahydrofuran are SHBG inhibitors found mainly in the plant root (Gansser 1995).

TONGKAT ALI

Dose: 100mg twice daily up to 400mg daily

Eurycoma longifolia, commonly known as tongkat ali or longjack, is an herb with a traditional use for infertility and for alleviating low T symptoms, although the studies are limited to one human study that showed tongkat to increase total T in hypogonadal men and reduce the Aging Males' Symptoms score (Tambi, 2012). Two studies showed tongkat to improve fertility, libido, and ED, yet these were evident without any changes in T (Ismail, 2012; Tambi, 2010). Yet other studies exhibit statistically significant improvements in total and free T levels (Henkel, 2014; Tambi, 2012). Although andropause includes both symptoms of low T and clinically low levels of T, there are men who are symptomatic with normal T levels who may benefit from tongkat supplementation. Therefore, *E. longifolia's* mechanism of action may be through its ergogenic effects; it does seem to improve andropausal symptoms and improve T levels in select cohorts.

ASHWAGANDHA

Dose: 1 g to 5 g daily (begin at the lower dose and modify according to clinical presentation)

Withania somnifera, a popular Ayurvedic herb, has countless indications ranging from anxiety and stress to fatigue and depression. However, one of its lesser known uses is for andropause-type symptoms including low T and infertility. The use of *Withania* in infertile men was shown to improve semen quality and normalize LH, PRL, and T levels in infertile men (Ahmad, 2010; Gupta, 2013; Mahdi, 2011). Mahdi et al. showed Ashwagandha not only improved sperm motility and T in infertile men, but increased benefits also were seen in stressed men (as evident by questionnaire responses and serum cortisol), and in fact it reduced cortisol levels while increasing T levels (Mahdi, 2011). The significance of lowered cortisol is discussed later in the section on pregnenolone steal.

Ashwagandha is an adaptogenic herb in that it works, or adapts, to the body's level of stress and helps normalize levels if the body is overexerting stress hormones (adrenaline, cortisol, insulin, glucagon) or insufficiently synthesizing these hormones. Although the specific mechanism(s) of action is unknown, it is suspected that *Withania* may work on the hypothalamus to improve PRL, LH, and FSH balance and reduce oxidative stress, which may contribute to andropause (Mahdi, 2011).

DAMIANA

Dose: *Turnera diffusa* 1:3 extract, 1 ml, three to four times daily

The Latin names of this herb, *Turnera diffusa* and *T. aphrodisiaca*, imply its traditional use as an aphrodisiac herb. Although human studies are limited, its traditional use as a sexual stimulant was tested in animal models. Sexually exhausted rats were shown to improve in achieving ejaculation along with reducing mounting time and intercopulatory interval, suggesting these rats had an increased sex drive or libido (Arletti, 1999; Estrada-Reyes, 2009). These studies support the traditional use of Damiana as an aphrodisiac, and it may be even more effective clinically in combination with other botanicals, showing promising benefit in patients in improving sexual desire and ability to perform sexually.

YOHIMBE

Dose: 6 mg of yohimbine HCl daily or 1–1.5ml of Yohimbe bark extract daily or prior to sexual activity

- L-arginine: 6 g prior to sexual activity

Pausinystalia yohimbe bark, with its main constituent yohimbine, has been linked to improved ED. Considering ED is a common symptom of andropause, whether it is psychogenic or organic, yohimbe bark is potentially beneficial. Yohimbe acts as an alpha-2 adrenergic antagonist to reduce peripheral sympathetic activity (Charney 1982; Goldberg 1983). A large meta-analysis found yohimbine, the main constituent of Yohimbe bark, to be safe; it may be effective in 34% to 73% of case of male impotence, with more benefit observed in psychogenic ED (Shamloul, 2010). Combination of yohimbine with arginine was shown to enhance the efficacy for ED (Lebret, 2002).

RED KOREAN GINSENG (RKG)

Dose: 1800 to 2700mg divided daily

Of the many ginseng varieties, *Panax ginseng* shows the most promise to benefit the andropausal male, not to be confused with other "ginsengs" (American ginseng *Panax quinquefolius,* Siberian ginseng *Eleutherococcus senticosus,* etc.). The rhizome and root of this herb are most commonly used as they contain high ginsenoside content; however, the root is also beneficial as it too contains ginsenosides (Nag, 2012).

RKG research has supported most its benefits on erectile function, with or without possible changes in T levels. IIEF scores improved in men consuming RKG, which included both psychogenic and organic ED. Although the mechanism of action is unknown, it is suspected that the ginsenosides are the active ingredient. Two particular ginsenosides have been shown to reduce 5-alpha-reductase activity by binding to the enzyme and reducing activity, thus reducing conversion T to DHT (Leung, 2010; Murata, 2012). *Panax* may increase T levels in hypogonadic or infertile men to normal levels, although without the benefit seen in normogonadic men, which may explain its benefits in ED (Hong, 2002; Salvati, 1996).

HORNY GOAT WEED

Dose: 50 to 150mg icariin daily

- *Epimedium spp.* supplementation must be standardized to icariin, which typically equates to 1,700 to 5,000mg of *Epimedium* standardized to 0.3% icariin.

Horny goat weed, *Epimedium spp.*, is this herb's common name as found in most supplement formulas. The main constituent studied is icariin, which has been shown to have phosphodiesterase-5 inhibitory effects, similar to common ED drugs (Agli, 2008; Liu, 2005; Ning, 2006; Shindel, 2010). Although not as strong as the common ED drug sildenafil, *Epimedium* did improve NOS and nNOS where sildenafil had no impact (Zhang, 2012). Improvements were more pronounced with repeated dosing, differentiated from pharmaceutical PDE-5 inhibitors, which are more for "on-demand" use (Liu, 2005; Shindel, 2010). The benefits of *Epimedium* on T levels have only been tested in animal models, with evidence showing the herb to improve T in chemically castrated male rats, yet at high dose and with no benefit in physically castrated rats (Liu, 2005; Shindel, 2010; Zhang, 2006). Theoretically *Epimedium* is proposed to work on the testes to improve T. Clinically *Epimedium's* ergogenic use is suspect, and although it is likely effective for ED, its use for clinically low T levels in andropausal men is with reservation.

Given the low risk of adverse effects from these botanicals, they are generally considered safe and effective at the suggested doses. In fact these herbs work synergistically to improve male sexual performance and androgen levels. Combined use is encouraged as a more profound response and benefit is noted when they are used together. At the prescribed levels these botanicals are otherwise deemed safe. However, *Pausinystalia yohimbe* must be used with caution in men with hypertension, and *Epimedium spp.* may interact with individuals on anticoagulation therapies.

Therapeutic Dietary Interventions

Before any drug interventions or major supplementation protocols, dietary modifications should be considered and may have a significant impact on andropause.

Hormonal levels are very much influenced by dietary intake, thus andropause as a hormone-related syndrome is also influenced by diet. Carbohydrate-restricted diets typically seem to be the most common recommendation given to individuals dealing with weight management concerns. Prescribing a low-carbohydrate diet does not necessarily mean the individual will be eating more healthfully or will be bettering his health.

Seen clinically but also supported by research, increased body fat is associated with decreased T levels and increased estrogen levels (Gautier, 2013; Vona-Davis, 2013), an andropausal man's worst nightmare. It is well known that adipocytes contain aromatase, which converts T to estradiol, thus indicating men with greater visceral adiposity have greater aromatization and lower T levels. However, increased adipocytes is not the only cause of increased aromatization, as visceral fat contains increased interleukin-6, IL-1 beta, tumor necrosis factor-alpha, and cyclooxygenase-2 (COX-2) activity, all of which induce aromatase gene transcription and activity (Gautier, 2013). Additionally, increased COX-2 activity of arachidonic acid results in increased prostaglandin-E2 activity. These eicosanoids have been shown to increase aromatase activity (Davies, 2002).

A key way to reduce adiposity, reduce inflammatory markers, and decrease COX-2 activity of arachidonic acid would be through a low-carbohydrate diet. Evidence shows delta-5 desaturase, the rate-limiting enzyme in the production of anti-inflammatory eicosapentaenoic acid and docosahexaenoic acid and pro-inflammatory arachidonic acid, is upregulated by insulin (Brenner, 2003). With the consumption of a diet high in arachidonic acid (AA), precursor omega-6 fatty acids and high glycemic foods, the increased activity of delta-5 desaturase will promote AA metabolism to proinflammatory eicosanoids including prostaglandin E2, thus increasing aromatase and lowering T levels.

To combat such effects, it would be prudent to increase polyunsaturated fatty acids, including omega-3 oils, and monounsaturated fatty acids, to decrease the production of proinflammatory eicosanoids and improve anti-inflammatory eicosanoids 1-series and 3-series production.

To do so one would reduce consumption of high glycemic foods (foods greater than a 55 glycemic index [GI]) and increase consumption of omega-3 essential oils. Yet the balance becomes tricky here, because in these situations practitioners would typically suggest a low-carbohydrate diet, but this may not be best diet for the andropausal male.

THE LOW-CARB PREDICAMENT

A low-carbohydrate diet can reduce insulin response, thus reducing AA metabolism. However, strict low-carb diets, compared to >50% low-carb diets, have been shown to reduce T levels, increase cortisol levels (Anderson, 1987; Lane, 2010; Tsai, 1993), and reduce thyroid hormones (Jaya Kumar, 1990; Meikle, 2004; Spaulding 1976). Low T3 and T4 may reduce LH response to GnRH, resulting in reduced Leydig cell activity (Kumar, 2014).

DIETARY RESOLUTION

The predicament is: Do we lower carb intake to reduce adiposity, decrease aromatization, and decrease proinflammatory markers but then risk lowering T and thyroid hormone? The happy medium is to not limit carbohydrate intake but to limit carb intake to foods with a GI of less than 55. These foods do not spike blood glucose levels, limiting insulin response, but still provide adequate carbohydrate intake to maintain an optimal T to cortisol ratio. Protein does not elicit a strong blood sugar increase and, along with fat, reduces the glycemic load of a meal. Therefore, including a fat and protein with each meal is essential in stabilizing blood sugar and insulin and supporting anabolic hormone production. Fibrous foods have a low GI, and the addition of fiber to meals can further lower a meal's GI. It would be prudent to have patients increase fiber intake to more than 30g daily. The addition of flax meal, psyllium husk, wheat bran, or chia seeds prior to or with meals would be ideal to help regulate blood sugar levels. It must not be forgotten, these measures will also promote a lifestyle of healthful eating and likely will result in decreased body fat, thus improving multiple health measures and outcomes.

Simple Lifestyle Changes: Sleep and Stress

The importance of sleep is often overlooked in health. Physicians typically might tell their patients they should be getting seven to eight hours of sleep, yet it is often said without emphasis. Be assured, sleep is *very* important, especially so for optimal androgen levels and sexual health.

Several studies have shown sleep deprivation, from less than four hours to four to six hours, reduces total and free T levels (Goh, 2009; Wittert 2014a, 2014b). Men who slept less than four hours and between four and six hours have a 39.0% and 9.8% decrease in total T and a 29.0% and 7.0% decrease in free T, respectively, with also a decreased sexual motivation evident by decreased masturbation (Goh, 2009). Improving sleep in androgen-deficient men may improve endogenous T production up to 40%, which matches or even supersedes testosterone replacement therapy (TRT), as sleep recovers proper episodic T increases, rather than in a tonic pattern seen with TRT (Goh, 2009). Sleep deprivation also interrupts proper hypothalamic-pituitary-adrenal axis function, thus resulting in increased total cortisol levels, which also may contribute to disrupted androgens (Leproult, 1997).

The stress hormone, cortisol (C), is often compared to the total T ratio. Total T and C reference ranges must be considered as well as compared to the T:C ratio, as a high cortisol level can negatively inhibit GnRH and Leydig T production (Terburg, 2009) indicating overtaxed adrenals and suboptimal gonadal function. It is important to manage stress levels whether it is chronic stress or acute stress such as overtraining (Urhausen 1995).

Guided meditation is a simple and easy way for men to reduce their stress, particularly as a consequence of a chaotic lifestyle. These sessions may last from 5 to 20 minutes with multiple resources available, including mobile applications, CDs, and even online programs. This has shown to be very effective clinically. Additionally, particular herbs such as Ashwagandha (Ahmad, 2010; Ambiye 2013; Gupta 2013; Mahdi, 2011) and *Rhodiola rosea* (Shevtsov 2003; Spasov, 2000) act as adaptogens to help reduce cortisol and balance physiology in high-stress situations and lifestyles.

EXERCISE

Physical activity is one of the more underappreciated therapies physicians can prescribe. Most Americans do not exercise, and only about 20% regularly participate in aerobic or weight-resistant exercise (National Center for Health Statistics, 2014); yet engaging in exercise to mitigate symptoms in the andropausal male has been shown to be very important. However, telling men to just exercise is not enough; more specifically, high-intensity exercise with weights over endurance training has been shown to be most effective in improving androgen levels and reducing cortisol levels.

High-intensity training and weight training have been shown in multiple studies to improve total and free T levels and GH and reduce total cortisol levels and SHBG in men (Goh, 2009; Grandys, 2009; Kraemer, 1989, 1991). By contrast, endurance training including long-distance running/cycling has been shown to reduce T levels and increase cortisol levels (Perna 1995, Lac 2000). In the case of inexperienced men hiring a personal trainer, it may be best at first to properly instruct them in technique but also to develop a legitimate high-intensity program with weight-bearing exercise and to educate them about its significance for health and sexual function. Exercises to emphasize would be large compound movements involving larger muscles such as squats, deadlifts, bench presses, pull-ups, back rows, and shoulder presses. This type of training elicits an increase in lactic acid, which has been shown to increase activity of cAMP and cP450SCC particularly in Leydig cells, thus increasing T production (Kraemer, 2005; Lin, 2001).

The Pregnenolone Steal

As evident in the previous sections, cortisol seems to be the anti-testosterone hormone. Seen in multiple studies, when cortisol levels are elevated T levels seem to drop. This is something that has been evident among many integrative physicians and is called the "pregnenolone steal." The synthesis of both androgens and corticosteroids all share the common precursor pregnenolone. The determination of which hormone is produced is organ- and enzyme specific, as male androgens are mostly made in the testes whereas corticosteroids are mainly made in the cortex of the adrenal glands. Under unique situations when chronic stress is prevalent, cortisol steals the pregnenolone from the androgens. Therefore cortisol synthesis may be ramped up while androgen synthesis may be diminished, thus increasing activity of 11-beta-hydroxylase and decreasing 17-beta-hydroxysteroid dehydrogenase and 17,20 lyase activity.

This is an explanation for why and how chronic stress and elevated cortisol results and is related to diminished T and androgens (including DHEA). In addition to andropause symptoms, other key clues that a pregnenolone steal may be occurring include low vitamin D3 levels, chronic fatigue, low cholesterol levels, inability of the pupils to maintain miosis (pupillary contraction), and orthostatic hypotension.

Environmental Endocrine Disruption

Genistein from soy and other foods, pharmaceuticals such as diethylstilbestrol (DES), and chemicals such as bisphenol A (BPA), dichlorodiphenyltrichloroethane (DDT), polychlorinated biphenyls (PCBs), dioxins, and nonpersistent insecticides are a short list of the common contaminants found in our environment that are known endocrine disruptors. They are found in plastics, landfill runoff, insecticides, pesticides, and foods (meat, dairy, fish), just to name a few. These chemicals have been shown to lower total and free T levels in healthy men, change sperm motility and sperm count, and alter hormones such as TSH, T4, T3, estrogens, and SHBG (Bonde, 2008; Meeker, 2006; Richthoff, 2003; Turyk, 2006; Zawatski, 2013). In addition, lead also has been shown to damage Leydig cells (Rodamilans, 1988; Ronis, 1996; Thoreux-Manlay, 1995).

These substances are often overlooked as possible etiological causes of low libido, low T levels, and all other symptoms of andropause, possibly because the research studies have been small and flawed due to incomplete exposure and assessment endpoints. However, awareness is necessary and exposure *must* be minimized, particularly in the younger population. One may argue that many contaminants have been banned and are no longer used; however, considering their long half-life, lipophilic nature, and tendency to bioaccumulate, they are *still* detectable in our environment and in ourselves (Diamanti-Kandarakis, 2009). In the younger population the latency between exposure and outcome makes it difficult to measure any effect, and as food is a common source of endocrine disruptors, studies correlating adolescent body mass index with pubertal outcomes may overlook the impact of these environmental toxins (Zawatski, 2013).

Late onset hypogonadism, as we have seen clinically, is occurring much earlier with some individuals presenting in their 20s and 30s, not in middle age; andropause typically occurs in middle-aged men. We must question if the sexual dysfunctions men today are experiencing in their 20s, 30s, and 40s are to some extent a result of early life endocrine disruptor exposure, or due to genetics, nutrition and lifestyle, or advancing age. Therefore, to reduce exposure, patients should be advised to follow local fish and wildlife advisories about eating fish and game that may have high levels of environmental contaminants and to favor consumption of organic produce and grass-fed, hormone-free, free-range animal products, thus avoiding, for example, increased consumption of DES and rBST.

Case Study: 22-Year-Old Male with Low Libido

J.R., a 22-year-old male student, presented to clinic with a chief complaint of low libido, feeling as if he does not have any sexual drive, and with decreased sexual pleasure. He claims to be having sex approximately two times per month. Patient complains of fatigue, difficulty studying, and high irritability but without ED.

Patient has been in a monogamous relationship for the past year. The patient's medical history and current health is unremarkable, with a BP is 110/75 and a BMI of 26.1 kg/m^2. Patient is technically overweight but does weight-bearing exercise five to six times weekly and has played soccer competitively since age five. Patient is a high-strung student and puts great stress upon himself for top grades, resulting in only five to six hours of sleep nightly. He does not present with any symptoms of estrogen dominance and is negative for gynecomastia. Testicles are descended bilaterally, nonnodular, nontender to palpation; the epididymis and spermatic cord are palpable and within normal limits.

Patient is not currently taking any pharmaceutical drugs and denies history of any anabolic steroid use. Patient supplements with creatine monohydrate, whey protein, and a multivitamin. Most recent laboratory tests were unremarkable with no makers in the abnormal range. Total cholesterol is 190, HDL 70, LDL 105, TG 110, and fasting blood glucose 75. Follow-up laboratory testing was ordered for early morning (before 10 AM) resulting in total T 287 ng/dL (L), free T 4.8 ng/dL (L), SHBG 14 nmol/L, estradiol 9 pg/mL (L), TSH 2.8 mIU/L, total T4 7.2 mcg/dL, free T3 3.1 pg/dL, DHEA-S 178 mcg/dL (L), PRL 10 ng/mL. Additional testing that included a chest x-ray, ultrasound of abdomen, pelvis, and testicles, and MRI of pituitary were all negative.

Patient has been dealing with symptoms for almost two years without resolution. The patient was offered testosterone gel and intramuscular TRT by his endocrinologist yet he refused because of long-term complications and his young age. A second endocrinologist suggested clomiphene citrate, which the patient was considering after two years of unresolved symptoms.

The patient was put on a protocol, which included 100mg DHEA before bed and a 8-oz tincture of Ashwagandha (40%), *Damiana* (28%), *Panax ginseng* (28%), *Yohimbe* (2%) dosed at 1 tsp t.i.d. After four weeks the patient reported his irritability, fatigue, and difficulty studying completely resolved yet his libido was still an issue. After retesting his blood work, results were similar to his previous lab work yet now his DHEA-S was 380 mcg/dL and his total T and free T were in the low-normal range at 315 ng/dL and 5.1 ng/dL, respectively.

Further modifications to his current protocol included maca root ¾ tsp daily along with increasing weight-bearing exercises (compound movements of squats, deadlifts, bench press, back row, and shoulder press) but limiting to only four days per week. He was advised to sleep seven to eight hours minimum nightly. If he had to decide between an extra hour of sleep or weight-training he was advised to choose the extra hour of sleep.

Following this treatment the patient returned in six weeks in a much better mood, with complete resolution of all symptoms and a slight improvement in libido. However, he claimed this was still not normal for him, as he was traditionally more sexually driven and his decreased sex drive was "unlike" him. A final modification was ½ tsp *M. pruriens* twice daily with food.

Eight weeks following the addition of *M. pruriens* the patient claimed his sex drive was back to normal with sexual activity up to two to three times per week with increased sexual desire. The patient was very content. He was then retested; his total T was 698 ng/dL and free T was 7.9 ng/dL and all laboratory tests were within normal limits. J.R. was removed from his herbal formula and *Mucuna*, his DHEA was reduced to 50mg nightly, and he continued the maca. He continued with his sleep and exercise regime. Because of his high-stress lifestyle as a student he found his sexual health suffered when he steered away from his sleep, exercise, and supplements but otherwise he is no longer hypogonadic or symptomatic.

This case might have presented in analogous form in a midlife, andropausal man, in which case the clinical strategy would have been much the same: counseling to improve lifestyle in such key areas as sleep, exercise, stress reduction, optimum nutrition, and avoiding environmental chemicals as much as possible. Furthermore, implementing a personalized and precise regime of botanicals and nutrients, all with excellent modern and traditional evidence for safety and efficacy, allows for a foundational and functional approach in addressing sexual dysfunction. In fact, many cases in the authors' practices have successfully been resolved in this way.

Professionals should become accustomed to educating men from their teen and early adult years on to adopt optimum lifestyle practices, which will help them experience better health and more fulfilling sexual lives, from youth into middle age and beyond. Finally, given the considerable evidence presented on mechanism of action, efficacy, and safety of botanicals and nutraceuticals, both professionals and patients alike should consider these therapeutics as viable treatment options, with patients seeking out a well-qualified provider who may guide them in the modern clinical use of botanicals and nutraceuticals.

REFERENCES

Age-Related Eye Disease Study Research Group (2001). A randomized, placebo-controlled,clinical trial of high-dose supplementation with vitamins C and E, beta carotene, and zinc for age-related macular degeneration and vision loss: AREDS report no. 8. *Arch Ophthalmol.* 119(10):1417–1436.

Agli MD, Galli G V, Cero ED, et al. (2008). Potent inhibition of human phosphodiesterase-5 by icariin derivatives potent inhibition of human phosphodiesterase-5 by icariin derivatives. *J Nat Prod.* 71:5–10.

Ahmad MK, Mahdi AA, Shukla KK, et al. (2010). Withania somnifera improves semen quality by regulating reproductive hormone levels and oxidative stress in seminal plasma of infertile males. *Fertil Steril.* 94:989–996.

Ambiye VR, Langade D, Dongre S, Aptikar P, Kulkarni M, Dongre A (2013). Clinical evaluation of the spermatogenic activity of the root extract of Ashwagandha (Withania somnifera) in oligospermic males: a pilot study. *Evid Based Complement Altern Med.* 2013.

Andersson KE (2011). Mechanisms of penile erection and basis for pharmacological treatment of erectile dysfunction. *Pharmacol Rev.* 63:811–859.

Anderson KE, Rosner W, Khan MS, et al. (1987). Diet-hormone interactions: protein/carbohydrate ratio alters reciprocally the plasma levels of testosterone and cortisol and their respective binding globulins in man. *Life Sci.* 40(18):1761–1768.

Araujo AB, Wittert GA (2011). Endocrinology of the aging male. *Best Pract Res Clin Endocrinol Metab.* 25:303–319.

Arletti R, Benelli A, Cavazzuti E, Scarpetta G, Bertolini A (1999). Stimulating property of Turnera diffusa and Pfaffia paniculata extracts on the sexual behavior of male rats. *Psychopharmacology (Berl).* 143(1):15–19.

Ben-Jonathan N, LaPensee CR, LaPensee EW (2008). What can we learn from rodents about prolactin in humans? *Endocr Rev.* 29:1–41.

Bhasin S, Cunningham GR, Hayes FJ, et al. (2010). Testosterone therapy in men with androgen deficiency syndromes: an Endocrine Society clinical practice guideline. *J Clin Endocrinol Metab.* 95:2536–2559.

Blomberg Jensen M, Nielsen JE, Jørgensen A, et al (2010). Vitamin D receptor and vitamin D metabolizing enzymes are expressed in the human male reproductive tract. *Hum Reprod.* 25(5):1303–1311.

Bonde JP, Toft G, Rylander L, et al. (2008). Fertility and markers of male reproductive function in Inuit and European populations spanning large contrasts in blood levels of persistent organochlorines. *Environ Health Perspect.* 116:269–277.

Brenner RR (2003). Hormonal modulation of Δ6 and Δ5 desaturases: case of diabetes. *Prostaglandins, Leukot Essent Fat Acids.* 68(2):151–162.

Charney DS, Heninger GR, Sternberg DE (1982). Assessment of alpha-2 adrenergic autoreceptor function in humans: effects of oral yohimbine. *Life Sci.* 30(23):2033–2041.

Chung KW, Kim SY, Chan WY, Rennert OM (1986). Androgen receptors in ventral prostate glands of zinc deficient rats. *Life Sci.* 38:351–356.

Cicero AFG, Bandieri E, Arletti R (2001). Lepidium meyenii Walp. improves sexual behaviour in male rats independently from its action on spontaneous locomotor activity. *J Ethnopharmacol.* 75(2–3):225–229.

Cormio L, De Siati M, Lorusso F, et al. (2011). Oral L-citrulline supplementation improves erection hardness in men with mild erectile dysfunction. *Urology.* 77:119–122.

Corona G, Isidori AM, Buvat J, et al. (2014). Testosterone supplementation and sexual function: a meta-analysis study. *J Sex Med.* 11:1577–1592.

Corona G, Rastrelli G, Maggi M (2013). Diagnosis and treatment of late-onset hypogonadism: systematic review and meta-analysis of TRT outcomes. *Best Pract Res Clin Endocrinol Metab.* 27:557–579.

Cunningham GR, Stephens-Shields AJ, Rosen RC, et al. (2014). Association of sex hormones with sexual function, vitality, and physical function of symptomatic older men with low testosterone levels at baseline in the testosterone trials. *J Clin Endocrinol Metab.* 100(3):1146–1155.

Cynober L, Moinard C, De Bandt JP (2010). The 2009 ESPEN Sir David Cuthbertson. Citrulline: A new major signaling molecule or just another player in the pharmaconutrition game? *Clin Nutr.* 29:545–551.

Davies G, Martin L-A, Sacks N, Dowsett M (2002). Cyclooxygenase-2 (COX-2), aromatase and breast cancer: a possible role for COX-2 inhibitors in breast cancer chemoprevention. *Ann Oncol.* 13(5):669–678.

Denti L, Pasolini G, Sanfelici L, et al. (2000). Aging-related decline of gonadal function in healthy men: correlation with body composition and lipoproteins. *J Am Geriatr Soc.* 48:51–58.

Diamanti-Kandarakis E, Bourguignon J-P, Giudice LC, et al (2009). Endocrine-disrupting chemicals: an Endocrine Society scientific statement. *Endocrine Reviews.* 30:293–342.

Dording CM, Fisher L, Papakostas G, et al. (2008). A double-blind, randomized, pilot dose-finding study of maca root (L. Meyenii) for the management of SSRI-induced sexual dysfunction. *CNS Neurosci Ther.* 14(3):182–191.

Dumoulin SC, Perret BP, Bennet AP, Caron PJ (1995). Opposite effects of thyroid hormones on binding proteins for steroid hormones (sex hormone-binding globulin and corticosteroid-binding globulin) in humans. *Eur J Endocrinol.* 132: 594–598.

Estrada-Reyes R, Ortiz-López P, Gutiérrez-Ortíz J, Martínez-Mota L (2009). Turnera diffusa Wild (Turneraceae) recovers sexual behavior in sexually exhausted males. *J Ethnopharmacol.* 123(3):423–429.

Eyles DW, Smith S, Kinobe R, Hewison M, McGrath JJ (2005). Distribution of the vitamin D receptor and 1 alpha-hydroxylase in human brain. *J Chem Neuroanat.* 29(1):21–30.

Feldman HA, Goldstein I, Hatzichristou DG, Krane RJ, McKinlay JB (1994). Impotence and its medical and psychosocial correlates: results of the Massachusetts Male Aging Study. *J Urol.* 151:54–61.

Feldman HA, Longcope C, Derby CA, et al. (2002). Age trends in the level of serum testosterone and other hormones in middle-aged men: longitudinal results from the Massachusetts Male Aging Study. *J Clin Endocrinol Metab.* 87(2):589–598.

Gansser D, Spiteller G (1995). Plant constituents interfering with human sex hormonebinding globulin: evaluation of a test method and its application to Urtica dioica root extracts. *Zeitschrift für Naturforsch.* 50c:98–104.

Ganßer D, Spiteller G (1995). Aromatase inhibitors from Urtica dioica roots. *Planta Med.* 61(2):138–140.

Gautier A, Bonnet F, Dubois S, et al. (2013). Associations between visceral adipose tissue, inflammation and sex steroid concentrations in men. *Clin Endocrinol (Oxf).* 78:373–378.

Goh VHH, Tong TYY (2009). The association of exercise with sleep, sex steroids, sexual activities and aging in Asian men. *J Mens Health.* 6(4):317–324.

Goldberg MR, Robertson D (1983). Yohimbine: a pharmacological probe for study of the alpha 2-adrenoreceptor. *Pharmacol Rev.* 35(3):143–180.

Gonzales GF, Córdova A, Vega K, et al. (2002). Effect of Lepidium meyenii (MACA) on sexual desire and its absent relationship with serum testosterone levels in adult healthy men. *Andrologia.* 34:367–372.

Grandys M, Majerczak J, Duda K, Zapart-Bukowska J, Kulpa J, Zoladz JA (2009). Endurance training of moderate intensity increases testosterone concentration in young, healthy men. *Int J Sports Med.* 30(7):489–495.

Gupta A, Mahdi AA, Ahmad MK, et al. (2011). A proton NMR study of the effect of Mucuna pruriens on seminal plasma metabolites of infertile males. *J Pharm Biomed Anal.* 55:1060–1066.

Gupta A, Mahdi AA, Shukla KK, et al. (2013). Efficacy of Withania somnifera on seminal plasma metabolites of infertile males: A proton NMR study at 800 MHz. *J Ethnopharmacol.* 149:208–214.

Hammond GL, Hogeveen KN, Visser M, Coelingh Bennink HJ (2008). Estetrol does not bind sex hormone binding globulin or increase its production by human HepG2 cells. *Climacteric.* 11(Suppl 1):41–46.

Harman SM, Metter EJ, Tobin JD, Pearson J, Blackman MR (2001). Longitudinal effects of aging on serum total and free testosterone levels in healthy men. Baltimore Longitudinal Study of Aging. *J Clin Endocrinol Metab.* 86:724–731.

Henkel RR, Wang R, Bassett SH, et al. (2014). Tongkat Ali as a potential herbal supplement for physically active male and female seniors—a pilot study. *Phyther Res.* 28(4):544–550.

Hong B, Ji YH, Hong JH, Nam KY, Ahn TY (2002). A double-blind crossover study evaluating the efficacy of Korean red ginseng in patients with erectile dysfunction: a preliminary report. *J Urol.* Nov;168(5):2070–2073.

Islam N, Trainer PJ (1998). The hormonal assessment of the infertile male. *Br J Urol.* 82(1):69–75.

Ismail SB, Wan Mohammad WMZ, George A, Nik Hussain NH, Musthapa Kamal ZM, Liske E (2012). Randomized clinical trial on the use of PHYSTA freeze-dried water extract of Eurycoma longifolia for the improvement of quality of life and sexual well-being in men. *Evid Based Complement Altern Med.* 2012.

Jaya Kumar B, Khurana ML, Ammini AC, Karmarkar MG, Ahuja MM (1990). Reproductive endocrine functions in men with primary hypothyroidism: effect of thyroxine replacement. *Horm Res.* 34(5–6):215–218.

Kelly MP, Kight MA, Castillo S (1998). Trophic implications of altered body composition observed in or near the nails of hemodialysis patients. *Adv Ren Replace Ther.* 5(3):241–251.

Kiecolt-Glaser JK, Belury MA, Andridge R, Malarkey WB, Glaser R (2011). Omega-3 supplementation lowers inflammation and anxiety in medical students: a randomized controlled trial. *Brain Behav Immun.* 25:1725–1734.

Kinuta K, Tanaka H, Moriwake T, Aya K, Kato S, Seino Y (2000). Vitamin D is an important factor in estrogen biosynthesis of both female and male gonads. *Endocrinology.* 141(4):1317–1324.

Koch E (1995). Pharmacology and modes of action of extracts of Palmetto fruits (Sabal Fructus), stinging nettle roots (Urticae Radix) and pumpkin seed (Cucurbitae Peponis Semen) in the treatment of benign prostatic hyperplasia. In D Loew, N Rietbrock (Eds.), *Phytopharmaka in Forschung Und Klinischer Anwendung* (pp. 55–57). Heidelberg: Steinkopff.

Kraemer WJ, Gordon SE, Fleck SJ, et al. (1991). Endogenous anabolic hormonal and growth factor responses to heavy resistance exercise in males and females. *Int J Sports Med.* Apr;12(2):228–235.

Kraemer WJ, Patton JF, Knuttgen HG, et al. (1989). Hypothalamic-pituitary-adrenal responses to short-duration high-intensity cycle exercise. *J Appl Physiol.* 66(1):161–166.

Kraemer WJ, Ratamess NA (2005). Hormonal responses and adaptations to resistance exercise and training. *Sports Med.* 35:339–361.

Kumar A, Shekhar S, Dhole B (2014). Thyroid and male reproduction. *Indian J Endocrinol Metab.* 18(1):23.

Lac G, Berthon P (2000). Changes in cortisol and testosterone levels and T/C ratio during an endurance competition and recovery. *J Sports Med Phys Fitness.* 40(2):139–144.

Lane AR, Duke JW, Hackney AC (2010). Influence of dietary carbohydrate intake on the free testosterone: cortisol ratio responses to short-term intensive exercise training. *Eur J Appl Physiol.* 108(6):1125–1131.

Langdahl BL, Stenkjaer L, Carstens M, Tofteng CL, Eriksen EF (2003). A CAG repeat polymorphism in the androgen receptor gene is associated with reduced bone mass and increased risk of osteoporotic fractures. *Calcif Tissue Int.* 73:237–243.

Lebret T, Hervé J-M, Gorny P, Worcel M, Botto H (2002). Efficacy and safety of a novel combination of L-arginine glutamate and yohimbine hydrochloride: a new oral therapy for erectile dysfunction. *Eur Urol.* 41(6):608–613.

Lee DM, Tajar A, Pye SR, et al. (2012). Association of hypogonadism with vitamin D status: the European Male Ageing Study. *Eur J Endocrinol.* 166(1):77–85.

Lei KY, Abbasi A, Prasad AS (1976). Function of pituitary-gonadal axis in zinc-deficient rats. *Am J Physiol.* 230:1730–1732.

Leproult R, Copinschi G, Buxton O, Van Cauter E (1997). Sleep loss results in an elevation of cortisol levels the next evening. *Sleep.* Oct;20(10):865–870.

Leung KW, Wong AS-T (2010). Pharmacology of ginsenosides: a literature review. *Chin Med.* 5:20.

Libè R, Barbetta L, Dall'Asta C, et al. (2004). Effects of dehydroepiandrosterone (DHEA) supplementation on hormonal, metabolic and behavioral status in patients with hypoadrenalism. *J Endocrinol Invest.* 27:736–741.

Lin H, Wang SW, Wang RY, Wang PS (2001). Stimulatory effect of lactate on testosterone production by rat Leydig cells. *J Cell Biochem.* 83:147–154.

Liu TC, Lin CH, Huang CY, Ivy JL, Kuo CH (2013). Effect of acute DHEA administration on free testosterone in middle-aged and young men following high-intensity interval training. *Eur J Appl Physiol.* 113:1783–1792.

Liu W-J, Xin Z-C, Xin H, Yuan Y-M, Tian L, Guo Y-L (2005). Effects of icariin on erectile function and expression of nitric oxide synthase isoforms in castrated rats. *Asian J Androl.* 7:381–388.

Mahdi AA, Shukla KK, Ahmad MK, et al. (2011). Withania somnifera improves semen quality in stress-related male fertility. *Evid Based Complement Altern Med.* 2011:1–9.

Meeker JD, Ryan L, Barr DB, Hauser R (2006). Exposure to nonpersistent insecticides and male reproductive hormones. *Epidemiology.* 17:61–68.

Meikle AW (2004). The interrelationships between thyroid dysfunction and hypogonadism in men and boys. *Thyroid.* 14(Suppl 1):S17–S25.

Moinard C, Cynober L (2007). Citrulline: a new player in the control of nitrogen homeostasis. *J Nutr.* 137:1621S–1625S.

Muller M, Van Den Beld AW, Van Der Schouw YT, Grobbee DE, Lamberts SWJ (2006). Brief report: effects of dehydroepiandrosterone and atamestane supplementation on frailty in elderly men. *J Clin Endocrinol Metab.* 91:3988–3991.

Murata K, Takeshita F, Samukawa K, Tani T, Matsuda H (2012). Effects of ginseng rhizome and Ginsenoside RO on testosterone 5?—reductase and hair re-growth in testosterone-treated mice. *Phyther Res.* 26(1):48–53.

Nag SA, Qin J-J, Wang W, Wang M-H, Wang H, Zhang R (2012). Ginsenosides as anticancer agents: in vitro and in vivo activities, structure-activity relationships, and molecular mechanisms of action. *Front Pharmacol.* 3(February):25.

Nangia AK, Hill O, Waterman MD, Schwender CEB, Memoli V (2007). Testicular maturation arrest to testis cancer: spectrum of expression of the vitamin D receptor and vitamin D treatment in vitro. *J Urol.* 178(3):1092–1096.

National Center for Health Statistics (2014). With special feature on prescription drugs. http://www.ncbi.nlm.nih.gov/books/NBK209224/.

Neuzillet Y, Hupertan V, Cour F, Botto H, Lebret T (2013). A randomized, double-blind, crossover, placebo-controlled comparative clinical trial of arginine aspartate plus adenosine monophosphate for the intermittent treatment of male erectile dysfunction. *Andrology.* 1(2):223–228.

Ning H, Xin ZC, Lin G, Banie L, Lue TF, Lin CS (2006). Effects of icariin on phosphodiesterase-5 activity in vitro and cyclic guanosine monophosphate level in cavernous smooth muscle cells. *Urology.* 68:1350–1354.

Om AS, Chung KW (1996). Dietary zinc deficiency alters 5 alpha-reduction and aromatization of testosterone and androgen and estrogen receptors in rat liver. *J Nutr.* 126:842–848.

Ostojic SM, Calleja J, Jourkesh M (2010). Effects of short-term dehydroepiandrosterone supplementation on body composition in young athletes. *Chin J Physiol.* 53:19–25.

Pérez-Fernandez R, Alonso M, Segura C, Muñoz I, García-Caballero T, Diéguez C (1996). Vitamin D receptor gene expression in human pituitary gland. *Life Sci.* 60(1):35–42.

Perna FM, McDowell SL (1995). Role of psychological stress in cortisol recovery from exhaustive exercise among elite athletes. *Int J Behav Med.* 2(1):13–26.

Pilz S, Frisch S, Koertke H, et al. (2011). Effect of vitamin D supplementation on testosterone levels in men. *Horm Metab Res.* 43(3):223–225.

Rajfer J (2000). Relationship between testosterone and erectile dysfunction. *Rev Urol.* 2:122–128.

Reiter WJ, Pycha A, Schatzl G, et al. (1999). Dehydroepiandrosterone in the treatment of erectile dysfunction: a prospective, double-blind, randomized, placebo-controlled study. *Urology.* Mar;53(3):590–594; discussion 594–595.

Richthoff J, Rylander L, Jönsson BAG, et al. (2003). Serum levels of 2,2',4,4',5,5'-hexachlorobiphenyl (CB-153) in relation to markers of reproductive function in young males from the general Swedish population. *Environ Health Perspect.* 111:409–413.

Rodamilans M, Martinez-Osaba MJ, To-Figueras J, et al. (1988). Inhibition of intratesticular testosterone synthesis by inorganic lead. *Toxicol Lett.* 42:285–290.

Romero MJ, Platt DH, Caldwell RB, Caldwell RW (2006). Therapeutic use of citrulline in cardiovascular disease. *Cardiovasc Drug Rev.* 24:275–290.

Ronis MJ, Badger TM, Shema SJ, Roberson PK, Shaikh F (1996). Reproductive toxicity and growth effects in rats exposed to lead at different periods during development. *Toxicol Appl Pharmacol.* 136:361–371.

Salvati G, Genovesi G, Marcellini L, et al. (1996). Effects of Panax Ginseng C.A. Meyer saponins on male fertility. *Panminerva Med.* 38(4):249–254.

Samman S, Robert DCK (1987). The effect of zinc supplements on plasma zinc and copper levels and the reported symptoms in healthy volunteers. *Med J Aust.* 146(5):246–249.

Sarne DH, Refetoff S, Rosenfield RL, Farriaux JP (1988). Sex hormone-binding globulin in the diagnosis of peripheral tissue resistance to thyroid hormone: the value of changes after short term triiodothyronine administration. *J Clin Endocrinol Metab.* 66(4):740–746.

Schmidt K (1983). The effect of an extract of Radix Urticae and various secondary extracts on the SHBG of blood plasma in benign prostatic hyperplasia. *Fortschr Med.* 101:713–716.

Schöttner M, Ganßer D, Spiteller G (1997). Lignans from the roots of Urtica dioica and their metabolites bind to human sex hormone binding globulin (SHBG). *Planta Med.* 63(06):529–532.

Selva DM, Hammond GL (2009). Thyroid hormones act indirectly to increase sex hormone-binding globulin production by liver via hepatocyte nuclear factor-4? *J Mol Endocrinol.* 43:19–27.

Shamloul R (2010). Natural aphrodisiacs. *J Sex Med.* 7(1 Part 1):39–49.

Shevtsov VA, Zholus BI, Shervarly VI, et al. (2003). A randomized trial of two different doses of a SHR-5 Rhodiola rosea extract versus placebo and control of capacity for mental work. *Phytomedicine.* 10(2):95–105.

Shindel AW, Xin ZC, Lin G, et al. (2010). Erectogenic and neurotrophic effects of icariin, a purified extract of horny goat weed (Epimedium spp.) in vitro and in vivo. *J Sex Med.* 7:1518–1528.

Shiota A, Hotta Y, Kataoka T, Morita M, Maeda Y, Kimura K (2013). Oral l-Citrulline supplementation improves erectile function in rats with acute arteriogenic erectile dysfunction. *J Sex Med.* 10:2423–2429.

Shukla KK, Mahdi AA, Ahmad MK, Jaiswar SP, Shankwar SN, Tiwari SC (2010). Mucuna pruriens reduces stress and improves the quality of semen in infertile men. *Evid Based Complement Altern Med.* 7(1):137–144.

Shukla KK, Mahdi AA, Ahmad MK, Shankhwar SN, Rajender S, Jaiswar SP (2009). Mucuna pruriens improves male fertility by its action on the hypothalamus-pituitary-gonadal axis. *Fertil Steril.* 92:1934–1940.

Sobrinho LG (2003). Prolactin, psychological stress and environment in humans: Adaptation and maladaptation. *Pituitary.* 6:35–39.

Spasov AA, Wikman GK, Mandrikov VB, Mironova IA, Neumoin VV (2000). A double-blind, placebo-controlled pilot study of the stimulating and adaptogenic effect of Rhodiola rosea SHR-5 extract on the fatigue of students caused by stress during an examination period with a repeated low-dose regimen. *Phytomedicine.* 7(2):85–89.

Spaulding SW, Chopra IJ, Sherwin RS, Lyall SS (1976). Effect of caloric restriction and dietary composition on serum T3 and reverse T3 in man. *J Clin Endocrinol Metab.* 42(1):197–200.

Stone M, Ibarra A, Roller M, Zangara A, Stevenson E (2009). A pilot investigation into the effect of maca supplementation on physical activity and sexual desire in sportsmen. *J Ethnopharmacol.* 126:574–576.

Tambi MIBM, Imran MK (2010). Eurycoma longifolia Jack in managing idiopathic male infertility. *Asian J Androl.* 12:376–380.

Tambi MIBM, Imran MK, Henkel RR. Standardised water-soluble extract of Eurycoma longifolia, Tongkat ali, as testosterone booster for managing men with late-onset hypogonadism? *Andrologia.* 2012;44:226–230.

Terburg D, Morgan B, van Honk J (2009). The testosterone-cortisol ratio: a hormonal marker for proneness to social aggression. *Int J Law Psychiatry.* 32(4):216–223.

Thoreux-Manlay A, Vélez de la Calle JF, Olivier MF, Soufir JC, Masse R, Pinon-Lataillade G (1995). Impairment of testicular endocrine function after lead intoxication in the adult rat. *Toxicology.* 100:101–109.

Tsai L, Karpakka J, Aginger C, Johansson C, Pousette Å, Carlström K (1993). Basal concentrations of anabolic and catabolic hormones in relation to endurance exercise after short-term changes in diet. *Eur J Appl Physiol Occup Physiol.* 66(4):304–308.

Turyk ME, Anderson HA, Freels S, et al. (2006). Associations of organochlorines with endogenous hormones in male Great Lakes fish consumers and nonconsumers. *Environ Res.* 102:299–307.

Urhausen A, Gabriel H, Kindermann W (1995). Blood hormones as markers of training stress and overtraining. *Sports Med.* 20(4):251–276.

Vona-Davis L, Rose DP (2013). The obesity-inflammation-eicosanoid axis in breast cancer. *J Mammary Gland Biol Neoplasia.* 18:291–307.

Wehr E, Pilz S, Boehm BO, März W, Obermayer-Pietsch B (2010). Association of vitamin D status with serum androgen levels in men. *Clin Endocrinol (Oxf).* 73(2):243–248.

Weiss EP, Shah K, Fontana L, Lambert CP, Holloszy JO, Villareal DT (2009). Dehydroepiandrosterone replacement therapy in older adults: 1- and 2-y effects on bone. *Am J Clin Nutr.* 89:1459–1467.

Wijnands KAP, Vink H, Briedé JJ, et al. (2012). Citrulline a more suitable substrate than arginine to restore no production and the microcirculation during endotoxemia. *PLoS One.* 7.

Wittert G (2014a). The relationship between sleep disorders and testosterone. *Curr Opin Endocrinol Diabetes Obes.* 21(3):239–243.

Wittert G (2014b). The relationship between sleep disorders and testosterone in men. *Asian J Androl.* 16(2):262–265.

Wu FCW, Tajar A, Beynon JM, et al. (2010). Identification of late-onset hypogonadism in middle-aged and elderly men. *N Engl J Med.* 363:123–135.

Zawatski W, Lee MM (2013). Male pubertal development: Are endocrine-disrupting compounds shifting the norms? *J Endocrinol.* 218:R1–R12.

Zhang ZB, Yang QT (2006). The testosterone mimetic properties of icariin. *Asian J Androl.* 8(5):601–605.

Zhang J, Wang YB, Ma CG, et al. (2012). Icarisid II, a PDE5 inhibitor from Epimedium wanshanense, increases cellular cGMP by enhancing NOS in diabetic ED rats corpus cavernosum tissue. *Andrologia.* 44(Suppl.1):87–93.

Zheng BL, He K, Kim CH, et al. (2000). Effect of a lipidic extract from Lepidium meyenii on sexual behavior in mice and rats. *Urology.* 55(4):598–602.

Zitzmann M (2009). The role of the CAG repeat androgen receptor polymorphism in andrology. *Front Horm Res.* 37:52–61.

Zitzmann M, Nieschlag E (2003). The CAG repeat polymorphism within the androgen receptor gene and maleness. *Int J Androl.* 26:76–83.

6

Integrative Treatment of Male Sexual Dysfunction

JEFFREY A. ALBAUGH

Chapter 4 covered the common types of male sexual dysfunction including premature ejaculation, delayed ejaculation, erectile dysfunction, lack of libido, and Peyronie's disease. Normal sexual function and dysfunction were described. In addition, the traditional medical approaches to sexual dysfunction were also discussed. This chapter focuses on other approaches including behavioral and complementary approaches to treating sexual dysfunction.

An important aspect of the integrative approach to sexual dysfunction involves the approach to patients and/or partners. It is imperative to use a nonjudgmental, open therapeutic approach to establish a working relationship with the patient and/or partner. Some clinicians use the PLISSIT model (Permission, Limited Information, Specific Suggestions, and Intensive Therapy) with their patients to approach sexual problems (Annon, 1974). The approach begins with getting and giving permission (the P in the model) to talk about sex, intimacy, and sexual problems. Giving the patient/partner permission opens communication and acknowledges the patient's concerns. In addition, the clinician is asking the patient permission to educate him or her on sex. The next step involves giving limited information (the LI in the model) about sex and sexual problems. Information may include describing sex and sexual function, addressing and correcting misinformation about sexuality, and discussing common male sexual problems. After giving limited information, the clinician may provide specific suggestions (the SS in the model) that may be helpful in addressing sexual problems. The clinician must fully understand the patient's and/or partner's full medical/psychological/social/sexual history. It is also important to gain a complete perspective on the nature of the sexual problem and the patient/partner sexual practices to understand how treatment options might be incorporated into lovemaking practices. The course of treatment will be guided by the patient/partner's sexual health goals.

The final step in the PLISSIT model is to provide or refer the patient for intensive therapy (the IT of the model). Throughout the process the clinician should be cognizant of the patient's and partner's goals, feelings, and expectations.

Education is an essential element of a successful sexual dysfunction program. Patients will not continue to use proven therapies if they do not have an understanding of how these treatments work and feel comfortable using the treatment. The clinician should approach the subject with openness and honesty. All patients and their partners should be made to feel accepted and acknowledged in their personal struggles with sexual dysfunction. Discussion should center on factual information and focus on resolving the patient's problem. Teaching strategies should include written, video, model reference, and verbal communication. The key to patient success with the therapy lies in the clinician's ability to effectively work one on one with the patient and/or the partner to incorporate the techniques/treatment into loveplay with their partner. The clinician should be comfortable talking about all aspects of the various therapies.

Healthy lifestyle choices including a low-fat, low-cholesterol plant-based diet, regular cardiovascular exercise, eliminating nicotine, and tightly controlling medical issues that impact sexual function are all essential to sexual health. These factors were discussed in some detail in Chapter 4, and these lifestyle changes are among the ways a man can try to reverse sexual dysfunction. Part of managing sexual dysfunction involves addressing and minimizing possible underlying etiologies or contributing factors. Controlling underlying medical conditions such as diabetes, high blood pressure, hyperlipidemia, obesity, and kidney disease is important. If a particular medication is problematic to sexual function, it may be possible to work with the other prescribers to change or discontinue that medication and resolve the erectile problem. In determining how to move forward with the first and subsequent treatments, the clinician and couple must explore preferences for treatment, potential tolerability of techniques or treatments, convenience, and/or concomitant medications and medical conditions.

Premature Ejaculation

Alternative treatments for premature ejaculation include layered condoms, herbal remedies, and behavioral techniques. Layered condoms offers a technique to desensitize the penis by adding multiple condoms over the penis. The patient can add or subtract condoms if he feels too much or not enough during sex. It can be helpful to have the condoms ready to go next to the bed so they are easily accessible during sex. The layered condoms can be used alone or in conjunction with behavioral techniques and/ or a numbing spray such as Promescent. Behavioral techniques can be effective.

Distraction can sometimes be helpful and can be accomplished by changing the focus, when sexual sensation gets too close to climax, to something nonsexual such as the parts of a car or running through mathematical equations. In addition, some men can last longer on the bottom during sex verses on top, and then once they are ready to climax they could change positions if they want. Most men can last longer after they have recently climaxed, so they may masturbate to climax within 24 to 48 hours before sex knowing they want to last longer during sex with their partner. The stop-start method can also be used to delay climax. As a man begins to feel close to climax, he stops all stimulation on the penis or of himself until the feeling subsides (this generally takes about 30 seconds). These behavioral techniques can be used together or separately to delay climax.

Acupuncture has also been used to treat premature ejaculation. In a randomized, placebo-controlled trial acupuncture worked better than placebo to improve the Premature Ejaculation Diagnostic Tool scores (from 16 to 11, $p = 0.001$) and the intravaginal ejaculation latency times (Sunay, 2011). It did not work as well as paroxetine but was better than placebo (Box 6.1).

Box 6.1 Integrative Sexual Medicine Case Study

John is a 29-year-old male with a lifelong history of premature ejaculation with inability to control ejaculation, which happens within one minute of intravaginal penetration, and this is causing him great distress. He is otherwise healthy and reports no other medical issues. He has a normal BMI, and free and total testosterone and prolactin blood levels are all normal. He has tried delaying techniques such as distraction and the squeeze technique to no avail. He is frustrated and upset. His physical exam is completely normal. He and his wife are both willing to work on improving sex and intimacy. After a full discussion of all possible treatment options he decides to try using behavioral techniques first, and he and his wife begin working on sensate focus (regular planned loveplay to improve intimacy). They begin using behavioral techniques with stop/start method, distraction, male on bottom positioning, and recent climax with sensate focus sessions. Once the couple moves into the fourth phase of sensate focus and are allowed to move toward penetrative sex, John begins using Promescent spray beginning with 1 spray applied to the penis 10 minutes prior to sex, layered condoms, or both the spray and condoms. Both the spray and the condoms help him to last longer, and, within a few sessions focused in phase 4 of sensate focus, he is able to delay climax to four to five times longer. Sometimes the numbness related to the spray is problematic, so he prefers the layered condoms, but the combination of the spray and a condom work best for him. Both John and his wife report improved sexual satisfaction with the longer intravaginal latency time.

Delayed Ejaculation

Treatment of delayed ejaculation is aimed at relieving any underlying anxiety or distress and helping a man to focus on reaching climax. Depending on any underlying psychological anxiety or distress, it may be helpful to treat these underlying conditions with cognitive-behavioral therapy with a trained therapist (Althof, 2012). In some cases the couple may need to have therapy as well. Performance anxiety and loss of sexual self-confidence need to be addressed. Working with men to overcome performance anxiety and build sexual confidence can be accomplished through psychotherapy as well as teaching men some relaxation techniques and working with them on repeated exposure to positive sexual experience without emphasis on performance. It may be beneficial to work with partners as well since partner support and cooperation is essential to success.

Mindfulness and anchoring to stay focused on sex during sexual relations can be helpful. Mindfulness is a process of teaching a man to stay present during sex, focusing on the sensations of sex and not letting his mind drift to anything but sex. A man can also use anchoring to stay present during sex by focusing on the sensations of sex (such as the warmth of the vagina around the penis).

If insufficient stimulation is the problem, adding vibratory stimulation, enhancing mental stimulation, and/or increasing sensations with more vigorous thrusting during sex can help (Althof, 201). Vibratory stimulation taps into a different part of the neurologic system and may improve ability to reach climax. Erotica can be used to enhance mental stimulation and improve ability to climax. Fantasy can also increase arousal and help lead to climax (McMahon, 201). In addition, more vigorous thrusting or positioning for increased sensation may improve the ability to orgasm.

Some men are able to climax with masturbation but not with a partner. It may be helpful to incorporate techniques used during masturbation to reach successful orgasm during sex with a partner. Masturbatory retraining may be helpful in learning to climax/orgasm. It may also help to move toward techniques that assimilate intercourse during masturbation such as using a sleeve or tube to thrust the penis within to reach climax. Once a man is able to climax with masturbation, he can work on incorporating those techniques into sex with his partner to help him climax.

Korean ginseng (panax ginseng) is an herb that has been shown to improve sexual function in limited research but including multicenter, placebo-controlled, double-blind clinical trials (Choi, 2012; Jang, 2008). In the most recently published multicenter placebo-controlled, double-blind study not only were the primary endpoints of improving erectile function met, but the authors also noted improvement in the orgasmic function domain of the International Index of Erectile Function at four ($p = 0.02$) and eight ($p = 0.002$) weeks (Choi, 2012). This botanical may be helpful for enhancing orgasm in patients with delayed orgasm, but further research is needed looking at that particular indication.

Erectile Dysfunction

There is consistent positive data supporting red ginseng (Panax ginseng) and heart-healthy lifestyle changes for erectile dysfunction, and an article recently criticized the current erectile dysfunction guidelines for not including the evidence for both in clinical guidelines (Moyad, 2012). In a 2008 systematic review of red ginseng for treating erectile dysfunction, seven randomized controlled trials were examined and the collective evidence supported the effectiveness of red ginseng in the treatment of erectile dysfunction. However, the total sample size from all trials was too low and the methodology quality too weak to draw definitive conclusions; more rigorous studies are necessary (Jang, 2008). A more recent multicenter, placebo-controlled, double blind study of 119 men receiving four 350mg tablets of red ginseng versus placebo showed improved overall International Index of Erectile Function scores mean of 40.95 to a mean of 46.19 (p = 0.002) and the erectile function domain at eight weeks (mean of 17.17 to mean of 18.59, p = 0.05), and no safety issues with changes in hormonal or cholesterol markers were noted (Choi, 2012). Only one case of mild gastrointestinal upset was reported at four weeks, and no side effects were reported at eight weeks (Choi, 2012).

The Erektor is a nonsurgical, external penile support device that fits on the flaccid penis using a ring at the base of the penis attached by a thin, hard splint system that runs the length of the underside of the penis down to the head where it attaches to the underside of a second ring, which fits snugly behind the corona of the glans (Figure 6.1). This splint system is worn during sex to hold the penis during penetrative sex. Although it has no formal research to support efficacy, it was described in *Urology Times* as a new product in 2009 as well as the *Spine Cord Injury Life Newsletter*. The website has many testimonials from men who have used it, but research is needed to determine further the efficacy and safety of the device.

Peyronie's Disease

Although there is limited evidence to support oral therapies for Peyronie's disease, there are some options that may be helpful with this disease. Colchicine 1.2 mg twice daily for three to five months decreased plaque size and improved penile curvature in a pilot study (Akkus, 1994). The study included 24 men, and the side effects reported were mild to severe gastrointestinal problems (most commonly diarrhea); four of the patients discontinued it due to the side effects. Another

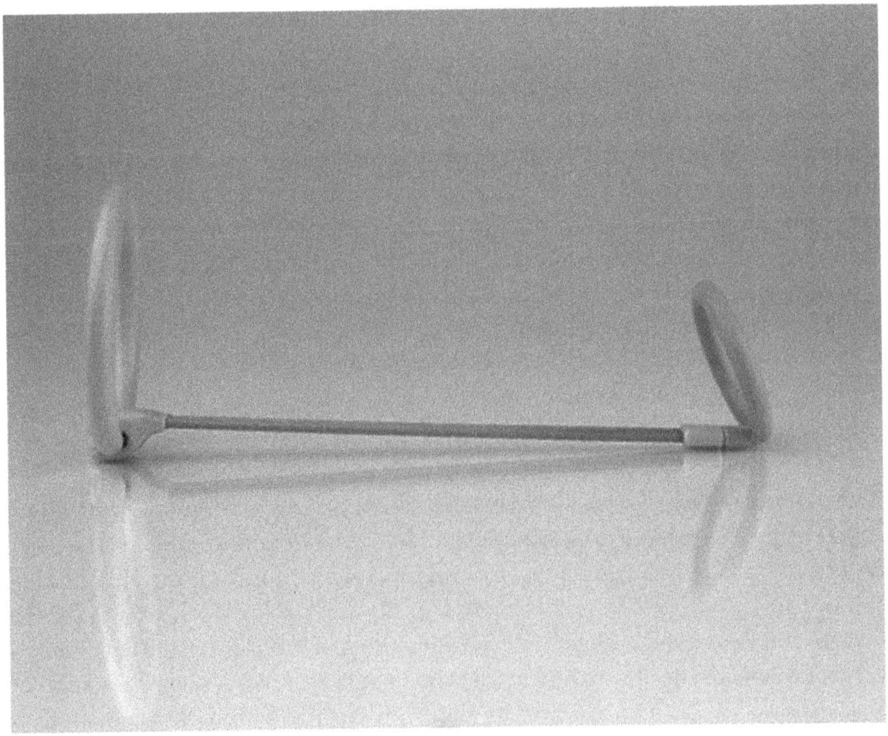

FIGURE 6.1. Erektor Device.
Reprinted with permission of Global Life Technologies.

randomized, double-blind controlled study concluded that four months of therapy of up to 2.5 mg colchicine daily did not show benefit over placebo (Safarinejad, 2004). It may also be used in combination with Vitamin E. Acetyl-L carnitine is a short-chain derivative of carnitine that has antiproliferative effects endothelial cells and may decrease pain and curvature and/or inhibit disease progression with Peyronie's disease when taken (1 g) twice daily (Biagiotti, 2001), but another study did not show efficacy alone or in combined with Vitamin E when compared with placebo (Safarinejad, 2007). Coenzume Q10 is a potent antidoxiant and was studied in a double-blind, placebo-controlled trial demonstrating a statistically significant reduction in plaque size, decreased penile curvature, and improved erectile function (Safarinejad, 2010). Research has not revealed favorable results for either Vitamin E or omega-3 fatty acids (Safarinejad, 2007, 2009).

The penile traction device is used for mechanotransduction and/or expansion of tissue by traction to form new connective tissue. The Fast Size Penile Extender (Fast Size Medical, Aliso Viefo, California, USA) and the Andro-Penis (Andromedical, Madrid, Spain) traction devices have been studied in men with

Peyronie's disease. The Fast Size Penile extender was worn for two to eight hours a day for six months and the rods were lengthened every two weeks by 0.5 cm of penile length by 10 men with a 33% mean reduction (10- to 45-degree decrease) in curvature reported in all men (Levine, 2008). The Anro-Penis was used for a minimum of four hours a day by 15 men, and researchers found a non-significant mean 4-degree improvement in overall curvature and average 1.2-cm gain in flaccid penile length (Gontero, 2009). The vacuum device (Osbon ErecAid) was used for 10 minutes twice daily without any tension rings for 12 weeks, and men had a clinically and statistically significant improvement in penile length and angle of curvature and pain (Raheem, 2010).

Lack of Libido

Maca is one of the most commonly cited herbs on the Internet for sexual dysfunction and has been used as an aphrodisiac. There are only a few small studies to support this claim. Maca root may improve sexual function including libido (Shin, 2010). Four randomized controlled trials met the inclusion criteria for the review, and two of them suggested a positive effect of maca on sexual dysfunction or sexual desire in healthy menopausal women or healthy adult men while another study showed slight improvement in men with erectile dysfunction and a fourth trial failed to show an effect on sexual function in healthy male cyclists. None of the studies assessed adverse effects of maca. Although maca did not significantly improve mild erectile function when compared with placebo, the overall sense of well-being was improved in the maca group.

In this volume, see in particular Chapter 24 for a larger discussion of traditional botanical medicine for sexual problems.

Sexual Dysfunction after Cancer Treatment

The negative impact of sexual dysfunction in men after cancer treatment can be devastating for some men. Cancer treatments, including chemotherapy, surgery, and radiation, often impact sexual function. The surgical and radiation treatments for the more common cancers in men, such as prostate and colorectal cancer, typically have an impact on sexual function. A cross-sectional study of men with prostate cancer in the United States revealed that 78% of men at two years and 72% of men at five years were struggling with erectile dysfunction after radical prostatectomy (Penson, 2008). Erectile dysfunction is also common after radiation therapy with similar prevalence of erectile dysfunction in men undergoing prostatectomy versus external beam radiation at two years and a little higher prevalence of erectile dysfunction at five years in men who underwent radical prostatectomy versus external beam radiation (79.3% vs. 63.5%; Potosky, 2004). Sexual side effects are commonly reported regardless of treatment with nerve sparing (robotic or open) radical prostatectomy, radiation therapy, or androgen deprivation therapy. Erectile dysfunction occurred in 54% to 65.5% of men after colorectal surgery (Den Oudsten, 2012; Milbury, 2003). In addition, chemotherapy and/or hormone deprivation treatments can impact sexual desire and function. Since sexual side effects are common, it is important to try and preserve sexual function in men who have undergone treatment for cancer, especially cancers in the pelvic area such as prostate or colorectal cancer.

Sexual function preservation is dependent on preserving maximal overall health. As discussed previously, healthy lifestyle choices including regular cardiovascular exercise; a low-fat, low-cholesterol, high-fiber plant-based diet; proper sleep hygiene; and controlling other medical conditions is important with all men including those who have undergone cancer treatment. Penile rehabilitation strategies have been mostly implored for men after prostate cancer treatment, but the same strategies may help preserve function in men after colorectal surgery (which also involves peripheral nerve damage). Although research is needed to further determine efficacy of various penile rehabilitation strategies, men may find these strategies beneficial. Penile rehabilitation is aimed at preserving pretreatment function by promoting regular blood flow to the penis. In the most basic sense, it means regular stimulation of the penis to promote blood flow on a daily basis. This may also involve using various treatments such as PDE-5 inhibitors, regular stretching inside the vacuum device, regular use of penile injections or intraurethral suppository, or some of the previously mentioned integrative treatments from this chapter that may enhance blood flow response within the penis during stimulation. There is limited research to support penile rehabilitation, and most of it has been done with men post-prostatectomy. The theory behind penile rehabilitation is that the treatments may provide improved blood flow and oxygenation to the penis thereby improving function of the penis.

Some urology healthcare providers encourage patients to take phosphodiesterase type 5 (PDE-5 inhibitors such as sildenafil, tadalafil, vardenafil, or avanafil) after removal of the prostate, whether open radical prostatectomy or robotic radical prostatectomy or radiation therapy for penile rehabilitation. Providers may instruct the patient to take the PDE-5 inhibitors either daily or three times a week for penile rehabilitation, but this can be very costly. Conflicting research continues emerge about the early use of these medications for penile rehabilitation to promote better erectile function, making the decision and expense of this strategy challenging. Several studies have been done to date, and men who regularly take oral erectile dysfunction medications reported better return of their own erectile function (Incrocci, 2006; McCullough, 2008; Montorsi, 2004; Mulhall, 2005; Nehra, 2005; Padma-Nathan, 2003; Raina, 2008; Schwartz, 2004; Valicenti, 2001; Wilder, 2000). The studies looked at using medications or the vacuum device in the first three to six months after surgery. The dosage, agent, and frequency of these medications have not been clearly determined from the research. Probably the most compelling research for PDE-5 inhibitors was done by evaluating erectile function in 76 men with normal preoperative erectile function who underwent bilateral nerve sparing radical prostatectomy (Padma-Nathan, 2003). The men took sildanfil 50 to 100mg every night for 36 weeks beginning four weeks after surgery and then stopped the medication for a washout period. They found that 27% of the men taking sildenafil versus 4% in the placebo group had a return of spontaneous normal erections. Another study by McCullough and associates showed that men using 50 to 100mg of sildanfil nightly had improved nocturnal erections compared with men treated with placebo (McCullough, 2008). In a blinded, placebo-controlled study using vardenafil researchers ultimately did not find a difference between taking the medication on demand versus daily and even placebo after the washout period off the medications (Montorsi, 2008). This has added to confusion about efficacy of penile rehabilitation with PDE-5 inhibitors and with rising costs of these agents and more rigid research showing a lack of efficacy, penile rehabilitation with PDE-5 inhibitors remains challenging. Regular use of the vacuum device without tension rings for daily stretching may improve return of a man's spontaneous erections after prostate cancer treatment and minimize penile shrinkage (Kohler, 2007; Raina, 2006). The combination of an oral agent and the vacuum device resulted in 30% of men after radical prostatectomy reporting return of spontaneous erections (Raina, 2005). Although it is unclear exactly how much and how often the vacuum device should be used, in studies the device was used for approximately 10 minutes a day, pumping up the penis several times and allowing it to stand erect inside the vacuum device full of blood for a few minutes then releasing the suction and starting over. The penis stretches are done inside the pump to promote blood flow. Penile injections were the first medical treatment to be used for penile rehabilitation after prostatectomy to improve erectile function (Montorsi, 1997). In this small study, the penile injections during the first six months after radical prostatectomy had a 67% return of spontaneous erections sufficient for intercourse versus a 20% return of spontaneous erections for the control group. Raina and colleagues reported that 56% (10 of 18 patients) who had used a combination of oral agents and injections of alprostadil had return of partial erections at approximately six months, yet the men still needed treatment to be able to have sexual intercourse (Raina, 2008). The researchers also used trimix injections with four additional patients and reported that of the 22 total patients on either alprostadil or trimix, 50% of the patients had return of natural erections at six months. In another study using oral PDE-5 inhibitors for penile rehabilitation, patients who failed the oral agents were treated with intracavernosal injections (Mulhall, 2005). After 18 months the group undergoing penile rehabilitation contained a greater percentage of men who were able to engage in intercourse unassisted by medication (52% vs. 19%, $p < 0.001$).

Regardless of the method of enhancing blood flow to the penis for penile rehabilitation after cancer treatment, getting regular erections and blood flow may improve function. Penile rehabilitation should be started at the onset of symptoms during or after treatment. Although research is limited and more research is needed to understand the role and efficacy of penile rehabilitation for erectile dysfunction after cancer treatment (especially regarding some of the integrative approaches outlined in this chapter), erectile function may be improved using these strategies. If adverse reactions are minimal with a treatment and the individual has improved response using it, it may be worth exploring regular use of the treatment to promote blood flow and erectile function.

Pelvic floor muscle exercises bring more blood to the genital area and therefore may improve sexual function. In one study 122 men with erectile dysfunction participated in 20 sessions (30 minutes each session of voluntary contractions coupled with electrical stimulation) with a pelvic floor physical therapists to increase ishiocavernosus muscle strength; 99% of participants presented a positive trend with improved change in intracavernous pressure during the contraction (Lavoisier, 2015). Although the authors state pelvic muscle rehabilitation was found to be beneficial in erectile dysfunction, the pressure change was really the only measure of improved function (there were no other reported measures of erectile function). Another previous study showed pelvic floor exercises were effective treatment for erectile dysfunction, and authors identified them as a first-line approach for prevention and treatment of erectile dysfunction (Dorey, 2004a, 2004b). Although further research is needed, with no real negative effects of pelvic floor exercises/rehabilitation and a possible benefit, this treatment offers a safe way to possibly improve erectile function in men with erectile dysfunction from various etiologic causes including cancer treatment. Although further research is also needed to determine optimal approaches to penile rehabilitation, if the treatment has minimal untoward effects and may be beneficial, it might be prudent for a man to try various treatments that lead to improved blood flow into the penis and potential improved erectile function.

Conclusion

Although there are a myriad of treatments for male sexual dysfunction available to men, readily available herbal remedies often lack sufficient research to support safety and efficacy. Some of the diverse educational and supportive patient- and partner-oriented behavioral strategies, integrative medicine treatment options, and medical devices to address male sexual dysfunction have some limited research to support them and have been presented in this chapter. Each individual has to evaluate the pros and cons of each treatment option and the available research supporting that treatment to determine if it is something he wants to try.

Providing patients with information about the various treatment options and lifestyle modifications empowers them with the information and knowledge to make more informed choices in their journey toward better sexual health.

REFERENCES

Akkus E, Carrier S, Rehman J, Breza J, Kadioglu A, Lue TF (1994). Is colchicine effective in Peyronie's disease? A pilot study. *Urology*. 44(2):291–295.

Althof SE (2012). Psychological interventions for delayed ejaculation/orgasm. *Int J Impot Res*. 24(4):131–136.

Annon J (1974). *The behavioral treatment of sexual problems*. Honolulu, HI: Enabling Systems.

Biagiotti G, Cavallini G (2001). Acetyl-L-carnitine vs tamoxifen in the oral therapy of Peyronie's disease: a preliminary report. *Br J Urol Int*. 88(1):63–67.

Choi YD, Park CW, Jang J, et al. (2012). Effects of Korean ginseng berry extract on sexual function in men with erectile dysfunction: a multicenter, placebo-controlled, double-blind clinical study. *Int J Impot Res*. 25(2):45–50.

Den Oudsten BL, Traa MJ, Thong MS, et al. (2012). Higher prevalence of sexual dysfunction in colon and rectal cancer survivors compared with the normative population: a population-based study. *Eur J Cancer*. 48(17):3161–3170.

Dorey G (2004a). Pelvic floor exercises as a treatment for men with erectile dysfunction. *Nurs Times*. 100(12): 65–67.

Dorey G, Speakman M, Feneley R, Swinkels A, Dunn C, Ewings P (2004b). Randomised controlled trial of pelvic floor muscle exercises and manometric biofeedback for erectile dysfunction. *Br J Gen Prac*. 54(508):819–825.

Gontero P, Di Marco M, Giubilei G, et al. (2009). Use of penile extender device in the treatment of penile curvature as a result of Peyronie's disease: results of a phase II prospective study. *J Sex Med*. 6(2):558–566.

Incrocci L, Slagter C, Slob AK, Hop WC (2006). A randomized, double-blind, placebo-controlled, cross-over study to assess the efficacy of tadalafil (Cialis) in the treatment of erectile dysfunction following three-dimensional conformal external-beam radiotherapy for prostatic carcinoma. *Int J Radiation Oncol Biol Physics*. 66(2):439–444.

Jang DJ, Lee MS, Shin BC, Lee YC, Ernst E (2008). Red ginseng for treating erectile dysfunction: a systematic review. *Br J Clin Pharmacol*. 66(4):444–450.

Kohler TS, Pedro R, Hendlin K, et al. (2007). A pilot study on the early use of the vacuum erection device after radical retropubic prostatectomy. *Br J Urol Int*. 100(4):858–862.

Lavoisier P, Roy P, Dantony E, Watrelot A, Ruggeri J, Dumoulin S (2015). Pelvic-floor muscle rehabilitation in erectile dysfunction and premature ejaculation. *Phys Ther*. 94(12):1731 43.

Levine LA, Newell M, Taylor FL (2008). Penile traction therapy for treatment of Peyronie's disease: a single-center pilot study. *J Sex Med*. 5(6):1468–73.

McCullough AR, Levine LA, Padma-Nathan H (2008). Return of nocturnal erections and erectile function after bilateral nerve-sparing radical prostatectomy in men treated nightly with sildenafil citrate: subanalysis of a longitudinal randomized double-blind placebo-controlled trial. *J Sex Med.* 5(2):476–484.

McMahon CG, Jannini E, Waldinger M, Rowland D (2012). Standard operating procedures in the disorders of orgasm and ejaculation. *J Sex Med.* 10:204–209.

Milbury K, Cohen L, Jenkins R, Skibber JM, Schover LR (2013). The association between psychosocial and medical factors with long-term sexual dysfunction after treatment for colorectal cancer. *Support Care Cancer.* 21(3):793–802.

Montorsi F, Guazzoni G, Strambi LF, et al. (1997). Recovery of spontaneous erectile function after nerve-sparing radical retropubic prostatectomy with and without early intracavernous injections of alprostadil: results of a prospective, randomized trial. [see comment]. *J Urol.* 158(4):1408–1410.

Montorsi F, Nathan HP, McCullough A, et al. (2004). Tadalafil in the treatment of erectile dysfunction following bilateral nerve sparing radical retropubic prostatectomy: a randomized, double blind, placebo controlled trial. *J Urol.* 172(3):1036–1041.

Montorsi F, Brock G, Lee J, et al. (2008). Effect of nightly versus on-demand vardenafil on recovery of erectile function in men following bilateral nerve-sparing radical prostatectomy. *Eur Urol.* 54(4):924–931.

Moyad MA, Park K (2012). What do most erectile dysfunction guidelines have in common? No evidence-based discussion or recommendation of heart-healthy lifestyle changes and/or Panax ginseng. *Asian J Androl.* 14(6):830–841.

Mulhall JP, Land S, Parker M, Waters WB, Flanigan RC (2005). The use of an erectogenic pharmacotherapy regimen following radical prostatectomy improves recovery of spontaneous erectile function. *J Sex Med.* 2(4):532–540; discussion 540–542.

Nehra A, Grantmyre J, Nadel A, Thibonnier M, Brock G (2005). Vardenafil improved patient satisfaction with erectile hardness, orgasmic function and sexual experience in men with erectile dysfunction following nerve sparing radical prostatectomy. *J Urol.* 173(6):2067–2071.

Padma-Nathan H, McCullough AR, Giuliano F, et al. (2003). Postoperative nightly administraton of sildenafil citrate significantly improved the return of normal spontaneous erectile function after bilateral nerve-sparing radical retropubic prostatectomy with and without early intracavernous injections of alprostadil: results of a prospective, randomized trial. *J Urol.* 169(4 Suppl):375–376.

Penson DF, McLerran D, Feng Z, et al. (2008). 5-year urinary and sexual outcomes after radical prostatectomy: results from the Prostate Cancer Outcomes Study. *J Urol.* 179(5 Suppl):S40–S44.

Potosky AL, Davis WW, Hoffman RM, et al. (2004). Five-year outcomes after prostatectomy or radiotherapy for prostate cancer: the prostate cancer outcomes study. *J Natl Cancer Inst.* 96(18):1358–1367.

Raheem AA, Garaffa G, Raheem TA, et al. (2010). The role of vacuum pump therapy to mechanically straighten the penis in Peyronie's disease. *Br J Urol Int.* 106(8):1178–1180.

Raina R, Agarwal A, Allamaneni SS, Lakin MM, Zippe CD (2005). Sildenafil citrate and vacuum constriction device combination enhances sexual satisfaction in erectile dysfunction after radical prostatectomy. *Urology*. 65(2):360–364.

Raina R, Agarwal A, Ausmundson S, et al. (2006). Early use of vacuum constriction device following radical prostatectomy facilitates early sexual activity and potentially earlier return of erectile function. *Int J Impot Res*. 18(1):77–81.

Raina R, Pahlajani G, Agarwal A, Zippe CD (2008). Early penile rehabilitation following radical prostatectomy: Cleveland Clinic experience. *Int J Impot Res*. 20(2):121–126.

Safarinejad MR (2004). Therapeutic effects of colchicine in the management of Peyronie's disease: a randomized double-blind, placebo-controlled study. *Int J Impot Res*. 16(3):238–243.

Safarinejad MR (2009). Efficacy and safety of omega-3 for treatment of early-stage Peyronie's disease: a prospective, randomized, double-blind placebo-controlled study. *J Sex Med*. 6(6):1743–1754.

Safarinejad MR (2010). Safety and efficacy of coenzyme Q10 supplementation in early chronic Peyronie's disease: a double-blind, placebo-controlled randomized study. *Int J Impot Res*. 22(5):298–309.

Safarinejad MR, Hosseini SY, Kolahi AA (2007). Comparison of vitamin E and propionyl-L-carnitine, separately or in combination, in patients with early chronic Peyronie's disease: a double-blind, placebo controlled, randomized study. *J Urol*. 178(4 Pt 1):1398–1403 [discussion 1403].

Schwartz EJ, Wong P, Graydon RJ (2004). Sildenafil preserves intracorporeal smooth muscle after radical retropubic prostatectomy. *J Urol*. 171(2 Pt 1):771–774.

Shin BC, Lee MS, Yang EJ, Lim HS, Ernst E (2010). Maca (L. meyenii) for improving sexual function: a systematic review. *BMC Complement Altern Med*. 10:44.

Sunay D, Sunay M, Aydoğmuş Y, et al. (2011). Acupuncture versus paroxetine for the treatment of premature ejaculation: a randomized, placebo-controlled clinical trial. *Eur Urol*. May;59(5):765–771.

Valicenti RK, Choi E, Chen C, et al. (2001). Sildenafil citrate effectively reverses sexual dysfunction induced by three-dimensional conformal radiation therapy. *Urology*. 57(4):769–773.

Wilder RB, Chou RH, Ryu JK, et al. (2000). Potency preservation after three-dimensional conformal radiotherapy for prostate cancer: preliminary results. *Am J Clin Oncol*. 23(4):330–3.

SECTION II

Lifestyle and Environmental Impacts on Sexual Function

7

Optimum Macronutrition for Sexual Health

RONALD L. HOFFMAN AND LEYLA MUEDIN

"One cannot think well, love well, sleep well, if one has not dined well."
—Virginia Woolf

"Ask your doctor if you're healthy enough for sex."
—Viagra commercial

Popular magazines and websites often entice readers with breezy articles like "Top Ten Aphrodisiac Foods" ("Does your romantic life need a little more spark? Before you bust out the candles and lingerie—try out these top aphrodisiacs!" *Optiming*, December 23, 2012).

Is there any substance to the claim that diet—or specific foods—can influence sexual performance? It is no secret that good health and vitality can do wonders for libido and sexual function. Therefore, the importance of nutrition in sexual health cannot be overemphasized. With the prevalence of obesity in America and its associated morbidities such as metabolic syndrome, type 2 diabetes, and cardiovascular disease (CVD), it is not surprising that sexual dysfunction is on the rise. But what of oft-touted foods like oysters ("eat fish, live longer, eat shellfish love longer"), chocolate, and others?

It is easy to dispense with claims that rhinoceros horn, asparagus, bananas, even carrots can reboot flagging desire. These beliefs harken back to the ancient "doctrine of signatures," which posits that form follows function when it comes to plant properties.

Less easy to dispatch are the reputed benefits of shellfish. Their slimy texture surely evokes the doctrine of signatures, but pop science claims of their aphrodisiac properties are based on their high content of zinc. While zinc deficiency can result in infertility, there is no evidence that zinc supplementation will promote sexual performance in zinc-replete persons.

For that matter, why not eat Prairie Oysters—a euphemism for bull testicles—to boost sexual potency? While they contain actual testosterone, it has not been shown to be absorbed orally, which is why bioavailable oral testosterone preparations are synthetic, chemically modified forms of the natural hormone.

Chocolate makes most lists of sex-promoting foods, partially because it makes a delicious, precious gift—a sensuous delight that beats flowers or a greeting card as an invitation to dalliance. But chocolate deserves serious consideration as a mild sexual stimulant on several counts. It contains phenyethylamine (PEA), a brain neurotransmitter associated with "limerance"—the feeling of being infatuated. It also delivers theobromine, a substance that produces feelings of relaxation and well-being. Additionally, chocolate contains healthy antioxidant polyphenols that promote endothelial responsiveness and nitric oxide activation (Afoakwa, 2008). Nonetheless, research that supports an actual effect of chocolate on sex drive is scant.

Despite the implausibility of specific "aphrodisiac foods," the impact of diet on sexual performance remains indisputable. It is mostly mediated via its influence on metabolic syndrome, a lifestyle disease and a risk factor for CVD and type 2 diabetes in both men and women (Wilson, 2005). The atherogenic dyslipidemia associated with CVD is also associated with erectile dysfunction (ED). In fact, many healthcare practitioners view ED as a clue that coronary artery disease may be present or will develop in the not-too-distant future (Menezes, 2011; Miner, 2015).

What makes us "sexy"? Sexual behavior is governed by a complex combination of physical and emotional factors. Among the major dampers on sexual performance are conditions that require medications. The diseases themselves, or the side effects engendered by their treatment, are to a greater or lesser extent responsible for decreased libido or ED. This is true especially for depression, for which selective serotonin reuptake inhibitors (SSRIs) are prescribed, associated with decreased sexual response, particularly in women, and for hypertension, where antihypertensives are broadly associated with adverse impacts on male sexual performance. In fact, a disproportionate percentage of men who report ED suffer from diabetes or hypertension; for these men, the use of phosphodiesterase inhibitors is often not completely satisfactory (Lee, 2015). Since depression, diabetes, and hypertension may have, at least in part, nutritional antecedents, this suggests a plausible pathway by which a healthy diet could help avert the debilitation caused by these conditions as well as the sexual side effects of the pharmaceutical treatments they invoke.

The Standard American Diet, which consists of processed foods, refined carbohydrates, simple sugars, factory-farmed meats, poor-quality vegetable oils, and trans fats and is bereft of fiber and fresh fruits and vegetables, is an invitation to metabolic syndrome. It is estimated to affect up to 60 percent of American adults.

Abdominal obesity, typically associated with hyperinsulinism and insulin resistance, is not just a fat depot but behaves as an endocrine gland. Spikes in insulin can create menstrual irregularities in women, acne, hirsutism, and fertility problems associated with polycystic ovary syndrome (Hyman, 2015). In men, excess adipose tissue sequesters testosterone, reducing circulating levels of testosterone; fat cells also aromatize circulating testosterone, effectively converting portions of it to estrogen. Therefore, the prevalence of obesity in our society may at least in part explain the increasing incidence of hypogonadism in men. This often becomes a vicious cycle as low testosterone promotes decreased muscle mass, lack of motivation, sedentary behavior, and increased obesity. Low testosterone not only diminishes libido and leads to ED, it is also responsible for decreased feelings of well-being and optimism in men as well as women (Hyman, 2015). Undoubtedly this impacts self-perceived attractiveness.

It is often stated in integrative medicine and nutrition that "food is information." This information has the ability to switch genes on or off, propagate inflammation and hormonal dysregulation leading to disease, or, alternatively, reduce inflammation and eliminate risk factors for dysfunction and disease. There is mounting evidence that a lower carbohydrate, whole-foods Mediterranean-style diet is the antidote to chronic diseases such as metabolic syndrome, CVD, and diabetes (Rees, 2013; Volek, 2015). A recent European study showed that, in men with heart disease suffering from ED, a Mediterranean diet rich in olive oil and dietary polyphenols helped to improve heart function (Angelis, 2014).

One novel way to approach the relationship between food and sexual functioning is via the admittedly now-discredited perspective of 19th-century health crusader Dr. John Harvey Kellogg. Dr. Kellogg was an avowed crusader against "sexual excesses" and a notorious foe of masturbation. It may be said, without exaggeration, that the inventions of corn flakes and granola were intended to stem the tide of "licentiousness" and "unchastity." In his 1877 treatise *Plain Facts for Old and Young: Embracing the Natural History and Hygiene of Organic Life* (Kellogg, 1887), he enunciated his theories about the relationship between diet and lewd behavior:

> He who would keep in subjection his animal nature, must carefully guard the portal to his stomach. . . . Only the most simple and wholesome food should be eaten, and only in such moderate quantities as are required to replenish the tissues . . . Disgust for simple food is one of the traits that a victim of this vice [masturbation] is likely to possess. . . . The science of physiology teaches that our very thoughts are born of what we eat. . . . Eat fruits, grains, milk, and vegetables. There is a rich variety of these kinds of food, and they are wholesome and unstimulating. Graham flour, oatmeal, and ripe fruit are the indispensables of a dietary for those who are suffering from sexual excesses.

If there were any truth to Kellogg's fanciful assertions, then pursuit of optimal sexual performance would dictate scrupulously *avoiding* his prohibitions. Is there any scientific proof that, say, caloric restriction, eschewal of animal protein, or eating bland starchy foods impacts sexual performance?

One study evaluated the effect of a vegetarian diet on plasma prolactin, testosterone, and dehydroepiandrosterone in men and women. It found that, indeed, diet restriction could produce hormonal changes, specifically a decrease in circulating sex hormones (Hill, 1979). This is thought to be salutary from the standpoint of prevention and treatment of hormone-responsive cancers, and an inhibitory effect on sexual performance might be inferred, but, as with similar studies that mirror these findings, it is not clear whether the changes in sex hormone levels seen with a vegetarian diet are of sufficient magnitude to have behavioral consequences.

Another study examined the impact of diet on sex hormone binding globulin (SHBG) in men (Longcope, 2000). It is known that, because SHBG binds to circulating testosterone, there is an inverse relationship between SHBG and bioavailable testosterone. The intakes of overall calories, fat (animal or vegetable), and carbohydrates were not associated with changes in SHBG, but a protein-restricted diet prompted higher levels of SHBG, which, in turn, lowered free testosterone. The authors conclude: "The decrease in bioavailable testosterone can then result in declines in sexual function" (Longcope, 2000).

By contrast, a study examined the effects of a low-cholesterol, low-fat, high-carbohydrate diet (Pritikin diet) on serum lipids and hormones in male subjects. While it found a decline in estradiol levels, testosterone was unaffected (Monroe, 1985).

Could protein supplementation boost testosterone production in men? It turns out that it depends on the protein source. Soy protein, rich in isoflavones, is often thought to be estrogenic, hence its reputed therapeutic application to prostate cancer. One study examined the effects of short-term soy protein supplementation in men (Kraemer, 2013). It found a decline in testosterone levels after 14 days of soy protein use, despite no impact on serum estradiol. By contrast, whey protein did not impact testosterone but facilitated exercise recovery. In yet another trial, administration of branched chain amino acids attenuated endurance exercise-induced declines in testosterone (Carli, 1992).

What about the effects of dietary fat and fiber? It may be that dietary fat and cholesterol are substrates for steroidogenesis, while fiber sequesters sex hormones and promotes their enterohepatic elimination. A 1996 study compared a low-fat, high-fiber diet to a high-fat, low-fiber diet in men. It found a 13% higher level of testosterone excretion in men consuming the high-fat, low-fiber diet (Dorgan, 1996).

What about the effects of caloric restriction? A study performed in lean men showed increases in SHBG and decreases in free and total testosterone after an average of 7.4 years of caloric restriction (Cangemi, 2010). But while full implementation of hormone-altering caloric restriction entails limitation of up to 40%

to 60% of energy substrates, caloric restriction of just 20% was not found to alter testosterone concentrations in experimental animals (Rocha, 2007).

Another mechanism by which Kellogg's animal protein–restricted regimen might stifle sex drive is via iron deficiency. A high percentage of premenopausal women are prone to it, especially when on vegetarian diets. A recent study specifically highlighted reversal of sexual dysfunction in women suffering from iron deficiency anemia when corrected with iron supplementation (Gulmez, 2014).

Therefore, it could safely be said that Dr. Kellogg might have been on to something when he advocated an abstemious, low-protein, low-fat, starchy, fiber-replete, iron-poor diet to dampen desire. But while extreme and prolonged caloric deprivation may undermine sexual desire, a practice called intermittent fasting (IF) has the potential to stoke it.

Gaining popularity among bodybuilders to consolidate their muscle gains, IF involves meal timing with short (14 to 24 hrs) periods of fasting or light eating to "reboot" bodily repair and regenerative systems. Among the endocrine effects of IF is a transient surge in testosterone in nonobese men (Rojdmark, 1989). Fasting appears to reduce levels of the appetite-regulating hormone leptin (Weigle, 1997); leptin levels, in turn, have an inverse relationship with testosterone (Behr, 1997).

Testosterone levels are not the only determinants of sexual performance. Another way some have conceptualized the effects of diet on sexuality is via its impact on nitric oxide (NO) synthesis. NO is responsible for endothelial function and vasodilation, a critical determinant in male sexual performance; NO also enables engorgement of vaginal tissues in women (Figure 7.1). Phosphodiesterase-inhibitor drugs for ED invoke the NO principle to enhance penile blood flow. In their 2015 comprehensive review titled "Lifestyle Modifications and Erectile Dysfunction: What Can Be Expected?" (Maiorino, 2015) the authors write:

> Modifiable risk factors for ED include smoking, lack of physical activity, wrong diets, overweight or obesity, metabolic syndrome, and excessive

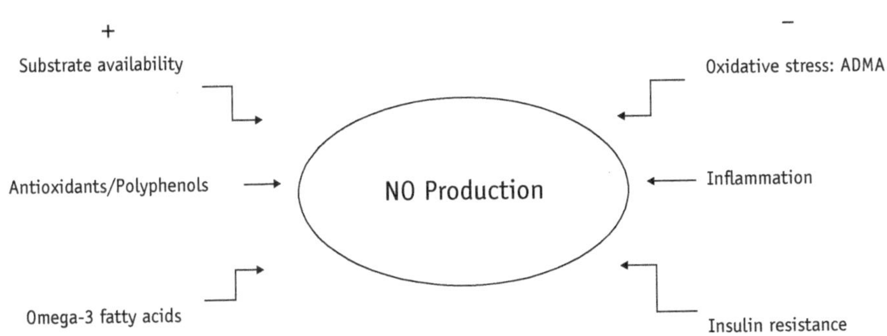

FIGURE 7.1. Diet and lifestyle-mediated effects on nitric oxide production.

> alcohol consumption. Quite interestingly, all these metabolic conditions are strongly associated with a pro-inflammatory state that results in endothelial dysfunction by decreasing the availability of nitric oxide (NO), which is the driving force of the blood genital flow.

Is there such a thing as a proinflammatory diet? The inclusion of high levels of omega-6 vegetable oils in the Standard American Diet bereft of healthy omega-3 oils from fish and nuts is a recipe for ill health. Vegetable oils such as corn, soybean, cottonseed, safflower, and even much-vaunted canola oil are promoters of inflammation. Factory-farmed animals are fed a grain-heavy diet consisting of corn and soy in feedlots. This changes the ratio of omega 6:3 in the animal flesh to more like 20:1 instead of a healthy 2:1 ratio. The animal is much less healthy and prone to fat storage or "marbling"—which is the objective of the commercial meat industry. We are what we eat eats: the net result for human health is to promote synthesis of proinflammatory prostanoids; these, in turn, negatively impact endogenous nitric acid production (Meldrum, 2012).

Another consequence of inflammation is generation of free radicals that further deplete NO. An anti-inflammatory diet, rich in antioxidant vitamins and polyphenols, can mitigate free radical suppression of NO (Meldrum, 2010, 2012). Increased intake of antioxidants vitamins C, and E and folate potentiate endothelial NO synthesis and prevent its catabolism (Meldrum, 2010, 2012).

Further, in addition to the direct effects of obesity via testosterone sequestration in adipose tissue and aromatization, as well as documented indirect effects of insulin resistance on testosterone bioavailability discussed previously, metabolic syndrome is known to suppress NO synthesis (Meldrum, 2010, 2012). Therefore, reversal of insulin resistance via weight loss and a low-carb diet might reinvigorate sexual performance via mutually reinforcing mechanisms—simultaneously raising testosterone and boosting NO.

Finally, endogenous nitric oxide synthesis is also impacted by substrate availability. Nitrate-rich foods are biotransformed into NO. Foods rich in nitrates include celery, beets, and other vegetables (Box 7.1). Foods high in citrulline like watermelon are also thought to support arginine pathway-dependent NO synthesis.

Mood impacts sexuality. Depression itself is characterized by loss of interest in sex. It is often accompanied by vegetative symptoms like insomnia, which is notorious for depleting testosterone (Challem, 2007). The hormonal correlates of depression include amplification of cortisol and suppression of testosterone (Challem, 2007).

Additionally, many antidepressant medications interfere with sexual function. Frequently used SSRIs are notorious for producing ejaculatory delay in men, anorgasmia in women, and diminished libido in both sexes.

Therefore, another pathway by which diet can influence sexual performance is via its impact on mood. Much has been made of the effects of a proinflammatory

Box 7.1 Vegetables Moderate to High in Nitrates

Arugula
Beets
Cabbage (regular, Chinese, and savoy)
Carrots
Celeriac
Celery
Chervil
Dill
Endive
Fennel
Kohlrabi
Leek
Lettuce
Parsley
Radish
Spinach
Turnip
Watercress

diet on depression (Challem, 2007) and the ameliorative effects of omega-3 fatty acids (Challem, 2007) and spices like turmeric (Deguchi, 2015).

Mood disorders are often observed in individuals whose diets are compromised with antinutrients such as sugar, refined carbohydrates, and manmade trans fats. A diet high in sugar depletes B vitamins and magnesium, necessary micronutrients for a healthy metabolism, mood, and brain function.

Moreover, a high refined carbohydrate diet can cause blood sugar imbalances such as hypoglycemia, resulting in physiological stress. Hypoglycemia sets off a physiologic alarm that mobilizes the stress response of the adrenal glands, secreting catecholamines and cortisol in an effort to restore normal blood sugar levels.

This sustained adrenal hyperstimulation can cumulatively elevate cortisol levels, which leads to anxiety, sleep disturbances, and shrinkage of the hippocampus; the latter is associated with short-term memory lapses and "brain fog." Persistently high cortisol levels will also promote fat deposition and retention, particularly abdominal fat, and will reinforce metabolic resistance to any weight loss efforts on the part of the individual.

To compound matters, unhealthy trans fats are thought to replace healthy omega-3 fats in phospholipid cell membranes in the brain, causing rigidity of fluid bilayers and reducing the efficient firing and functioning of neurotransmitters. This further potentiates mood disorders such as anxiety and depression.

Often underappreciated is the role of hidden or masked food intolerance in physical and emotional health. These are food reactions that can be delayed by hours or even several days and are not mediated via classical IgE pathways (Gaby, 1998). These food intolerances create symptoms and conditions as varied as fatigue, migraine, attention deficit hyperactivity disorder, irritable bowel syndrome, emotional lability, aphthous ulcers, and recurrent infections. To further complicate matters, there is the concept of allergic addiction. The allergist Dr. Doris Rapp asserts that this occurs when individuals experience short-term relief after ingesting foods that are later demonstrated to be the cause of their chronic symptoms (www.DrRapp.com). Food substances in susceptible individuals can provoke inappropriate levels of exorphins, opiate-like chemicals that could induce fatigue, brain fog, depression, aggression and violence, or even a drug-like haze. For example, gluten contains five distinct exorphins while dairy contains eight distinct exorphins. In many cases, it is a favorite or highly craved food or beverage that is likely to cause symptoms. The use of elimination diets in clinical practice can prove helpful in reducing and eliminating symptoms as well as reversing otherwise intractable medical conditions (Sarkola, 2003).

Dr. Melvyn R. Werbach is quoted as saying, "It is nutrition that feeds both body and mind. The food we eat supplies the biochemical basis for everything that happens (or should happen) in our bodies" (Challem, 2007). In essence, dietary improvements and enhanced nutrition status help relieve anxiety and depression by fostering the biochemical changes necessary to enhance mood. For example, serotonin production is dependent upon adequate intakes of the amino acid L-tryptophan. Tryptophan is an essential amino acid found in high biological value proteins like beef, poultry, and eggs. Orthomolecular psychiatry is premised on correcting nutritional deficiencies or overcoming genetic susceptibilities with targeted diet and supplements.

Researchers found that 80 percent of people with mood disorders knew that the foods they ate affected how they felt (Challem, 2007). Sugary foods and alcohol were found to contribute to mood problems, whereas fish, vegetables, and fruit were associated with better moods. We need ample amounts of high-quality fats and proteins and some good-quality, low–glycemic index carbohydrates in order to obtain the nutrients necessary to make neurotransmitters to support mood and brain function. All these factors may combine to explain the demonstrated efficacy of the Mediterranean diet in combating ED and enhancing sexual performance.

But the vaunted Mediterranean diet makes allowance for moderate consumption of alcoholic beverages—particularly polyphenol-rich red wine. What evidence is there for alcohol's impact on sexual performance?

Research shows that moderate drinkers (one to two drinks/day) have slightly higher hormone levels than teetotalers (Sarkola, 2003). But while alcohol may seem to grease the skids for romance via its disinhibitory actions, a reverse effect kicks in with greater than two drinks per day. Alcoholism is notorious for depleting testosterone in men (Emanuele, 1998). Sustained high levels of alcohol consumption are neurotoxic, with potentially irreversible deleterious effects on male genital innervation.

Clearly how we live, eat, drink, and sleep has a dramatic effect on sexual health. The foods and beverages we choose impact hormones, signaling molecules like NO, neurotransmitters, moods, digestion, and nutrition status, all of which translate to how well we feel and function sexually.

REFERENCES

Afoakwa EO (2008). Cocoa and chocolate consumption—are there aphrodisiac and other benefits for human health? *S Afr J Clin Nutr*. 21(3):107–113.

Angelis, A, et al. (2014). Atheromatosis, arteriosclerosis and deterioration of cardiac structure and performance in erectile dysfunction patients: A pivotal contribution of the Mediterranean diet in cardiovascular health. *EuroEcho-Imaging*. Abstract P185. http://www.medpagetoday.com/Endocrinology/GeneralEndocrinology/48979.

Behr HM, Simoni M, Nieschlag E (1997). Strong association between serum levels of leptin and testosterone in men. *Clin Endocrinol (Oxf)*. 47(2):237–240.

Cangemi R, Friedmann AJ, Holloszy JO (2010). Long-term effects of calorie restriction on serum sex-hormone concentrations in men. *Aging Cell*. 9(2):236–242.

Carli GI, Bonifazi M, Lodi L, et al. (1992). Changes in the exercise-induced hormone response to branched chain amino acid administration. *Eur J Appl Physiol Occup Physiol*. 64(3):272–277.

Challem J (2007). *The food-mood solution*. Hoboken, NJ: John Wiley.

Deguchi A (2015). Curcumin targets in inflammation and cancer. *Endocr Metab Immune Disord Drug Targets*. 15(2):88–96.

Dorgan JF, Judd JT, Loncope C, et al. (1996). Effects of dietary fat and fiber on plasma and urine androgens and estrogens in men: a controlled feeding study. *Am J Clin Nutr*. 64(6):850–855.

Emanuele MA, Emanuele NV (1998). Alcohol's effects on male reproduction. *Alcohol Health Res World*. 22(3):195–201.

Gaby AR (1998). The role of hidden food allergy/intolerance in chronic disease. *Alt Med Rev*. 3(2):90–100.

Gulmez H, Akin Y, Savas M, et al. (2014). Impact of iron supplementation on sexual dysfunction of women with iron deficiency anemia in short term: a preliminary study. *J Sex Med*. 11(4):1042–1046.

Hill PB, Wynder EL (1979). Effect of a vegetarian diet and dexamethasone on plasma prolactin, testosterone and dehydroepiandrosterone in men and women. *Cancer Lett*. 7(5):273–282.

Hyman M (2015). Killing your sex drive one bite at a time: 5 surprising ways sugar lowers libido. http://drhyman.com/blog/2014/08/29/killing-sex-drive-one-bite-time-5-surprising-ways-sugar-lowers-libido/.

Kellogg, JH (1887). *Plain facts for old and young: embracing the natural history and hygiene of organic life.* Burlington, IA: I.F. Segner.

Kraemer WJ, Solomon-Hill G, Volk BM, et al. (2013). The effects of soy and whey protein supplementation on acute hormonal responses to resistance exercise in men. *J Am Coll Nutr.* 32(1):66–74.

Lee DM, Nazroo J, Pendleton N (2015). Erectile dysfunction and phosphodiesterase type 5 inhibitor use: associations with sexual activities, function and satisfaction in a population sample of older men. *Int J Impot Res.* Jul;27(4):146–151.

Longcope C, Feldman HA, McKinlay JB, et al. (2000). Diet and sex-hormone-binding globulin. *J Clin Endocrinol Metab.* 85(1):293–296.

Maiorino MI, Bellastella G, Esposito K (2015). Lifestyle modifications and erectile dysfunction: What can be expected? *Asian J Androl.* 17(1):5–10.

Meldrum DR, Gambone JC, Morris MA, et al. (2010). A multifaceted approach to maximize erectile function and vascular health. *Fertil Steril.* 94(7):2514–2520.

Meldrum DR, Gambone JC, Morris MA, et al. (2012). Lifestyle and metabolic approaches to maximizing erectile and vascular health. *Int J Impot Res.* 24(2):61–68.

Menezes A, Artham S, Lavie CJ, et al. (2011). Erectile dysfunction and cardiovascular disease. *Postgrad Med.* 123(3):7.16.

Miner M, Kim ED (2015). Cardiovascular disease and male sexual dysfunction. *Asian J Androl.* 17:3–4.

Monroe B, Rosenthal MD, R. Barnard RJ, et al. (1985). Effects of a high-complex-carbohydrate, low-fat, low-cholesterol diet on levels of serum lipids and estradiol. *Am J Med.* 78(1):23–27.

Rees K, Hartley L, Flowers N, et al. (2013). Mediterranean dietary pattern for the primary prevention of cardiovascular disease. *Cochrane Database Syst Rev.* 12:8.

Rocha JS, Bonkowski MS, Franca LR, et al. (2007). Mild calorie restriction does not affect testosterone levels and testicular gene expression in mutant mice. *Exp Biol Med.* 232(8):1050–1063.

Rojdmark S, Asplund A, Rossner S (1989). Pituitary-testicular axis in obese men during short-term fasting. *Acta Endocrinol (Copenh).* 121(5):727–732.

Sarkola T, Eriksson CJ (2003). Testosterone increases in men after a low dose of alcohol. *Alcohol Clin Exp Res.* 27(4):682–685.

Volek JS, Feinman RD (2005).Carbohydrate restriction improves the features of metabolic syndrome: metabolic syndrome may be defined by the response to carbohydrate metabolism. *Nutr Metab.* 2:31.

Weigle DS, Duell PB, Connor WE, et al. (1997). Effect of fasting, refeeding, and dietary fat restriction on plasma leptin levels. *J Clin Endocrinol Metab.* 82(2):561–565.

Wilson PWF, D'Agostino RB, Parise H, et al. (2005). Metabolic syndrome as a precursor of cardiovascular disease and type 2 diabetes mellitus. *Circulation.* 112:3066–3072.

8

Sexual Dysfunction Secondary to Micronutrient Deficiency: A Common But Unrecognized Adverse Effect of Many Pharmaceuticals

BARBARA BARTLIK, JANET MINDES, AND SHEILA SPERBER HAAS

Introduction

Sexual dysfunction is caused by a wide range of factors. Clinicians and even many lay people are very familiar with causes such as psychological and interpersonal relationship problems, physical illness, endocrine imbalances, and the known adverse effects of medications (Neel, 2012). Healthcare providers are generally not aware, however, of the significant role played by nutritional deficiencies. This comprises both macronutrients—protein, fats, and carbohydrates—and micronutrients, i.e., those needed in minute quantities (the spectrum of vitamins and minerals).

One of us (Dr. Bartlik) has been treating sexual dysfunction for 25 years and developed early on an awareness of the fundamental contributions of adequate nutrition to various aspects of healthy sexual function. Dr. Bartlik owes her initial recognition of the significant connection between micronutrient deficiencies and sexual function to her mentor, sex therapy pioneer Helen Singer Kaplan, MD, PhD. It began with Dr. Kaplan sharing her observations of the effects of vitamin B12 deficiency and the restoration of good sexual function with B12 repletion. Dr. Bartlik's initial knowledge of the pivotal role of micronutrients continued to expand. Her subsequent experience over the years with assessing micronutrient levels of new patients, addressing deficiencies, and then observing many positive outcomes continually suggested to her the importance of key micronutrients in the etiology of some sexual dysfunction.

The final piece fell unexpectedly into place as Dr. Bartlik began to attend meetings delineating the functional medicine approach to health and wellness. There she learned about medication-induced micronutrient deficiencies involving a large number of commonly prescribed drugs, and her greater recognition of this complex interaction emerged. Dr. Bartlik incorporated this knowledge into her practice as well, to very good result for her patients.

This chapter represents the distillation of Dr. Bartlik's cumulative knowledge and years of clinical experience, as supported by scientific evidence. It describes the dimensions of this seriously underrecognized and undertreated problem and familiarizes clinicians with interactions between medications and micronutrients that can lead to micronutrient deficiency and subsequent sexual dysfunction. The micronutrients with the greatest impact on healthy sexual functioning are discussed and their respective mechanisms of action explained. Practitioners are also given guidance for determining when and how to assess micronutrient status and on methods of administering micronutrients to patients with sexual disorders with the goal of reducing or eliminating these micronutrient deficiencies and mitigating the sexual problem.

Medications Can Affect Micronutrients to Produce Sexual Dysfunction

Drug-induced side effects fall into two categories. Direct manifestations of the pharmacologic activity underlying a drug's therapeutic effects are unavoidable. Adverse effects due to vitamin and mineral deficiencies caused by a drug, however, are amenable to preventive and therapeutic intervention. They are avoidable when appropriate micronutrients are given before symptoms have appeared and reversible when these micronutrients are replenished after symptoms have developed. Vitamin and mineral deficiencies are prevalent. Dr. Mercola states that the majority of Americans are not getting adequate amounts of micronutrients due to causes that include impoverished soil, digestive issues, and inadequate intake (the most common cause; Mercola, 2015a). Exposure to environmental toxins also can contribute to micronutrient deficiency. The greatest micronutrient deficiencies, in order of severity, are vitamin D, vitamin K2, magnesium, vitamin B12, vitamin E, vitamin A, and iodine (Mercola, 2015a). Vitamin deficiencies contribute to endocrine imbalances and metabolic disorders that underlie a large number of sexual complaints.

Micronutrient deficiency's role in sexual dysfunction, though common, is minimally recognized clinically and has received little attention by investigators. For example, the standard of care for most male patients who consult their primary care doctor or urologist for sexual dysfunction (typically, erectile dysfunction [ED]) is a prescription for either testosterone or a sexually enhancing pharmaceuticals such as sildenafil (Viagra®), without first being evaluated for nutritional deficiencies. Female patients may be given a prescription for flibanserin (Addyi®), Food and Drug Administration (FDA) approved in 2015 for premenopausal women with hypoactive sexual desire disorder, or for estrogen–progesterone hormone replacement therapy, all without first being tested for nutritional deficiencies.

Although suboptimal diet is the most prevalent and acknowledged cause of these nutritional deficiencies, many of the medications that people take can cause or contribute to them. Nearly half of all Americans take one or more prescription drugs (WebMD, 2016a). Yet few physicians and even fewer lay people are aware that many common medications deplete vitamins and minerals. Even fewer recognize that such depletion, in turn, can impair sexual function or that this depletion can be prevented or reversed.

Addressing the Problem

In Dr. Bartlik's clinical experience, when the initial step involves detecting and treating micronutrient deficiencies, a significant proportion of sexual problems improve or resolve completely without any need for pharmaceuticals. As one example, diuretics are known to deplete magnesium (Duarte, 1968). Thus an individual taking a diuretic may develop symptoms of magnesium deficiency. The recognized consequences of magnesium deficiency are muscle cramps, arrhythmia, headache, irritability, depression, and constipation (Rosanoff, 2012). But sexual dysfunction can also result from magnesium deficiency (see later discussion) yet is typically not recognized as a consequence. Replenishing magnesium often relieves sexual dysfunction as well as the commonly recognized symptoms, and without impairing the diuretic's therapeutic effects.

For male patients already on pharmaceutical treatment for ED, correcting micronutrient deficiencies may well enable use of lower dosages or even stopping this drug treatment altogether. Even reducing the dosage would attenuate the unpleasant and potentially severe side effects associated with these medications (e.g., flushing, headache, dyspepsia, nasal congestion, diarrhea, visual disturbances, and, rarely, sudden blindness, heart attack, and stroke). In addition, many patients who are initially placed on standard treatments for sexual disorders (pharmaceuticals and psychotherapy) and fail to respond, or respond poorly, tend to respond more adequately once micronutrient deficiencies are corrected.

A common situation involves patients who are aware that their medication—for medical conditions such as hypertension or hyperlipidemia—diminishes their sexual capabilities. Their response is often noncompliance with an often critical medication, frequently without telling their physician. Here again, identifying and addressing micronutrient deficiencies may reduce or eliminate the sexual side effects and thus improve compliance and health. This situation highlights the need for physicians regularly to ask their patients on pharmaceuticals if they are experiencing any disturbance in sexual function, and—if they are—to discuss noncompliance issues and also assess the patient's micronutrient status. Then any deficiencies can be treated. This direct dialogue helps the patient in multiple ways and is likely to make medical management by the physician easier and more productive.

Culprit Medications

The list of common health conditions treated with medications that can deplete micronutrients is long. They include pain, hypertension, cardiac arrhythmias, birth control, menopausal symptoms, rhinitis, gastric reflux, ulcer, mood disorders, psychosis, asthma, diabetes, hormone deficiencies, hyperlipidemia, infections, and baldness (Cass, 2007, 2013). And patients often suffer from more than one condition and thus take multiple medications, which is likely to magnify any adverse effects due to medication-induced micronutrient deficiency.

Table 8.1 lists medications reported to be associated with sexual dysfunction and—when known—the micronutrients they deplete. In most cases, the mechanisms underlying micronutrient depletion remain to be identified. It is certainly possible that genetic alleles and/or mutations affecting metabolic factors determine whether a given patient will develop a particular drug-related side effect. Some of the other reasons that medications deplete micronutrients are that they may (a) alter patients' appetites so that, for example, they crave carbohydrates and nutrient-poor foods; (b) raise the pH of the upper gastrointestinal tract, impairing absorption of specific micronutrients; and (c) reduce absorption by binding with key nutrients (Cass, 2013).

The sexual side effects of these pharmaceuticals are often underrecognized because patients do not typically self-report them during clinical trials, as opposed to such obvious symptoms as headache, nausea, and dizziness. In addition, because sexuality is complex and affected by multiple issues, including mood and partner problems, patients may not realize that their emerging sexual symptoms are actually secondary to their new medication unless they are asked targeted questions. Illustrative is a study that assessed the rate of sexual dysfunction in subjects on selective serotonin reuptake inhibitors. When it was left to the subjects to report sexual symptoms spontaneously, only 14% indicated it. But when subjects were questioned directly about changes in sexual activity, the incidence of reported sexual dysfunction rose to 58% (Montejo-Gonzalez, 1997).

Table 8.1. Common Medications Reported to Affect Micronutrient Levels and/or Sexual Function with Long-Term Use

Medication	*Micronutrient Depleted or Diminished*	*Effect on Sexual Functioning*
Antihypertensive/Antiarrhythmic		
Loop diuretics *Furosemide* (Lasix) *Bumentanide* (Bumex)	Potassium, magnesium, zinc, calcium, pyridoxine, thiamin, vitamins B6, and C (Duarte, 1968; FDA, 2012; Iezhitsa, 2005; Lameire, 1988; Mydlik, 1999; Pelton, 1999; Zenuk, 2003)	Impotence (Medlineplus.gov)
Thiazide diuretics *Chlorthiazide* (Diuril) *Hydrochlorthiazide* (Esiddrix)	Potassium, magnesium, sodium, zinc (Braun, 2013; Clayton, 2006; FDA, 2012; Iezhitsa, 2005; Pak, 2000)	Impotence (Medlineplus.gov); ↓ Libido, erection, ejaculation difficulty (Chang, 1991)
Beta blockers *Propranolol* (Inderal) *Metoprolol* (Lopressor) *Atenolol* (Tenormin)	CoQ10 (Cass, 2013; Kishi, 1977) Decrease melatonin (Stoschitzky, 1999)	↓ Frequency of sexual activity and ↓ Erectile function (Fogari, 2001) ↓ Sexual desire (Suzuki, 1988) ↑ Ejaculation problem (Suzuki, 1988) Erectile dysfunction (Steinke, 2016)
Angiotensin-converting-enzyme inhibitors *Benazepril* (Lotensin) *Captopril* (Capoten)	Zinc (Braun, 2013) and sodium (Kishi, 1977; Schilling, 1988)	Erectile dysfunction (Shiri, 2007) ↓ Sexual activity (Fogari, 2001) ↓ Testosterone levels (Fogari, 2001) ↑ Impotence (Pierpont, 1981)
Angiotensin II receptor antagonists *Eprosartan* (Teveten) *Olmesartan* (Banicar)	Zinc, sodium, potassium (Braun, 2013)	Erectile dysfunction (Ekman, 2010) May improve erectile activity (Nicolai, 2014)

(*continued*)

Table 8.1. Continued

Medication	*Micronutrient Depleted or Diminished*	*Effect on Sexual Functioning*
Calcium channel blockers *Amlodipine* (Norvasc) *Diltiazem* (Cardizem)	Potassium (Korn, 2016)	↓ Vaginal pulse amplitude responses (Meston, 2007) Erectile dysfunction (Shiri, 2007)
Cardiac glycoside *Digoxin* (Lanozix)	Potassium, magnesium (Iezhitsa, 2005)	Adversely affects male sexual function (Gupta, 1998)
Vasodilators *Hydralazine* (Apresoline)	Vitamin B6 (Vidrio, 1990)	Impotence (Medlineplus.gov)
Acid Suppressing Medication and Antacids		
Proton pump inhibitors *Omeprazole* (Prilosec, Nexium)	Vitamin B12 (Lam, 2012; Saltzman, 1994), protein, minerals (Pitchumoni, 2012), vitamin D, magnesium, calcium, zinc, folic acid, iron, vitamin C (Cass, 2013; FDA, 2011; Hutchinson, 2007; McColl, 2009; Sorenson, 2012)	↓ Erectile function (Lam, 2013) ↑ Gynaecomastia in males (Carvajal, 2007)
H2-receptor antagonists *Famotidine* (Pepcid)*Ranitidine* (Zantac)*Cimetidine* (Tagamet)	Vitamin B12 (Hechtman, 2012; Lam, 2013), folic acid, vitamin D, iron (Cass, 2013), zinc (Aymard, 1988; Meletis, 2005; Sturniolo, 1991), calcium (Skikne, 1981), vitamin D, beta carotene (Pitchumoni, 2012), vitamin C (Hechtman, 2012)	↓ Sexual desire and ability Erectile dysfunction (WebMD, 2016a) Impotence (Medlineplus.gov)
Antacids *Aluminum and Magnesium mixtures* (Tums and Rolaids)	Vitamin B12, vitamin D, folic acid, calcium, iron, zinc (Korn, 2016)	None reported
Oral Hypoglycemic/Antidiabetic Agents		
Biguanides *Metformin* (Glucophage)	Vitamin B12 (de Jager, 2010; Heidelbaugh, 2013) folic acid (Wulffele, 2003; Samaras, 2013)	↑ Erectile function in men on sildenafil (Viagra) (Rey-Valzacchi, 2012)

Table 8.1. Continued

Medication	Micronutrient Depleted or Diminished	Effect on Sexual Functioning
Sulfonylureas *Glyburide* (Diabeta, Micronase) *Glipizide* (Glucotrol)	CoQ10 (Kishi, 1977)	Erectile dysfunction (WebMD.com)
Hormones		
Hormone replacement therapy *Conjugated equine estrogen and medroxyprogesterone* (Provera, Prempro) *Combined oral contraceptives*	B6, B12, folic acid (Cass, 2013), magnesium (Meletis, 2005; Seelig, 1993)	
Pharmaceutical contraceptives	Folic acid, B1, B2, B3, B6, B12, vitamin C, zinc, selenium, magnesium (Seelig, 1993), calcium (Palmery, 2013; Webb, 1980)	↓ Desire (Caruso, 2004), arousal, pleasure (Frellick, 2016) Sexual pain (Battaglia, 2012)
Neuro-Psychiatric Medications		
Benzodiazepines *Lorazepam* (Ativan) *Diazepam* (Valium)	Melatonin, biotin, folic acid, vitamins D, K, calcium (Cass, 2007)	Erectile dysfunction (Balon, 1993)
Lithium (Eskalith)	Folic acid, inositol (Cass, 2013)	Erectile Dysfunction ↓ Sexual thoughts, satisfaction, testosterone (Elnazer, 2015) Female sexual dysfunction (Phillips, 2000)
Antipsychotics (Traditional)		
Haloperidol (Haldol) *Fluphenazine* (Prolixin) *Chlorpromazine* (Thorazine)	Vitamin B2, CoQ10, melatonin, manganese (Cass, 2007)	↑ Erectile dysfunction ↓ Libido ↑ Prolactin (Park, 2012)
Antipsychiotics (Atypical)		
Risperidone *(Risperdal)* Clozapine *(Clozaril)* Olanzapine *(Zyprexa)*	Potassium, sodium (Laake, 2016) B6 (Stargrove, 2008)	↑ Prolactin, sexual dysfunction ↓ Libido (Park, 2012) Impotence (Boyarsky, 2000) Abnormal ejaculations (Boyarsky, 2000)

(*continued*)

Table 8.1. Continued

Medication	*Micronutrient Depleted or Diminished*	*Effect on Sexual Functioning*
Antidepressants		
Tricyclic antidepressants *Amitriptyline* (Elavil)	Zinc (Hechtman, 2012), riboflavin, sodium, CoQ10 (Korn, 2016), vitamin B2 (Laake, 2016)	↓ Desire, erection, orgasm lubrication (Boyarsky, 2000)
Selective Serotonin Reuptake Inhibitors (SSRIs) *Sertraline* (Zoloft) *Fluoxetine* (Prozac) *Escitalopram* (Lexapro)	Betaine, folic acid, vitamin B12 (Korn, 2016), folate (Stargrove, 2008)	↓ Desire, arousal and orgasm (Balon, 1993) ↑ Sexual dysfunction (Jacobsen, 2015; Lahon, 2011)
Monoamine Oxidase Inhibitor (MAOIs)		
Phenelzine (Nardil)	B6 (Stargrove, 2008)	↑ Impotence (Medlineplus.gov)
Anticonvulsants		
Gabapentin (Neurontin)	Vitamin D (Pitchumoni, 2012), folate (Linnebank, 2011)	Sexual dysfunction (Kaufman, 2011)
Barbiturates Phenobarbitol *(Luminol, Nembutol)* Seconal sodium *(Seconal)*	Folic acid (Linnebank, 2011), vitamin D (Davie, 1983), vitamin K, B12, B6, carnitine, calcium (Shafer, 1975), biotin (Mock, 1998; Schmitt, 1984)	Impotence (Medlineplus.gov)
Phenytoin (Dilantin)	Vitamin D (Davie, 1983), vitamin B6, B12, B1, calcium (Shafer, 1975), folic acid (Linnebank, 2011; Morrell, 2002; Sener, 2006), glutathione, carnitine (Pitchumoni, 2012; Rodriguez-Segade, 1989; Schmitt, 1984)	Sexual dysfunction (Bóné, 2006) ↑ Erectile dysfunction (Verrotti, 2016)
Valproic acid (Depakote)	Carnitine (Rodriguez-Segade, 1989), folic acid (Morrell, 2002) Biotin, vitamin D (Pitchumoni, 2012)	↓ Libido (Calabrò, 2012) Less likely to cause sexual dysfunction than others in class (Feltner, 2003) Erectile dysfunction (Calabrò, 2009)

Table 8.1. Continued

Medication	*Micronutrient Depleted or Diminished*	*Effect on Sexual Functioning*
Levitiracepam (Keppra) *Pregabalin* (Lyrica)	Vitamin B6 (Mintzer, 2012) Vitamin B12 (Linnebank, 2011)	Erectile dysfunction
Topiramate (Topomax)	Vitamin B12 (Linnebank, 2011)	↓ Free androgen index (Reis, 2012)
Carbamazepine (Tegretol) *Oxcarbazine* (Trileptyl)	Biotin (Mock, 1998), B12 (Frenkel, 1973), calcium (Bouillon, 1975), vitamin D (Bouillon, 1975), folate (Linnebank, 2011), vitamin E (Higashi, 1980), L-carnitine (Hiraoka, 1997)	↓ Libido, arousal, erection, ejaculation (Boyarsky, 2000)
Other Medications		
Antibiotics *Penicillin, cephalosporins, macrolides, quinolones*	Probiotics, vitamins B1, B2, B3, B6, B12, C, K, iron, and Magnesium (Pitchumoni, 2012; Meletis, 2005)	None reported (short term)
Antibiotics *Tetracycline*	Vitamins B1, B2, B6, B12, K zinc (Mapp, 1976), probiotics, calcium, magnesium, selenium, B2, CoQ10 (Neuvonen, 1976)	↑ Erectile Dysfunction (Balon, 1993)
Aminoglycosides *Gentamycin*	Magnesium, calcium, and potassium imbalances (Landau, 1997)	None reported (short term)
Corticosteroids *Prednisone* (Deltasone, Rayos, Prednisone Intensol)	Folic acid, potassium (Widmer, 1995), zinc, copper (Peretz, 1987), vitamin C (Levine, 1983), vitamin D (Pitchumoni, 2012; Hechtman, 2012), calcium (Lems, 1998; Hechtman, 2012), selenium (Peretz, 1987), magnesium (Rolla, 1990), chromium, selenium (Hechtman, 2012)	↓ Testosterone levels ↓ Estrogen levels (Weisman, 2001) Negative effects on sexual activity (e.g., ED) (Kohler, 2016) ↓ Sexual function in women (Christensen, 2014)

(*continued*)

Table 8.1. Continued

Medication	*Micronutrient Depleted or Diminished*	*Effect on Sexual Functioning*
Nonsteroidal Anti-Inflammatory Drugs (NSAIDs)		
Ibuprofen (Advil, Motrin) *Naproxen sodium* (Aleve)	Vitamin C, iron, folic acid, potassium, and vitamin C (Pitchumoni, 2012)	Erectile dysfunction (Gleason, 2011) Impotence (Medlineplus.org)
Acetylsalicylic acid (Aspirin, Ecotrin)	Vitamin C (Basu, 1982; Das, 1992; Pitchumoni, 2012), iron (via blood loss) (Palme, 1978), folic acid, potassium, sodium (Pitchumoni, 2012), zinc (Hechtman, 2012)	Erectile disorder (Mercola, 2015b)
Statins *Rosuvastatin* (Crestor) *Atorvastatin* (Lipitor) *Simvastatin* (Zocor)	CoQ10 (Langsjoen, 2003; Samaras, 2013)	Impotence (Crane, 2001; Folkers, 1990; Langsjoen, 2005) Affects sperm parameters (Pons-Rejraji, 2014) ↓ Testosterone levels (Corona, 2010)
Antihistamines *Diphenhydramine* (Benadryl, Sominex) *Fexofenadine* (Allegra) *Cetirizine* (Zyrtec) *Chlorphenniramine* (Chlor-Trimeton) *Desloratidine* (Clarinex) *Hydroxyzine* (Atarax)	Vitamin C (Hechtman, 2012)	↑ Impotence (Medlineplus.org)

Micronutrient effects and effects on sexual functioning are listed if there was one or more report(s) in the literature. Some, but not all of the medications in each class above, demonstrated the side effects listed here.

Another difficulty in gaining an accurate picture of a drug's sexual impact is that ED is the sole condition commonly addressed and treated. Among Dr. Bartlik's patients, however, diminished erectile ability usually does not exist in isolation but is often associated with impaired desire and arousal and diminished, delayed, or absent orgasm. ED is the sole dysfunction typically reported, however, because it is the most obvious and the easiest to measure.

Dr. Bartlik has observed that medications causing adverse sexual effects in men are just as likely to cause them in women, but women are much less likely to tell their doctors about sexual symptoms (WebMD, 2016b). In addition to a range of attitudinal, behavioral, and emotional reasons, a major factor is that the absence of pharmaceutical treatments for women—until very recently—would tend to discourage reporting the problem. And the one FDA-approved medication, flibanserin, is rarely used because the prescribing physician must register it through the risk and mitigation strategy service despite the relative absence of dangerous side effects. Although flibanserin can cause dizziness and fainting when taken with certain medications and alcohol, these are fairly commonplace adverse reactions. Because women's sexual dysfunction is inherently more challenging to study and to treat, the great majority of studies are limited to male subjects, and most of those limit their focus to erectile ability. In this vacuum, clinicians tend to generalize data gathered from studies on men to include women. This strategy can be seriously misleading.

Micronutrients Required for Healthy Sexual Function

The following highlights the micronutrients that, in Dr. Bartlik's experience, have the greatest impact on sexuality. We indicate why they are important, point to representative research, and note common medications that induce deficiency. Table 8.1 contains an extensive list of drugs with reported effects on sexual function and on micronutrient levels. We also present information on determining individual patient laboratory values and provide guidance for replenishing low levels. Space limitations prevent inclusion of every micronutrient and botanical supplement that may improve sexual function. (See other chapters for further discussion.)

THE B VITAMINS

These water-soluble vitamins are absorbed in the small intestine, then travel to the liver where they are metabolized into their active coenzyme forms. A key role of B vitamins is producing energy from the breakdown of carbohydrates, protein, and fat. They are essential to the functioning of the Krebs citric acid cycle and the production of adenosine triphosphate (ATP), the energy currency of the cell. In addition, B vitamins—particularly folate, B6, and B12—are critical for proper methylation, the biochemical process that helps convert the amino acid metabolite homocysteine into the amino acids methionine and cysteine. B vitamins also maintain the integrity of the nerves and circulation. B6, B12, and folate are required for synthesis of the neurotransmitters dopamine, serotonin, epinephrine, and norepinephrine. These are also the B vitamins most commonly involved in sexual dysfunction and thus a primary focus in this chapter. B3 (niacin) and B5 (pantothenic acid) also contribute to good sexual function and are also included.

B12, Folate, B6, and Sexual Dysfunction

It has long been known that B12 deficiency is associated with infertility in men and women, which can be reversed with B12 supplementation (Bennett, 2001; Isoyama, 1986). In addition, one case report of a patient with ED—who was homozygous for the CT677 mutation in methylation and resistant to treatment with sildenafil—responded to sildenafil after receiving folic acid and vitamin B12 for two months (Tsamatropoulos, 2008). When vitamin B12 and folate were discontinued, he once again failed to respond to sildenafil. B12 directly supports synthesis of the neurotransmitter acetylcholine, which is critical for genital engorgement (Sahelian, 2003). It appears to do so by increasing the activity of choline acetyltransferase (demonstrated in cats by Nadeau, 1988), which is directly responsible for acetylcholine synthesis. B12 deficiency also leads to erectile disorder by inducing neuropathy of the autonomic nerves that innervate the penis (Saenz de Tejada, 1988). Folate is essential for the synthesis of dopamine, the catecholamine neurotransmitter most strongly associated with sexuality. Folate deficiency is associated with both erectile dysfunction and premature ejaculation (Yan, 2014). Vitamin B6 aids dopamine and norepinephrine production (Mayo Clinic Women's HealthSource, 2013), both of which are integral to sexual function (Nagaraj, 2009). Dopamine is involved in all phases of the sexual response cycle (Melis, 1995). Norepinephrine levels rise with arousal and erection, and levels peak during orgasm (Nagaraj, 2009).

Causes of Deficiency

Methylenetetrahydrofolate Reductase Mutation

It is extremely important to recognize that a significant number of people with adequate B6, B12, and folate in their bloodstream and on laboratory testing are nonetheless unable to utilize these critical vitamins because they are not in their methylated—that is, activated—form. Methylation is accomplished by the methylenetetrahydrofolate (MTHFR) reductase enzyme, but a large portion of the population carries one or more of several mutations in the MTHFR gene that cause the synthesis of an enzyme unable to accomplish sufficient methylation. These individuals suffer from a chronic insufficiency of B12, B6, and folate that can impair sexual function, hormone metabolism, cardiovascular health, red blood cell (RBC) formation, and energy production (and which also play a role in mental illness and cancer). Individuals with an MTHFR mutation typically benefit significantly from taking supplements with the methylated forms of these three B vitamins. Most commercial laboratories test for MTHFR polymorphisms, as do laboratories that specialize in genetic testing. It is important to bear in mind that patients with multiple polymorphisms may have atypical reactions to methylating agents.

Diet

Some people are deficient in folate and B12 because their diet lacks adequate amounts of these micronutrients. Vegans and vegetarians are at particular risk because animal products are the only foods that contain B12, and they are often advised to supplement this essential vitamin. People over age 65 whose diet has lost variety may have eliminated foods that are rich in vitamin B12.

Aging

B12 absorption diminishes with aging (Baik, 1999), and thus deficiency is more common in older populations. This is exacerbated if a diet has become low in B12. The low B12 levels found more commonly among people over 65 may help to explain the impaired sexual function that typically occurs as people age.

Medications

Many medications are suspected of interfering with vitamin B12 absorption. It has been established that metformin (an oral medication for diabetes) and proton pump inhibitors (PPIs; oral medications that reduce gastric acid) cause B12 deficiency (Heidelbaugh, 2013).

Metformin, used to lower high blood sugar in type II diabetes, is now also used increasingly as preventive therapy in prediabetic patients and off-label to treat polycystic ovary syndrome. Documented adverse effects include B12 deficiency and increased homocysteine (de Jager, 2010). The study authors suggest that patients taking metformin be routinely screened for B12 deficiency and given supplementation where indicated (de Jager, 2010). The effect of metformin on folic acid, however, is equivocal. Some studies have shown effects similar to those on B12 (Wulffele, 2003), while others do not (Pongchaidecha, 2004). Studies assessing the effect of metformin on ED have been limited to observation of benefit when metformin is given to nondiabetic but insulin-resistant men who had been responding poorly to sildenafil (Rey-Valzacchi, 2012; Seftel, 2012).

Both *PPI drugs* and *histamine H2-receptor antagonists* (i.e., H2-blockers) suppress the production of gastric acid in patients with gastroesophageal reflux disease and are among the most commonly used pharmaceuticals in the United States. Yet gastric acid is necessary for cleaving vitamin B12 from ingested dietary protein. Thus these drugs impair digestion and absorption of B12 from food, putting patients at risk for deficiency and ensuing—potentially severe—complications (Heidelbaugh, 2013). This can include the development of sexual dysfunction. Earlier attempts to document this drawback with PPIs and H2-blockers generated inconclusive results because patient samples were extremely small and highly skewed to geriatric patients. Population data were needed.

This was provided by use of the vast and comprehensive Kaiser Permanente Northern California patient database, which made the association clear (Lam, 2013). Results enabled the investigators to conclude that both "previous and current gastric acid inhibitor use was significantly associated with the presence of vitamin B12 deficiency" (Lam, 2013). Other work has identified patients at particularly increased risk for this impact of PPIs on vitamin B12—those over age 60 (Valuck, 2004) and those with certain genetic polymorphisms in cytochrome P450 2C19 (CYP2C19), one of the liver's main drug-metabolizing enzymes (Sagar, 1999). PPIs have also been reported to cause ED (Lam, 2013), perhaps due to multiple micronutrient deficiencies caused by this group of medications (FDA, 2011).

Determining Deficiency of B Vitamins

Homocysteine

The normal range for homocysteine is 4 to 15 μmol/L. Ideally, it should be under 8.0 μmol/L (Lombardo, 2010). Elevated homocysteine is a sign of possible deficiency in B12, B6, and folate (Kado, 2005). More common causes are a methylation defect and/or use of gastric acid-reducing drugs. Homocysteine has a strong inhibitory effect on endothelium-dependent nitric oxide (NO) production, which in turn will impair sexual functioning (Cheng, 2009). It is a marker for ED (Demir, 2006). Lombardo found an association between the C677T MTHFR mutation and ED and hypothesized that elevated homocysteine in these patients may be responsible (Lombardo, 2010). He also found that patients with ED who did not respond to sildenafil had high homocysteine and low folic acid levels. After treatment with vitamin B6 and folic acid, 89% of this group showed significantly reduced homocysteine levels and also became responsive to sildenafil.

Mean Corpuscular Volume

Elevated MCV—the average volume of red cells—is another clue to a pathologic cause of low B12 and folate levels, possibly a problem with methylation. Although the normal range is 80 to 96 fL/red cell, a level of 91 is not ideal. The patient with an MCV >91 may need methylated B12 and/or folate even with normal or elevated serum levels.

Restoring Vitamin B Levels

The richest sources of vitamin B6 include fish (especially tuna and salmon), beef liver and other organ meats, potatoes and other starchy vegetables, legumes (especially chick peas), and fruit (other than citrus; National Institutes of Health [NIH], 2016b) Vitamin B12 is found naturally only in animal-derived foods. The best sources are beef liver (and other organ meats), trout, salmon, clams, beef, ham, milk, yogurt, eggs, and cheese. In addition, some foods—especially breakfast cereals and some soy products—have been fortified (NIH, 2016c). Folate is found in beets, celery, peanuts, spinach, avocado, and sunflower seeds (Institute of Medicine, 1998).

When a nutritional supplement is determined to be the most reliable way to replenish these deficiencies, the appropriate dose is very individual. When large doses are needed, it is highly important to keep in mind that the B vitamins work in concert, and thus it is best to supplement them all simultaneously via a B-complex blend. They are frequently referred to as B-50 and B-100 supplements (referring to the predominant dosage). When an MTHFR mutation is responsible for B vitamin deficiencies, then supplementation—but *only* with the methylated forms—is essential. The terms indicating this are *methylcobalamin* for B12, *pyridoxal 5'-phosphate* for B6, and *L-5-methyltetrahydrofolate* for folate.

Vitamin B5 (Pantothenic Acid)

As with B12, B5 is a cofactor for enzymatic reactions needed for the synthesis of acetylcholine (Kennedy, 2016). B5 is also a cofactor in mitochondrial energy production (Depeint, 2006) and is necessary for the synthesis of steroid hormones (e.g., testosterone and estrogen) needed for healthy sexual functioning (Hever, 2014). Vitamin B5 is widely available in many foods, most notably avocados, mushrooms, lentils, broccoli, sweet potatoes, and cauliflower (Hever, 2014). Some authors suggest taking 0.3 to 1 gm of B5 with 1 to 3 gm of choline 20 to 30 minutes before sexual activity (Mercola, 2008) as a less expensive alternative to such erectile-enhancing medications as sildenafil.

Vitamin B3 (Niacin)

Niacin plays a role in sex hormone production (Page, 2011). In a double-blind placebo-controlled study of patients with ED, it was shown to improve dysfunction significantly in moderate to severe ED but had no benefit in patients with mild to moderate ED (Ng, 2011). Patients should be warned that niacin causes temporary flushing, which can be uncomfortable (Kamanna, 2009; Svedmyr, 1969).

Vitamin C

Vitamin C can have a positive effect on sexual function through its ability to boost NO production (Nathan, 2010). Vitamin C is concentrated in both the cortex and the medulla of the adrenal glands, where it is a cofactor for steroidogenesis and catecholamine synthesis (Patak, 2004). Vitamin C is thus needed for the synthesis of sex hormones, including estrogen (Wu, 2008) and testosterone (Getoff, 1993), and it supports adrenal function overall (Patak, 2004), which is critical for sexuality. In women of reproductive age, the adrenals produce half of the body's testosterone (Shifren, 2004). Androgens promote sexual function in women (Shifren, 2004) as well as in men, and in postmenopausal women the androgens of adrenal origin are essential to maintaining sexual function (Traish, 2002). As stated previously, the catecholamines dopamine and norepinephrine are integral to sexual function. Vitamin C also aids in the uptake, transport, and mobilization of iron, which—as will be noted later—is also important for sexual function. Thus medications that deplete vitamin C can result in low iron and consequent anemia. Acetylsalicylic acid (ASA)—aspirin—taken regularly by a great many patients for its anti-inflammatory and cardioprotective effects—blocks both the absorption of vitamin C (Basu, 1982) and its uptake by leukocytes (Das, 1992). Thus it may be advisable to take vitamin C supplements during ASA use. Serum or plasma vitamin C levels are adequate measures of clinical status.

Nonsteroidal anti-inflammatory drugs (NSAIDs) increase urinary excretion of vitamin C. Patients on NSAIDs are more likely to develop moderate to severe ED, even after controlling for comorbidities (Gleason, 2011), and this may actually reflect a vitamin C deficiency. PPIs—by virtue of their ability to increase gastric pH, that is, make the stomach less acid (Heidelbaugh, 2013)—diminish the absorption of vitamin C as well as iron and vitamin B12. Patients with *H. pylori* infections who take PPIs are at increased risk of vitamin C depletion (McColl, 2009).

Significant amounts of vitamin C are found in citrus fruits, berries, melons, and certain vegetables—especially peppers, broccoli, brussels sprouts, cabbage, snow peas, and cauliflower.

Vitamin D

Vitamin D is necessary for promoting the activity of tyrosine hydroxylase, the rate-limiting enzyme in catecholamine synthesis (Jiang, 2014), and, as stated previously, the catecholamines—dopamine and norepinephrine—are critical for sexual functioning. Because vitamin D receptors and vitamin D-metabolizing enzymes are richly present in human testicular tissue, ejaculatory tract, and mature spermatozoa (Blomberg-Jensen, 2011), vitamin D deficiency can also have significant direct effects on sexual function. Vitamin D also appears to be important in maintaining adequate testosterone levels. Inadequate catecholamine levels due to deficient vitamin D can result in depression, thus contributing indirectly to sexual dysfunction (Jorde, 2008). Depression is known to be associated with sexual dysfunction (Baldwin, 2001). Vitamin D deficiency also impairs the absorption of magnesium, which plays multiple roles in sexual health (discussed later; Brannan, 1976).

Supporting vitamin D's significant role in sexual function, Barassi showed that deficiency is common in patients with ED and that patients with arteriogenic ED were more likely to be deficient (Barassi, 2014). In a small controlled study that found an association between low vitamin D and low free and total testosterone levels, supplementing men at the lower testosterone levels with vitamin D increased their testosterone levels in a dose-dependent manner, approaching a plateau at higher levels of supplementation (Pilz, 2011).

Medications that deplete vitamin D include antacids, PPIs, H2-receptor agonists, barbiturates, and corticosteroids. Patients with vitamin D levels of $\leq$ 20 ng/ml are deficient. The Endocrine Society recommends vitamin D levels between 40 and 60 ng/ml (Holick, 2011), and the normal range at most laboratories is 30 to 100 ng/ml. Very few foods naturally contain vitamin D (see later discussion), and it is primarily synthesized in the skin in response to sun exposure. Patients who are deficient may be supplemented orally with vitamin D3 in doses from 1,000 to 10,000 IU per day. The dose is determined, and then periodically adjusted, according to the patient's 25-hydroxy vitamin D level. When depletion is severe, intramuscular injections are available for rapid repletion, and adequate levels are maintained via oral supplements.

It is important to take vitamin K2 along with vitamin D3 supplements, as it is essential in insuring that calcium goes to bone metabolism, which is helpful, rather than to the arteries, where it is harmful because of calcification (Mercola, 2014). Vitamin D aids absorption of calcium from the digestive tract, and vitamin K2 activates osteocalcin, which binds calcium so that it can contribute to bone formation. Healthy bones are important to maintaining lifelong sexual activity. Osteoporosis, in addition to the more obvious health consequences for affected women and men, can interfere with sexual activity. Vitamin A is advisable to take along with vitamin D, as it activates vitamin D receptors. Optimal vitamin D function also requires magnesium and zinc (Mercola, 2015b). These micronutrients are best taken in conjunction with food, particularly fatty foods. The few foods with any significant vitamin D content are fatty fish (e.g., salmon, tuna, mackerel) and fish liver oils, which also provide vitamin A. Small amounts are found in beef liver, cheese, and egg yolks (NIH, 2016d) and mushrooms exposed to UV light.

Magnesium

Magnesium affects sexual function through multiple important pathways. It relaxes blood vessels, allowing for improved circulation and decreased resistance to blood flow. This is important to erectile capability. Magnesium is required for production of NO (Northcott, 2004) and acetylcholine (Miller, 2004), both of which are needed for smooth muscle relaxation in erectile tissue. Magnesium plays a role in dopaminergic, noradrenergic, and serotonergic neurotransmission (Cardoso, 2008). Dopamine is important for sexual desire, arousal, and orgasm (Kruger, 2003). Norepinephrine is involved in sexual arousal (Meston, 2000, 2007). Serotonin is necessary for maintaining positive mood and for the ability to sleep, both necessary for healthy sexual functioning. Magnesium also modulates tissue sensitivity to vitamin D (Zofkova, 1995), which plays a role in sexual function (discussed earlier). Magnesium deficiency can cause insomnia (Watt, 1988). In a vicious cycle, insomnia causes a reduction in testosterone (Luboshitzky, 2001), which exacerbates the insomnia in men and the negative impact on testosterone. Low testosterone is well known to diminish sexual function (Kirby, 2015).

Magnesium deficiency is common in the general population (Rosanoff, 2012). One major factor is dietary. Approximately half of Americans consume less than the recommended daily allowance (RDA) of this essential mineral (Moshfegh, 2005). Suboptimal dietary habits that deplete magnesium include overconsumption of alcohol, sugar, refined or processed foods, and salt (Blomberg-Jensen, 2010). Obese individuals are especially likely to be magnesium deficient, and it is well known that central adiposity and metabolic syndrome are associated both with low levels of testosterone and with sexual dysfunction (Wang, 2011).

Certain chronic diseases impair magnesium levels. Because excess glucose induces marked renal excretion of magnesium, poorly controlled diabetes is frequently associated with magnesium deficiency, and diabetes—both types 1 and 2—is strongly associated with sexual dysfunction (Barbagallo, 2015; Pontiroli, 2013). Magnesium deficiency plays a major role in the development of atherosclerosis (Nadler, 1995), and there is a strong association between cardiovascular disease and sexual dysfunction (Nascimento, 2013). Hyperlipidemic subjects given magnesium supplementation have shown significantly improved lipid profiles (reduced total and LDL cholesterol, increased HDL cholesterol; Fox, 2001).

The other major cause of magnesium deficiency is the use of medications—especially those treating hypertension—that exacerbate an already reduced magnesium level. Hypertensive patients are typically prescribed antihypertensive medication—for example, angiotensin-converting enzyme (ACE) inhibitors, angiotensin II receptor blockers (ARBs), beta blockers, calcium channel blockers, and thiazide or loop diuretics—all of which affect multiple factors contributing to sexual function. Thiazide diuretics deplete magnesium, potassium (Iezhitsa,

2005), and zinc (Brannan, 1976). In addition, the loop diuretics also deplete, calcium, thiamine, vitamin C, and pyridoxine (Duarte, 1968; FDA, 2012; Lameire, 1988; Iezhitsa, 2005; Mydlik, 1999; Zenuk, 2003). A vicious cycle can occur in which hypomagnesemia develops and leads to hypertension, then a diuretic is prescribed that increases magnesium loss and ultimately requires a greater amount of antihypertensive medication. Sexual dysfunction is a major consequence of this vicious cycle.

Although it is generally believed that older-generation antihypertensive medications, such as diuretics and beta blockers, are more likely to cause sexual dysfunction than newer-generation drugs such as calcium antagonists, ACE inhibitors, and ARBs (WebMD, 2016a), some studies (Ekman, 2010) find them not so dissimilar. The exception is ARBs, which are considered the least likely to cause sexual dysfunction and may even improve erectile activity (Doumas, 2006). Beta blockers also deplete CoQ10 levels (Kishi, 1977), and are associated with ED and reduced sexual activity (Steinke, 2016). Calcium channel blockers also deplete CoQ10 and are reported to cause sexual dysfunction (Shiri, 2007). CoQ10 has profound cardiac benefits, all of which support sexual functioning. It reduces blood pressure (Rosenfeldt, 2007), improves ejection fraction (DiNicolantonio, 2015), improves endothelial function (Dai, 2011), protects mitochondrial function (Deichmann, 2010), and augments NO (Pepe, 2007).

Finally, cardiac glycosides, such as digoxin, deplete magnesium and potassium (Iezhitsa, 2005) and are also known to affect male sexual function adversely (Gupta, 1998). The responsible mechanism may involve digoxin's inhibitory effect on the sodium/potassium ATP pump, which regulates smooth muscle tone. The authors proposed that digoxin could be an effective treatment for recurrent priapism.

Measuring Magnesium

A convenient way to measure magnesium is by obtaining RBC magnesium concentration levels by atomic absorption spectrometry (Abraham, 1981). Although not ideal, this is usually more accurate than serum levels because serum contains only 1% of the body's magnesium stores. (RBC magnesium levels may be more sensitive at detecting deficiency, but they do not correlate perfectly with other measures of total body magnesium or magnesium status; Hansen, 2014). Ideally, RBC magnesium levels should be in the higher end of the 4.7 to 7.0 mg/dl range. Serum levels are homeostatically regulated and thus become low only once an individual is severely depleted. Serum levels should be no less than 2.0 mg/dl, or 1.7 mEg/L. Although magnesium can also be measured in white blood cells, this test is not currently available in all states.

In practice, as long as renal function is normal most clinicians simply choose to give a therapeutic trial of oral magnesium. Excess magnesium is excreted by the kidney, so when renal function is normal it is extremely difficult to overdose on magnesium, and toxicity is rare.

Replenishing Magnesium

Foods high in magnesium include green leafy vegetables, almonds, cashews, peanuts, black beans, avocado, oatmeal, and edamame (green soy beans; NIH, 2016a). Intermediate amounts of magnesium are found in chocolate, fruits, meat, and fish. Cooking—especially boiling—destroys magnesium content. As a dietary supplement, magnesium may be taken as magnesium glycinate, magnesium taurate, or other chelated forms. The RDA for elemental magnesium is 420 mg/day for men and 320 mg/day for women over the age of 31 (NIH, 2016a). If a laxative effect is also desired, the patient can take 8.75 to 25 gm or 290 ml liquid magnesium citrate for acute effect, or approximately 30 ml per day preventively. Alternatively, 400 to 1200 mg of magnesium oxide may be taken up to four times a day. Patients with sexual dysfunction in whom magnesium depletion is a likely factor should take the highest dose that does not cause uncomfortably loose bowel movements. It is ideal for patients to determine their daily dose of magnesium based on bowel tolerance. In addition, daily amount of magnesium can vary depending on a given day's diet, amount of exercise, and water consumption. Patients can choose to alternate between magnesium citrate or magnesium oxide and one of the chelated forms of magnesium. It is also possible to absorb magnesium through the skin by applying creams or oils or by taking Epsom salt baths or foot baths.

Blood pressure is reduced when magnesium is taken whether or not a patient is on antihypertensive medication (Itoh, 1997). For hypertensive patients, restoring a sufficient magnesium level is often enough to reduce blood pressure to acceptable levels and minimize or avoid the need for antihypertensive medication (Yogi, 2010). Although it can take months for a depleted individual to fully replenish magnesium stores, most patients experience some therapeutic benefit within just a few days.

ON THE HORIZON

Several micronutrients with logical involvement in sexual dysfunction have not yet attracted significant research activity, but existing findings are suggestive.

Boron

Boron is a trace mineral that aids the function and metabolism of estrogen and may also enhance testosterone levels. In a study of eight male subjects, one week of supplementation with boron at 10 mg a day resulted in significant decreases in sex hormone-binding globulin and in mean plasma estradiol, while mean plasma free testosterone increased. TNF-α, cortisol, and high-sensitivity C-reactive protein—all biomarkers of inflammation—were diminished (Naghii, 2011). In a study of 12 postmenopausal women, boron supplementation of 3 mg/day markedly elevated serum concentrations of estradiol and testosterone, which are both critical for good sexual function (Nielsen, 1987).

Vitamin E

Vitamin E improves blood flow (Takasaki, 2010) and aids in sex hormone production (Mumford, 2016). Foods with a high concentration of vitamin E include broccoli, spinach, asparagus, sunflower seeds, almonds, hazlenuts, peanuts, salmon, egg yolks, and sweet potato (NIH, 2016e). Serum and plasma levels of vitamin E can be measured by most commercial laboratories.

Iron

Oxygen-rich blood is one of the most important components for erectile health, and the amount of oxygen present is dependent on iron. Thus there is a well-documented association of iron deficiency and anemia with sexual dysfunction in both men and women (Gulmez, 2014). In an uncontrolled study of reproductive-aged women with iron-deficiency anemia (IDA), a course of iron supplementation resulted in significant improvement on the Female Sexual Function Index (Rosen, 2000) and in reduced anxiety (Gulmez, 2014). Adequate iron levels clearly contribute to good sexual function directly via effects on tissue. This study suggests an indirect effect as well, although results of this single study must be confirmed in a placebo-controlled trial to see if the degree to which reduction in anxiety is a significant placebo effect. If this study proves to be accurate, then adequate iron levels affect sexual function indirectly through reduced anxiety via an improved behavioral status potentially more conducive to intimacy.

Unfortunately, the widely used PPIs interfere significantly with iron absorption. Illustrating the impact is an interesting study. Patients homozygous for the gene causing hemochromatosis (a disease of excess iron) who underwent frequent phlebotomy to reduce excess RBCs required the procedure less frequently while taking PPIs (Hutchinson, 2007). Vegans and vegetarians are classically at risk of developing IDA if they do not make the effort to consume sufficient iron, especially if on PPIs.

Iron-containing foods include beef, chicken, liver, clams, mussels, oysters, beans, broccoli, and spinach. It should be noted that vegetables contain nonheme iron, which is less well absorbed than the heme iron found in meat.

Zinc

We know that zinc is necessary for the production of both testosterone and sperm (Netter, 1981). It is thought to improve prostate health (Franklin, 2007) and increase libido, despite conflicting results so far from studies that have used zinc to treat men with sexual disorders (Ebisch, 2007). Zinc depletion is associated with hypogonadism (Karaca, 2007) and hypothyroidism (Betsy, 2013).

A meta-analysis assessed eight earlier studies of antihypertensive medications that had reported zinc measurements. It concluded that ACE inhibitors, ARBs, and thiazide diuretics have the potential to reduce zinc levels (Braun, 2013). One complicating factor is that patients taking antihypertensive medications often have comorbidities—for example, heart failure and diabetes—that are also associated with problems in zinc metabolism. Another is that, not uncommonly, a thiazide diuretic—which blocks the reabsorption of zinc in the kidney—is often taken together with one of the other medications in this zinc-depleting group, intensifying the problem.

Because zinc competes with copper in intestinal absorption, a copper supplement may be needed when supplemental zinc is taken. Levels of both minerals can be measured in serum or in RBCs. Zinc–copper combination products are available. It is important to inform patients that nausea is a common side effect of zinc and that it can be minimized by splitting the dose into several smaller doses a day, by taking it with meals, and/or by taking it as zinc picolinate. Zinc-containing foods include oysters (one reason for their reputation as an aphrodisiac), beef, pork, chicken, kidney beans, and brown rice (NIH, 2016f). (Brown rice should be sourced carefully to make sure it is not contaminated with arsenic.)

OTHER NUTRIENTS TO NOTE

Although not micronutrients, these other nutrients carry therapeutic benefits in the context of sexual dysfunction and thus are included in this section.

Omega-3 Fatty Acids and the Mediterranean Diet

Most people consume far too little of the essential long-chain polyunsaturated omega-3 fatty acids (FA)—the n-3 PUFAs eicosapentaenoic acid (EPA) and docosahexaenoic acid (DHA). Although they are macronutrients rather than micronutrients, among their many health benefits are two in particular that make them appropriate to include in the context of sexual dysfunction caused by medication-induced micronutrient deficiencies. The Mediterranean diet (MeDi) includes particularly rich and varied sources of the omega-3 FAs. When this traditional dietary pattern—the most highly associated with optimal consumption of omega-3 FAs in the West—is diligently followed, it is associated with significantly higher micronutrient levels (Castro-Quezada, 2014; Simopoulos, 2001, 2004). It is also associated with a significantly lower risk for the chronic health problems—for example, cardiovascular disease, hypertension, and diabetes—that are treated with medications carrying a significant probability of inducing micronutrient depletions. Gaining and maintaining good omega-3 FA levels may enable many patients on these medications to eliminate them or reduce their dose.

Despite their unique biochemical properties and biological importance, omega-3 FAs are not endogenous, and food sources are fairly limited. Especially abundant are fish from cold northern waters, which are rich in EPA and DHA. Plant-sourced oils contain omega-3 FAs in the form of alpha-linolenic acid, which the body must transform into EPA and DHA (Kim, 2014). These sources include seeds (especially flax and chia seeds), nuts and the oils derived from them, olives and olive oil, and certain vegetables (including kale, spinach, cauliflower, and cabbage). Additional options are nutritional supplements. These are fish oil and krill oil. The vegetarian/vegan option is algae-derived omega-3 FAs.

Deficiencies reflect dietary inadequacy. The Standard American Diet of processed foods, meat and dairy from grain-fed animals, minimal vegetable consumption, and high sugar intake yields minimal quantities of omega-3 FAs and high amounts of the proinflammatory omega-6 FAs (Bourre, 2007; James, 2000; Simopoulos, 2001). Vegans, vegetarians, and others who do not regularly eat fatty fish are also highly likely to be deficient in EPA and DHA.

Among their many biological effects, the omega-3 FAs support primary neuronal function, reduce inflammation, support cardiovascular health and positive mood, and mitigate anxiety disorders (Green, 2006; Liu, 2013, Mazza, 2007; Petersson, 2016; Pusceddu, 2016; Seo, 2005), all of which are directly or indirectly relevant to sexual functioning. EPA and DHA in particular reduce proinflammatory PG2 prostaglandins, which constrict blood vessels (Bouree, 2007; Mantzioris, 2000; Saldeen, 2004), and omega-3 FAs in an animal model enhanced microcirculation by increased vasodilator production, among other mechanisms (Conde, 2007). The omega-3 FAs also increase RBC deformability, making blood less viscous and thus improving circulation and increasing oxygen delivery to tissues (Simopoulos, 1999).

The MeDi is highly documented as an effective dietary approach to omega-3 FA adequacy and includes dietary elements that carry multiple important health benefits; there are limited data on dietary change and sexual function. One trial addressing this found that among men and women with newly diagnosed type 2 diabetes, the MeDi reduced deterioration of sexual function over the eight-year study period more than a low-fat diet did (Maiorino, 2016). Recent epidemiologic studies found that type 2 diabetic men and women most adherent to the MeDi had the lowest prevalence of ED or female sexual dysfunction (FSD) and were more likely to be sexually active (Esposito, 2010 a, b). A review of the sparse literature noted clinical trials showing the MeDi more effective than a control diet in ameliorating ED or restoring absent ED in men with obesity or metabolic syndrome (Esposito, 2010a). And as stated, the MeDi ensures higher levels of micronutrients that are critical to sexual function.

Changing one's diet so comprehensively is not easy to accomplish for many people. EPA/DHA supplements are the option for those who have failed or refuse to try. Those on vegan or vegetarian diets can take algae-derived omega-3 supplements. Supplements have the added advantage of being acceptably free of the mercury contamination that can occur in fish. The recommended daily dosage of omega-3 FAs is 2 to 4 g/day (Miller, 2011).

The Carnitines

Carnitine is an amino acid derivative found in nearly every cell of the body that transports FAs to the mitochondria to be used as fuel for energy production. It exists in three variants, each with its own mechanism of action, and they work synergistically together. The two of interest here are acetyl-L-carnitine and propionyl-L-carnitine. Acetyl-L-carnitine is chiefly involved in moving fat molecules across the mitochondrial membrane. Propionyl-L-carnitine plays an important role in protecting the function and health of endothelial cells, protecting vascular cells from damage and optimizing endothelial function by increasing the production of vasodilating nitric oxide. Dietary carnitine is derived primarily from meat, fish, and dairy products.

Propionyl-L-carnitine has been shown effective in treating ED. In a study of diabetic men with poor circulation and ED who had failed monotherapy with sildenafil (Gentile, 2004), subjects responded substantially more adequately when sildenafil was taken together with propionyl-L-carnitine. In a six-month study in older men, propionyl-L-carnitine was most effective when used together with acetyl-L-carnitine in improving erectile function, desire, energy, mood, and general well-being (Cavallini, 2004). The combination outperformed testosterone, which itself showed benefit compared to placebo. Men with postprostatectomy ED responded better to sildenafil when taken together with this carnitine combination (Cavallini, 2005). Because the carnitines have an indirect pro-oxidant effect, they are best taken with the antioxidant alpha-lipoic acid.

Conclusion

Sexual dysfunction is a common adverse effect of many medications that occurs as a direct and recognized side effect of a particular drug. Also established—but little recognized—is the fact that micronutrient deficiencies are commonly caused by many widely used pharmaceuticals. That these micronutrient deficiencies can impair sexual function reflects the fact that many of them are vital to endocrine and metabolic balance in ways that play significant roles in sexual function. Because micronutrient deficiency secondary to long-term pharmaceutical use is common but little recognized, it is typically untreated and thus the impact on sexual dysfunction remains untreated as well. This chapter provides guidance on the specific micronutrients involved, the large variety of commonly prescribed drugs involved in producing deficiencies, how to identify a given patient's micronutrient deficiencies, how to resolve them, and how—when possible—to prevent them.

Patients on medications who have developed sexual dysfunction should be evaluated for relevant micronutrient deficiencies and given nutritional supplements where appropriate. It may also be advisable for patients on pharmaceuticals to take nutritional supplements preventively, based on the nutrients that those pharmaceuticals are likely to deplete. To the degree that restoring micronutrient levels reduces the burden of sexual side effects, medication compliance should improve. Finally, in addition to preserving sexual function, optimum micronutrient status benefits overall health and vitality.

REFERENCES

Abraham GE, Lubran MM (1981). Serum and red cell magnesium levels in patients with premenstrual tension. *Am J Clin Nutri.* 34(11):2364–2366.

Aymard JP, Aymard B, Netter P, Bannworth B, Trechot P, Streiff F (1988). Haematological adverse effects of histamine H2-receptor antagonists. *Med Toxicol Adverse Drug Exp.* 3:430–448.

Baik HW, Russell RM (1999). Vitamin B12 deficiency in the elderly. *Annu Rev Nutr.* 19:357–377.

Baldwin DS (2001). Depression and sexual dysfunction. *Br Med Bull.* 57(1): 81–99.

Balon R, Yeragani VK, Pohl R, Ramesh C (1993). Sexual dysfunction during antidepressant treatment. *J Clin Psych.* 54(6):209–212.

Barassi A, Pezzilli R, Colpi GM, Corsi-Romanelli MM, Melzi d'Eril GV (2014). Vitamin D and erectile dysfunction. *J Sex Med.* 11(11):2792–800.

Barbagallo M, Dominguez JL (2015). Magnesium and type 2 diabetes. *World J Diabet.* 6(10):1152–1157.

Basu TK (1982). Vitamin C–aspirin interactions. *Int J Vitam Nutr Res Suppl.* 23:83–90.

Battaglia C, Battaglia B, Mancini F, Busacchi P, Paganotto MC et al. (2012). Sexual behavior and oral contraception: a pilot study. *J Sex Med.* Feb;9(2):550–557.

Bennett M (2001). Vitamin B12 deficiency, infertility and recurrent fetal loss. *J Reprod Med.* 46(3):209–212.

Betsy A, Binitha M, Sarita S (2013). Zinc deficiency associated with hypothyroidism: an overlooked cause of severe alopecia. *Int J Trichology*. Jan 5(1):40–42.

Blomberg-Jensen M, Nielsen JE, Jørgensen A, Rajpert-De Meyts E, Kristensen DM, et al. (2010). Vitamin D receptor and vitamin D metabolizing enzymes are expressed in the human male reproductive tract. *Hum Reprod*. 25(5):1303–1311.

Blomberg-Jensen M, Bjerrum PJ, Jessen TE, Neilsen JE, Joensen UN, Olesen IA, Petersen JH, Dissing S, Jørgensen N (2011). Vitamin D is positively associated with sperm motility and increases intracellular calcium in human spermatozoa. *Hum Reprod*. 26(6):1307–1317.

Bóné B, Janszky J (2006). [Epilepsy and male sexual dysfunction: etiology, diagnosis and therapy]. *Ideggyogy Sz*. May 20;59(5–6):148–152.

Bouillon R, Reynaert J, Claes JH, et al. (1975). The effect of anticonvulsant therapy on serum levels of 25-hydroxy-vitamin D, calcium, and parathyroid hormone. *J Clin Endocrinol Metab*. 41:1130–1135.

Bourre JM (2006). Effects of nutrients (in food) on the structure and function of the nervous system: update on dietary requirements for brain. Part 2: macronutrients. *J Nutr Health Aging*. Sep-Oct;10(5):386–399.

Bourre JM (2007). Dietary omega-3 fatty acids for women. *Biomed Pharmacother*. Feb-Apr;61(2–3):105–112.

Boyarsky KB, Hirschfeld MR (2000). The management of medication-induced sexual dysfunction. *Essent Psychopharmacol*. 3(2):151–170.

Brannan PG, Vergne-Marini P, Pak CY, Hull AR, Fordtran JS (1976). Magnesium absorption in the human small intestine: results in normal subjects, patients with chronic renal disease, and patients with absorptive hypercalciuria. *J Clin. Invest*. 57(6): 1412–1418.

Braun LA, Rosenfeldt F (2013). Pharmaco-nutrient interactions—a systematic review of zinc and antihypertensive therapy. *Int J Clin Pract*. Aug;67(8):717–725.

Calabrò RS, Bramanti P, Italiano D, Ferlazzo E (2009). Topiramate-induced erectile dysfunction. *Epilepsy Behav*. Mar;14(3):560–561.

Calabrò RS, Italiano D, Militi D, Bramanti P (2012). Levetiracetam-associated loss of libido and anhedonia. *Epilepsy Behav*. 24:283–284.

Cardoso CC. Lobato KR, Binfaré RW, Ferreira PK, Rosa AO, Santos AR, Rodrigues AL (2009). Evidence for the involvement of the monoaminergic system in the antidepressant-like effect of magnesium. *Prog Neuropsychopharmacol Biol Psychiatry*. Mar 17;33(2):235–242.

Caruso S, Agnello C, Intelisano G, Farina M, Di Mari L, Cianci A (2004). Sexual behavior of women taking low-dose oral contraceptive containing 15 microg ethinylestradiol/60 microggestodene. *Contraception*. Mar; 69(3):237–240.

Carvajal A, Macias D, Gutiérrez A, Ortega S, Sáinz M, Martín Arias LH, Velasco A (2007). Gynaecomastia associated with proton pump inhibitors: a case series from the Spanish Pharmacovigilance System. *Drug Saf*. 30(6):527–531.

Cass, H (2007). *Supplement your prescription: what your doctor doesn't know about nutrition*. Laguna Beach, CA: Basic Health Publications.

Cass, H (2013). A practical guide to avoiding drug-induced nutrient depletion. Nutrition Review. http://nutritionreview.org/2013/04/practical-guide-avoiding-drug-induced-nutrient-depletion/

Castro-Quezada I, Román-Viñas B, Serra-Majem L (2014). The Mediterranean Diet and nutritional adequacy: a review. *Nutrients*. 6(1):231–248.

Cavallini G, Caracciolo S, Vitali G, Modenini F, Biagiotti G (2004). Carnitine versus androgen administration in the treatment of sexual dysfunction, depressed mood, and fatigue associated with male aging. *Urology*. 63:641–646.

Cavallini G, Modenini F, Vitali G, Koverech A (2005). Acetyl-L-carnitine plus propionyl- L-carnitine improve efficacy of sildenafil in treatment of erectile dysfunction after bilateral nerve-sparing radical retropubic prostatectomy. *Urology*. 66:1080–1085.

Chang WS, Fine R, Siegal D, Chesney M, Black D, Hulley BS (1991). The impact of diuretic therapy on sexual function. *Arch Intern Med*. Dec;151:2402–2408.

Cheng Z, Yang X, Wang H (2009). Hyperhomocysteinemia and endothelial dysfunction. *Curr Hypertens Rev*. 5(2):158–165.

Chi-Fai Ng, Chui-Ping Lee, Ho LA, Lee WYV (2011). Effect of niacin on erectile function in men suffering erectile dysfunction and dyslipidemia. *J Sex Med*. 8(10):2883–2893.

Christensen B (2014). Inflammatory bowel disease and sexual dysfunction. *Gastroenterol Hepatol*. Jan;10(1):53–55.

Clayton JA, Rodgers S, Blakey J (2006). Thiazide diuretic prescription and electrolyte abnormalities in primary care. *Br J Clin Pharmacol*. 61:87–95.

Conde CM, Cyrino FZ, Bottino DA, Gardette J, Bouskela E (2007). Longchain n-3 polyunsaturated fatty acids and microvascular reactivity: observation in the hamster cheek pouch. *Microvasc Res*. May;73(3):237–247.

Corona G, Boddi V, Balercia G, Rastrelli G, De Vita G, Sforza A, Forti G, Mannucci E, Maggi M (2010). The effect of statin therapy on testosterone levels in subjects consulting for erectile dysfunction. *J Sex Med*. Apr;7(4 Pt 1):1547–1556.

Crane FL (2001). Biochemical functions of coenzyme Q10. *J Am Coll Nutr*. 20:591–598.

Dai YL, Luk TH, Yiu KH, Wang M, Yip PM, et al. (2011). Reversal of mitochondrial dysfunction by coenzyme Q10 supplement improves endothelial function in patients with ischaemic left ventricular systolic dysfunction: a randomized controlled trial. *Atherosclerosis*. Jun, 216(2):395–401.

Das N, Nebioglu S (1992). Vitamin C aspirin interactions in laboratory animals. *J Clin Pharm Ther*. 17:343–346.

Davie MW, Emberson CE, Lawson DE, et al. (1983). Low plasma 25-hydroxy vitamin D and serum calcium levels in institutionalized epileptic subjects: associated risk factors, consequences and response to treatment with vitamin D. *Q J Med*. 52:79–91.

Deichmann R, Lavie C, Andrews S (2010). Coenzyme Q10 and statin-induced mitochondrial dysfunction. *Ochsner J*. (1):16–21.

de Jager J, Kooy A, Lehert P, Wulffele MG, van der Kolk J, Bets D, Kerburg J, Donker AJ, Stehouwer CD (2010). Long term treatment with metformin in patients with type 2 diabetes and risk of vitamin B-12 deficiency: randomised placebo controlled trial. *BMJ*. 340:2181.

Demir T, Comlekçi A, Demir O, Gülcü A, Calypkan S, Argun L, et al. (2006). Hyperhomocysteinemia: a novel risk factor for erectile dysfunction? *Metabolism*. 55:1564–1568.

Depeint F, Bruce WR, Shangari N, Mehta R, O'Brien PJ (2006). Mitochondrial function and toxicity: role of the B vitamin family on mitochondrial energy metabolism. *Chem Biol Interact*. Oct 27;163(1–2):94–112.

DiNicolantonio, JJ Bhutani J, McCarty M, O'Keefe J, et al. (2015). Coenzyme Q10 for the treatment of heart failure: A review of the literature. *Open Heart* 2.1: e000326.

Doumas M, and Douma, S (2006). The effect of antihypertensive drugs on erectile function: a proposed management algorithm. *J Clin Hyperten*. 8:359–363.

Duarte GC (1968). Effects of ethacrynic acid and furosemide on urinary calcium, phosphate and magnesium. *Metab J: Clin Exp*. 17(10): 867–876.

Ebisch IM, Thomas CM, Peters WH, Braat DD, Steegers-Theunissen RP (2007). The importance of folate, zinc and antioxidants in the pathogenesis and prevention of subfertility. *Hum Reprod Update*. Mar-Apr;13(2):163–174.

Ekman E, Hägg S, Sundström A, Werkström V (2010). Antihypertensive drugs and erectile dysfunction as seen in spontaneous reports, with focus on angiotensin II type 1 receptor blockers. *Drug Healthc Patient Saf*. 2:21–25.

Elnazer HY, Sampson A, Baldwin D (2015). Lithium and sexual dysfunction: an under-researched area. *Hum Psychopharmacol*. Mar;30(2):66–69.

Esposito K, Giugliano F, Maiorino MI, Giugliano D (2010a). Dietary factors, Mediterranean diet and erectile dysfunction. *J Sex Med*. Jul;7(7):2338–2345.

Esposito K, Maiorino M, Bellastella G, Giugliano F, Romano M, et al (2010b) Determinants of female sexual dysfunction in type 2 diabetes. *International Journal of Impotence Research*. May-Jun;22(3):179–84.

Food and Drug Administration (2011). FDA Drug Safety Communication: Low magnesium levels can be associated with long-term use of proton pump inhibitor drugs (PPIs). http://www.fda.gov/Drugs/DrugSafety/ucm245011.htm. 3/2/2011.

Food and Drug Administration (2012). Drug Safety Communication: Diovan HCT (valsartan/hydrochlorothiazide) tablets. http://www.fda.gov/Safety/MedWatch/SafetyInformation/ucm212125.htm.

Feltner DE, Crockatt JG, Dubovsky SJ, Cohn CK, Shrivastava RK, Targum SD, Liu-Dumaw M, Carter CM, Pande AC (2003). A randomized, double-blind, placebo-controlled, fixed-dose, multicenter study of pregabalin in patients with generalized anxiety disorder. *J Clin Psychopharamacol*. Jun;23(3):240–299.

Fogari R, Zoppi A, Poletti L, Marasi G, Mugellini A, Corradi L (2001). Sexual activity in hypertensive men treated with valsartan or carvedilol: a crossover study. *Am J Hypertens*. 14(1):27–31.

Folkers K, Langsjoen P, Willis R, et al. (1990). Lovastatin decreases coenzyme Q levels in humans. *Proc Natl Acad Sci U S A*. 87:8931–8934.

Fox C, Ramsoomair D, Carter C (2001). Magnesium: its proven and potential clinical significance. South Med J. 94:1195–1201.

Franklin RB, Costello LC (2007). Zinc as an anti-tumor agent in prostate cancer and in other cancers. *Arch Biochem Biophys.* Jul;463(2):211–217.

Frellick M (2016). Oral contraceptives don't impair sexual function, at least overall. *Medscape Med News.* August 19.

Frenkel EP, McCall MS, Sheehan RG (1973). Cerebrospinal fluid folate, and vitamin B12 in anticonvulsant-induced megaloblastosis. *J Lab Clin Med.* 81:105–115.

Gentile V, Vicini P, Prigiotti G, Koverech A, Di Silverio F (2004). Preliminary observations on the use of propionyl-L-carnitine in combination with sildenafil in patients with erectile dysfunction and diabetes. *Curr Med Res Opin.* 20:1377–1384.

Getoff N, Hartmann J, Schittl H, Gerschpacher M, Quint RM (2011). Photo-induced regeneration of hormones by electron transfer processes: potential biological and medical consequences. *Radiat Phys Chem (Oxf.)* 1993. Aug;80(8):890–894.

Gleason MJ, Slezak MJ, Jung H, Reynolds K, Stephen K et al. (2011). Regular nonsteroidal anti-inflammatory drug use and erectile dysfunction. *J Urol.* Apr;185: 1388–1393.

Green P, Hermesh H, Monselise A, Marom S, Presburger G, Weizman A (2006). Red cell membrane omega-3 fatty acids are decreased in nondepressed patients with social anxiety disorder. *Eur Neuropsychopharmacol.* Feb;16(2):107–113.

Gulmez H, Akin Y, Savas M, Gulum M, Ciftci H, Yalcinkaya S, Yeni E (2014). Impact of iron supplementation on sexual dysfunction of women with iron deficiency anemia in short term: a preliminary study. *J Sex Med.* 11(4):1042–1046.

Gupta S, Salimpour P, Saenz de Tejada I, Daley J, Gholami S, Daller M, Krane RJ, Traish AM, Goldstein I (1998). A possible mechanism for alteration of human erectile function by digoxin: inhibition of corpus cavernosum sodium/potassium adenosine triphosphates activity. *J Urol.* 159(5):1529–36.

Hansen KE, Nabak AC, Johnson RE, et al. (2014). Isotope concentrations from 24-h urine and 3-h serum samples can be used to measure intestinal magnesium absorption in postmenopausal women. *J Nutrit.* 144(4):533–537.

Hechtman L (2012). *Clinical naturopathic medicine.* New York: Elsevier.

Heidelbaugh JJ (2013). Proton pump inhibitors and risk of vitamin and mineral deficiency: evidence and clinical implications. *Ther Adv Drug Saf.* Jun;4(3):125–133.

Hever J (2014). *The vegiterranean diet: the new and improved Mediterranean eating plan—with deliciously satisfying vegan recipes for optimal health.* Boston: Da Capo Lifelong Books.

Higashi A, Tamari H, Ikeda T, et al. (1980). Serum vitamin E concentration in patients with severe multiple handicaps treated with anticonvulsants. *Pediatr Pharmacol.* 1:129–134.

Higgins A, Nash M, Lynch MA (2010). Antidepressant-associated sexual dysfunction: impact, effects and treatment. *Drug Healthc Patient Saf.* 2:141–150.

Hiraoka A, Arato T, Tominaga I (1997). Reduction in blood free carnitine levels in association with changes in sodium valproate (VPA) disposition in epileptic patients treated with VPA and other anti-epileptic drugs. *Biol Pharm Bull.* 20:91–93.

Holick MF, Binkley NC, Bischoff-Ferrari HA, et al. (2011). Evaluation, treatment, and prevention of vitamin D deficiency: an Endocrine Society clinical practice guideline. *Endocrine Society J Clin Endocrinol Metab.* Jul;96(7):1911–1930.

Hutchinson C, Geissler CA, Powell JJ, Bomford A (2007). Proton pump inhibitors suppress absorption of dietary non-haem iron in hereditary haemochromatosis. *Gut.* Sep;56(9): 1291–1295.

Iezhitsa IN (2005). Potassium and magnesium depletions in congestive heart failure—pathophysiology, consequences and replenishment. *Clin Calcium.* Nov;15(11):123–133.

Institute of Medicine (1998). *Food and Nutrition Board. Dietary Reference Intakes: Thiamin, Riboflavin, Niacin, Vitamin B6, Folate, Vitamin B12, Pantothenic Acid, Biotin, and Choline.* Washington, DC: National Academy Press.

Isoyama R, Baba Y, Harada H, et al. (1986). Clinical experiences of methylcobalamin (CH3-B12)/clomiphene citrate combined treatment in male infertility. *Hinyokika Kiyo.* 32(8):1177–1183.

Itoh K, Kawasaka T, Nakamura M (1997). The effects of high oral magnesium supplementation on blood pressure, serum lipids and related variables in apparently healthy Japanese subjects. *Br J Nutr.* Nov;78(5):737–750.

Jacobsen PL, Mahableshwarkar AR, Chen Y, Chrones L, Clayton AH (2015). Effect of Vortioxetine vs. Escitalopram on sexual functioning in adults with well-treated major depressive disorder experiencing SSRI-induced sexual dysfunction. *J Sex Med.* Oct;12(10):2036–2048.

James MJ, Gibson RA, Cleland LG (2000). Dietary polyunsaturated fatty acids aninflammatory mediator production. *Am J Clin Nutr.* Jan;71(1 Suppl):343S–348S.

Jiang P, Zhang LH, Cai HL, Li HD, Liu YP, Tang MM, He X (2014). Neurochemical effects of chronic administration of calcitriol in rats. *Nutrients.* 6(12):6048–6059.

Jorde R, Sneve M, Figenschau Y, Svartberg J, Waterloo K (2008). Effects of vitamin D supplementation on symptoms of depression in overweight and obese subjects: randomized double blind trial. *J Intern Med.* 264(6):599–609.

Kado DM, Karlamangla AS, Huang MH, Troen A, Rowe JW, Selhub J, Seeman TE (2005). Homocysteine versus the vitamins folate, B6, and B12 as predictors of cognitive function decline in older high-functioning adults: MacArthur Studies of Successful Aging. *Am J Med.* Feb;118(2):161–167.

Kamanna VS, Ganji SH, Kashyap ML (2009). The mechanism and mitigation of niacin-induced flushing. *Int J Clin Pract.* Sep;63(9):1369–1377.

Kaufman KR, Struck, PJ (2011). Gabapentin-induced sexual dysfunction. *Epilepsy Beh.* Jul;21(3):324–326.

Karaca Z, Tanriverdi F, Kurtoglu S, Tokalioglu S, Unluhizarci K, Kelestimur F (2007). Pubertal arrest due to Zn deficiency: the effect of zinc supplementation. *Hormones (Athens).* Jan-Mar;6(1):71–74.

Kennedy DO (2016). B Vitamins and the Brain: Mechanism, Dose and Efficacy—a review. *Nutrients.* 8(2),68.

Kim KB, Nam YA, Kim HS, Hayes AW, Lee BM (2014). α-Linolenic acid: nutraceutical, pharmacological and toxicological evaluation. *Food Chem Toxicol.* Aug;70:163–178.

Kirby EW, Verges D, Matthews J, Carson CC, Coward RM (2015). Low testosterone has a similar prevalence among men with sexual dysfunction due to either Peyronie's disease or erectile dysfunction and does not correlate with Peyronie's disease severity. *J Sex Med.* Mar;12(3):690–696.

Kishi T, Watanabe T, Folkers K (1977). Bioenergetics in clinical medicine XV: inhibition of coenzyme Q10-enzymes by clinically used adrenergic blockers of beta-receptors. *Res Commun Chem Pathol Pharmacol.* 17:157–164.

Kohler ST, McVary TK (2016). *Contemporary treatment of erectile dysfunction: a clinical guide,* 2nd ed. (Contemporary Endocrinology). New York: Springer, Humana Press.

Korn L, Lake J (2016). *Nutrition essentials for mental health: a complete guide to the food-mood connection.* New York: W. W. Norton.

Krüger THC, Haake P, Chereath D, et al. (2003). Specificity of the neuroendocrine response to orgasm during sexual arousal in men. *J Endocrinol.* 177:57–64.

Laake D, Compart P (2016). *The ADHD and autism nutritional supplement handbook: the cutting-edge biomedical approach to treating the underlying deficiencies and symptoms of ADHD and autism.* Beverly, MA: Quarto Publishing Group, Fair Winds Press.

Lahon K, Shetty MH, Paramel A, Sharma G (2011). Sexual dysfunction with the use of antidepressants in a tertiary care mental health setting—a retrospective case series. *J Pharmacol Pharmacother.* 2(2):128–131.

Lam JR, Schneider JL, Zhao W, Corley DA (2013). Proton pump inhibitor and histamine 2 receptor antagonist use and vitamin B12 deficiency. *JAMA.* 310(22):2435–2442.

Lameire N, Dodion L (1988). Acute and chronic effects of torasemide in healthy volunteers. *Arzneimittelforschung.* 38(1A):167–171.

Landau D, Kher KK (1997). Gentamicin-induced Bartter-like syndrome. *Pediatr Nephrol.* 11:737–740.

Langsjoen PH, Langsjoen AM (2003). The clinical use of HMG CoA-reductase inhibitors and the associated depletion of coenzyme Q10: A review of animal and human publications. *Biofactors.* 18(1–4):101–111.

Langsjoen PH, Langsjoen JO, Langsjoen AM, Lucas LA (2005). Treatment of statin adverse effects with supplemental coenzyme Q10 and statin drug discontinuation. *Biofactors.* 25(1–4):147–152.

Lems WF, Van Veen GJ, Gerrits MI, et al. (1998). Effect of low-dose prednisone (with calcium and calcitriol supplementation) on calcium and bone metabolism in healthy volunteers. *Br J Rheumatol.* 37:27–33.

Levine MA, Pollard HB (1983). Hydrocortisone inhibition of ascorbic acid transport by chromaffin cells. *FEBS Lett.* 158:134–138.

Linnebank M, Moskau S, Sammler A, et al. (2011). Antiepileptic drugs interact with folate and vitamin B12 serum levels. *Ann Neurol.* Feb;69(2):352–359.

Liu JJ, Galfalvy HC, Cooper TB, Oquendo MA, Grunebaum MF, Mann JJ, Sublette ME (2013). Omega-3 polyunsaturated fatty acid (PUFA) status in major depressive disorder with comorbid anxiety disorders. *J Clin Psychiatry.* Jul;74(7):732–738.

Lombardo F, Tsamatropoulos P, Piroli E, Culasso F, Jannini EA (2010). Treatment of erectile dysfunction due to C677T mutation of the MTHFR gene with vitamin B6 and folic acid in patients non responders to PDE5i. *J Sex Med.* 7(1):216–223.

Luboshitzky R, Zabari Z, Shen-Orr Z, Herer P, Lavie P (2001). Disruption of the nocturnal testosterone rhythm by sleep fragmentation in normal men. *J Clin Endocrin Metab.* 86(3):1134–1139.

Maiorino MI, Bellastella G, Caputo M, et al. (2016). Effects of Mediterranean diet on sexual function in people with newly diagnosed type 2 diabetes: The MÈDITA trial. *J Diabet Complications.* Nov–Dec;30(8):1519–1524.

Mantzioris E, Cleland LG, Gibson RA, Neumann MA, Demasi M, James MJ (2000). Biochemical effects of a diet containing foods enriched with n-3 fatty acids. *Am J Clin Nutr.* Jul;72(1):42–48.

Mapp RK, McCarthy TJ (1976). The effect of zinc sulphate and of bicitropeptide on tetracycline absorption. *S Afr Med J.* 50:1829–1830.

Mazza M, Pomponi M, Janiri L, Bria P, Mazza S (2007). Omega-3 fatty acids and antioxidants in neurological and psychiatric diseases: an overview. *Prog Neuropsychopharmacol Biol Psychiatry.* Jan;31(1):12–26.

Mayo Clinic (2013). Vitamin B6 (Pyridoxine) background. Mayo Foundation for Medical Education and Research. http://www.mayoclinic.org/drugs-supplements/vitamin-b6/background/hrb-20058788.

McColl KE (2009). Effect of proton pump inhibitors on vitamins and iron. *Am J Gastroenterol.* Mar;104(Suppl 2):S5–S9.

Meletis C, Jacobs T (2005). *Interactions between drugs and natural medicines: a simple guide to a complex subject,* 2nd ed. Sandy, OR: Eclectic Medical Publications.

Melis MR, Argiolas A (1995). Dopamine and sexual behavior. *Neurosci Biobehav Rev.* 19(1):19–38.

Mercola J (2014). 7 signs you may have a vitamin D deficiency http://articles.mercola.com/sites/articles/archive/2014/05/28/vitamin-d-deficiency-signs-symptoms.aspx May.

Mercola J (2015a). 11 most common nutrient deficiencies. http://articles.mercola.com/sites/articles/archive/2015/10/19/most-common-nutrient-deficiencies.aspx Oct.

Mercola J (2015b). Four important fat-soluble vitamins. http://articles.mercola.com/sites/articles/archive/2015/04/20/important-fat-soluble-vitamins.aspx Apr.

Meston CM, Frohlich PF (2000). The neurobiology of sexual function. *Arch Gen Psychiatry.* 57(11):1012–30.

Meston CM, Bradford A (2007). Sexual dysfunctions in women. *Annu Rev Clin Psychol.* 3:233–256.

Miller D (2004). *The Salts of the Earth and Synthetic Insanity: The Bipolar Theory: A Physical Explanation of Mental Illness.* Lincoln, NE: iUniverse, Inc, 35.

Miller M, Stone NJ, Ballantyne C, et al. (2011). Triglycerides and cardiovascular disease: a scientific statement from the American Heart Association. *Circulation.* 123:2292–333.

Mintzer S (2012). Plaques on the wall: inducing anticonvulsant use and atherogenesis. *Epilepsy Curr.* Nov;12(6):229–231.

Mock DM, Mock NI, Nelson RP, Lombard KA (1998). Disturbances in biotin metabolism in children undergoing long-term anticonvulsant therapy. *J Pediatr Gastroenterol Nutr.* 26:245–250.

Montejo-Gonzalez AL, Llorca G, Izquierdo JA, et al. (1997). SSRI-induced sexual dysfunction: fluoxetine, paroxetine, sertraline, and fluvoxamine in a prospective, multicenter, and descriptive clinical study of 344 patients. *J Sex Marital Ther.* 23(3):176–194.

Morrell JM (2002). Folic acid and epilepsy. *Epilepsy Curr.* 2(2):31–34.

Moshfegh A, Goldman J, Cleveland L (2005). What we eat in America, Nhanes 2001–2002: usual nutrient intake from food compared to dietary reference intakes. US Department of Agriculture, Agricultural Research Service. https://www.ars.usda.gov/ARSUserFiles/80400530/pdf/0102/usualintaketables2001-02.pdf.

Mumford SL, Browne RW, Schliep KC, et al. (2016). Serum antioxidants are associated with serum reproductivehormones and ovulation among healthy women. *J Nutr.* Jan;146(1):98–106.

Mydlik M, Derzsiova K, Zemberova E (1999). Influence of water and sodium diuresis and furosemide on urinary excretion of vitamin B6, oxalic acid and vitamin C in chronic renal failure. *Miner Electrolyte Metab.* 25(4–6):352–356.

Nadeau A, Roberge AG (1988). Effects of vitamin B12 supplementation on choline acetyltransferase activity in cat brain. *Int J Vit Nutrition Res.* 58(4):402–406.

Nadler JL, Rude RK (1995). Disorders of magnesium metabolism. *Endocrinol Metab Clin North Amer.* 24:623–641.

Nagaraj AK, Pai NB, Rao S, Rao TS, Goyal N (2009). Biology of sexual dysfunction. *Online J Health Allied Sci.* 8(1):1–7.

Naghii MR, Mofid M, Asgari AR, Hedayati M, Daneshpour MS (2011). Comparative effects of daily and weekly boron supplementation on plasma steroid hormones and proinflammatory cytokines. *J Trace Elem Med Biol.* 25(1):54–58.

Nascimento ER, Maia ACO, Pereira V, Soares-Filho G, Nardi AE, Silva AC (2013). Sexual dysfunction and cardiovascular diseases: a systematic review of prevalence. *Clinics.* 68(11):1462–1468.

Nathan BS, Zund J, Gottlieb B (2010). *The nitric oxide (NO) solution: how to boost the body's miracle molecule to prevent and reverse chronic disease.* Austin, TX: Neogenis.

National Institutes of Health, Office of Dietary Supplements. (2016a). Magnesium: dietary supplement fact sheet. https://ods.od.nih.gov/FactSheets/magnesium-HealthProfessional.

National Institutes of Health, Office of Dietary Supplements. (2016b). Vitamin B6: dietary supplement fact sheet. https://ods.od.nih.gov/factsheets/VitaminB6-Health Professional.

National Institutes of Health, Office of Dietary Supplements. (2016c). Vitamin B12: dietary supplement fact sheet. https://ods.od.nih.gov/factsheets/VitaminB12-Health Professional.

National Institutes of Health, Office of Dietary Supplements. (2016d). Vitamin D: dietary supplement fact sheet. https://ods.od.nih.gov/factsheets/VitaminD-Health Professional.

National Institutes of Health, Office of Dietary Supplements. (2016e). Vitamin E: dietary supplement fact sheet. https://ods.od.nih.gov/factsheets/VitaminE-HealthProfessional/#h3.

National Institutes of Health, Office of Dietary Supplements (2016f). Zinc Fact Sheet for Health Professionals. https://ods.od.nih.gov/factsheets/Zinc-HealthProfessional/#h3, updated: February 11, 2016.

Neel AB (2012). 7 meds that can wreck your sex life. AARP. Apr. 25. http://www.aarp.org/health/drugs-supplements/info-04-2012/medications-that-can-cause-sexual-dysfunction.html.

Netter A, Hartoma R, Nahoul K (1981). Effect of zinc administration on plasma testosterone, dihydrotestosterone, and sperm count. *Arch Androl.* Aug;7(1):69–73.

Neuvonen PJ (1976). Interactions with the absorption of tetracyclines. *Drugs.* 11:45–54.

Nicolai MPJ, Liem SS, Both S, et al. (2014). A review of the positive and negative effects of cardiovascular drugs on sexual function: a proposed table for use in clinical practice. *Neth Heart J.* 22(1):11–19.

Nielsen FH, Hunt CD, Mullen LM, Hunt JR (1987). Effect of dietary boron on mineral, estrogen, and testosterone metabolism in postmenopausal women. *FASEB J.* 1(5):394–397.

Northcott CA, Watts SW (2004). Low [Mg2+]e enhances arterial spontaneous tone via phosphatidylinositol 3-kinase in DOCA-salt hypertension. *Hypertension.* 43(1):125–129.

Page L (2011). *Healthy healing.* 14th ed. Eden Prairie, MN: Healthy Healing Publications.

Pak CY (2000). Correction of thiazide-induced hypomagnesemia by potassium-magnesium citrate from review of prior trials. *Clin Nephrol.* 54:271–275.

Palme G, Koeppe P (1978). Comparative experimental studies in animals and humans on gastrointestinal blood loss following antirheumatic pharmacotherapy. *Arzneimittelforschung.* 28:426–428.

Palmery M, Saraceno A, Vaiarelli A, Carlomagno G (2013). Oral contraceptives and changes in nutritional requirements. *Eur Rev Med Pharmacol Sci.* Jul;17(13):1804–13.

Park WY, Kim Y, Lee HJ (2012). Antipsychotic-induced sexual dysfunction and its management. *World J Mens Health.* Dec;30(3):153–159.

Patak P, Willenberg HS, Bornstein SR (2004). Vitamin C is an important cofactor for both adrenal cortex and adrenal medulla. *Endocr Res.* Nov;30(4):871–875.

Pelton R, LaValle JB, Hawkins EB, Krinsky DL (1999). *Drug-induced nutrient depletion handbook.* Wolters Kluwer Clinical Drug Information, Lexi-Comp. http://www.chiro.org/nutrition/ABSTRACTS/Nutrient_Depletion_Checklist_Diuretics.shtml.

Pepe S, Marasco SF, Haas SJ, Sheeran FL, Krum H, Rosenfeldt FL (2007). Coenzyme Q10 in cardiovascular disease. *Mitochondrion.* 7(Suppl 1):S154–S167.

Peretz A, Neve J, Vertongen F, et al. (1987). Selenium status in relation to clinical variables and corticosteroid treatment in rheumatoid arthritis. *J Rheumatol.* 14:1104–1107.

Petersson SD, Philippou E (2016). Mediterranean diet, cognitive function, and dementia: A systematic review of the evidence. *Adv Nutr.* Sep;7(5):889–904.

Phillips NA (2000). Female sexual dysfunction: evaluation and treatment. *Am Fam Physician.* Jul;62(1):127–136.

Pierpont GL, Francis GS, Cohn JN (1981). Effect of captopril on renal function in patients with congestive heart failure. *Br Heart J.* 46:522–527.

Pilz S, Frisch S, Koertke H, Kuhn J, Dreier J, Obermayer-Pietsch B, Wehr E, Zittermann A (2011). Effect of vitamin D supplementation on testosterone levels in men. *Horm Metab Res.* 43:223–225.

Pitchumoni CS, Dharmarajan, T (2012). Geriatric gastroenterology. In GS Gunturu TS Dharmarajan (Eds.), *Drug-nutrient interaction* (pp. 89–98). New York: Springer.

Pongchaidecha M, Srikusalanukul V, Chattananon, et al. (2004). Effect of metformin on plasma homocysteine, vitamin B12 and folic acid: a cross-sectional study in patients with type 2 diabetes mellitus. *J Med Assoc Thai.* 87(7):780–787.

Pons-Rejraji H, Brugnon F, Sion B, et al. (2014). Evaluation of atorvastatin efficacy and toxicity on spermatozoa, accessory glands and gonadal hormones of healthy men: a pilot prospective clinical trial. *Reprod Biol Endocrinol.* 12:65.

Pontiroli AE, Cortelazzi D, Morabito A (2013). Female sexual dysfunction and diabetes: a systematic review and meta-analysis. *J Sex Med.* Apr;10(4):1044–1051.

Pusceddu MM, Kelly P, Stanton C, Cryan JF, Dinan TG (2016). N-3 polyunsaturated fatty acids through the lifespan: implication for psychopathology. Review. *Int J Neuropsychopharmacol.* 19(12).

Reis LO (2012). Variations of serum testosterone levels in prostate cancer patients under LH-releasing hormone therapy: an open question. *Endocr Relat Cancer.* May 24;19(3):R93–R98.

Rey-Valzacchi GJ, Costanzo PR, Finger LA, et al. (2012). Addition of metformin to sildenafil treatment for erectile dysfunction in eugonadal nondiabetic men with insulin resistance: a prospective, randomized, double-blind pilot study. *J Androl.* Jul-Aug;33(4):608–614.

Rodriguez-Segade S, de la Pena CA, Tutor JC, et al. (1989). Carnitine deficiency associated with anticonvulsant therapy. *Clin Chim Acta.* 181:175–181.

Rolla G, Bucca C, Bugiani M, et al. (1990). Hypomagnesemia in chronic obstructive lung disease: effect of therapy. *Magnes Trace Elem.* 9:132–136.

Rosanoff A, Weaver Cm, Rude RK (2012). Suboptimal magnesium status in the United States: are the health consequences estimated? *Nutr Rev.* 70(3):153–164.

Rosanoff A (2005). Magnesium and hypertension. *Clin Calcium.* 12(2):225–260.

Rosen R, Brown C, Heiman J, et al. (2000). The Female Sexual Function Index (FSFI): a multidimensional self-report instrument for the assessment of female sexual function. *J Sex Marital Ther.* 26:191–208.

Rosenfeldt FL, Haas SJ, Krum H, et al. (2007). Coenzyme Q10 in the treatment of hypertension: a meta-analysis of the clinical trials. *J Hum Hypertens.* Apr;21(4):297–306.

Saenz de Tejada I, Goldstein I (1988). Diabetic penile neuropathy. *Urol Clin North Am.* Feb;15(1):17–22.

Sagar M, Janczewska I, Ljungdahl A, Bertilsson I, Seensalu R (1999). Effect of CYP2C19 polymorphism on serum levels of vitamin B12 in patients on long-term omeprazole treatment. *Aliment Pharmacol Ther.* 13:453–458.

Sahelian R (2003). *Natural sex boosters: supplements that enhance stamina, sensation, and sexuality for men and women*. New York: Square One.

Saldeen P, Saldeen T (2004). Women and omega-3 fatty acids. *Obstet Gynecol Surv.* Oct;59(10):722–730.

Saltzman JR, Kemp JA, Golner BB, Pedrosa MC, Dallal GE, Russell RM (1994). Effect of hypochlorhydria due to omeprazole treatment or atrophic gastritis n protein-bound vitamin B12 absorption. *J Am Call Nutr.* 13:584–591.

Samaras D, Samaras N, Lang PO, Genton L, Frangos E, Pichard C (2013). Effects of widely used drugs on micronutrients: a story rarely told. *Nutrition.* 29(4):605–610.

Schilling H, Scheler F (1988). Angiotensin-converting enzyme inhibition: side effects and risks. *Z Kardiol.* 77(Suppl 3):47–54.

Schmitt BP, Nordlund DJ, Rodgers LA (1984). Prevalence of hypocalcemia and elevated serum alkaline phosphatase in patients receiving chronic anticonvulsant therapy. *J Fam Pract.* 18:873–877.

Seelig MS (1993). Interrelationship of magnesium and estrogen in cardiovascular and bone disorders, eclampsia, migraine and premenstrual syndrome. *J Am Coll Nutr.* 12:442–458.

Seftel AD (2012). Re: addition of metformin to sildenafil treatment for erectile dysfunction in eugonadal non-diabetic men with insulin resistance: a prospective, randomized, double blind pilot study. *J Urol.* May;187(5):1787–1788.

Sener U, Zorlu Y, Karaguzel O, et al. (2006). Effects of common anti-epileptic drug monotherapy on serum levels of homocysteine, vitamin b12, folic acid and vitamin B6. *Seizure.* 15:79–85.

Seo T, Blaner WS, Deckelbaum RJ (2005). Omega-3 fatty acids: molecular approaches to optimal biological outcomes. *Curr Opin Lipidol.* Feb;16(1):11–18.

Shafer RB, Nuttall FQ (1975). Calcium and folic acid absorption in patients taking anticonvulsant drugs. *J Clin Endocrinol Metab.* 41:1125–1129.

Shifren JL (2004). The role of androgens in female sexual dysfunction. *Mayo Clin Proc.* Apr;79: S19–S24.

Shiri R, Koskimaki J, Hakkinen J, Auvinen A, Tammela TL, Hakama M (2007). Cardiovascular drug use and the incidence of erectile dysfunction. *Int J Impot Res.* Mar-Apr;19(2):208–212.

Simopoulos AP (1999). Essential fatty acids in health and chronic disease. *Am J Clin Nutr.* Sep;70(3): 560S–569S.

Simopoulos AP (2001). The Mediterranean diets: What is so special about the diet of Greece? The scientific evidence. *J Nutr.* Nov;131(11):3065S–73S.

Simopoulos AP (2004). Omega-3 fatty acids and antioxidants in edible wild plants. *Biol Res.* 37(2):263–277.

Skikne BS, Lynch SR, Cook JD (1981). Role of gastric acid in food iron absorption. *Gastroenterology.* 81:1068–1671.

Sorenson M, Grant WB (2012). Does vitamin D deficiency contribute to erectile dysfunction? *Dermato-Endocrinology.* 4(2):128–136.

Stargrove BM, Treasure J, Mckee LD (2008). *Herb, nutrient, and drug interactions: clinical implications and therapeutic strategies*. New York: Mosby.

Steinke EE, Hill TJ, Mosack V (2016). Medication use and predictors of sexual activity in men and women with CVD. *J Am Assoc Nurse Pract.*;28(2):91–97.

Stoschitzky K, Sakotnik A, Lercher P, Zweiker R, Maier R, Liebmann P, Lindner W (1999). Influence of beta-blockers on melatonin release. *Eur J Clin Pharmacol.* 55(2):111–115.

Sturniolo GC, Montino MC, Rossetto L, et al. (1991). Inhibition of gastric acid secretion reduces zinc absorption in man. *J Am Coll Nutr.* 10:372–375.

Suzuki H, Tominaga T, Kumagai H, Saruta T (1988). Effects of first-line antihypersive agents on sexual function and sex hormones. *J Hypertens Suppl.* Dec;6(4):S649–S651.

Svedmyr N, Harthon L, Lundholm L (1969). The relationship between the plasma concentration of free nicotinic acid and some of its pharmacologic effects in man. *Clin Pharmacol Ther.* Jul-Aug;10(4):559–570.

Takasaki A, Tamura H, Miwa I, Taketani T, Shimamura K, Sugino N (2010). Endometrial growth and uterine blood flow: a pilot study for improving endometrial thickness in the patients with a thin endometrium. *Fertil Steril.* Apr;93(6):1851–1858.

Traish A, Kim N, Min K, Munarriz R, Goldstein I (2002). Role of androgens in female genital arousal: receptor expression, structure, and function. *Fertil Steril.* 77(4):11–18.

Valuck RJ, Ruscin JM (2004). A case-control study on adverse effect: H2 blockersor proton pump inhibitor use and risk of vitamin B12 deficiency in older adults. *J Clin Epidemiol.* 57:422–428.

Verrotti A, Mencaroni E, Cofini M, Castagnino M, Leo A, Russo E, Belcastro V (2016). Valproic acid metabolism and its consequences on sexual functions. *Curr Drug Metab.* 17(6): 573–581.

Vidrio H (1990). Interaction with pyridoxal as a possible mechanism of hydralazine hypotension. *J Cardiovasc Pharmacol.* 15:150–156.

Wang C, Jackson G, Jones HT, Matsumoto AM, Nehra A (2011). Low testosterone associated with obesity and the metabolic syndrome contributes to sexual dysfunction and cardiovascular disease risk in men with type 2 diabetes. *Diabetes Care.* Jul; 34(7):1669–1675.

Watt DL (1988). The nutritional relationships of magnesium. *J Orthomol Med.* 3(4):197–201.

Webb JL (1980). Nutritional effects of oral contraceptive use: a review. *J Reprod Med.* 25:150–156.

WebMD (2016a). Erectile Dysfunction Health Center. http://www.webmd.com/erectile-dysfunction/guide/blood-pressure-medication-and-ed.

WebMD (2016b). Female sexual dysfunction. http://www.webmd.com/women/guide/sexual-dysfunction-women#1.

Weisman HM, Weinblatt EM, Louie SJ, Saunders WB (2001). *Treatment of the rheumatic diseases: a companion to Kelley's textbook of rheumatology*, 2nd ed. New York: Elsevier Health, Saunders.

Widmer P, Maibach R, Kunzi UP, et al. (1995). Diuretic-related hypokalaemia: the role of diuretics, potassium supplements, glucocorticoids and beta 2- adrenoceptor agonists. Results from the Comprehensive Hospital Drug Monitoring programme, Berne (CHDM). *Eur J Clin Pharmacol.* 49(1–2):31–36.

Wu X, Iguchi T, Itoh N, Okamoto K, Takagi T, Tanaka K, Nakanishi T (2008). Ascorbic acid transporter by sodium-dependent vitamin C transporter 2 stimulates steroidogenesis in human choriocarcinoma cells. *Endocrinology.* 149:73–83.

Wulffele MG, Kooy A, Lehert P, Bets D, Ogterop JC (2003). Effects of short-term treatment with metformin on serum concentrations of homocysteine, folate and vitamin B-12 in type 2 diabetes: randomised placebo controlled trial. *J Intern Med.* 254:455–463.

Yan WJ, Yu N, Yin TL, Zou YJ, Yang J (2014). A new potential risk factor in patients with erectile dysfunction and premature ejaculation: folate deficiency. *Asian J Androl.* Nov-Dec; 16(6):902–906.

Yogi A, Callera GE, Antunes TT, et al. (2010). Vascular biology of magnesium and its transporters in hypertension. *Magnes Res.* 23:207–215.

Zenuk C, Healey J, Donnelly J, Vaillancourt R, Almalki Y, Smith S (2003). Thiamine deficiency in congestive heart failure patients receiving long term furosemide therapy. *Can J Clin Pharmacol.*10:184–188.

Zofkova I, Kancheva RL (1995). The relationship between magnesium and caliotropichormones. *Magnes Res.* 8(1):77–84.

9

Birth Control Pills and Hormonal Contraception: Effects on Female Sexual Desire and Pleasure

BARBARA BARTLIK, AMANDEEP KAUR,
CHELSEA SCHOEN, AND JULIE KOLZET

Literature Review

First and foremost, we present basic information regarding hormonal contraception (HC) and its uses. Oral contraception (OC) can be viewed as a component of the larger category of HC. In addition to an oral delivery method/pill, HC can be delivered by way of a vaginal ring, transdermal patch, subcutaneous implant, and intrauterine device. OC, or, "the pill," as it is commonly referred to, is the most widely used form of HC. With regard to its mechanism of action, HC works to prevent pregnancy by delivering synthetic estrogen and progestin to the body, which inhibit luteinizing hormone and follicle stimulating hormone elevation, prevent the ovary from releasing an egg, thicken the cervical mucus impeding the travel of sperm to the egg, and alter the lining of the uterus, making implementation difficult.

Research over the past several decades has demonstrated the pill to be a highly effective contraceptive method, one that has led to successful fertility control for millions of women worldwide (Golobof, 2016). Sonfield (2013) discussed how the pill has also lead to social and economic benefits for women, ranging from freedom to pursue education, to personal autonomy and increased relationship satisfaction. Research suggests that 62% of women of reproductive age are currently using OC (Jones, 2012). The pill also is effective in preventing unwanted pregnancy but only when used as directed. Failure to do so, including both incorrect and inconsistent use, can result in unwanted pregnancy. Moreover, the pill does not offer protection against sexually transmitted infections, and the financial cost associated with monthly use can make OC a prohibitive option for some women.

Oral contraceptives have changed since they were first introduced in the 1960s. Today, oral contraceptives are prescribed in many formulations—synthetic estrogen combined with different types of progestin or progestin-only preparations. The new oral contraceptives have a lower dose of active ingredients than their predecessors, and there is data to suggest that these new birth control pills offer benefits to women in addition to preventing pregnancy. According to Kaunitz (2002), modern oral contraceptives have been linked to noncontraceptive benefits, including cycle control, relief of dysmenorrhea and menorrhagia, treatment of endometriosis, treatment of premenstrual syndrome or dysphoric disorder, reduced occurrence of acne and hirsutism, increased bone mass, ovarian cyst suppression, and decreased risk of ovarian and endometrial cancers. Jones (2011) has documented that 14% of pill users—1.5 million women—rely on oral contraceptives for exclusively noncontraceptive purposes. Data also reveal negative side effects of the pill, including breast tenderness, nausea, headaches, stroke, heart attack, venous thrombosis, pulmonary embolism, gallstones, hypertension, increased frequency of respiratory and gastrointestinal infections, yeast overgrowth, vulvodynia, weight gain, increased risk of breast and cervical cancer, depression, and disturbances in sexual functioning (Box 9.1; Lidegaard, 1999; Mercola, 2004; Palmery, 2013).

With respect to sexual functioning, there is data to suggest that the reported negative effects of OC on sexual function affect primarily desire and sexual

Box 9.1 Adverse Effects of Hormonal Contraceptives

- Bloating/swelling
- Impaired muscle gains
- Reduced total DNA methylation
- Breast tenderness
- Increased arterial plaque
- Nausea
- Cramps
- Increased atherosclerosis risk
- Sleep disturbance
- Depression/mood changes
- Increased oxidative stress
- Thromboembolism
- Elevations in C-reactive protein
- Increased risk of breast cancer
- Vaginal infections
- Fetal abnormalities
- Increased risk of cervical cancer
- Venous thrombosis
- Gall bladder disease
- Irregular bleeding
- Weight gain
- Headaches/migraines
- Ischemic stroke
- Yeast overgrowth
- Higher cortisol levels
- Low libido
- Higher triglycerides
- Low bone density
- Hypertension
- Miscarriages

(Campesi, 2012; Lidegaard, 1990; Mercola, 2004; Meulenberg, 1987; Palmery, 2013; Scholes, 2010; prepared with assistance of Turano V).

thoughts (Caruso, 2003; Smith, 2014). Battaglia (2012) noted additional negative side effects of oral contraception on sexual functioning (Battaglia, 2014). Findings indicated that oral contraceptive use may lead to decreased vaginal lubrication and the presence of vaginal pain. Similarly, Bohm-Starke (2004) noted that oral contraceptives may increase vulvar mucosal sensitivity, which is associated with pain and inflammation of the vulva. Furthermore, there is evidence to suggest that that some women discontinue the use of the pill or use it incorrectly as a result of sexual side effects. Of an estimated 45 million women ages 15 to 44 who ever used OC, there is data that indicate 8.6 million or 19% discontinued use because of side effects, and, of these, 5.4% stated that they discontinued use due to a decrease in sexual pleasure (Jones, 2010). Consistent with this research are findings by Sanders (2001), who points out that emotional and sexual side effects of the pill may be the best predictors of discontinuation.

In contrast, there is a body of research demonstrating that not all women experience negative sexual side effects when taking oral contraceptives, which suggests that more research is needed on the subject (Davis, 2004; Davison, 2008). With respect to sexual functioning and use of OC, Burrows (2012) provided a comprehensive discussion; the authors reiterated that the sexual side effects of hormonal contraceptives are understudied, particularly with regard to the impact on libido. Ultimately, their review of the literature revealed mixed effects of oral contraceptives on libido, with a small percentage of women experiencing a decrease or increase and the majority being unaffected. Large-scale, placebo-controlled studies on oral contraceptives are needed. Frellick (2016) provides support for this argument and cites the works of both Graham (1993) and Graham (1995) when pointing out that only two small-scale placebo-controlled trials on oral contraceptives and sexuality have been conducted between 1978 and 2011—both with inconclusive findings. With respect to these findings, the authors indicated that the lack of knowledge on the subject might be due to lack of interest from the pharmaceutical industry to perform large-scale, placebo-controlled trials due to the associated financial costs. To add to current knowledge regarding the relationship between use of OC and sexual functioning, Frellick (2016) conducted a placebo-controlled trial on combined OC (an estrogen and a progestin) in 340 young, healthy women. The authors found no significant reduction in overall sexual function by a combined OC, in comparison to placebo. However, they documented impairment on three subscales: sexual desire, arousal, and pleasure. The results suggest that OC use may negatively impact some aspects of sexual function but not all. Frellick (2016) is significant because it was the first assessing the impact of OC on sexual functioning with a large number of participants and a placebo arm. However, this study is not without limitations. One limitation outlined by the authors is that the duration of treatment with OC was limited to three months. Thus it is possible that some of the effects recorded may change over time in either direction. Patients may not develop sexual side effects until months or years after starting OC. Additionally, the authors emphasized some of the complexities involved in conducting research on female sexuality, which involves the interplay of many factors including physical, emotional, psychosocial, cognitive, and cultural.

Oral Contraceptives and Their Effect on Sexual Functioning: Mechanisms Involved

One mechanism that explains how oral contraceptives affect sexual function is well documented, namely the effects they have on androgen levels. An additional mechanism that may also affect sexual function is related to vitamin and mineral deficiencies, which can affect liver function and the metabolism of estrogen and other hormones. The scientific community has not adequately researched this mechanism. Nevertheless, there is some data to suggest that the adverse effects of contraceptives on sexuality may be related, in part, to a decreased levels of certain vitamins and minerals (Bennett, 2001; Hammed, 2001; Isoyama, 1986; Palmery, 2013; Pelton, 2001; Prasad, 1975, 1976; Webb, 1980). As further discussed later, restoring these levels may lead to improved sexual functioning.

MECHANISMS RELATED TO DIMINISHED ANDROGENS

A recent meta-analysis by Zimmerman (2014) offers insight into the mechanisms thought to influence the association between combined oral contraceptives and sexual functioning. Their data points to the role of androgen levels, specifically, reduced testosterone, which is integral to female sexual functioning (Kaplan, 1993). In concurrence with these findings, causal pathways between lower testosterone and combined oral contraceptives include the inhibition of testosterone production by the ovaries and adrenals, along with increased concentrations of sex hormone binding globulin (SHBG). Notably, SHBG binds to the hormones considered integral to sexual functioning, mainly estrogen and testosterone. SHBG binding results in less circulating estrogen and testosterone to the brain and other tissues. Moreover, SHBG attaches to testosterone more strongly than it does to estrogen, further reducing the body's supply of free testosterone.

In a separate yet related study, Panzer (2006) found that women with a history of oral contraceptive use have four times the level of SHBG in their bloodstream compared with women who have never used OC. These data suggest that SHBG elevation may persist even after oral contraceptives are discontinued. This could have longstanding effects on sexual function.

MECHANISMS RELATED TO VITAMIN AND MINERAL DEFICIENCIES

A study by Palmery (2013) revealed a positive association between OC use and deficiencies in vitamins and minerals. These findings echo those from an earlier study by Webb (1980), suggesting that OC use may be associated with depletions in vitamins B2 (riboflavin), B6 (pyridoxine), B9 (folate), B12 (cobalamin), C, and zinc. Additionally, Palmery (2013) identified diminished levels of magnesium, selenium, and vitamin B1 (thiamine). Collectively, these results indicate that one way to bolster overall health and well-being in women using OC may be to attend to any vitamin and mineral deficiencies that may affect the metabolism of estrogen and other hormones in the liver and adversely impact women's sexual functioning.

Vitamin and Mineral Deficiencies

FOLATE AND VITAMIN B12

It is important to emphasize that our current understanding of how vitamin and mineral deficiencies impact sexual functioning is limited. Given the dearth of literature, it is difficult to compare the data from the clinical arena with any relevant scientific research. Nevertheless—and as described in detail later—there is reason to believe that OC may be associated with sexual problems via micronutrient depletion. Illness (conceptualized here in its broadest sense as physical and mental illness), along with inadequate nutrition, is a well-known predictor of negative sexual health outcomes. It is well known that oral contraceptives deplete folate (vitamin B9). We know that some manufacturers of oral contraceptives add a form of folate called levomefolate to their contraceptive pill (Bayer, 2010). This is done in an effort to prevent serious birth defects (e.g., spina bifida and cleft palate) associated with folate deficiency. It may seem paradoxical to add folate to oral contraceptives, however, given the rate of contraceptive failure, this measure is well advised. Furthermore, taking folate along with contraceptives may reduce the risk of birth defects in women who conceive shortly after discontinuing contraceptives.

There is reason to believe that if oral contraceptives reduce folate levels to such an extent that measures to offset these changes are taken, they may also be capable of compromising mental and sexual health. Indeed, folate is integral to neurotransmitter formation and hormone metabolism. Taking it one step further, there is data to suggest that folate deficiencies have been associated with malabsorption of vitamins and minerals, along with increased urinary excretion and accelerated metabolism by microsomal enzyme induction (Shojania, 1982).

It is worth mentioning that folate and B12 deficiency may also cause megaloblastic anemia (Oh, 2003), resulting in red blood cells that are larger than normal, immature, and less efficient at transporting oxygen. This is an important consideration because anemia may result in decreased oxygenation of sexually important tissues.

As mentioned, some literature demonstrates that OC use is associated with vitamin B6 deficiency (Palmery, 2013; Pelton, 2001; Prasad, 1976). Vitamin B6 is needed to make melatonin from serotonin. Thus a lack of vitamin B6 can be responsible for decreased sleep quality. Sleep deficiencies have been shown to be associated with reduced testosterone levels in men (Warner, 2011), and there is a very substantial body of evidence that supports the association between low testosterone in men and sexual problems. The same may be true for women; however, the possible effect of sleep deprivation on testosterone or other hormone levels in women is not yet elucidated (Krapf, 2009). Research is needed. Notably, B6 is needed by both women and men to make serotonin from the amino acid tryptophan. Adequate levels of serotonin are necessary for normal sleep and positive mood, both of which are conducive to healthy sexual functioning.

With regard to vitamin B12, there is data to suggest that it can be reduced by OC (Palmery, 2013; Pelton, 2001; Wertalik, 1972). This is an important consideration as Vitamin B12 deficiency has been associated with depression, dementia, fatigue, pernicious anemia, neurological damage (Briani, 2013), and sexual dysfunction (Bennett, 2001; Isoyama, 1986). Deficiencies in these three B vitamins (B6, B12, and folate) is known to increase homocysteine levels, which may be pertinent regarding the effect of OC on sexuality. Elevated homocysteine has negative effects on cardiovascular and mental health; it is a proposed neurotoxin, and high homocysteine has been associated with sexual dysfunction (Demir, 2006), depression, dementia, coronary artery disease, stroke, peripheral vascular disease, and more (Kuo, 2005).

Methylation and the Role of Folate and Vitamin B12

Methylation is a vital metabolic process, and folate and vitamin B12 are cofactors for methylation. In this context, folate and vitamin B12 play a key role in detoxification, neurotransmitter formation, gene expression, enzyme and antibody production, and hormone synthesis and metabolism (Bottiglieri, 1996). Consistent with other data presented, deficiencies in folate and vitamin B12 due to oral contraceptive use may lead to problems in methylation. This in turn may put women taking oral contraceptives at higher risk for impaired metabolic functioning. Further, individuals with polymorphisms of the methyltetrahydrofolate reductase (MTHFR) enzyme might, theoretically, be at an elevated risk of developing sexual side effects as a result of taking oral contraceptives due to impaired neurotransmitter and hormone function. There is data to suggest that women who are carriers of the MTHFR 677TT and Factor V Leiden polymorphisms may be at an elevated risk of stroke. Slooter (2005) concluded that, to prevent stroke, women starting oral contraceptives should be tested for these polymorphisms. It is notable that smoking tobacco also impairs methylation (Anttila, 2003), and smoking is linked to serious prothrombotic conditions in OC users (Lidegaard, 1999). Other researchers suggested that progesterone-only contraceptives may be a safer choice than combined oral contraceptives for women with thrombocytic genetic polymorphisms (Gomes, 2004). It is the estrogen component of contraceptives that traditionally has been associated with coagulopathies. Estrogen receptors in the liver are involved in the manufacture of clotting factors (Gebel Berg, 2015).

VITAMIN B1 (THIAMINE) AND B2 (RIBOFLAVIN)

Oral contraceptives have been linked to deficiencies in thiamine and riboflavin. There is data that shows thiamine deficiency to be associated with depression, memory loss, muscle weakness (Pelton, 2001), and sudden infant death syndrome (Jeffrey, 1985). As an extension of these findings, and because thiamine deficiency has been linked to impairments in nerve transmission, it is reasonable to hypothesize that sexual functioning may be impaired by thiamine deficiency. Additionally, riboflavin deficiency has been associated with weakness, dermatitis, and anemia. While there are no reports that riboflavin deficiency is linked to sexual dysfunction, it is possible that riboflavin may indirectly be linked to such problems. This is because B vitamins work best in concert with one another. More specifically, it is generally advised that individuals take a blend of all the B vitamins that contain the high-quality, methylated forms of vitamin B6, B12, and folate. Taking too much of one B vitamin can result in an imbalance of the others.

VITAMIN C

There is some evidence, albeit scant, that suggests that vitamin C levels should be considered when examining the impact of vitamin and mineral deficiencies on a woman's sexual health and performance in the wake of oral contraceptives. Zal Mostafavi-Pour (2012) explained that a reduction in vitamin C levels has been found in the plasma and urine of women on oral contraceptives. These researchers hypothesized that estrogen contained in some oral contraceptives may facilitate the breakdown of vitamin C. In accordance with this, Gottleib (2010) found that women on low-dose oral contraceptives show signs of increased oxidative stress and that vitamin C and E supplements may counteract this. As they explain, vitamin C may play a role in promoting sexual function by aiding hormone synthesis and metabolism and by boosting nitric oxide formation, which promotes local blood circulation (Bryan, 2010). Significantly, vitamin C supports the uptake and transport of iron, which is involved in sexual function. Vitamin C is also a cofactor in catecholamine synthesis for the conversion of dopamine to norepinephrine. Both dopamine and norepinephrine play roles in sexual function (Bornstein, 2003).

ZINC

Zinc is thought to play an important role in both synaptic and sexual functioning. Although definitive research on zinc and sexuality is lacking, there is some evidence to suggest that women taking oral contraceptives have lower plasma zinc levels compared to women who are not taking oral contraceptives (Halsted, 1968; Hess, 1977; King, 1978; Palmery 2013).

VITAMIN E

Vitamin E in plasma was reduced in OC users, and at the same time clotting activity increased. These changes were notably reversed with vitamin E administration (Renaud, 1987). Vitamin E improves circulation and on that basis may have some sexual benefit (Mumford, 2016; Takasaki, 2010), though this has not been shown directly.

SELENIUM

In one study by Heese (1988), selenium levels were lower in women on low-dose triphasic contraceptives, compared to women not on these contraceptives. This is an important consideration because selenium is a cofactor for thyroid enzymes and integral to the function of antioxidant enzymes. Proper thyroid function is essential to overall health, as well as sexual health.

MAGNESIUM

The vasoconstriction associated with chronic magnesium deficiency (Rosanoff, 2005) may have a negative impact on sexual function. Magnesium deficiency is associated with hypertension, muscle cramps, headache, psychiatric disorders, diabetes, insulin resistance (Huerta, 2005), and constipation. Furthermore, magnesium is integral to energy production and is present in the molecule magnesium adenosine triphosphate (Mg-ATP). We could therefore hypothesize that a diminution of magnesium from oral contraceptive use could adversely affect a woman's energy production and, thereby, sexual functioning. In addition, low magnesium levels may increase the risk of thrombosis from oral contraceptives (Olatunbosun, 1974).

When micronutrients are depleted and affect processes relevant to sexual functioning, this does not necessarily mean that those individuals cannot function sexually. However, to the extent that many complain of sexual difficulties and unsatisfactory sexual experiences, whose cause may be difficult to identify, we feel increased attention to the possible impacts of oral contraceptives on micronutrients is warranted. Routinely assessing micronutrients and repleting them may help many to function better. Table 9.1 lists foods that contain a high concentration of key nutrients, such as glutathione, zinc, magnesium, and B vitamins.

Recommendations from Clinical Practice

RESTORING HORMONE BALANCE AND DETOXIFICATION

Micronutrient deficiencies associated with contraceptives, such as those described in previous sections of this chapter, can negatively impact hormone metabolism and other functions performed by the liver. Hormonal contraceptives alter bile transport and increase the risk of gall stone formation (Kern, 1987). They also affect the synthesis of clotting factors in the liver, which are responsible for thrombotic side effects. This is thought to be a function of the estrogen component, ethinyl estradiol in these medications. In addition, combined oral contraceptives adversely affect lipid and carbohydrate metabolism by increasing serum triglycerides, low density lipoprotein cholesterol, and insulin levels (Godsland, 1990). These metabolic effects did not occur with progesterone-only contraceptives.

It is important to understand the pathways by which estrogen in contraceptives is metabolized. This will shed light on how vitamin and mineral deficiencies may be associated with impaired sexual health in HC users. Armed with this information, clinicians might be more likely to recommend diet and lifestyle changes.

Ethinyl estradiol, the main estrogen in contraceptives, is more potent than natural estradiol and more difficult to eliminate from the body (Berg, 2015). Estrogens, including ethinyl estradiol, are detoxified and metabolized by the liver through Phase I (hydroxylation) and Phase II (methylation, sulfation, and

Table 9.1. Foods Containing High Concentrations of Key Nutrients

Glutathione	*B12*	*Zinc*	*Magnesium*	*B6*	*B1*
Spinach	Beef liver	Lobster	Peanuts	Beef liver	Beef liver
Potatoes	Tuna	Oysters	Almonds	Turkey	Asparagus
Avocado	Salmon	Crab	Spinach	Tuna	Avocado
Asparagus	Trout	Beef	Soymilk	Salmon	Spinach
Zucchini	Beef	Poultry	Black beans	Chickpeas	Brussels sprouts
Grapefruit	Poultry	Pork	Brown rice	Potatoes	Black-eyed peas
Peaches	Egg	Beans	Potato	Banana	
Strawberries	Yogurt	Chickpeas	Edamame	Squash	
	Breakfast cereals	Cashews	Banana	Nuts	
	(fortified)	Peas		Watermelon	
		Kidney Beans			
		Oatmeal			
		Milk			

National Institutes of Health, 2016, prepared with the assistance of Turano V.

glucuronidation) detoxification (Hyman, 2007). Ethinyl estradiol is metabolized primarily through intestinal sulfation and hepatic oxidation, glucuronidation, and sulfation (Zhang, 2007).

Phase I detoxification is aided by antioxidants, which are listed in Box 9.2, as well as cruciferous vegetables, sulfurophane, and indole-3-carbinol (IC3). IC3 aids the metabolism of estrogen in the liver, directing it toward more benign pathways and away from known carcinogenic ones. Phase II makes estrogen metabolites easier to excrete. Methylation is dependent upon an adequate supply of methylating agents such as methylfolate, folinic acid, methylcobalamin (vitamin B12), and S-adenosyl-l-methionine (SAM-e). Estrogens also are sulfated to ready them for excretion, so a good supply of sulfur is necessary, as is found in egg yolk, garlic, onion, and cruciferous vegetables. Sulfur may be taken as a nutritional supplement as organic sulfur or methysulfonylmethane (MSM). If there is not enough sulfur present, it may be extracted from joint cartilage, which can contribute to arthritis. Estrogens also are glucaronidated by the attachment of glucaronyl groups, which render the complex ready for excretion into the bile.

Depending on the bacterial milieu in the intestines, the glucaronyl group can either remain attached to the estrogen, in which case the complex is eliminated

Box 9.2 Nutrients to Enhance Hormone Balance and Liver Detoxification

- Antioxidants: vitamins A, E, resveratrol, curcumin, N-acetyl-cysteine, green tea, and alpha-lipoic acid
- Methylcobalamin (B12); methyl B6 (P5P); methylfolate; folinic acid
- Magnesium
- Multivitamin (containing vitamins C, E, selenium, zinc, among others)
- Phosphatidylcholine
- Taurine, L-methionine, L-lysine, L-threonine
- Dandelion root, milk thistle
- Omega-3 fatty acids
- Isoflavonoids (red clover, kudzo root, and genistein and diadzein for soy)
- Organic sulfur, methylsulfonylmethane
- Chaste berry fruit extract
- Chinese herbs: ginger, mint leaf, licorice root, Chinese peony root, fenugreek, and dong quai
- Glutathione
- Flax, chia, and other seeds/oils
- Psyllium husk
- Calcium D-glucarate

(Hyman, 2007; prepared with the assistance of Turano V).

in the stool, or it can be cleaved off, in which case the estrogen is free to be reabsorbed into the body. Certain intestinal bacteria produce an enzyme that cleaves off the glucaronyl group. The nutritional supplement calcium D-glucarate renders that enzyme inactive and thereby promotes excretion of hormones.

If stool remains in the intestines too long, as occurs with constipation, hormonal metabolites are more likely to be reabsorbed. Therefore, treating constipation is essential when detoxifying from hormones. Magnesium citrate and buffered vitamin C are very helpful for constipation, as is dietary fiber as found in flax seed, psyllium husk, chia seed, and legumes. Fiber improves the balance of intestinal flora and reduces the deconjugation of estrogens. Probiotics found in fermented foods also inhibit beta glucuronidase-producing bacteria. A high-quality daily probiotic is recommended.

Glutathione, the body's main endogenous antioxidant, also plays a role in Phase II detoxification and the elimination of hormones. L-methionine is a sulfur-containing amino acid that aids glutathione production. It is also an antioxidant, that is necessary for the production of endogenous SAMe, which supports positive mood. Glutathione is necessary for the metabolism of alcohol, which is why alcohol should not be consumed by those attempting to achieve better hormone balance. Caffeine and nicotine also compete for glutathione, so these should be eliminated as well. Glutathione supplementation may be given orally, topically, intranasally, or intravenously. Vitamin C promotes the production of glutathione.

Other ways to reduce the burden on the liver and enhance hormone balance are to eliminate high-fat foods, processed foods, additives and preservatives, refined flour and sugar (Hyman, 2007), transfat, and high fructose corn syrup. A modified elimination diet often is recommended, excluding the previously mentioned foods plus gluten and soy. In addition, meat, eggs, and dairy products from animals given exogenous hormones also are eliminated (Chighzola, 2012; Hyman, 2007). Moreover, the casein found in dairy products can lead to autoimmune problems in certain individuals (Otaegui, 2007). Organic meats and vegetables are recommended to minimize exposure to hormones, antibiotics, xenoestrogens, pesticides, and herbicides. Also eliminated are xenoestrogens from plastics, cosmetics, and household products. Eating adequately spaced, high-protein, low-carbohydrate meals throughout the day is important to reduce insulin resistance and balance glucose levels. High-protein diets enhance estrogen metabolism, whereas low-protein diets do the opposite (Hyman, 2007).

It is important to drink plenty of pure, clean water daily. The water supply can be contaminated with xenoestrogens, hormonal contraceptives, chlorine, fluoride, and other toxins. Investing in a high-quality water filter may be advisable. Daily exercise is highly recommended, as it helps regulate hormone function, as well as promote positive mood and decrease stress levels. Techniques to reduce stress include yoga, meditation, and hot baths. Epsom salt baths provide some extra magnesium, which aids detoxification.

Botanicals and nutrients that aid hormone balance and detoxification are listed in Box 9.2. These include taurine, glycine, dandelion root, omega-3 fatty acids, milk thistle, and isoflavonoids. Soy is a good source of isoflavonoids, which direct estrogen metabolism toward more benign metabolites. As most American soy is genetically modified, those concerned with genetically modified organisms (GMOs) may wish to consume only organic non-GMO soy products. Aside from soy, isoflavonoids may be found in kudzu root and red clover (Hyman, 2007). Omega-3 fatty acids also direct estrogen metabolism to more benign, less carcinogenic metabolites (Lord, 2002). Resveratrol, rosemary, oregano, thyme, sage, and turmeric aid the metabolism of hormones in the liver. Certain traditional Chinese medicine herbs may help balance hormones. In one controlled study, fenugreek increased free testosterone and estradiol levels in women; scores on measures of sexual desire and arousal also were higher for the fenugreek versus a control group (Rao, 2015). Fenugreek is known to lower blood sugar in diabetics (WebMD, 2016).

Women with estrogen dominance, a condition where there is an inadequate amount of progesterone to balance the amount of estrogen in a women's body, may need the support of exogenous progesterone. This may be taken orally or topically in the evening, because it has a mild sedative effect (Hyman, 2007). In cycling women, progesterone may be given during the last two weeks of the menstrual cycle.

LABORATORY TESTS

Several laboratory tests may be useful in evaluating patients who have problems with hormone balance. These tests are listed in Box 9.3. In brief, it is important that the patient's vitamin D (Holick, 2008) and thyroid function be checked. Reverse T3 gives an indication of the degree of inflammation present. It also is useful to measure markers of inflammation, such as highly sensitive-C reactive protein to monitor progress. Gamma glutamyltransferase (GGT) is a liver enzyme that when elevated suggests alcohol consumption and liver disease. GGT reconstitutes glutathione, the body's main antioxidant. When GGT is elevated, it is a sign that glutathione is deficient and that supplementation may be needed. Ideally, GGT should be in the lower levels of the normal range (Koenig, 2015).

Serum vitamin B12 and folate levels may not be adequate to reveal deficiencies. Some patients with methylation defects have normal or high plasma levels of both of these vitamins, when they actually are quite deficient. This is because they have difficulty converting the inactive forms of vitamin B12 and folate into the active methylated forms. Thus the inactive forms cannot be utilized and build up to high levels in the bloodstream. If homocysteine levels are elevated, there is likely to be a need for supplementation of vitamin B12, folate, and other

Box 9.3 Laboratory Tests for Patients with Hormonal Imbalance

4 point cortisol
2:4:16 alpha-OH estrogen ratios
SHBG, albumin
Free and total testosterone
Progesterone
Estradiol, estrone, estriol
DHEA
Fasting insulin, fasting glucose
Hemoglobin A1c
Luteinizing hormone (LH) and follicle stimulating hormone (FSH)
Homocysteine
Whole blood histamine
CBC with differential
Ferritin
Comprehensive metabolic panel
Highly sensitive-reactive protein (Hs-CRP)
Gamma glutamyltransferase (GGT)
TSH, Free T4, Free T3, T3, T4,
TPO antibodies, anti-thyroglobulin antibodies
Reverse T3
25-Hydroxy vitamin D
Serum vitamin C, vitamin E, and selenium levels
Single nucleotide polymorphisms involved in hormone metabolism and detoxification:
  5, 10- methylenetrahydrofolate reductase (MTHFR)
  CYP IBI
  Catechol-o-methyltransferase (COMT)
  Glutathione S-transferase MI (GSTMI)
  Factor V Leiden
Activated protein C (for factor V Leiden)
Serum zinc and copper levels
RBC and serum magnesium levels
Serum B12 and methylmalonic acid
RBC and serum folate, formiminoglutamate

(Hyman, 2007; prepared with assistance of Turano V).

methylating factors. Other laboratory tests that may be helpful include red blood cell folate levels, which are more accurate than serum folate levels; however, there is controversy about this. Both homocysteine and methylmalonic acid (MMA) are elevated with B12 deficiency, but only homocysteine and not

MMA is elevated in folate deficiency (Bjorke Monsen, 2003). It is important to distinguish between the two conditions because folate supplementation will correct anemia but will not address B12 deficiency, which can lead to irreversible neurological damage (Herrmann, 2008; Oh, 2003). Physicians planning to start a patient on oral contraceptives should consider laboratory testing for MTHFR and other polymorphisms listed in Box 9.3. Patients positive for the MTHFR mutation may be at increased risk of developing blood clots and other adverse effects (Slooter, 2005).

Case Study

Ms. M was a 29-year-old-woman who presented with symptoms of depression, constipation, and sexual dysfunction. She had been on oral contraceptives for nine years, which she was taking to prevent pregnancy. She reported low libido and the complete inability to achieve orgasm for several years. She had been orgasmic in her teens and early 20s, prior to starting oral contraceptives and during the first few years on them, but her orgasmic ability gradually diminished. At the time of presentation, she had no interest in sex and participated in it only for the sake of her fiancé. She also noted that she did not lubricate as well as she used to and that intercourse was uncomfortable, which it had not been prior to taking oral contraceptives. She had been through several failed, lengthy, monogamous relationships that were adversely affected by her sexual problems and was determined to "make this one work." She reported that her relationship with her fiancé was loving and satisfying, except for her sex life.

During the first visit, the patient was advised to stop her OC at the start of her next cycle, which she did, and to obtain some blood tests. Laboratory tests showed that she was heterozygous for both the C6771 and the A1298 MTHFR genes. Her SHBG was elevated, her free testosterone was undetectable, and her vitamin D was below range. She was prescribed a commercial protein shake specifically designed to enhance the metabolism of female sex hormones, to be consumed two times a day. It contained high-quality multivitamins, methylcobalamin (vitamin B12), methyl B6, methylfolate, flaxseed, choline, vitamin A, turmeric, isoflavones from kudzu root extract, lysine, threonine, and taurine, among other ingredients. She was placed on a modified elimination diet of no alcohol, sugar, gluten, dairy, nightshades, or unhealthy oils and fats. Non-GMO soy products were permitted. She was prescribed the following: a probiotic twice a day, magnesium citrate dosed to bowel tolerance; omega-3 fatty acids (2 g per day); resveratrol (200 mg twice daily); vitamin D3 (5000 units a day); a combination pill designed to enhance hormone metabolism containing IC3 (150 mg) and rosemary leaf extract (50 mg) twice daily, a combination pill intended to improve female sexual function containing fenugreek (300 mg), and a blend of methylating agents (methylcobalmin, methylfolate, zinc, B6, and betaine HCI) twice daily.

The patient returned three months later and said her sexual interest had improved, and she was able to achieve orgasm about 50% of the time that she engaged in sexual activity. She felt encouraged, and her fiancé was pleased and supportive. Six months into treatment, her orgasmic ability improved to 80% of the time, and her libido was notably stronger. She initiated sex with her partner fairly often, which she never did before. The patient reported that she had positive feelings of anticipation about her sexual experiences and spontaneous sexual fantasies, which she had not felt for years while on contraceptives. Her mood also improved and she no longer experienced depressive symptoms. She commented that, compared to how she felt before, she now felt "sexually alive" for the first time in many years.

The case described demonstrates how oral contraceptives can impact a woman's sexual interest, sexual functioning, mood, relationship, and overall quality of life. Notably, the patient did not make the connection between sexual dysfunction and her oral contraceptive use. She had experienced multiple failed relationships in the past, in part due to her sexual dysfunction. Her SHBG was elevated, contributing to low free testosterone levels. The protein shake, resveratrol, rosemary, IC3, and methylating agents helped her system to process and eliminate undesirable estrogens. Magnesium citrate aided with the detoxification process, relieved constipation, and enhanced the elimination of estrogen metabolites through the bowel. The probiotic provided healthier bowel flora, which reduced reabsorption of estrogen metabolites.

Conclusion

Research on oral contraceptives' effects on sexuality show varied results with reports of both improved, unchanged, and worsened sexual function. As such, it is difficult to make sense of the complex and conflicting findings. In cases where sexual problems have been associated with OC use, one mechanism is well documented. This mechanism is related to the effects oral contraceptives have on free testosterone levels. An additional mechanism that may also affect sexual function is related to vitamin and mineral deficiencies. Vitamin and mineral deficiencies impair the metabolism of estrogen and other hormones, which may adversely affect sexual function. Researchers would do well to explore this mechanism of micronutrient deficiency further. Indeed, it is worth taking into account that one way to bolster overall health and well-being in women using HC may be to attend to vitamin and mineral deficiencies arising from HC use, as well as from other causes, that may impact sexual functioning. This is particularly important for patients with genetic defects in methylation. Vitamin and mineral supplements, along with diet modifications, can be therapeutic.

When a woman on HC presents with sexual dysfunction, it is important that the possibility of a medication side effect be considered. Unfortunately, many patients are not aware of the potential negative side effects of contraceptives on sexuality. In addition, some clinicians admit that they deliberately do not mention psychiatric or sexual side effects, because to do so might bring on those very symptoms.

Regrettably, the information that clinicians receive from HC manufacturers tends to downplay the adverse mental and sexual health effects. Were the biological basis (hormonal and nutritional) for such adverse effects better recognized, clinicians would be more likely to have informative discussions with their patients about these issues.

Currently, the decision to use HC is taken very lightly by both patients and healthcare providers, without adequate consideration of the risks. We would argue that HC use should be regarded with greater caution. Patients and providers should take note of side effects and consider contraception alternatives, such as condoms, other barrier methods, and nonhormonal intrauterine devices. Patients need education especially on the correct use of condoms. Tech-savvy young women also are turning to digital ovulation-tracking apps (software applications) with names like DaysyView, Kindara, and Clue (Bennett, 2016).

Given that many women will continue to use HC for their convenience, more widespread awareness of the concerns and data set out in this chapter would allow many women to develop self-care and provider-based strategies to buffer against the problematic impact of HC use.

We also would strongly suggest that all patients, especially those taking HC, be counseled about the important role that optimal nutrition plays in sexual functioning, mental health, and health in general. Patients using HC should consider eating foods high in methylating agents, (e.g., meat, beets, and leafy greens). They also should consider taking nutritional supplements containing methylating agents, such as SAMe and methylated B vitamins. Finally, more research regarding the safety of such supplements needs to be done.

REFERENCES

Anttila S, Hakkola J, Tuominen P, Elovaara E, Husgafvel-Pursiainen K, Karjalainen A, Hirvonen A, Nurminen (2003). Methylation of cytochrome P4501A1 promoter in the lung is associated with tobacco smoking. *Cancer Res.* Dec 15;63(24):8623–8628.

Battaglia C, Morotti E, Persico N, Battaglia B, Busacchi P, Casadio P, Venturoli S (2014). Clitoral vascularization and sexual behavior in young patients treated with drospirenone-ethinyl estradiol or contraceptive vaginal ring: a prospective, randomized, pilot study. *J Sex Med.* 11:471–480.

Bayer HealthCare Pharmaceuticals (2010) U.S. FDA approves second Bayer oral contraceptive that also raises folate levels. August 16. https://www.prnewswire.com/news-releases/us-fda-approves-second-bayer-oral-contraceptive-that-also-raises-folate-levels-112057124.html

Bennett J (2016). Chill, pill. *Vogue Magazine*. Sept;606:608.

Bennett M (2001). Vitamin B12 deficiency, infertility and recurrent fetal loss. *J Reprod Med*. 46(3):209–212.

Bjorke Monsen AL, Ueland PM (2003). Homocysteine and methylmalonic acid in diagnosis and risk assessment from infancy to adolescence. *Am J Clin Nutr*. Jul;78(1):7–21.

Bohm-Starke N, Johannesson U, Hilliges M, Rylander E, Torebjörk E (2004). Decreased mechanical pain threshold in the vestibular mucosa of women using oral contraceptives: A contributing factor in vulvar vestibulitis? *J Reprod Med*. Nov;49(11):888–892.

Bottiglieri T (1996). Folate, vitamin B12, and neuropsychiatric disorders. *Nutr Rev*. Dec;54(12):382–390.

Bornstein SR, Yoshida-Hiroi M, Sotiriou S, Levine M, Hartwig HG, Nussbaum RL, Eisenhofer G (2003). Impaired adrenal catecholamine system function in mice with deficiency of the ascorbic acid transporter (SVCT2). *FASEB J*. Oct;17(13), 1928–1930.

Briani C, Dalla Torre C, Citton V, Manara R, Pompanin S, Binotto G, Adami F (2013). Cobalamin deficiency: clinical picture and radiological findings. *Nutrients*. 5(11):4521–4539.

Bryan N, Gottleib B, Zand J (2010). *The nitric oxide (NO) solution: how to boost the body's miracle molecule*. Kindle Edition.

Burrows LJ, Basha M, Goldstein A (2012). The effects of hormonal contraceptives on female sexuality: a review. *J Sex Med*. 9:2213–2223.

Campesi I, Sanna M, Zinellu A, et al. (2012). Oral contraceptives modify DNA methylation and monocyte-derived macrophage function. *Biol. Sex Diff*. January 27;2:4.

Caruso S, Agnello C, Intelisano G, Farina M, Di Mari L, Cianci A (2003). Sexual behavior of women taking low-dose oral contraceptive containing 15 microg ethinylestradiol/60 microggestodene. *Contraception*. Mar;69(3):237–240.

Chighizola C, Menori PL (2012). The role of environmental estrogens and autoimmunity. *Autoiminn Rev*. May;11(6–7):A493–501.

Davis AR, Castano PM (2004). Oral contraceptives and libido in women. *Ann Rev Sex Res*. 15:297–320.

Davison SL, Bell RJ, LaChina M, Holden SL, Davis SR (2008). Sexual function in well women: stratification by sexual satisfaction, hormone use, and menopause status. *J Sex Med*. May;5(5):1214–1222.

Demir T, Comlekçi A, Demir O, Gülcü A, Calýpkan S, Argun L, Seçil M, Yepil S, Esen A (2006). Hyperhomocysteinemia: a novel risk factor for erectile dysfunction. *Metab Clin Exper*. 55(12):1564–1568.

Frellick M (2016). Oral contraceptives don't impair sexual function, at least overall. *Medscape Med News*, August 19. https://www.medscape.com/viewarticle/867671

Gebel Berg E (2015). The chemistry of the pill. *ACS Cent Sci*. Mar 25;1(1):5–7.

Godsland IF, Crook D, Simpson R, Proudler T. Felton C, Lees B, Anyaoku V, Devenport M, Wynn V (1990). The effects of different formulations of oral contraceptive agents on lipid and carbohydrate metabolism. *N Engl J Med.* Nov 15;323(20):1375–1381.

Golobof A, Kiley J (2016). The current status of oral contraceptives: progress and recent innovations. *Semin Reprod Med.* May;34(3):145–151.

Gomes MP, Deitcher SR (2004). Risk of venous thromboembolic disease associated with hormonal contraceptives and hormone replacement therapy: a clinical review. *Arch Intern Med.* 164(18):1965–1976.

Graham C, Sherwin BB (1993). The relationship between mood and sexuality in women using an oral contraceptive as a treatment for premenstrual symptoms. *Psychoneuroendocrinology.* 18(4):273–281.

Graham CA, Ramos R, Bancroft J, Maglaya C, Farley TM (1995). The effects of steroidal contraceptives on the well-being and sexuality of women: a double blind, placebo-controlled, two-centre study of combined and progestogen-only methods. *Contraception.* Dec;52(6):363–369.

Halsted JA, Haddey BM, Smith JC (1968). Plasme zinc and copper in pregnancy and after oral contraceptive. *Lancet.* 292 (7562):278–279.

Hammed A, Majeed T, Rauf S, Ashraf M, Jalil MA, Nasrullah M, Hussan A, Noreen R (2001). Effect of oral and injectable contraceptives on serum calcium, magnesium and phosphorus in women. *J Ayub Med Coll Abbottabad.* 13:24–25.

Heese HD, Lawrence MA, Dempster WS, Pocock F (1988). Reference concentrations of serum selenium and manganese in healthy nulliparas. *S Afr Med J.* Feb 6;73(3):163–165.

Herrmann W, Obeid R (2008). Causes and early diagnosis of vitamin B12 deficiency. *Dtsch Arztebl Int.* Oct;105(40):680–685.

Hess FM, King JC, Margen S, (1977). Effect of low zinc intake and oral contraceptive agents on nitrogen utilization and clinical findings in young women. *J. Nutr.* 107:2219–2227.

Holick MF, Chen TC (2008). Vitamin D deficiency: a worldwide problem with health consequences. *Am J Clin Nutr* 87:1080S–1086S.

Huerta MG, Roemmich JN, Kington ML, Bovbjerg VE, Weltman AL, et al. (2005). Magniesum deficiency is associated with insulin resistance in obese children. *Diabetes Care.* May;28(5):1175–1181.

Huerta MG, Roemmich JN, Kington ML, Bovbjerg VE, Weltman AL, et al. (2005). Magniesum deficiency is associated with insulin resistance in obese children. *Diabetes Care.* May;28(5):1175–1181.

Hyman, AM (2007). The life cycles of women: restoring balance. *Altern Ther Health Med.* 13(3):10–16.

Isoyama R, Baba Y, Harada H, Kawai S, Shimizu Y, Fuji M, Fujisawa S, Takihara H, Koshido Y, Sakatoku J (1986). Clinical experiences of methylcobalamin (CH3-B12)/clomiphene citrate combined treatment in male infertility. *Hinyokika Kiyo.* 32(8):1177–1183.

Jeffrey HE, McCleary BV, Hensley WJ, Read DJ (1985). Thiamine deficiency—a neglected problem of infants and mothers—possible relationships to sudden infant death syndrome. *Aust N Z J Obstet Gynaecol.* Aug;25(3):198–202.

Jones J, Mosher W, Daniels K (2010). *Use of contraception in the United States: 1982–2008.* Hyattsville, MD: US Department of Health and Human Services, Centers for Disease Control and Prevention, National Center for Health Statistics.

Jones RK (2011). *Beyond birth control: the overlooked benefits of oral contraceptive pills.* New York: Guttmacher Institute.

Kaplan HS, Owett T (1993). The female androgen deficiency syndrome. *J Sex Martial Ther.* 19(1):3–24.

Kaunitz AM (2002). Noncontraceptive health benefits of oral contraceptives. *Rev Endocr Metab Disord.* 3(3):277–283.

Kern F, Everson TG (1987). Contraceptive steroids increase cholesterol in bile: mechanisms of action. *J Lipid Res.* 28:828–839.

King JC, Raynolds WL, Margen S (1978). Absorption of stable isotopes of iron, copper, and zinc during oral contraceptive use. *Am J Clin Nutr.* 31:1198–1203.

Koenig G, Seneff S (2015). Gamma-glutamyltransferase: a predictive biomarker of cellular antioxidant inadequacy and disease risk. *Dis Markers.* 2015;2015:818570.

Krapf JA, Simon M (2009). The role of testosterone in the management of hypoactive sexual desire disorder in postmenopausal women. *Maturitas.* Jul 20;63(3):213–219.

Kuo HK, Sorond FA, Chen JH, Hashmi A, Milberg WP, Lipsitz LA (2005). The role of homocysteine in multisystem age-related problems: a systematic review. *J Gerontol A Biol Sci Med Sci.* Sep;60(9):1190–1201.

Lidegaard O (1999). Smoking and use of oral contraceptives: impact on thrombotic diseases. *Am J Obstet Gynecol.* Jun;180(6 Pt 2):S357–3S63.

Lord RS, Bongiovanni B, Bralley JA (2002). Estrogen metabolism and the diet-cancer connection: rationale for assessing the ratio of urinary hydroxylated estrogen metabolites. *Altern Med Rev.* Apr;7(2):112–129.

Mercola. (2004). Just say "no" to birth control pills. http://articles.mercola.com/sites/articles/archive/2004/10/27/birth-control-part-two.aspx.

Meulenberg PM, Ross HA, Swinkels LM, Benraad TJ (1987). The effect of oral contraceptives on plasma-free and salivary cortisol and cortisone. *Clin Chim Acta.* Jun 15;165(2–3):379–385.

Mumford SL, Browne RW, Schliep KC, et al. (2016). Serum antioxidants are associated with serum reproductive hormones and ovulation among healthy women. *J Nutr.* Jan;146(1):98–106.

National Institutes of Health, Office of Dietary Supplements. (2016). Vitamin B6: dietary supplement fact sheet. https://ods.od.nih.gov/factsheets/VitaminB6-HealthProfessional/;

Olatunbosun DA, Adeniyi FA, Adadevoh BK (1974). Effect of oral contraceptives on Serum magnesium levels. *Int J Fertil.* 19(4):224–226.

Oh R, Brown DL (2003). Vitamin B12 deficiency. *Am Fam Physician.* 167(5):979–986.

Otaegui D, Mostafavi S, Bernard CC, Lopez de Munain A, Mousavi P, Oksenberg JR, Baranzini SE (2007). Increased transcriptional activity of milk-related genes following the active phase of experimental autoimmune encephalomyelitis and multiple sclerosis. *J Immunol.* Sept;179(6):4074–4082.

Palmery M, Saraceno A, Vaiarelli A, Carlomagno G (2013). Oral contraceptives and changes in nutritional requirements. *Eur Rev Med Pharmacol Sci.* Jul;17(13):1804–1813.

Panzer C, Wise S, Fantini G, Kang D, Munarriz R, Guay A, Goldstein I (2006). Impact of oral contraceptives on sex hormone-binding globulin and androgen levels: a retrospective study in women with sexual dysfunction. *J Sex Med.* Jan;3(1):104–113.

Pelton R (2001). *Drug-induced nutrient depletion handbook.* Hudson, OH: LexiComp.

Prasad AS, Oberleas D, Moghassi KS, Stryker JC, Lei KY (1975). Effects of oral contraceptives agents on nutrients II. *Am J Clin Nutr.* Apr;28(4):385–391.

Prasad, AS, Lei KY, Moghissi KS, Stryker JC, Oberleas D (1976). Effects of oral contraceptives on nutrients III: vitamins B6, B12, and folic acid. *Am J Obstet Gynecol.* Aug 15;125(8):1063–1069.

Rao A, Steels E, Beccaria G, Inder WJ, Vitetta L (2015). Influence of a Specialized Trigonella foenum-graecum Seed Extract (Libifem), on Testosterone Estradiol and Sexual Function in Healthy Menstruating Women, a Randomised Placed Controlled Study. *Phytothe Res.* 29(8):1123–1130.

Renaud S, Ciavatti M, Perrot L, Berthezene F, Dargent D, Condamin P (1987). Influence of vitamin E administration on platelet functions in hormonal contraceptive users. *Contraception.* 36:347–358.

Rosanoff A (2005). Magnesium and hypertension. *Clin Calcium.* 12(2):225–260.

Sanders SA, Graham CA, Bass JL, Bancroft J (2001). A prospective study of the effects of oral contraceptives on sexuality and well-being and their relationship to discontinuation. *Contraception.* Jul;64(1):51–8.

Scholes D, Ichikawa L, LaCroix A.Z, Spangler L, Beasley JM, Reed S, Ott SM (2010). Oral contraceptive use and bone density in adolescent and young adult women. *Contraception.* 81(1):35.

Shojania AM (1982). Oral contraceptives: effect of folate and vitamin B12 metabolism. *Can Med Assoc J.* Feb;126(3):244–247.

Slooter AJ, Rosendaal FR, Tanis BC, Kemmeren JM, van der Graaf Y, Algra A (2005). Prothrombotic conditions, oral contraceptives, and the risk of ischemic stroke. *J Thromb Haemost.* Jun;3(6):1213–1217.

Smith NK, Jozkowski KN, Sanders SA (2014). Hormonal contraception and female pain, orgasm and sexual pleasure. *J Sex Med.* Feb;11(2);462–470.

Sonfield A, Pollack HA (2013). The Affordable Care Act and reproductive health: potential gains and serious challenges. *J Health Polit Policy Law.* Apr;38(2):373–391.

Takasaki A, Tamura H, Miwa I, Taketani T, Shimamura K, Sugino N (2010). Endometrial growth and uterine blood flow: a pilot study for improving endometrial thickness in the patients with a thin endometrium. *Fertil Steril.* Apr;93(6):1851–1858.

Warner J (2011). Sleep loss may lower testosterone. http://www.webmd.com/men/news/20110603/sleep-loss-may-lower-testosterone.

Webb JL (1980). Nutritional effects of oral contraceptive use: a review. *J Reprod Med.* 25:150–156.

WebMD.(2016). Fenugreek. http://www.webmd.com/vitamins-supplements/ingredientmono- 733-fenugreek.aspx?activeingredientid=733&.

Wertalik LF, Metz EN, LoBuglio AF, Balcerzak SP (1972). Decreased serum B 12 levels with oral contraceptive use. *JAMA*. 221:1371–1374.

Zal F, Mostafavi-Pour Z, Amini F, Heidari A (2012). Effect of vitamin A and C supplements on lipid peroxidation and GSH-dependent antioxidant enzyme status in the blood of women consuming oral contraceptives. *Contraception*. 8:62–66.

Zhang H, Cui D, Wang B (2007). Pharmacokinetic drug interactions involving 17α-ethinylestradiol: a new look at an old drug. *Clin Pharmacokinet*. 46(2):133–157.

Zimmerman Y, Eijkemans MJ, Coelingh Bennink HJ, Blankenstein MA, Fauser BC (2014). The effect of combined oral contraception on testosterone levels in healthy women: a systematic review and meta-analysis. *Hum Reprod Update*. Jan-Feb;20(1),76–105.

10

Toxins: A Pervasive 21st-Century Problem for Health and Sexuality

ADRIENNE SPROUSE

"It is no longer a question of whether the patient in front of you has an accumulation (body burden) of toxic chemicals. The question is 'How large is the burden and how are these chemicals contributing to your patient's illness?' "

—Sprouse (2015)

Sources of Toxins in Daily Life

More than 80,000 chemicals are in commercial use in the United States. Most were released to the market before any studies were conducted on their low-dose, long-term effects on human health. These chemicals are ubiquitous in our culture and environment.

Beginning in the early 1960s, the National Health and Nutrition Examination Survey, a program of studies conducted by the US Centers for Disease Control and Prevention (CDC), has conducted a series of surveys focusing on different population groups or health topics. In 1999, the survey became a continuous program, examining a nationally representative sample of about 5,000 persons each year who are located in counties across the United States. Environmental exposures is one of the health topics studied.

Table 10.1 represents data from national samples collected beginning 1999–2000 and as recently as 2011–2012 and lists the categories of environmental chemicals identified in the American population (CDC, 2015). More than 300 chemicals were recorded, albeit not all chemicals were measured in each national sample. Common exposures for each chemical or category are listed.

Table 10.1. Simplified Listing of Chemicals in the Fourth Report Updated Tables, February, 2015

Chemical Name or Category	*Common Exposures*
Tobacco Smoke	Firsthand smoke, secondhand smoke, thirdhand smoke: breathing residual smoke on clothing, porous items, hair when actual smoking is not present
Bisphenol A	Plastics containers, thermal paper used as sales receipts, CDs, DVDs, coatings on many food and beverage cans, linings of water pipes, baby bottles and sipping cups, manufacture of flame retardants, fungicide
Other Environmental Phenols	Soaps, detergents, toys, surgical cleaners
Disinfectant By-Products	Drinking chlorinated water, breathing chlorinated water vapors released into the air inside your home or outside a factory or laboratory, chlorinated swimming pools
Fungicides and Metabolites	Disinfectants, pesticides, produce, canned beer, preservation of construction products, metal working fluids, slurries, pigments, stains, used for sapstain control of freshly cut lumber
Herbicides and Metabolites	Lawncare products, soil, crops
Sulfonylurea Herbicides	Lawncare products, produce, soil
Carbamate Pesticides and Metabolites	Insecticides, produce
DDT	Produce, soil, travel to countries where DDT is currently used as insecticide
Other Organochlorine Pesticides and Metabolites	Insecticides, exterminated buildings
Organophosphorus Insecticides and Metabolites	Fruits, vegetables, farm associated soil, air, dust, pets
Pyrethroid Pesticides and Metabolites	Insecticides, exterminated buildings
Dichlorophenol Pesticides and Metabolites	Pesticides, chlorination of water
Metals and Metalloids	Mercury—dental amalgams, tuna, swordfish, thermometer, vaccines; Cadmium—cigarette smoke; Arsenic—paint pigments, chemotherapy medicines, pesticides, herbicides, fungicides, contaminated water supplies; Aluminum—pots and pans, deodorant

Table 10.1. Continued

Chemical Name or Category	*Common Exposures*
Parabens	Sunscreen products, shampoo, cosmetics, foods, beverages, pharmaceuticals
Perchlorate and Other Anions	Fertilizers, rodenticides, food, tobacco smoke
Perfluorinated Compounds: Surfactants	Stain repellant, used in manufacture of Gore-Tex and Teflon
Phthalate and Phthalate Alternative Metabolites	Plastic food containers, nail polish, cosmetics, air fresheners, toys, flooring tiles, medical devices, cosmetics, vinyl, adhesives, and inks
Phytoestrogens and Metabolites	Soybeans, flaxseed; during digestion of whole grains, seeds, some fruits, and vegetables
Polycyclic Aromatic Hydrocarbon Metabolites	Incomplete combustion of coal, oil, gas, garbage, heavy oils, coal tar, roofing tar, creosote
Volatile Organic Compounds and Metabolites	Mold contamination, printing inks, paints, varnishes, adhesives, petrochemicals, carpets, composite wood products, sealing caulks, solvents, upholstery fabrics, vinyl floors, air fresheners, air cleaners that produce ozone, cleaning and disinfecting chemicals, cosmetics, moth balls, gas stoves, cleaning products, newspapers, non-electric space heaters, photocopiers, smoking, wood-burning stoves
Polybrominated Diphenyl Ethers and PBB 153	Flame retardants, mattresses, foods
Polychlorinated Biphenyls (PCBs)	Banned worldwide but redistribution of PCBs already present in soil, water, and food

Source: Sprouse A (2015). *The evaluation and treatment of the chemically exposed patient—a primer for healthcare professionals and their patients*, 2nd ed. New York: Adrienne Sprouse.

Mold contamination from water-damaged buildings produces a wide variety of toxic chemicals including mycotoxins, which are chemicals specifically emitted by molds such as trichothecenes, aflatoxins, ochratoxins, zearlenones, and others (Croft, 1966; Kotsonis, 1975, 939). Genital involvement in animals including uterotrophy, vulvar hypertrophy, preputial enlargement in castrated males, and prominent mammary glands in both sexes has been described from exposure to metabolites of *Fusarium graminearum* (*Gibberella zeae*) (Stob, 1962).

Certain mold species also emit volatile organic compounds (VOCs), which include benzene, hexane, acetone, toluene, styrene, furan, and others (Ahearn, 1996; Ezenou, 1994). Exposure to each of these chemicals can produce illness

(Budavari, 1996). Table 10.2 lists elected mold metabolites and their adverse health effects. Simultaneous exposure to multiple chemicals can be synergistic and cause heightened illness as these chemicals may combine to form secondary compounds and may compete for the same clearance pathways in the body.

Table 10.2. Selected Mold Metabolites and Their Adverse Health Effects

Genre	*Mold Metabolite*	*Adverse Health Effects*
Stachybotrys	Trichothecenes	Vomiting, diarrhea, blurred vision, nasal irritation, cough, weakness, ataxia, hypotension, coagulopathy, burning pain, redness, blisters and necrosis of the skin, nervous disorders, meningeal hemorrhage of the brain, hemorrhage of the mucosal epithelium of the stomach and intestine, immune system dysfunction, apoptosis, death
Penicillium	Penicillic acid	Carcinogenic
Penicillium	Verrucosidin	Neurotoxic and tremorogenic
Penicillium	Ochratoxin A	Nephrotoxic, immunotoxic, carcinogenic
Aspergillus	Ochratoxin A	Nephrotoxic, immunotoxic, carcinogenic
Aspergillus	Ethanol	Nausea, vomiting, mental excitation or depression, drowsiness, impaired perception, incoordination, coma, death
Aspergillus	Acetone	Irritation of the eye and nose, drowsiness, weakness, unsteady gait, defatting dermatitis
Aspergillus	Toluene	Irritation of the eye, nose, respiratory system, and skin, CNS excitation or depression, nausea, hilarity, lassitude, drowsiness, impaired balance, paresthesias, vision disturbances, headache, narcosis, collapse, death from respiratory failure or sudden ventricular fibrillation, hepatotoxic, nephrotoxic
Aspergillus	Benzene	Dizziness, headache, vomiting, visual disturbances, staggering gait, hilarity, fatigue, CNS depression, loss of consciousness, respiratory arrest, bone marrow depression, leukemia, irritation of the eye, nose, respiratory system, and skin
Aspergillus	Gliotoxin	Numerous immunosuppressive functions
Aspergillus	Patulin	Immunotoxic

Note: CNS = central nervous system.

Source: Sprouse A (2015). *The evaluation and treatment of the chemically exposed patient—a primer for healthcare professionals and their patients*, 2nd ed. New York: Adrienne Sprouse.

How Toxic Chemicals Enter the Body

Xenobiotics can enter the body via the air we breathe, the foods we eat, and the water we drink; through skin absorption or maternal-fetal transmission; or from the implantation of medical materials and devices. Once in the body, these toxicants enter the bloodstream and circulate to all organ systems. Many are small, nonpolar, lipophyllic molecules that can dissolve in fat tissue, making them resistant to biological clearance. Other internal storage depots are the bones, liver, kidneys, and plasma proteins. These internal reservoirs may serve as sources of ongoing chemical exposure and continue to damage tissue and/or dysregulate vital aspects of metabolism even after the external chemical sources have been removed.

Endocrine-Disrupting Chemicals Derail Sexual Function

Chemicals from different categories in Table 10.1 and Table 10.2 behave as endocrine disrupters, interfering with endogenous hormones and producing adverse developmental, reproductive, metabolic, neurological, and immunological effects in mammals. Many serve as obesagens and diabetagens, contributing to the obesity epidemic and increase in Type II diabetes (Legler, 2015). Endocrine disrupting chemicals (EDCs) with estrogenic activity can be active at concentrations far below those currently being tested in toxicological studies. Using only very high doses in those studies can dramatically underestimate bioactivity.

During the past decade a number of pesticides, industrial byproducts, manufactured products such as plastics, and natural chemicals have been shown to disrupt the endocrine system and to be potentially bioactive at low environmentally-relative doses (Colerangle, 1997). During fetal life, low levels of EDCs can provide the mechanism by which even slight alterations in hormonal activity during very brief critical developmental periods can potentially lead to irreversible changes in the course of cell differentiation. The cellular changes were associated with permanent alterations in gene activity and organ function (Li, 1997; Newbold, 1995).

A critical concern is the potential for species differences in response to EDCs with estrogenic activity. It is well known that the gene structure and ligand binding properties of the classical estrogen receptor (ER-α) have been highly conserved (have experienced relatively little change) among vertebrates after 300 million years of evolution. Thus the binding of an estrogenic chemical to an ER-α in fish, amphibians, reptiles, birds, and mammals including humans shows relatively little difference (Katzenellenbogen, 1979; Pakdel, 1989; White, 1994). All of the toxic effects of the chemicals in Table 10.1 cannot be reported in this chapter due to space constraints. Bisphenol A (BPA) is presented as an EDC to demonstrate basic mechanisms of actions, alterations in sexual development, and challenges in determining adverse health effects.

BPA

BPA is classified as a weak estrogen because it binds with a much lower affinity to membrane and nuclear estrogen receptor ER-α compared to 17-β estradiol. Yet studies demonstrate even weak estrogens are known to affect the mammary gland during early development. In animal models, prepubertal rats exposed to both low and relatively high doses of BPA for just 11 days (11ug/kg/d and 54 mg/kg/d, respectively) showed the mammary gland architecture was affected in both dose groups with an increase in the number of epithelial structures and altered structure suggesting advanced development (Jones, 2010). Fetal and neonatal exposure via maternal treatment of just 0.25 ug/kg/d produced alterations in tissue organization observed in puberty and throughout adulthood long after the exposures ended. Even adult exposure to BPA altered the tissue organization of the mammary gland in ways that are similar to the effects observed after perinatal exposure (Vandenberg, 2012). These results are also consistent with the effects of low doses of BPA exposure on mammary epithelial cells in culture (Wetherill, 2007). In reviewing the scientific literature and public policy, a weight of evidence approach has been used to assess low-dose EDC studies. A dose of 400 ug/kg/day or lower was selected for BPA although the Environmental Protection Agency (EPA) calculated a reference dose for BPA of 50 ug/kg/d based on a lowest observed adverse effect level (LOAEL) of 50 mg/kg/d (Welshons, 2003).

Because estrogen is implicated in a number of cancers both as an initiator and a promoter, environmental chemicals that mimic estrogen cannot be ruled out as carcinogens. In particular, researchers need to determine whether exposure to EDCs early in life is related to development of cancer later in life.

Exposure to BPA during fetal development also affects the androgen receptor binding activity in the prostate gland, changing its tissue organization and cytokeratin expression as well as the volume of the prostate and the number and size of dorso-lateral prostate ducts. Fetal exposure also influences the incidence of adult onset prostatic intraepithelial neoplasia lesions, especially when the BPA exposure is accompanied by a concurrent exposure of testosterone and estradiol, mimicking adulthood hormonal status and the endocrine changes associated with aging in men that also typically accompany the onset of prostate cancer (Wetherill, 2002). Very low doses (0.1–10 nM or 0.023–2.3 ng/mL) of BPA induces proliferation of human prostate cancer cells via binding to a mutant form of the androgen receptor found in some prostate tumors.

A concept related to low-dose is that of Non-Monotonic Dose Response Curve (NMDRC). In this kind of response curve, the observed defects may be linear or nonlinear, but the slope of the curve changes signs somewhere within the range of doses examined (Figure 10.1). NMDRCs are often U-shaped, with maximal responses measured at the endpoints observed at low and high doses, or inverted U-shapes, with maximum responses observed at intermediate doses. Some cases are more complicated with multiple points along the curve at which the slope of the curve reverses sign (Vandenberg, 2012).

Another example of nonmonotonic dose response to estrogens is the set of experiments showing that the lower doses of diethylstilbestrol (DES), another estrogen mimicker, resulted in an increase in prostate size. This dose was in the physiological range of estrogenic activity and falls within the low-dose range for exposure. In contrast, in the same studies, a 10,000-times higher dose of DES (200 μg/kg/day) resulted in gross abnormalities in the reproductive organs including a marked reduction in prostate size (Gupta 2000a, 2000b; vom Saal, 1997). This higher dose of DES falls within the toxicological dose range and represents a high-dose range of exposure. The currently accepted LOAEL dose of bisphenol A of 50 mg/kg/day was reported from high-dose toxicological studies (National Toxicology Program, 2001). Those studies are typical in using 50,000 to 500,000 times higher doses than the 2 to 20 μg/kg/day doses administered to pregnant mice, a dose accurately predicted to be bioactive in male mouse fetuses (Morrissey, 1987).

Nonmonotonicity is not synonymous with low dose because there are low-dose effects that follow monotonic dose response curves. Thus it is not required that a study include doses that span from the true low-dose range to the high toxicological range to detect nonmonotonicity. In a system that is responding nonmonotonically, it is not appropriate to use a high-dose test to predict low-dose effects. All current exposure standards employed by governmental agencies around the world, including the Food and Drug Administration and the EPA have been developed using an assumption of monotonicity. The low-dose range, which presumably is what the general public normally experiences, is rarely, if ever, tested directly (Nagel, 1997; Yang, 2015).

The standard procedure for regulatory testing typically involves a series of tests to establish the LOAEL, then a dose beneath that at which no adverse effect is observable (the NOAEL). Next a series of calculations is used to acknowledge uncertainty in the data, species differences, age differences, and so on, and those calculations beginning with the LOAEL or the NOAEL produce a reference dose that is presumed to be a safe exposure for humans. Typically, the reference dose is 3 to 1,000-fold lower than the NOAEL. That reference dose then becomes the allowable exposure and is deemed safe, even when it is *never examined directly*. For chemicals with monotonic linear dose response curves, this may be appropriate. But for chemicals that display nonmonotonic patterns, it is likely to lead to false

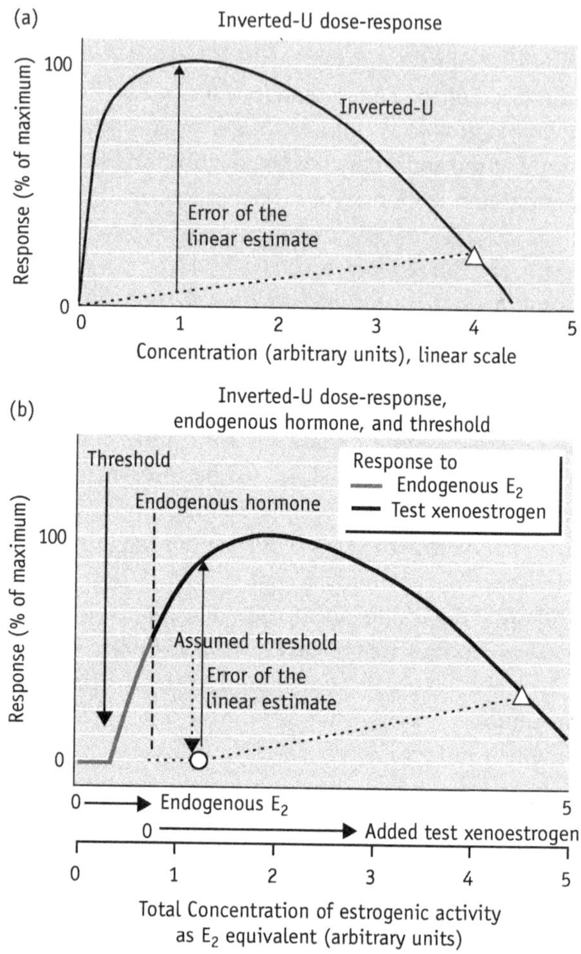

FIGURE 10.1. **(A)** The error due to assuming that the dose–response curve is linear (dotted line) when, in fact, the dose–response curve is nonmonotonic and forms an inverted U (solid black line). The error in estimating actual responses that will occur at doses below the test dose in a toxicological study increases as the concentration of the test dose increases relative to a test dose that would result in a maximum response. Figure 10.1A shows that the magnitude of the error in estimating responses at doses below the test dose for an EEDC (using linear extrapolation) is greater when the dose–response curve is nonmonotonic relative to the error when the dose–response curve is monotonic.
(B) The error associated with examining a single dose of a test chemical (triangle) with estrogenic activity, such as bisphenol A, that adds to an existing background level of endogenous estradiol, which is variable because of endogenous and exogenous factors. In the current model used in risk assessment, a linear extrapolation (dotted line) from the test dose (triangle) to an assumed threshold dose (circle) is used, based on the assumption that there will be an absence of response at this assumed threshold dose. In this figure, the assumption is that endogenous estrogen is already above threshold

negatives, concluding that exposure to the reference dose is safe when, in fact, it is not (Colerangle, 1997; Newbold, 2001; Welshons, 2003).

Mechanisms for NMDRCs include

1. Cytotoxicity
2. Cell and tissue specific receptors and cofactors
3. Receptor selectivity
4. Receptor downregulation and desensitization
5. Receptor competition
6. Endocrine negative feedback loops
7. Receptor presence in tissues other than target organs
8. Time of exposure

Although epidemiology studies examining the influence of BPA on breast cancer have proven to be inconclusive at best, epidemiologists must collect information about prenatal and neonatal exposures and relate them to adult breast cancer incidence. These types of studies should also take into consideration the effect of other xenoestrogens because their effect could be additive or even synergistic. Additionally, studies have shown that BPA exposure produces defects in the ovary (Newbold, 2007).

DICHLORODIPHENYLTRICHLOROETHANE

Dichlorodiphenyltrichloroethane (DDT), a chlorinated pesticide, came into widespread agricultural use in the United States in 1945. Rachel Carson, the biologist and author of *Silent Spring* (Carson, 1962), predicted its adverse effects on wildlife, especially birds. DDT was eventually banned for agricultural use in the United States in 1972. It is still used outside of the United States, and Americans become exposed by eating produce grown with DDT that is imported or by traveling to foreign countries where DDT is in use. A prospective study of young women's blood collected between 1959 and 1967 and stored by freezing reports that women exposed to relatively high levels of DDT prior to midadolescence are five times

---

FIGURE 10.1. Continued

for the estrogen receptor–mediated response to the EEDC (vertical dashed line). There can thus be no threshold for the response to the exogenous EEDC. The assumption of no response at the assumed threshold EEDC dose, when this is not the case, will result in a great error, potentially infinite, in estimating the response at this dose, if linear extrapolation from a high test dose is used instead of actually determining the shape of the dose–response curve.

Welshons WV, Thayer KA, Judy BM, Taylor JA, Curran EM, vom Saal FS. (2003). Large effects from small exposures: mechanisms for endocrine disrupting chemicals with estrogenic activity. *Environ Health Perspect.* 111(8):1003, Fig 4.

more likely to develop breast cancer later in life than women with lower exposures, but exposure after adolescence does not increase risk (Cohn, 2007). Combined with an analysis of the women's medical records, the median time to diagnosis of breast cancer after the sample was taken was 17 years. This study differs from the Long Island Breast Cancer Study (Gammon, 2002), which failed to find heightened breast cancer risk in response to exposure to environmental estrogens. The Long Island study concluded there was no link between adult exposures and adult cancer but did not consider developmental exposures when evaluating adult cancer risk.

Toxic Metals

Mercury, cadmium, arsenic, aluminum, and lead bind with proteins, forming toxic metal-protein complexes that have no biological functions, or they may compete with essential metals such as iron and calcium, displacing these from their biological activities. Cadmium and arsenic are carcinogens. Mercury is known to have a widespread distribution in aquatic ecosystems and to cause significant neurodevelopmental toxicity in humans. It also has been associated with low birth weight babies, which were more likely in South Carolina among women living in geographic areas of the upper quartile of predicted total mercury in fish compared to the lowest quartile. When stratified by race, the risk for low birth weight or preterm births (<37 weeks) was 10% to 18% more likely among African American mothers living in areas with the highest total fish mercury concentrations (Burch, 2014).

Reducing Exposure to Harmful Environmental Chemicals

FOOD AND WATER

Many chemicals enter our food supply: heavy metals, pesticides, herbicides, monosodium glutamate, and polybrominated biphenyl ethers (Schecter, 2006). Genetically modified organisms (or micro-organisms) are genetically engineered foods containing genetic sequencing that does not appear in nature.

People should choose organic foods whenever possible—those grown without toxic additives, shelf-life extenders, or synthetic pesticides. Integrated pest management utilizes current, comprehensive information on the life cycles of pests and their interaction with the environment to manage pests with the least possible hazard to people, property, and the environment (EPA, n.d.).

Food packaging is important. Glass, ceramic, cellophane, and paper food containers are preferable to plastics or aluminum foil, which can leach into food.

Foods packaged in cans may contain unwanted additives used in the can liners. Cooking in stainless steel, ceramic, glass, or cast iron pots and pans rather than aluminum or nonstick pans also reduces toxic exposures.

Natural or spring water bottled in glass rather than plastic is often preferable to tap water. When in doubt, people should have their water tested by the local health department or private company for heavy metals, solvents, and other contaminants. Home water purification systems are available and range from pitchers with special filters to whole-house systems.

PERSONAL CARE PRODUCTS

Surveys investigating the prevalence of fragrance sensitivity in the American population found 30.5% of the general population reported scented products on others were irritating (Caress, 2009). Fragrances are often added to soaps, shampoos, cosmetics, cleaning products, laundry detergent, dryer anti-cling sheets, and others. Ethanol, acetaldehyde, chloromethane, and 1, 4-dioxane are some of the VOCs found in fragrances and may not be disclosed on labels because they are considered trade secrets. Reading labels and researching unfamiliar ingredients is recommended. Often a reference phone number is provided that can be called for clarification, and sometimes a Material Safety Data Sheet is available. Many companies now make fragrance-free products.

HOUSEHOLD PRODUCTS

The study of VOCs emitted from several consumer and commercial products revealed sprayed products emit the highest amount of VOCs (~96 wt%) when compared to nonsprayed products and contain propane, iso-butane, and n-butane, which are components of liquefied petroleum gas (Dinh, 2015). Air fresheners, cleaning products containing toxic chemicals, fragranced laundry detergent, dish detergent, mildew-killing products used on tubs and showers, and oven cleaners may all contain chemicals that produce illness. Nontoxic products are available. Individuals should evaluate every chemical that touches their skin and wear impermeable gloves when washing dishes, windows, floors, and cleaning ovens.

MISCELLANEOUS EXPOSURES

Compared to electric stoves, gas stoves can emit higher levels of nitrous oxide, carbon monoxide, carbon dioxide, methane, non-methane hydrocarbons, total VOCs, and fine particulate matter (Wong, 2011). Mothproofing winter clothes with mothballs containing paradichlorobenzene can lead to central nervous system tissue damage and functional neurological decline (Dubey, 2014).

Excretion of Toxic Chemicals

More chemicals are eliminated from the body via the kidneys than by any other route. To a lesser extent, excretion is accomplished by the feces, the lungs, cerebral spinal fluid, sweat, saliva, and breast milk (Rozman, 1975). Many chemicals in Table 10.1 and Table.10.2 require biotransformation to render them more water-soluble before they can be excreted into the urine. This biotransformation is primarily accomplished through enzymatic pathways in the liver. Liver Phase I results in modification of reactive chemicals by oxidation, reduction, hydroxylation, sulfoxidation, deamination, dealkylation or methylation. These modifications usually involve mixed function oxidases, cytochromes b3 and P-450, and the glutathione Sacyltransferases. Liver Phase II may follow Phase I reactions or may proceed independently. During Phase II, the liver maximally converts fat-soluble substances to water-soluble substances, facilitating their excretion from the body (Buffaloe, 1996).

NUTRIENT SUPPLEMENTATION

Because Liver Phase II is nutrient dependent, those exposed to low levels of xenobiotics often have deficiencies in the nutrients directly involved in Phase II biotransformation (Anderson, 1991; Basu, 1977; Bidlack, 1984, 1986). Supplementing these deficiencies supports the detoxification process. Decisions about what specific nutrients to use and the schedule of treatments for supplementation are complex determinations and must consider the patient's ability to identify and remove the toxic chemicals from the environment; the patient's total toxic load, symptoms, age, liver, kidney, and vascular functions, and pH; and the patient's potential reactivity to the treatment itself. Because xenobiotic exposure can alter the immune system, hypersensitivity reactions can develop even to preservative-free treatment materials. Performing intradermal skin tests for all treatment materials identifies hypersensitivities, and then a neutralizing or desensitization dose can be administered to allow the patient to receive the treatments without significant reactions. Without this prior testing, it is difficult to determine if the patient is hypersensitive to the treatment or whether the treatment is producing a mild (but desired) symptom of detoxifying from the problematic chemical.

A common initial preservative-free intravenous infusion for a 70 kg man is

- Diluent: 0.9% sodium chloride (normal saline), 500 ml
- Sodium bicarbonate, 50 m equivalents

- Ascorbic acid (vitamin C), 12.5 grams
- Trace minerals-approximately chromium 10 micrograms; zinc 5 mg; copper 1 mg; selenium 60 micrograms; and manganese, 0.5 mg
- Magnesium sulfate, 5 grams
- B-complex, 25 mg each of B1, B2, B3, B5
- L-glutathione, 600 mg
- Taurine, 2 grams

Intravenous doses of Phase II nutrients coupled with sauna detoxification has been shown effective in eliminating toxicants from the body (Sprouse, 2013). Oral supplementation is usually added to the treatment program after several intravenous infusions have been given. Gastrointestinal intolerance requires oral doses lower than intravenous doses.

SAUNA DETOXIFICATION

The goal of the sauna detoxification program is the removal of solvents, pesticides, heavy metals, and other chemicals from the blood, fatty tissue, brain, and other target organs. Nutritional studies indicate that adequate vitamin and mineral intake is necessary for successful detoxification (Roehm, 1983; Schnare, 1984; Sprouse, 2013; Tretjak, 1990). Once mobilized, the toxicants enter the bloodstream and travel to the liver for processing or may be cleared from the body via sweat or lungs. The sauna temperature, length of time in sauna, and number of sessions are titrated for each patient.

HEAVY METAL CHELATION

In addition to the sauna, heavy metals may be removed by administering a chelating agent, a medication known to bind a particular heavy metal (Nurchi, 2013). The medication-heavy metal complex is excreted in the urine. Several rounds of chelation may be necessary to reduce heavy metal exposure. These treatments must also be titrated to the patient's condition and symptoms.

Common Chelating Medications include

- DMPS -2,3 dimercapto-1-propanesulfonic acid
    - Average adult oral dose 10 to 30 mg/kg a day in divided doses (check pediatric dose)
    - Common IV adult dose 50 mg diluted in 100 ml 0.9% Na Cl slow over 20 minutes (check pediatric dose)

- DMSA -2,3, dimercaptosuccinic acid (Chemet)
  - Average adult dose is 10 mg/kg orally three times a day for five days (15 doses) (check pediatric dose)
  - Common adult dose is 500 mg orally three times a day for five days (15 doses) (check pediatric dose)
  - Can produce worrisome side effects in sensitive patients
  - Can cause a rise in the liver enzyme ALT
- EDTA (not used by author) (Sprouse, 2015)

Case Study

SB was 53 years old when she first presented to my medical center in 2011 with multiple medical complaints including asthma, abdominal pain and bloating, interstitial cystitis, vulvo-vestibulitis, migraine, chronic fatigue, vasculitis, and cognitive deficits in word searching, memory, and concentration.

She was in her usual state of good health in 1984 when she moved into a new house, then slowly developed her severe symptoms. Treatment by a variety of specialists did not improve her health. Biopsy revealed interstitial cystitis unresponsive to Elavil, Elmiron, Valium, Urudus, and bladder distentions with dimethyl sulfoxide irrigations for two years. Although urine cultures were repeatedly negative, she was given Cipro without relief.

When tile was removed from her bathroom wall and black mold was observed, her breathing problems, gastrointestinal pain, interstitial cystitis, and vulvar pain all escalated. Air samples collected inside her home measured Alternaria, Cladosporium, Penicillium/Aspergillus, and several other molds. A complete mold abatement and renovation of her bathroom required her to leave the house and stay with a relative. SB eventually underwent a treatment program of avoidance of mold and mold toxins, infusions of intravenous nutrients, and sauna detoxification. Her bladder pain and other symptoms decreased.

Trichothecene mycotoxin was measured in her urine providing ongoing toxin exposure even when mold was not present. She further developed urethritis and vulvodynia. Continued IVs and sauna detoxifications decreased her multiple symptoms. Once out of the mold-contaminated house permanently, her interstitial cystitis entirely resolved, allowing her to enjoy sexual intercourse again without any bladder pain. However, her vulvodynia continued to be a problem. Upon re-exposure to chemicals, especially to mold toxins, her many other symptoms flared. Strict environmental controls, IV infusions, sauna detoxification, and eventually desensitization to molds, foods, and other factors gave SB her best results.

Conclusion

Toxicants are ubiquitous in our environment and can cause a wide range of illnesses including dysfunction of the sexual organs. Avoidance of toxic chemicals by choosing nontoxic products can help prevent further xenobiotic exposure. Treatment with nutrients, sauna, and/or heavy metal chelation can reduce the total toxic load.

REFERENCES

Ahearn DG, Crow SA, Simmons RB, et al. (1996). Fungal colonization of fiberglass insulation in the air distribution system of a multi-story office building: VOC production and possible relationship to a sick building syndrome. *J Indus Microbiol.* 16:280–285.

Anderson DE, Kappas A (1991). Dietary regulation of cytochrome p450. *Annu Rev Nutr.* 11:141–167.

Basu TK (1977). Interaction of drugs and nutrition. *J Human Nutr.* 31:449–458.

Bidlack WR, Brown RC, Mohan C (1986). Nutritional parameters that alter hepatic drug metabolism, conjugation, and toxicity. *Fed Proceed.* 46(2):142–148.

Bidlack WR, Smith CH (1984). The effect of nutritional factors on hepatic drug and toxicant metabolism. *J Am Diet Assoc.* 84(8):892–898.

Budavari S, O'Neil MJ, Smith A, Heckelman, PE, Kinneary JF, eds. (1996). *Merck Index*, 12th ed. Whitehouse Station, NJ: Merck Research Laboratories.

Buffaloe A (1996). *Chemicals in the environment and human sensitivity to them: their impact on health and procedures for clearing the body.* New York: Adrienne Buffaloe.

Burch JB, Wagner Robb S, et al. (2014). Mercury in fish and adverse reproductive outcomes: results from South Carolina. *Int J Health Geogr.* Aug;13–30.

Caress SM, Steinemann AC (2009). Prevalence of fragrance sensitivity in the American population. *J Environ Health.* Mar;71(7):46–50.

Carson R (1962). *Silent spring.* New York: Houghton Mifflin.

Centers for Disease Control and Prevention (2015). Fourth national report on human exposure to environmental chemicals: updated tables. http://www.cdc.gov/ biomonitoring/environmental_chemicals.html.

Cohn BA, Wolff MS, Cirillo PM, Scholtz RI (2007). DDT and breast cancer in young women: new data on the significance of age at exposure. *Environ Health Perspect.* Oct;115(10):1406–1414.

Colerangle JB, Roy D (1997). Profound effects of the weak environmental estrogen-like chemical bisphenol A on the growth of the mammary gland of Noble rats. *J Steroid Biochem Mol Biol.* 60:153–160.

Croft W (1966). Airborne outbreak of trichotecene toxicosis. *Atmos Environ.* 20(3):549–552.

Dinh TV, Kim SY, Son YS, et al. (2015). Emission characteristics of VOCs emitted from consumer and commercial products and their ozone formation potential. *Environ Sci Pollut Res Int.* Jun;22(12):9345–9355.

Dubey D, Sharma VD, Pass SE, Sawhney A, Stuve O (2014). Para-dichlorobenzene toxicity—a review of potential neurotoxic manifestations. *Ther Adv Neurol Disord.* May;7(3):177–187.

Environmental Protection Agency. (n.d.). Integrated pest management. http://www.epa.gov/agriculture/tipm.html.

Ezenou IM, Price DL, Simmons RB, Crow SA, Ahearn DG (1994). Fungal production of volatiles during growth on fiberglass. *App and Envir Microbiol.* Nov:4172–4173.

Gammon MD, Wolff MS, Neugut AI, et al (2002). Environmental toxins and breast cancer on Long Island. II. Organochlorine compound levels in blood. *Cancer Epidemiol Biomarkers Prev.* 11(8):686–697.

Gupta C (2000a). Reproductive malformation of the male offspring following maternal exposure to estrogenic chemicals. *Proc Soc Exp Biol Med.* 224:61–68.

Gupta C (2000b). The role of estrogen receptor, androgen receptor and growth factors in diethylstilbestrol-induced programming of prostate differentiation. *Urol Res.* 28:223–229.

Jones LP, Sampson A, Kang HJ, et al. (2010). Loss of BRCA1 leads to an increased sensitivity to bisphenol A. *Toxicol Lett.* 199:261–268.

Katzenellenbogen BS, Katzenellenbogen JA, Mordecai D (1979). Zearalenones: characterization of the estrogenic potencies and receptor interactions of a series of fungal beta-resorcylic acid lactones. *Endocrinology.* 105:33–40.

Kotsonis FN, Burdock, GA, Flamm WG (1975). Food toxicology. In CD Klassen, MO Amdur, J Doull (Eds.), *Casarett and Doull's toxicology: the basic science of poisons,* 5th ed. New York: Macmillan.

Legler J, Fletcher T, Govarts E, et al. (2015). Obesity, diabetes, and associated costs of exposure to endocrine-disrupting chemicals in the European union. *J Clin Endocrinol Metab.* Apr;100(4):1278–1288.

Li S, Washburn KA, Moore R, et al. (1997). Developmental exposure to diethylstilbestrol elicits demethylation of estrogen-responsive lactoferrin gene in mouse uterus. *Cancer Res.* 57:4356–4359.

Morrissey RE, George JD, Price CJ, Tyl RW, Marr MC, Kimmel CA (1987). The developmental toxicity of bisphenol A in rats and mice. *Fundam Appl Toxicol.* 8:571–582.

Nagel SC, vom Saal FS, Thayer KA, Dhar MG, Boechler M, Welsons WV (1997). Relative binding affinity-serum modified access (RBA-SMA) assay predicts the relative *in -vivo* bioactivity of the xenoestrogens bisphenol A and octylphenol. *Environ Health Perspect.* 105:70–76.

Newbold R (1995). Cellular and molecular effects of developmental exposure to diethylstilbestrol: implications for other environmental estrogens. *Environ Health Perspect.* Oct;103(Suppl 7):83–87.

Newbold RR, Banks EP, Bullock B, Jefferson WN (2001). Uterine adenocarcinoma in mice treated neonatally with genistein. *Cancer Res.* 61:4325–4328.

Newbold RR, Jefferson WR, Banks EP (2007). Long-term adverse effects of neonatal exposure to bisphenol A on murine female reproductive tract. *Reprod Toxicol.* 24:253–258.

National Toxicology Program (2001). Final report of the endocrine disruptors low-dose peer review panel. In *Endocrine disruptors low-dose peer review.* Research Triangle Park, NC: National Toxicology Program.

Nurchi VM, Alonso MC, Toso L, Lachowica JI, Crisponi G (2013). Chelation therapy for metal intoxication: comments from a thermodynamic viewpoint. *Mini Rev Med Chem.* Oct;13(11):1541–1549.

Pakdel F, Le Guellec C, Vaillant C, Le Roux MG, Valotaire Y (1989). Identification and estrogen induction of two estrogen receptors (ER) messenger ribonucleic acids in the rainbow trout liver: sequence homology with other ERs. *Mol Endocrinol.* 3:44–51.

Roehm DC (1983). Effects of a program of sauna baths and mega-vitamins on adipose DDE and PCBs and on clearing of symptoms of Agent Orange (Dioxin) toxicity. *Clin Res.* 31:243a.

Rozman KK, Klasen CD (1975). Absorption, distribution, and excretion of toxicants. In CD Klassen, MO Amdur, J Doull (Eds.), *Casarett and Doull's toxicology: the basic science of poisons*, 5th ed. (pp. 105–109). New York: Macmillan.

Schecter A, Haffner D, Colacino, et al. (2006). Polybrominated biphenyl ethers (PBDEs) levels in an expanded market basket survey of U.S. food and estimated PBDE dietary intake by age and sex. *Environ Health Perspect.* Oct; 114(10): 1515–1520.

Schnare DW, Ben M, Shields MG (1984). Body burden reductions of polychlorinated biphenyls, polybrominated biphenyls and chlorinated pesticides in human subjects. *Ambio.* 13:378–380.

Sprouse A, Curtis L, Bartlik B (2013). Organic solvent-induced bipolar disorder—a case report. *Adv Mind Body Med.* Summer; 27(3):19–23.

Sprouse A (2015). *The evaluation and treatment of the chemically exposed patient—a primer for healthcare professionals and their patients*, 2nd ed. New York: Adrienne Sprouse.

Stob M, Baldwin RS, Tuite J, Andrews FN, Gillette KG (1962). Isolation of an anabolic uterotropic compound from corn infected with Gibberella *zeae. Nature.* 196:1318.

Tretjak Z, Root DE, Tretjak A, et al. (1990). Xenobiotic reduction and clinical improvements in capacitor workers: a feasible method. *J Env Sci Health.* A25(7):731–751.

Vandenberg LN, Colborn T, Hayes TB, et al. (2012). Hormones and endocrine-disrupting chemicals: low-dose effects and nonmonotonic dose responses. *Endocr Rev.* Jun;33(3):378–455.

vom Saal FS, Timms BG, Montano MM, et al. (1997). Prostate enlargement in mice due to fetal exposure to low doses of estradiol or diethylstilbestrol and opposite effects at high doses. *Proc Natl Acad Sci U S A* 94:2056–2061.

Welshons WV, Thayer KA, Judy BM, et al. (2003). Large effects from small exposures: mechanisms for endocrine disrupting chemicals with estrogenic activity. *Environ Health Perspect.* 111(8):994–1006.

Wetherill YB, Akingbemi BT, Kanno J, et al. (2007). In vitro molecular mechanism of bisphenol A action. *Reprod Toxicol.* 24:178–198.

Wetherill YB, Petre CE, Monk KR, Puga A, Knudsen KE (2002). The xenoestrogen bisphenol A induces inappropriate androgen receptor activation and mitogenesis in prostatic adenocarcinoma cells. *Mol Cancer Ther.* 1:515–524.

White R, Jobling S, Hoare SA, Sumpter JP, Parker MG (1994). Environmentally persistent alkyl-phenolic compounds are estrogenic. *Endocrinology.* 135:175–182.

Wong TW, Wong AH, Lee FS, Qiu H (2011). Respiratory health and lung function in Chinese restaurant kitchen workers. *Occup Environ Med.* Oct;68(10):746–752.

Yang O, Kim HL, Weon, JI, Seo, YR (2015). Endocrine-disrupting chemicals: review of toxicological mechanisms using molecular pathway analysis. *J Cancer Prev.* Mar; 20(1):12–24.

11

Sexual Dysfunction and Exercise

KEONI (JOHN) TETA, RALPH ESPOSITO, AND JADE TETA

Overview

Sexual dysfunction in both men and women takes a toll on the enjoyment of sexual intimacy, putting undue strain on our amorous relationships. Impaired sexual function may stem from both physical and psychological causes or most often a combination of the two. No matter the cause, sexual dysfunction may be an indicator of poor health. Male sexual dysfunction is usually labeled as low libido and/or erectile dysfunction (ED) and for women as low libido, pain with intercourse (dyspareunia), or inability to orgasm (anorgasmia).

Whatever the origin, there is one often overlooked remedy that can productively address sexual dysfunction, and that is exercise. Exercise can impact most of the known causes of this disorder. The research is clear, sustaining a consistent exercise program that is individually tailored has whole-body benefits, thus improving cardiovascular health, diabetes, nerve damage, hormonal status, and psychological well-being. If any medicine can be considered an inexpensive but powerful cure-all, exercise can, and it is an excellent remedy for mitigating the myriad causes of sexual impairment.

This chapter discusses the physical and psychological causes of male and female sexual dysfunction (FSD) through the lens of exercise as a therapeutic treatment.

Male Sexual Dysfunction

Libido and healthy erectile function are instinctually part of a man's sexual health. An intricate communication among multiple organ systems that govern and impact sex drive results in sex-seeking behavior, arousal, and sexual activity. It has been known for centuries that a special male substance, that is, the "male hormone" testosterone (T), is one of the key factors in sex drive (Travison, 2006). Low libido is common in about 6% to 12% of older men exhibiting low libido from androgen deficiency (Araujo, 2011). Unfortunately only 15% of men will seek help for low libido (Corona, 2013). Symptoms associated with low libido include decreased spontaneous erections, depression/dysthymia, decreased motivation, increased body fat, and ED in intercourse (Bhasin, 2010).

Low libido and ED often, but not always, present together. ED impacts up to 18 million men older than 20 years of age in the United States (Selvin, 2007) with up to 60% of men above the age of 60 (Nicolosi, 2003), with a greater prevalence in men with cardiovascular risk and diabetes (Selvin, 2007). The International Index of Erectile Function helps measure the degree of ED, yet ED is commonly diagnosed clinically based on an inability to attain or maintain an erection sufficient for sexual activity.

The physician's goal must be to first understand why low libido and ED are manifesting. A holistic approach recognizing the relationships among the endocrine, cardiovascular, psychological, and neurological systems is necessary to truly elucidate the underlying cause(s) of the patient's aberrant presentation.

Exercise clearly has been shown to be effective in maintaining libido in men on androgen deprivation therapy (Cormie, 2013). In healthy men, exercise was shown to improve overall sexual function, arousal, and desire and even improve erectile function (Simon, 2015). Given that the current medical care for ED is primarily pharmaceutical, exercise at the very least can work to potentiate drug efficacy (Maio, 2010).

Exercise promotes health generally; more precisely, there are seven common, major factors in the etiology of ED and low libido that can be addressed with exercise: obesity, diabetes, nerve damage, hormonal imbalance, medications, cardiovascular dysfunction, and psychological imbalances.

OBESITY

Obesity may be one of the main obstacles for alleviating male and female sexual dysfunction. Given the phenomenon of age-related fat gain, age-related muscle loss, and the higher incidence of diabetes and heart disease with age, these risk factors predispose men and women to sexual dysfunction.

Obesity impairs endogenous T synthesis, which then causes more fat deposition in the viscera. As T levels are reduced there is a corresponding increase in waist circumference. About 40% of obese nondiabetic men and 50% of obese diabetic men aged above 45 years have low free T (Boloña, 2007). Together, obesity and diabetes further increase the prevalence of T deficiency by another 34%

(Boloña, 2007; Traish, 2014), further worsening libido. Although low T would be suspected to cause ED, the clearest association with ED is aging, as the correlation of total T with erectile functioning is weak (Isidori, 2005; Rhoden, 2002; Urgureanu, 2015).

Testosterone replacement therapy (TRT) has been shown to improve libido, bone mass, and insulin sensitivity and to reduce fat mass (Fong, 2015). Exercise, especially resistance training, has similar effects. Resistance training increases T and also improves all of the aforementioned conditions from low libido to obesity (Traish, 2006).

DIABETES

Diabetes mellitus (DM) significantly accelerates the aging process via glycation and chronic inflammation. Although Type I diabetes (DMI) is not typically known as a lifestyle disease, unlike Type II (DMII), both benefit from exercise. DMII is considered a major cause of sexual dysfunction. Exercise not only prevents DM; it also helps treat it along with its associated comorbidities, including sexual dysfunction.

Though DM is considered an endocrine disorder, it has a myriad of systemic effects on the cardiovascular and neurological system. When blood sugar levels are not managed properly, the production of advanced glycation end-products (AGEs) results in damage to endothelial and neuronal cells, contributing to endothelial dysfunction. AGEs are analogous to internal rusting of the body. As AGEs build up, the endothelium becomes more compromised. This alone contributes to neuropathy, but also AGEs prevent neuronal regeneration and plasticity, further exacerbating nerve damage.

The correlation with diabetes and male sexual dysfunction is inarguable. Diabetic ED is becoming a common term among physicians (Kamenov, 2015) as approximately 50% of men with DM also experience ED (Feldman, 2000). Low T and low libido are common in diabetic men, and this may be due to increased activity of aromatase converting T into estrogens (Kapoor, 2007). The "artery size" hypothesis proposes ED is an indicator of vascular and endothelial dysfunction, as increased adhesion factors along with decreased vasodilation decrease penile blood flow (Montorsi, 2003; Pegge, 2006). This hypothesis is magnified in diabetic men and thus explains the increased incidence of ED in that group. With decreased neurovascular supply and autonomic dysfunction, innervation to the penis is compromised, resulting in poor erections (Hecht, 2001).

Exercise is the common denominator to manage the endothelial, neurovascular, and hormonal aspects of diabetic sexual dysfunction by promoting a calorie deficit and balancing hormones that prime the body for lipolysis, which is one of the major risk factors for diabetes. Exercise works on a deeper level improving insulin sensitivity of glucose receptors thus improving sugar utilization.

NERVE DAMAGE

Somatic innervation to the penis via the pudendal nerve (S2-S4) is coordinated by the spinal cord sympathetic fibers resulting in the ejaculatory reflex. Improper function of peripheral sensory nerves and autonomics, which are commonly damaged in diabetes and cardiometabolic disease, can contribute to ED and ejaculatory issues (Fode, 2012).

Prolonged bicycling is one of the major risk factors for ED in men, with an incidence of 13% to 24% (Gulino, 2010). This is due in part to the compression of perineal arteries leading to chronic reduced penile perfusion. Bicycling and other exercises may cause nerve entrapment syndromes presenting as genitalia numbness (Leibovitch, 2005) and require caution in men with a compromised prostate given increased pressure on the perineal area (Grandys, 2009; Perna, 1995). Other less common symptoms include priapism, penile thrombosis, infertility, hematuria, torsion of spermatic cord, prostatitis, perineal nodular induration and elevated serum prostrate-specific antigen (Sommer, 2001). Men (and women) should use specially designed bicycle seats that are contoured to reduce pressure on the perineum.

LOW TESTOSTERONE

We know T is involved in libido (Traish, 2006), and although there is conflicting evidence with respect to T's role in erections, studies tend to agree that increasing T in hypogonadal men improves erections and sexual desire (Andersson, 2011; Cunningham, 2014; Rajfer, 2000; Wu, 2010). This is thought to be due to T's role in nitric oxide (NO) synthesis (Deenadayalu, 2010), along with maintaining the integrity and innervation of the pelvic floor (Puppo, 2016; Verhovshek, 2010). Yet one would expect T to improve ED, yet studies suggest TRT improves satisfaction and libido but not necessarily ED (Traish, 2006).

Exercise is the one intervention that can significantly improve T levels relative to other hormones, and it is much safer than any pharmaceutical intervention. The evidence tends to concur, high-intensity training (HIT) with weight resistance (body or free weight) is superior to endurance training, as HIT improves total T, free T, and human growth hormone (HGH) while reducing cortisol and sex hormone binding globulin (SHBG) (Goh, 2009; Grandys, 2009; Kraemer, 1989, 1991). Prolonged endurance training (cycling, marathon running, etc.) has shown less than desirable hormonal effects when compared to more anaerobic type exercise (Lac, 2000).

HIT with large compound movements including squats, bench press, deadlifts, rows, shoulder presses, and pull-ups elicit the most beneficial androgen response by increasing lactic acid, which activates Leydig cells' function in androgen production (Kraemer, 2005)

PHARMACEUTICAL-INDUCED SEXUAL DYSFUNCTION

Iatrogenic sexual dysfunction via pharmaceuticals is one lifestyle factor that can be managed quite effectively. The three major groups of drugs known to induce symptoms of sexual dysfunction are antidepressants, psychotropic drugs, and cardiovascular drugs. Antidepressants may be of little to no benefit in patients with mild to moderate depression (Fournier, 2010) and may even cause ED, anorgasmia and decreased libido (Brill, 2016; Werneke, 2006; Williams, 2006). Even common cardiovascular drugs such as beta-blockers cause ED in about 20% of men (Shabsigh, 2005), with diuretics and central alpha agonists also contributing to the issue (Düsing, 2005).

In the right situation and presentation there is no doubt that these drugs can be helpful. However, clinically we see these drugs being given to individuals who can easily rectify their depression or hypertension with exercise and lifestyle modifications.

Furthermore, clinicians should be more aware of pharmaceutical-induced sexual dysfunction (Conaglen, 2013). It is prudent (and also cost-effective) to consider exercise as a first-line therapy prior to a pharmaceutical intervention, as exercise can change or reduce a medication's use and mitigate these adverse effects.

CARDIOMETABOLIC DYSFUNCTION

Of all the diverse comorbidity risks with sexual dysfunction, cardiometabolic disease is most strongly associated (Chew, 2010; Conaglen, 2013; Lee, 2011). Cardiometabolic dysfunction is a mix of one or more of the following: diabetes, metabolic syndrome, obesity, and cardiovascular disease. Endothelium dysfunction, smooth muscle dysfunction, hormonal imbalance, and poor autonomic regulation are compromised, which may present clinically as hypertension, atherosclerosis, and the oft overlooked symptom of sexual dysfunction. As noted already, ED is a well-known indicator of cardiovascular disease (Montorsi, 2003).

Exercise addresses each of the four major contributors to cardiometabolic dysfunction; exercise enhances endothelial function by improving NO synthesis, blood pressure, and vascular remodeling induction (Joyner, 2009; Lu, 2016; Whelton, 2002); it improves autonomic vagal tone, thus improving heart rate variability (Lu, 2016); and it improves insulin sensitivity, glucose metabolism, and weight loss, thus improving hormonal balance (Thomas, 2006).

PSYCHOEMOTIONAL

Anxiety is the patient's anticipation of stress combined with his perception of the stressor. Sexual dysfunction has a strong connection to anxiety and depression, with both sexual dysfunction and psychoemotional stress being both a cause and a result of one another (Rosen, 2008; Tan, 2012). Clinically, this is evident more so in younger men who suffer with performance anxiety resulting in inability to perform in the bedroom. A one-time experience of ED or decreased sex drive may further worsen future experiences and ultimately exacerbate this cycle of sexual dysfunction and lead to avoiding intimacy altogether (Rosen, 2008).

When men and women lose sexual confidence, consistent exercise can be a significant tool to improve self-esteem, thus making intimacy and the sexual experience much more enjoyable (Lyons, 2015).

Female Sexual Dysfunction

FSD is a very frequent health problem and is observed in 25% to 92% of women (Laumann, 1999). However, unlike most men, women with sexual dysfunction may also experience pain with intercourse (dyspareunia), inability to orgasm (anorgasmia), and decreased arousal and desire. In women, sexual dysfunction risk factors include age, education, chronic diseases, pregnancy, and parity (Aslan, 2005; Çayan, 2004; Laumann, 1999).

Major causes of FSD may be based solely upon the fluctuations in hormones during various times in women's lives (e.g., pregnancy, breastfeeding, menopause). Surgical procedures may negatively influence sexual function as hysterectomies can lead to significant reductions in sex hormones. Other important factors in FSD include psychoemotional issues, unsatisfying interpersonal relationships, and even negative sociocultural influences, all of which can adversely affect one another. See Table 11.1 for the major causes of sexual dysfunction in women. As with men, androgen deficiency also plays an important role in women. Testosterone impacts desire, pleasure, and arousal and has been linked to hypoactive sexual desire (Nappi, 2009). Although the research tends to suggest that T is not solely responsible, its contribution is significant. Estrogen levels within the physiological range in menopausal women also aid in improving desire (Cappelletti, 2016). All

Table 11.1. Major Causes of Sexual Dysfunction in Women

Psychological	*Physical*	*Interpersonal Relationship*	*Sociocultural Influences*
Stress from work or family	Hormonal changes (pregnancy and menopause	Partner technique	Inadequate education
Sexual performance anxiety	Diabetes	Partner performance	Conflict w/ religious values
Depression/anxiety	Heart, liver, or kidney disease	Lack of privacy	Conflict w/ family values
Sexual orientation issues	Thyroid disease	Relationship conflict	Conflict w/ personal values
Traumatic sexual or physical experience	Neurological disorders	Relationship quality	Societal taboos
Body image and self-esteem problems	Pelvic injury		
	Medications, alcohol, or drugs		
	Fatigue		

Source: Elder (2010).

sex hormones (e.g., testosterone, dehydroepiandrosterone, estrogen, and progesterone) play a role and require relative balance for healthy sexual function. Given that the endocrine system communicates via multiple hormones systems (thyroid, adrenals), it would be prudent to assume balancing the endocrine system is key in assessing sexual function.

MENOPAUSE

Fluctuations in progesterone and estrogen in perimenopausal and postmenopausal women can explain multiple causes of FSD. Low estrogen, even in the low-normal range, may result in urogenital atrophy along with decreased lubrication, vaginal tone, elasticity, and genital engorgement.

The inadequate hormonal presentation can result in sexual pain including dyspareunia, vulvodynia, vaginal bleeding, pelvic floor pain, and consequently decreased sexual arousal and desire (Walsh, 2004). These negative sexual experiences may have a psychological effect, which can then create a mental block resulting in anorgasmia and reduced pleasure. This creates a cascade of further reducing sexual desire and arousal thus contributing to FSD and avoidance of intimacy.

PREGNANCY AND CHILDBIRTH

Several factors most common during pregnancy and postbirth have been noted as causes of FSD, such as maternal duties, sleep disorders, weakening of pelvic muscles, psychosocial changes, fatigue, changes in mental imaging of the body, and breastfeeding (Ahmadi, 2010; Nikpour, 2006). Breastfeeding lowers both estrogen and T while also increasing prolactin creating a hormonal environment that results in low libido. This decreased sexual desire can last for months even after nursing is stopped. Weakening of the pelvic floor muscles, especially true for vaginal deliveries, is likely the major cause of sexual dysfunction in postpartum women (Citak, 2010). Pelvic pain, especially pain with intercourse, is a sign of weakened pelvic floor muscles, with other signs such as urinary incontinence, urgency, incomplete bladder and bowel emptying, and heaviness in the perineal area.

MEDICATIONS

As with men, drug-induced sexual dysfunction may play a significant role in FSD. There are select classes of drugs that have been linked particularly to FSD including antihypertensives, antidepressants, anxiolytics, and illicit drugs. Exercise can be utilized in treating hypertension (Börjesson, 2016), depression (Schuch, 2015), and anxiety (Bartley, 2013; Wegner, 2014), which may allow removing or decreasing the medication dose, ultimately mitigating any possible pharmaceutical-induced sexual issues.

PSYCHOLOGICAL CAUSES OF FSD

The psychophysiology of the female sexual response was first described by Masters and Johnson as excitement, plateau, orgasm, and resolution (Masters, 1966). The newer circular model of female sexual response emphasizes emotional intimacy with both emotional and physical arousal (Kammerer-Doak, 2008). Multiple studies have clarified the strong relationship among depression/anxiety with arousal disorder, desire disorder, orgasmic dysfunction, and sexual pain disorder (Bradford, 2006; Cyranowski, 2004; Kuffel, 2006; Oliveira, 2013).

Exercise shines in tackling a variety of psychoemotional and physiological contributors to sexual problems, boosting mood, energy, and confidence, all of which can profoundly improve mind, body, and sexual function.

EXERCISE PRESCRIPTION

Exercise should be a first-line therapy in sexual dysfunction for men and women. Specific exercise recommendations that address the area of the body where the sexual dysfunction is occurring (i.e., the perineal area) and the ultimate cause (i.e., cardiovascular disease, menopause, diabetes) can be profoundly therapeutic.

Exercise can provide an immediate positive impact on the psychological causes of sexual dysfunction. Consistent exercise will also help to lower blood lipids, improve insulin resistance, decrease fasting blood sugar, and balance hormones, while also improving overall cardiovascular health. It is no secret that the majority of individuals that suffer from sexual dysfunction also tend to have an underlying physiological condition closely associated with poor lifestyle habits.

Male-Specific Studies

Specifically for men, exercise improves vascular endothelial function by increasing NO, enabling stronger erections. Lifestyle modifications that include physical activity and penile-focused exercise, like Kegels, should be part of any program geared to treating ED (Belardinelli, 2005; Meldrum, 2012). It is known that physical activity helps to alleviate ED and low libido and improve sexual satisfaction (Meldrum, 2012). In healthy men with increased cardiovascular risk, high-intensity exercise and stable relationships were associated with decreased risk of ED (Ettala, 2014). Additionally, Belardinelli (2005) showed that short-term moderate exercise on a cycle ergometer in heart failure patients also improved sexual function.

Niskanen (2004) conveyed that rapid weight loss via a low-calorie diet can be therapeutic in improving T and attenuating SHBG. Although exercise was not a factor in this study, it questions which improves T levels best, reduced energy intake or physical activity? Studies demonstrate increased aerobic physical activity has a greater effect than caloric restriction in inducing endogenous T production (Kumagai, 2016).

It is evident that any physical activity that fits the American College of Sports Medicine guidelines of 30 minutes a day most days has the potential to increase T levels (Hayes, 2013a, 2013b). But what is the effect seen with resistance training, and is it better than aerobic exercise?

Moradi (2015) showed that resistance training for 12 weeks in obese young men caused increases in both adiponectin and T levels. Levels of adiponectin are inversely correlated with body fat percentage in adults and closely associated with energy regulation in the body. Although more study is needed to determine which form of exercise is better for increasing T levels, the majority of evidence supports resistance training (Grandys, 2009; Perna, 1995).

Knowing that inactivity contributes to weight gain and worsening of ED, it is logical that losing extra fat improves endogenous T levels and sexual desire and most certainly ED (La Vignera, 2012).

Female-Specific Studies

Low libido in women is significantly correlated with waning T levels (Nappi, 2009). Marx (2001) showed that consistent resistance training in women for at least 12 weeks improves T levels and overall muscular performance with periodized high-volume multiple-sets workouts being better than low-volume, single-set circuit routines (Belardinelli, 2005).

Many disease states, and to some extent their associated treatments, have the potential to cause sexual dysfunction in women. Theses disease may include but are not limited to the following: breast cancer, polycystic ovary syndrome, multiple sclerosis, kidney disease, diabetes, depression and anxiety, and cardiovascular disease. Sexual dysfunction as a result of these diseases can benefit from perineal or vaginal-specific exercises, Kegels, yoga, and resistance training (Juraskova, 2013; Lara, 2015; Najafidoulatabad, 2014).

Exercise Specifics

The most significant consideration for the healthcare provider when prescribing exercise is consistency. Consistency can only be maintained by fitting the exercise regimens to the unique needs of the patient. Limitations must be expected as many patients will be limited due to injuries, fitness level, or even lack of enjoyment in exercise.

The healthcare provider should discuss the importance of exercising with regularity by employing a concept called structured flexibility. A rigid approach to prescribing, and fulfilling, an exercise regimen—that it has to be done in the same place, in the same way, or at only certain times, all the time—often is not possible for many reasons; life will always throw up obstacles that may hinder consistency with physical activity. Structured flexibility will give patients a psychological advantage in maintaining consistency with exercise by allowing them to skip sessions when needed, without feelings of guilt or failure. Giving patients alternative exercise regimens that fit their busy schedules is very helpful, including home workouts, or one-minute exercise bursts spread throughout the day.

HIGH-INTENSITY, MODERATE-INTENSITY, LOW-INTENSITY EXERCISE

Intensity is probably the most important consideration when recommending exercise. An exercise regimen that is too intense or not intense enough will compromise compliance and safety. A simple way to convey intensity to the patient is by using a modified Borg Scale, called the Metabolic Effect's Rate of Perceived Exertion or ME Intensity Scale (Table 11.2). Most people can get an idea of their intensity level during exercise based on how difficult it is to talk. This does generally correspond to percentage of maximum heart rate for most people.

High-intensity exercise is generally shorter in duration and has more exercise density (more exercises or repetitions are done per unit time) than either moderate intensity or low intensity. Higher intensity exercise may also be interspersed with short periods of rest (e.g., doing sprints or doing a set of squats) as opposed to long-duration jogging. This way of exercising tends to be more anabolic and better for fat loss. Think about the differences in body types of an sprinter and long-distance runner. The sprinter creates a more anabolic hormonal environment whereas the endurance runner creates a more catabolic hormonal milieu.

The stereotypical low-intensity exercise that is gaining traction is restorative yoga. The other is walking; both are easily implemented by most people. Restorative yoga, for many, may require finding a yoga studio, getting there, and scheduling. Yet walking can be done almost anytime and anywhere.

There really is not an upper limit of exercise-induced health benefits to exercise, as even the highest and lowest volume of moderate and vigorous intensity exercise reduces all-cause mortality and cardiovascular disease compared to a sedentary lifestyle (Eijsvogels, 2016), suggesting the more exercise one does, the better. The more an exercise routine fits an individual's unique needs and fitness level, the more consistent he or she will be and the easier it will be to fit it into a daily routine.

AEROBIC VERSUS ANAEROBIC

Aerobic exercise tends to benefit the cardiovascular system more than anaerobic exercise; whereas anaerobic exercise tends to benefit strength, fat loss, and muscle gain more. Aerobic exercise, as the name implies, depends on the oxygen energy-generating process in the body, whereas anaerobic exercise depends more on the non-oxygen energy-generating processes.

Table 11.2. Metabolic Effect's Rate of Perceived Exertion Scale

ME Intensity Scale (0–4)					
0	No activity at all, resting, meditation	Can talk with ease	Resting HR	No Intensity	Can be done long duration
1	Active rest, walking, restorative yoga, Tai Chi	Can talk with ease	Slightly elevated HR	Low Intensity	Can be done long duration
2	Fast walking, jogging, lifting weights	Talking is slightly difficult but can still be done	About 60% Max HR	Moderate Intensity	Can be done long duration
3	Fast jogging, running, cross-fit,	Difficult to talk	70%–85% Max HR	High Intensity	Cannot be sustained for more than 10–45 min
4	Sprinting, rest-based training	Need to stop and rest between bouts of exercise	85% Max HR	High Intensity	Cannot be sustained for more than 1–2 minutes

Note: HR = heart rate.

However, the whole notion of aerobic exercise and anaerobic exercise being two completely different things is a bit of a misnomer. The body is always using the aerobic energy generating systems whether one is sprinting or jogging or even sleeping. In other words, the proportion of which energy system one is using depends on the intensity of the exercise, which in turn helps determine the duration of exercise bout.

Knowing that both aerobic and anaerobic exercises benefit the body differently, it inclines us to use both types within one exercise session instead of doing primarily aerobic exercise one day and then resistance training the next.

EXERCISING BIG MUSCLES VERSUS SMALL MUSCLES

The hormonal cascade of effects from exercising big muscle groups as compared to small or single muscle groups is analogous to causing an avalanche or throwing a snowball. The difference in benefit is huge in terms of cardiovascular health, strength, lean mass gains, and weight loss. Doing just one set of lunges or squats will elevate heart rate quickly because muscles of the whole lower body are engaged and working. When polyset workouts are utilized, as they are in conventional body-building workouts, the exercise session will benefit the cardiovascular system and create a hormonal cascade that is more anabolic.

Resistance exercises that utilize big muscle groups (e.g., chest press or squats) as opposed to smaller muscles groups (e.g., tricep extension or leg extension) cause larger relative increases in anabolic hormones like T and HGH.

Furthermore, the more muscles that are exercised, the greater the potential there is to improve insulin sensitivity, as these muscles utilize glucose better and are more likely to halt and possibly reverse DMII.

REST-BASED TRAINING

Rest and exercise are complimentary and dependent upon one another. Rest is the single biggest determinant of exercise intensity. Without intense workouts, fitness results will be lacking, and, without rest, intensity will be compromised. Consider an Olympic sprinter. With no rest between races, the sprinter could not physically or mentally muster the same effort. By focusing attention on rest during training, it is possible to reset both mental and physical capacity to achieve more.

THE SYNERGY OF LOW-INTENSITY AND HIGH-INTENSITY EXERCISE

Low-intensity exercise (e.g., walking or restorative yoga) and high-intensity exercise (e.g., weight training or sprinting) are synergistic in their actions, just like sleep and activity. In order for the body to recover properly, age well, build muscle, burn fat, and optimize sexual function, it needs the correct balance of stress hormones (cortisol, adrenaline, noradrenaline) and growth-promoting hormones (T and HGH).

In the days and hours following an intense workout, it is most beneficial to maintain activity levels with low-intensity exercise in order to optimize stress hormones. Higher intensity exercise most days of the week is not sustainable for most and can lead to injuries, noncompliance with daily activity, and fatigue. This chronically elevates stress hormones, which is counterproductive for sexual dysfunction.

Yoga and walking help lower cortisol levels and perceived stress (Hill, 2008; Thirthalli, 2013; Wardwell, 2013). This effect appears to be magnified when the activity takes place in a natural setting—what the Japanese call Shinrin-yoku (Park, 2007). Walking in nature also has been shown to offer the added prosexual benefit of decreasing depressive rumination and lifting mood, which walking in an urban setting did not (Bratman, 2015). Other forms of low-intensity exercise are Taiji (ta'i chi) and Qigong (forms of exercise from traditional Chinese medicine [TCM]), both of which have been shown to lower perceived stress.

YOGA

Yoga, a restorative exercise, appears to enhance sexual performance in men and women (Dhikav, 2010; Sengupta, 2013). Yoga is effective for reducing anxiety and depression and improving core muscles, all effects that are useful for mitigating symptoms of sexual dysfunction. There is a long tradition of specific yoga poses for specific medical ailments, including for women's health (e.g., menstrual pain, infertility) and men's health (e.g., impotence, prostate problems; Iyengar, 2001).

Some yoga poses may help to improve blood flow and tone to the perineal area, such as Yoga Mudra, Lotus, and Moola (Mula) Bandha poses (Sengupta, 2013):

1. **Lotus Pose:** Is a cross-legged position in which the feet are placed on the opposite thighs. (see https://www.youtube.com/watch?v=S8WxwDmDvBI).
2. **Yoga Mudra Pose**: Seated in Lotus pose put both hands at your back. Grasp your left-hand wrist with your right hand and make a fist with

the left hand. Let your hands relax loosely on your back. Now bend forward while exhaling until your head touches the ground. If Lotus pose is too difficult, do this pose starting in regular cross-legged position. (See https://www.youtube.com/watch?v=qWp3fgZWq1c.)

3. **Moola/Mula Bandha**: Is also known as Root Lock and is similar to doing Kegels. Sitting in Lotus position or in a regular cross-legged position, imagine that you have to pee. The perineal muscles that you use to stop the flow of urine by contracting them is the Moola Bandha region. These are the muscles between you anus and genitals. (See https://www.youtube.com/watch?v=flWBFikeGAA.)

TRADITIONAL CHINESE MEDICINE: YIN AND YANG

TCM has its roots in one of the oldest medical systems known to humans, one in which yin (dark; female) and yang (bright; male) describe the complementarity and interconnectedness of seemingly opposite or contrary forces and how they interact dynamically to continually give rise to each other. This philosophical framework also is expressed in forms of exercise that exemplify these principles. TCM Qigong (chi kung) and Taiji (t'ai chi) are two forms of gentle TCM exercise that can improve overall health by creating a balance between resistance or force and fluidity and between the mind and the body. Now studies have shown these therapies to reduce cortisol levels (Chan, 2006; Lee, 2003), reduce stress (Griffith, 2008; Rogers, 2009), improve anxiety and depression (Chow, 2007), and improve heart rate variability (Fong, 2015), improving overall quality of life and well-being.

THERAPEUTIC PELVIC FLOOR EXERCISES AND SEXUAL DYSFUNCTION: IMPLICATIONS FOR MEN AND WOMEN

Pelvic floor muscles, in men and women, respond to neurological, endocrine, and cardiovascular input. Penile erection, ejaculation, and orgasm require autonomic and somatic innervation of the pelvic floor muscles, penile erectile tissue, arterial and venous structures, along with various hormonal influences. The pelvic floor muscles bulbocavernosus and ischiocavernosus contribute to the male sexual response (Rosenbaum, 2007). In fact, several studies have shown pelvic floor dysfunction as a major contributor to ED and anorgasmia, with improvements seen with Kegel exercises, thus preventing the need for pharmacological interventions (Claes, 1993; Dorey, 2004; Van Kampen, 2003). Kegel exercises, if done correctly, can be of great benefit for anyone with sexual dysfunction and urinary incontinence, especially following radical prostatectomy (Park, 2012).

The strength and tonicity of the female pelvic floor muscles have a significant role in female arousal, orgasm, lubrication, and vaginal tone. The contraction and tonicity of the pelvic muscles enhance sexual pleasure for both the male and female during coitus (Chambless, 1984; Shafik, 2000). In 1952, Dr. Kegel first identified the role of the pelvic floor muscles during sex and exposed hypotonic pelvic floor muscles as a cause of sexual dysfunction (Rosenbaum, 2007). Since 1952 multiple studies have identified Kegel exercises to improve vaginal tone, lubrication, female orgasm, and arousal in populations ranging from nonparous to multiparous women (Golmakani, 2015; Phillips, 2000).

Just as hypotonic pelvic floor muscles can contribute to sexual dysfunction, so too can hypertonic muscles. Vaginismus, dyspareunia, pelvic pain, and other sexual pain disorders have also seen significant improvement with pelvic floor exercises combined with biofeedback (Bergeron, 2001; Glazer, 1995; McKay, 2001), as these therapies have been shown to help relax the pelvic floor and reduce hypertonicity. Male chronic pelvic pain syndrome, often dismissed as chronic prostatitis, can also be attributed to hypertonic pelvic floor muscle, which too can contribute to ED, premature ejaculation, and anorgasmia, with biofeedback pelvic floor exercises showing improvement (Cornel, 2005; Ye, 2003).

HOW TO INCORPORATE KEGEL EXERCISES

Kegel exercises can be done anywhere or anytime and do not require any equipment.

To learn how to do Kegels try the following:

1. During urination, try to stop or at least slow down the flow of urine. If you can do this then you have successfully isolated the pubococcygeus muscles.
2. Try not to tense the muscles in legs, abdomen, or buttocks, and try not to hold your breath.

General recommendations for doing Kegels are the following:

1. Contract the pubococcygeus muscles to a count of 5 to 10 seconds and release the muscle to a count of 5 to10 seconds.
2. Can be done sitting, standing, or lying down.
3. Do at least one set of 10, three to five times per day every day.
4. Best if done daily.

POSSIBLE EXERCISE CONTRAINDICATIONS

Knowing that almost any type of exercise will give the body some benefit does not necessarily mean it will give benefit to the unique health needs of the sexual dysfunction patient. An example of a possible exercise contraindication for any man with ED or possibly a women with perineal pain is prolonged bicycling. Bicycling is one of the major risk factors for ED in healthy men, with incidence of 13% to 24% (Gulino, 2010), which may increase pressure on the perineal area, possibly compromising nerve function.

Exercise Specifics for Male and Female Patients

Any high-intensity exercise greater than 30 minutes will raise cortisol, yet this same exercise but shorter than 30 minutes also raises HGH and T. These hormones together promote fat loss and maintain muscle, whereas long-duration exercise fails to increase HGH and T while still elevating cortisol. This is especially true when adequate rest is not taken between workouts. Excess cortisol in the aging man contributes to weight gain, fatigue, and depression, all of which contribute to sexual dysfunction. For this reason, short-duration intense exercise should be part of most weekly exercise regimens combined with active rest like walking and yoga.

Also, combining rest into the intense exercise, as in rest-based training, makes the exercise more productive. "Go hard and rest hard" rest should be the most important component of high-intensity workouts. Individuals should rest as much as they need to go intense again while keeping the workout to between 10 to 30 minutes. Consistency is most important. If the exercise program is too intense, is too difficult, or does not take into account personalized needs (injury, gym access, etc.), then its consistency will be compromised.

General weekly guidelines for exercise specific for sexual dysfunction are as follows:

1. Daily rest-based activity—Walk daily (30 to 60 min) preferably in a nature setting. Yoga, Taiji, and Qigong can all be substitutes for walking.
2. High-intensity rest-based training—One to two times per week for no more than 30 minutes. This can include intense walking, sprinting, weight training or resistance bodyweight training. Exercises should primarily be multijoint movements that engage big muscle groups. Hybrid movements, using a combination of lower and upper body exercises, can significantly add to the intensity of the workout.
3. Kegels—5 to 10 minutes daily.

Next we describe a home workout and a workout using hybrid movements with dumbbells.

ASSISTED HOME BODYWEIGHT WORKOUT

Do three sets of 10 to 15 reps of each exercise or as many rounds as you can for 30 minutes. Do two to three times per week. Rest when you need to do and then start where you left off.

1. **Assisted squat**: Find a sturdy structure like a railing, door frame, or other suitably object to hold onto. While standing, set your feet about shoulder width apart. Look straight ahead or slightly up to help your spine stay relaxed and straight. Start squatting as if you were about to sit down while pushing your hips down and back, like you are putting your buttocks into the corner of a chair. Descend as far as you can with your back relaxed and straight and then stand back up pushing your feet into the floor. Use the support to ensure proper form throughout the entire squat. For a visual on this, see https://www.youtube.com/watch?v=OuR_Fp7ABoc.
2. **Assisted lunge:** Find a sturdy structure like a railing, door frame, or a chair or table. Keep your feet pointed forward throughout the entire movement. Begin the lunge by standing with your feet shoulder-width apart and then take an exaggerated step forward. When the front foot reaches a desired distance in front of you, begin to flex or bend the back knee while lowering your body toward the floor. Your front knee also has to bend to do this. Lower yourself to the floor until the back knee is just inches above the floor. To finish the lunge push back on your front foot in order to come back into standing position. Repeat with the other leg. Keep you back relaxed and straight through the entire movement. For a visual on this, see https://www.youtube.com/watch?v=jPftZUx_AZo.
3. **Glute bridge:** Lie on the floor with your knees bent and feet flat on the floor. Push your hips into the air toward the ceiling using your buttocks muscles. Squeeze your glutes or butt muscles at the peak of the movement. For a visual on this, see https://www.youtube.com/watch?v=8bbE64NuDTU.
4. **Back extensions**: Lie on the floor face down. You can put your arms in the following positions from easiest to most difficult: (a) at your sides, (b) perpendicular to your body, or (c) stretched out in front of you, or (d) put your hands on the back of your head. Now lift your arms and legs and head off the floor as far as you can by contracting all of your back muscles. Repeat. For a visual on this, see https://www.youtube.com/watch?v=Ikq3OCFoP7c.

5. **Assisted push-ups or regular push-ups:** Find a sturdy wall, bench, countertop, or bed. Doing an assisted or elevated push-up should be progressed to a regular push-up off the floor. From easiest to most difficult the assisted push-up can be done (a) off a wall (b) off a chair, bed, or countertop (the higher the height of structure you are doing it off of, the easier it is); or (c) from your knees. This is an exercise done with your face, palms, and toes facing toward the structure you are doing the push-up off of. Keep your legs and back straight and whole body in line, like a plank. Start with you arms extended straight and lower your body down until you chest touches the structure you are doing the push-ups off of. If you cannot lower you body all the way down then regress the exercise to a structure that is more elevated in order to make it easier. Push-ups are a great exercise that works the chest, shoulders, and arms. For a visual on this, see https://www.youtube.com/watch?v=IzsNqEP49ZM.

DUMBBELL WORKOUT

Do three sets of 10 to 15 reps of each exercise or as many rounds as you can for 20 to 30 minutes. Do two to three times a week.

1. **Dumbbell squat press:** Standing with dumbbells in each hand holding them at your shoulders with palms facing forward or toward your ears, do a squat, as explained in the "Assisted Home Bodyweight Workout" instructions. Once you go put to a standing position, press the dumbbells over your head. For a visual on this, see https://www.youtube.com/watch?v=MJa0907ROso.
2. **Dumbbell lunge curl:** Standing with dumbbells in each hand, holding them at your waist, do a lunge as explained instructions for "Assisted Home Bodyweight Workout." Once in the lunge position, do a biceps curl by bending your elbows or flexing your biceps so that the dumbbells come toward your shoulders. You can also do the curl simultaneously while your are lunging. Lower the dumbbell and then push back off of your front foot and come back to standing position. Switch legs and do it again. For a visual on this, see https://www.youtube.com/watch?v=VylJxgvj_Bo.
3. **Dumbbell push-up row:** Get in push-up position as explained in the "Assisted Home Bodyweight Workout" instructions. Have dumbbells in both hands on the floor. You can do this from your knees or from your toes (regular push-ups). Do a push-up (lower chest to the floor and push-up off of the dumbbells until you come back to plank position of the top of the push-up). Now raise one of the dumbbells to your ribs

and lower it to floor. Now raise the other dumbbell and lower it to the floor. You can also do a push-up between every row instead of doing an row in each arm before doing a push-up. Either way is good. Do another push-up and repeat. For a visual on this, see https://www.youtube.com/watch?v=gNdDIB90MVg.

4. **Bent over dumbbell row:** Standing with a dumbbell in each hand (palms facing your body), bend your knees slightly and bring your torso forward by bending at the waist. As you bring your torso forward be sure to keep your back straight until it is almost parallel to the floor. You may keep the head up or in line with your spine. The weights should be hanging in each arm above the floor. Stay in this position and bring both dumbbells to your ribs. Lower the dumbbells and repeat. For a visual on this, see https://www.youtube.com/watch?v=LktGPg-AkvY.
5. **Glute bridge with dumbbell:** Lie on the floor or use a bench with your shoulders on it. Your knees should be bent and feet flat on the floor. Lay a dumbbell or other weight across your hip line. If doing it off of the floor, raise your hips to the ceiling. If doing off of a bench, lower your hips to the floor or as low as you can and then push your hips to the ceiling. Repeat. For a visual on this, see https://www.youtube.com/watch?v=wlP-1qW3hog.

Putting It Together: Precise and Personalized Exercise

Exercise is a cost-effective, significant therapeutic tool in the treatment of male and female sexual dysfunction. The fitness community may always debate the best exercise, but, as set out in this chapter, the balance of exercise with a personalized approach including short-duration HIT and avoiding long-duration endurance training while incorporating active rest exercises can significantly mitigate sexual dysfunction, psychologically, neurologically and with respect to improved cardiovascular health.

It cannot be emphasized enough that the prescribed exercise *must* fit the patient. Consistency and participation is the key in any treatment plan regardless of how convincing the evidence for one exercise strategy or another, because without compliance the prescription has no way of influencing improvements. The best exercise is the one an individual can do on a consistent basis.

It also is crucial to not be sedentary, which decreases fitness, promotes weight gain, and negatively impacts overall health and sexual health. We now know that sitting for prolonged periods of time is associated with increased metabolic age and increased all-cause mortality (Rezende, 2016; Shadyab, 2017), effects that even regular exercise may not be able to override (Mercola, 2015). As emphasized in this chapter and others in this volume, multiple underlying medical conditions are associated both with poorer health status as we age and also with sexual dysfunction. Major components of any plan to improve general and mental health, and sexual function and fulfillment, should involve both a program of well-designed exercise and avoiding sedentary habits.

Often individuals are indoctrinated by societal fitness mottos such as "more is better" or "go hard or go home." We are led to believe we must work out as hard as possible for as long as possible to achieve optimal results, and then we wonder why most Americans give up their New Year's exercise resolutions by February. The goal is to assist the practitioner in developing a sustainable exercise program for the patient, to improve general health, mitigate specific medical issues, and provide therapeutic exercises to reverse, treat, and prevent sexual dysfunction.

REFERENCES

Ahmadi Shirvani M, Nesami M, Bavand M (2010). Maternal sexuality after child birth among Iranian women. *Pak J Biol Sci.* 13:385–389.

Andersson K-E (2011). Mechanisms of penile erection and basis for pharmacological treatment of erectile dysfunction. *Pharmacol Rev.* 63:811–859.

Araujo AB, Wittert GA (2011). Endocrinology of the aging male. *Best Pract Res Clin Endocrinol Metab.* 25:303–319.

Aslan G, Aslan D, Kizilyar A, Ispahi C, Esen A. (2005). A prospective analysis of sexual functions during pregnancy. *Int J Impot Res.* 17:154–157.

Bartley CA, Hay M, Bloch MH (2013). Meta-analysis: aerobic exercise for the treatment of anxiety disorders. *Prog Neuropsychopharmacol Biol Psychiatry.* 45:34–39.

Belardinelli R, Lacalaprice F, Faccenda E, Purcaro A, Perna G. (2005). Effects of short-term moderate exercise training on sexual function in male patients with chronic stable heart failure. *Int J Cardiol.* May 11;101(1):83–90.

Bergeron S, Binik YM, Khalifé S, et al. (2001). A randomized comparison of group cognitive-behavioral therapy, surface electromyographic biofeedback, and vestibulectomy in the treatment of dyspareunia resulting from vulvar vestibulitis. *Pain.* 91(3):297–306.

Bhasin S, Cunningham GR, Hayes FJ, et al. (2010). Testosterone therapy in men with androgen deficiency syndromes: an Endocrine Society clinical practice guideline. *J Clin Endocrinol Metab.* 95:2536–2559.

Bradford A, Meston CM (2006). The impact of anxiety on sexual arousal in women. *Behav Res Ther.* 44(8):1067–1077.

Bratman GN, Hamilton JP, Hahn KS, Daily GC, Gross JJ (2015). Nature experience reduces rumination and subgenual prefrontal cortex activation. *Proc Natl Acad Sci U S A.* Jul 14;112(28):8567–8572.

Brill M (2016). Antidepressants and sexual dysfunction. *Fertil Steril.* 81:35–40.

Boloña ER, Uraga MV, Haddard RM, et al. (2007). Testosterone use in men with sexual dysfunction: a systematic review and meta-analysis of randomized placebo-controlled trials. *Mayo Clin Proc.* Jan;82(1):20–28.

Börjesson M, Onerup A, Lundqvist S, Dahlöf B (2016). Physical activity and exercise lower blood pressure in individuals with hypertension: narrative review of 27 RCTs. *Br J Sport Med.* Mar.;50(6):356–361.

Cappelletti M, Wallen K (2016). Increasing women's sexual desire: the comparative effectiveness of estrogens and androgens. *Horm Behav.* 78:178–193.

Çayan S, Akbay E, Boslu M, Canpolat B, Acar D (2004). The prevalence of female sexual dysfunction and potential risk factors that may impair sexual function in Turkish women. *Urol Int.* 72:52–57.

Chambless DL, Sultan FE, Stern TE, O'Neill C, Garrison S, Jackson A (1984). Effect of pubococcygeal exercise on coital orgasm in women. *J Consult Clin Psychol.* 52(1):114–118.

Chan CLW, Tso IF, Ho RTH, et al. (2006) Short communication: the effect of a one-hour Eastern stress management session on salivary cortisol. *Stress Health.* 22(1):45–49.

Chew K-K, Finn J, Stuckey B, et al. (2010). Erectile dysfunction as a predictor for subsequent atherosclerotic cardiovascular events: findings from a linked-data study. *J Sex Med.* 7(1 Pt 1):192–202.

Chow YWY, Tsang HWH (2007). Biopsychosocial effects of qigong as a mindful exercise for people with anxiety disorders: a speculative review. *J Altern Complement Med.* 13(8):831–840.

Citak N, Cam C, Arslan H, Karateke A, Tug N, Ayaz R, Celik C (2010). Postpartum sexual function of women and the effects of early pelvic floor muscle exercises. *Acta Obstet Gynecol Scand.* 89:817–822.

Claes H, Baert L (1993). Pelvic floor exercise versus surgery in the treatment of impotence. *Br J Urol.* 71(1):52–57.

Conaglen HM, Conaglen JV (2013). Drug-induced sexual dysfunction in men and women. *Aust Prescr.* 36(2):42–46.

Corona G, Rastrelli G, Maggi M (2013). Diagnosis and treatment of late-onset hypogonadism: systematic review and meta-analysis of TRT outcomes. *Best Pract Res Clin Endocrinol Metab.* 27:557–579.

Cormie P, Newton RU, Taaffe DR, et al. (2013). Exercise maintains sexual activity in men undergoing androgen suppression for prostate cancer: a randomized controlled trial. *Prostate Cancer Prostatic Dis.* 16(2):170–175.

Cornel EB, van Haarst EP, Schaarsberg RWMB-G, Geels J (2005). The effect of biofeedback physical therapy in men with chronic pelvic pain syndrome type III. *Eur Urol.* 47:607–611.

Cunningham GR, Stephens-Shields AJ, Rosen RC, et al. (2014). Association of sex hormones with sexual function, vitality, and physical function of symptomatic older men with low testosterone levels at baseline in the testosterone trials. *J Clin Endocrinol Metab.* 100(3):1146–1155.

Cyranowski JM, Bromberger J, Youk A, Matthews K, Kravitz HM, Powell LH (2004). Lifetime depression history and sexual function in women at midlife. *Arch Sex Behav*. 33(6):539–548.

Deenadayalu V, Puttabyatappa Y, Liu AT, Stallone JN, White RE (2012). Testosterone-induced relaxation of coronary arteries: activation of BKCa channels via the cGMP-dependent protein kinase. *Am J Physiol Heart Circ Physiol*. 302(1):H115–H123.

Dhikav V, Karmarkar G, Verma M, Gupta R, Gupta S, Mittal D, Anand K. (2010). Yoga in male sexual functioning: a noncompararive pilot study. *J Sex Med*. Oct;7(10):3460–3466.

Dorey G, Speakman M, Feneley R, Swinkels A, Dunn C, Ewings P (2004). Randomised controlled trial of pelvic floor muscle exercises and manometric biofeedback for erectile dysfunction. *Br J Gen Pract*. 54(508):819–825.

Düsing R (2005). Sexual dysfunction in male patients with hypertension: influence of antihypertensive drugs. *Drugs*. 65(6):773–786.

Eijsvogels TM, Molossi S, Lee DC, Emery MS, Thompson PD (2016). Exercise at the extremes: the amount of exercise to reduce cardiovascular events. *J Am Coll Cardiol*. Jan 26;67(3):316–329.

Elder JA, Braver Y (2010). Female sexual dysfunction. Cleveland Clinic Center for Continuing Education. http://www.clevelandclinicmeded.com/medicalpubs/diseasemanagement/womens-health/female-sexual-dysfunction/.

Ettala OO, Syvänen KT, Korhonen PE, Kaipia AJ, Vahlberg TJ, Bostrom PJ, Aarnio PT (2014). High-intensity physical activity, stable relationship, and high education level associate with decreasing risk of erectile dysfunction in 1,000 apparently healthy cardiovascular risk subjects. *J Sex Med*. Sep;11(9):2277–2284.

Feldman HA, Johannes C, Derby A, et al. (2000). Erectile dysfunction and coronary risk factors: prospective results from the massachusetts male aging study. *Prevent Med*. 30(4):328–338.

Fode M, Krogh-Jespersen S, Brackett NL, Ohl DA, Lynne CM, Sønksen J (2012). Male sexual dysfunction and infertility associated with neurological disorders. *Asian J Androl*. 14(1):61–68.

Fong SSM, Wong JYH, Chung LMY, et al. (2015). Changes in heart-rate variability of survivors of nasopharyngeal cancer during Tai Chi Qigong practice. *J Phys Ther Sci*. 27(5):1577–1579.

Fournier JC, DeRubeis RJ, Hollon SD, et al. (2010). Antidepressant drug effects and depression severity: a patient-level meta-analysis. *JAMA*. 303(1):47–53.

Glazer HI, Rodke G, Swencionis C, Hertz R, Young AW (1995). Treatment of vulvar vestibulitis syndrome with electromyographic biofeedback of pelvic floor musculature. *J Reprod Med*. 40:283–290.

Goh VHH, Tong TYY (2009). The association of exercise with sleep, sex steroids, sexual activities and aging in Asian men. *J Mens Health*. 6(4):317–324.

Golmakani N, Zare Z, Khadem N, Shareh H, Shakeri MT (2015). The effect of pelvic floor muscle exercises program on sexual self-efficacy in primiparous women after delivery. *Iran J Nurs Midwifery Res*. 20(3):347–353.

Grandys M, Majerczak J, Duda K, Zapart-Bukowska J, Kulpa J, Zoladz JA (2009). Endurance training of moderate intensity increases testosterone concentration in young, healthy men. *Int J Sports Med.* 30(7):489–495.

Griffith JM, Hasley JP, Liu H, Severn DG, Conner LH, Adler LE (2008). Qigong stress reduction in hospital staff. *J Altern Complement Med.* 14(8):939–945.

Gulino G, Sasso F, D'Onofrio A, et al. (2010). [Sport, infertility and erectile dysfunction]. *Urologia.* 77(2):100–106. [Article in Italian]

Hayes LD, Grace FM, Sculthorpe N, Baker JS, Herbert P, Kilduff LP (2013a). Does chronic exercise attenuate age-related physiological decline in males? *Res Sports Med.* 21(4):343–354.

Hayes LD, Grace FM, Sculthorpe N, Herbert P, Ratcliffe JWT, Kilduff LP, Bakeret JS (2013b). The effects of a formal exercise training programme on salivary hormone concentrations and body composition in previously sedentary aging men. *Springerplus.* Dec;2(1):18.

Hecht MJ, Neundörfer B, Kiesewetter F, Hilz MJ (2001). Neuropathy is a major contributing factor to diabetic erectile dysfunction. *Neurol Res.* 23(6):651–654.

Hill EE, Zack E, Battaglini C, Viru M, Viru A, Hackney AC. (2008). Exercise and circulating cortisol levels: the intensity threshold effect. *J Endocrinol Invest.* Jul;31(7):587–591.

Isidori AM, Giannetta E, Gianfrilli D, et al. (2005). Effects of testosterone on sexual function in men: results of a meta-analysis. *Clin Endocrinol (Oxf).* Oct; 63(4):381–94.

Iyengar BKS (2001). *Yoga: the path to holistic health.* London, New York: Dorling Kindersley.

Joyner MJ, Green DJ (2009). Exercise protects the cardiovascular system: effects beyond traditional risk factors. *J Physiol.* 587(Pt 23):5551–5558.

Juraskova I, Mireskandari S, Jarvis S, Mok K, Peate M, Meiser B, Friedlander M, Cheah BC (2013). The acceptability, feasibility, and efficacy (phase I/II study) of the overcome (olive oil, vaginal exercise, and moisturizer) intervention to improve dyspareunia and alleviate sexual problems in women with breast cancer. *J Sex Med.* Oct;10(10):2549–2558.

Kamenov ZA (2015). A comprehensive review of erectile dysfunction in men with diabetes. *Exp Clin Endocrinol Diabetes.* 123(3):141–158.

Kammerer-Doak D, Rogers RG (2008). Female sexual function and dysfunction. *Obstet Gynecol Clin North Am.* 35(2):169–183, vii.

Kapoor D, Aldred H, Clark S, Channer KS, Jones TH (2007). Clinical and biochemical assessment of hypogonadism in men with type 2 diabetes: correlations with bioavailable testosterone and visceral adiposity. *Diabetes Care.* 30(4):911–917.

Kraemer WJ, Patton JF, Knuttgen HG, et al. (1989). Hypothalamic-pituitary-adrenal responses to short-duration high-intensity cycle exercise. *J Appl Physiol.* 66(1):161–166.

Kraemer WJ, Gordon SE, Fleck SJ, et al. (1991). Endogenous anabolic hormonal and growth factor responses to heavy resistance exercise in males and females. *Int J Sports Med.* Apr;12(2):228–235.

Kraemer WJ, Ratamess NA (2005). Hormonal responses and adaptations to resistance exercise and training. *Sports Med.* 35:339–361.

Kuffel SW, Heiman JR (2006). Effects of depressive symptoms and experimentally adopted schemas on sexual arousal and affect in sexually healthy women. *Arch Sex Behav.* 35(2):163–177.

Kumagai H, Zempo-Miyaki, Yoshikawa T, Tsujimoto T, Tanaka K, Maeda S (2016). Increased physical activity has a greater effect than reduced energy intake on lifestyle modification-induced increases in testosterone. *J Clin Biochem Nutr.* Jan;58(1):84–89.

La Vignera S, Condorelli R, Vicari E, D'Agata R, Calogero AE (2012). Physical activity and erectile dysfunction in middle-aged men. *J Androl.* 33(2):154–161.

Lac G, Berthon P (2000). Changes in cortisol and testosterone levels and T/C ratio during an endurance competition and recovery. *J Sports Med Phys Fitness.* 40(2):139–144.

Lara LA, Ramos FK, Kogure GS, et al. (2015). Impact of physical resistance training on the sexual function of women with polycystic ovary syndrome. *J Sex Med.* Jul;12(7):1584–1590.

Laumann EO, Paik A, Rosen RC (1999). Sexual dysfunction in the United States: prevalence and predictors. *JAMA.* 281:537–544.

Lee JC, Bénard F, Carrier S, Talwar V, Defoy I (2011). Do men with mild erectile dysfunction have the same risk factors as the general erectile dysfunction clinical trial population? *BJU Int.* 107(6):956–960.

Lee MS, Lee MS, Kim HJ, Moon SR (2003). Qigong reduced blood pressure and catecholamine levels of patients with essential hypertension. *Int J Neurosci.* 113(12):1691–1701.

Leibovitch I, Mor Y. The vicious cycling: Bicycling related urogenital disorders. *Eur Urol.* 2005;47(3):277–286.

Lu D-Y, Yang AC, Cheng H-M, et al. (2016). Heart rate variability is associated with exercise capacity in patients with cardiac syndrome X. *PLoS One.* 11(1):e0144935.

Lyons K, Winters-Stone K, Bennett J, Beer T (2015). The effects of partnered exercise on physical intimacy in couples coping with prostate cancer. *Heal Psychol.* May;35(5):509–513.

Maio G, Saraeb S, Marchiori A (2010). Physical activity and PDE5 inhibitors in the treatment of erectile dysfunction: results of a randomized controlled study. *J Sex Med.* 7(6):2201–2208.

Marx JO, Ratamess NA, Nindl BC, et al. (2001). Low-volume circuit versus high-volume periodized resistance training in women. *Med Sci Sports Exerc.* Apr;33(4): 635–643.

Masters WH, Johnson VE (1966). *Human sexual response.* New York: Little, Brown.

McKay E, Kaufman RH, Doctor U, Berkova Z, Glazer H, Redko V (2001). Treating vulvar vestibulitis with electromyographic biofeedback of pelvic floor musculature. *J Reprod Med.* 46(4):337–342.

Meldrum DR, Gambone JC, Morris MA, Esposito K, Giugliano D, Ignarro LJ (2012). Lifestyle and metabolic approaches to maximizing erectile and vascular health. *Int J Impot Res.* Mar-Apr;24(2):61–68.

Mercola J (2015). Even if you exercise, sitting eight hours a day significantly increases your risk of dying from any cause. February 6. http://fitness.mercola.com/sites/fitness/archive/2015/02/06/effects-prolonged- sitting.aspx.

Montorsi P, Montorsi F, Schulman CC (2003). Is erectile dysfunction the "tip of the iceberg" of a systemic vascular disorder? *Eur Urol.* 44(3):352–354.

Moradi F (2015). Changes of serum adiponectin and testosterone concentrations following twelve weeks resistance training in obese young men. *Asian J Sports Med.* Dec;6(4):e23808.

Najafidoulatabad S, Mohebbi Z, Nooryan K (2014). Yoga effects on physical activity and sexual satisfaction among the Iranian women with multiple sclerosis: a randomized controlled trial. *Afr J Tradit Complement Altern Med.* Aug 23;11(5):78–82.

Nappi RE, Lachowsky M (2009). Menopause and sexuality: prevalence of symptoms and impact on quality of life. *Maturitas.* 63(2):138–141.

Nicolosi A, Glasser DB, Moreira ED, Villa M (2003). Prevalence of erectile dysfunction and associated factors among men without concomitant diseases: a population study. *Int J Impot Res.* 15(4):253–257.

Nikpour S, Javaheri I, Yadavar Nikravesh M, Jamshidi R (2006). Study of sexual problems resulting from delivery in primiparous women referred to outpatient clinics in west of Tehran. *Razi J Med Sci.* 13:189–196.

Niskanen L, Laaksonen DE, Punnonen K, Mustajoki P, Kaukua J, Rissanen A (2004). Changes in sex hormone-binding globulin and testosterone during weight loss and weight maintenance in abdominally obese men with the metabolic syndrome. *Diabetes Obes Metab.* May;6(3):208–215.

Oliveira C, Nobre PJ (2013). The role of trait-affect, depression, and anxiety in women with sexual dysfunction: a pilot study. *J Sex Marital Ther.* 39(5):436–452.

Park BJ, Tsunetsugu Y, Kasetani T, Kagawa T, Miyazaki Y (2007). Physiological effects of Shinrin-yoku (taking in the atmosphere of the forest)—using salivary cortisol and cerebral activity as indicators. *J Physiol Anthropol.* Mar;26(2):123–128.

Park SW, Kim TN, Nam J-K, et al. (2012). Recovery of overall exercise ability, quality of life, and continence after 12-week combined exercise intervention in elderly patients who underwent radical prostatectomy: a randomized controlled study. *Urology.* Aug;80(2):299–305.

Pegge NC, Twomey AM, Vaughton K, Gravenor MB, Ramsey MW, Price DE (2006). The role of endothelial dysfunction in the pathophysiology of erectile dysfunction in diabetes and in determining response to treatment. *Diabet Med.* 23(8):873–878.

Perna FM, McDowell SL (1995). Role of psychological stress in cortisol recovery from exhaustive exercise among elite athletes. *Int J Behav Med.* 2(1):13–26.

Phillips NA (2000). Female sexual dysfunction: evaluation and treatment. *Am Fam Physician.* 62(1):127–136, 141–142.

Puppo V, Puppo G (2016). Comprehensive review of the anatomy and physiology of male ejaculation: premature ejaculation is not a disease. *Clin Anat.* Jan;29(1):111–119.

Rajfer J (2000). Relationship between testosterone and erectile dysfunction. *Rev Urol.* 2:122–128.

Rezende LFN, Sá TH, Mielke GI, Viscondi JYK, Rey-López JP, Garcia LMT (2016). All- cause mortality attributable to sitting time: analysis of 54 countries worldwide. *Amer J Prevent Med.* Aug;51(2):253–263.

Rhoden EL, Telöken CJ, Sogari PR, Vargas Souto CA (2002). The relationship of serum testosterone to erectile function in normal aging men. *J Urol.* Apr;167(4):1745–1748.

Rogers C, Larkey LK, Keller C (2009). A review of clinical trials of tai chi and qigong in older adults. *West J Nurs Res.* 31(2):245–279.

Rosen RC, Althof S (2008). Impact of premature ejaculation: the psychological, quality of life, and sexual relationship consequence. *J Sex Med.* 5(6):1296–1307.

Rosenbaum TY (2007). Pelvic floor involvement in male and female sexual dysfunction and the role of pelvic floor rehabilitation in treatment: a literature review. *J Sex Med.* 4(1):4–13.

Schuch FB, Deslandes AC, Stubbs B, Gosmann NP, da Silva CTB (2015). Neurobiological effects of exercise on major depressive disorder: a systematic review. *Neurosci Biobehav Rev.* 61:1–11.

Selvin E, Burnett AL, Platz EA (2007). Prevalence and risk factors for erectile dysfunction in the US. *Am J Med.* 120(2):151–157.

Sengupta P, Chaudhuri P, Bhattacharya K (2013). Male reproductive health and yoga. *Int J Yoga.* Jul;6(2):87–95.

Shabsigh R, Perelman M, Lockhart DC, Lue TF, Broderick GA (2005). Health issues of men: prevalence and correlates of erectile dysfunction. *J Urol.* 174(2):662–667.

Shadyab AH, Macera CA, Shaffer RA, et al. (2017). Associations of accelerometer-measured and self-reported sedentary time with leukocyte telomere length in older women. *Am J Epidemiol.* 185(3):172–184.

Shafik A (2000). The role of the levator ani muscle in evacuation, sexual performance and pelvic floor disorders. *Int Urogynecol J Pelvic Floor Dysfunct.* 11(6):361–376.

Simon RM, Howard L, Zapata D, Frank J, Freedland SJ, Vidal AC (2015). The association of exercise with both erectile and sexual function in black and white men. *J Sex Med.* 12(5):1202–1210.

Sommer F, König D, Graf C, et al. (2001). Impotence and genital numbness in cyclists. *Int J Sports Med.* 22(6):410–413.

Tan HM, Tong SF, Ho CCK (2012). Men's health: sexual dysfunction, physical, and psychological health—Is there a link? *J Sex Med.* 9(3):663–671.

Traish AM (2014). Testosterone and weight loss: the evidence. *Curr Opin Endocrinol Diabetes Obes.* Oct;21(5): 313–322.

Traish AM, Guay AT (2006). Are androgens critical for penile erections in humans? Examining the clinical and preclinical evidence. *J Sex Med.* 3(3):382–404; discussion 404–407.

Travison TG, Morley JE, Araujo AB, O'Donnell AB, McKinlay JB (2006). The relationship between libido and testosterone levels in aging men. *J Clin Endocrinol Metab.* 91(7):2509–2513.

Thomas DE, Elliott EJ, Naughton GA (2006). Exercise for type 2 diabetes mellitus. *Cochrane Database Syst Rev.* 3(3):CD002968.

Thirthalli J, Naveen GH, Rao MG, Varambally S, Christopher R, Gangadhar BN (2013). Cortisol and antidepressant effects of yoga. *Indian J Psychiatry.* Jul;55 (Suppl 3):S405–408.

Ungureanu MC, Costache II, Preda C, Mogoş V, Vulpoi C Leuştean L (2015). Myths and controversies in hypogonadism treatment of aging males. *Rev Med Chir Soc Med Nat Iasi.* Apr-Jun;119(2):325–333.

Van Kampen M, De Weerdt W, Claes H, Feys H, De Maeyer M, Van Poppel H (2003). Treatment of erectile dysfunction by perineal exercise, electromyographic biofeedback, and electrical stimulation. *Phys Ther.* 83(6):536–543.

Verhovshek T, Cai Y, Osborne MC, Sengelaub DR (2010). Androgen regulates brain-derived neurotrophic factor in spinal motoneurons and their target musculature. *Endocrinology.* 151(1):253–261.

Walsh KE, Berman JR (2004). Sexual dysfunction in the older woman: an overview of the current understanding and management. *Drugs Aging.* 21(10):655–675.

Wardwell KK, Focht, BC, Devries CA, O'Connell AA, Buckworth J (2013). Affective responses to self-selected and imposed walking in inactive women with high stress: a pilot study. *J Sports Med Phys Fitness.* Dec;53(6):701–12.

Wegner M, Helmich I, Machado S, Nardi AE, Arias-Carrion O, Budde H (2014). Effects of exercise on anxiety and depression disorders: review of meta- analyses and neurobiological mechanisms. *CNS Neurol Disord Drug Targets.* 13(6):1002–1014.

Werneke U, Northey S, Bhugra D (2006). Antidepressants and sexual dysfunction. *Acta Psychiatr Scand.* 114(6):384–397.

Whelton SP, Chin A, Xin X, He J (2002). Effect of aerobic exercise on blood pressure: a meta-analysis of randomized, controlled trials. *Ann Intern Med.* 136(7):493–503.

Williams VSL, Baldwin DS, Hogue SL, Fehnel SE, Hollis KA, Edin HM (2006). Estimating the prevalence and impact of antidepressant-induced sexual dysfunction in 2 European countries: a cross-sectional patient survey. *J Clin Psychiatry.* 67(2):204–210.

Wu FCW, Tajar A, Beynon JM, et al. (2010). Identification of late-onset hypogonadism in middle-aged and elderly men. *N Engl J Med.* 363:123–135.

Ye Z-Q, Cai D, Lan R-Z, et al. (2003). Biofeedback therapy for chronic pelvic pain syndrome. *Asian J Androl.* 5(2):155–158.

12

The Obesity Epidemic and Sexual Health

STEVEN LAMM AND JONATHAN BEKISZ

Sexual Function

Sexual function, sexual performance, and sexual pleasure are neither precisely definable nor quantifiable. We cannot physiologically assess, other than by crude and limited parameters, the human physiologic sexual response. There is no test for sexual response that is equivalent to pulmonary function tests, cardiac output, ejection fraction, or glomerular filtration rate, which provide measures that are precise in assessing lung, heart, and kidney status. At present, we are forced to rely on validated questionnaires, such as the International Index of Erectile Function for men and the Female Sexual Function Index (FSFI) for women, to elicit information about libido, arousal, erection quality, lubrication, and orgasm when we attempt to glean an understanding of the interplay between human sexual function and experience (Rosen, 2000; Rosen, 1999). There is a general understanding, within the field of sexual research, that improved methods are necessary to further our understanding of this complex topic. Maseroli (2016) recognized this need for a deeper exploration of sexuality in a 2015 study published in the *Journal of Endocrinological Investigation*. This study attempted to improve the six-question version of the FSFI by creating the Female Sexual Dysfunction Index-6. Nonetheless, our ability to accurately evaluate measures relating to human sexual behavior remains primitive at best.

Moreover, gender differences complicate our understanding even further. The male sexual response is historically depicted as depending largely on a man's arousal and performance alone, while the female sexual response is portrayed as a function of both her own biology as well as the nature of her relationship with her sexual partner. Such stereotypes are obviously not universally applicable, and the degree to which they may hold any truth is unknown. Nonetheless, we remain aware that gender differences in sexuality exist and the unique elements of the male and female sexual responses prevent adequately characterizing, by proper sexual function, a singular definition (Petersen, 2010).

It is a widely accepted fact that the human sexual response requires the elegant coordination of multiple organ systems, including the vascular, endocrine, immunologic, central and peripheral nervous, and, of course, gynecologic and urologic systems. Given the precise integration of so many of the body's systems that are necessary for an individual to attain proper sexual function, it is reasonable to conceive that it could be easily and profoundly disrupted by many human conditions, whether physiologic, psychologic, or social in nature.

Obesity

Obesity is undoubtedly among the most important medical problems of our time. When attempting to understand obesity, it is critical to recognize that it is not due to a single physiologic disorder. Although it is widely accepted that people with a body mass index (BMI) of greater than 30 are obese, this parameter is often a poor measure of true obesity. Because variations in body composition are not accounted for in its calculation, BMI fails to differentiate an individual with a high BMI due to increased muscle mass from one whose BMI is due to an increased percentage of body fat.

What we do know about obesity is that it represents a pathologic increase in body mass that can develop at almost any age. Obesity may be further subdivided into three classes based on the severity of BMI elevation. Class I is defined as 30.0 to 34.9 kg/m^2, class II is defined as 35.0 to 39.9 kg/m^2, and class III, also known as extreme or severe obesity (Bray, 2015), is defined as any index greater than or equal to 40.0 kg/m^2.

As the prevalence of obesity rises at an alarmingly rapid rate, effects of obesity are borne out in clinics and hospitals across the nation and the world. While its cost to the healthcare system, which is beyond the scope of this discussion, is unprecedented, the cost to individuals suffering from it is immeasurable. Obesity is debilitating at best and a deadly disease at worst. Much attention is often devoted to the widely known adverse health effects brought about by and associated with obesity. Among these are coronary artery disease, the metabolic syndrome, diabetes, hypertension, dyslipidemia, insomnia, restless leg syndrome, sleep apnea, obesity hypoventilation syndrome, circadian dysregulation, nonalcoholic fatty liver disease, peripheral vascular disease, and various musculoskeletal issues (Hargens, 2013; Klish, 2015).

While the physical effects of obesity are frequently addressed, the psychosocial issues associated with obesity are an often overlooked topic. Patients who are overweight and obese, particularly those who are young, can suffer from significant emotional trauma. They can be excluded from participating in certain activities, often because of the physical limitations imposed by their body habitus. Obese individuals may become socially isolated because of a lack of self-esteem. They may find themselves the victims of stigma, even at the hands of healthcare providers (Hinman, 2015). This can further compound the morbidity faced by those with obesity, as they may become averse to the idea of seeking care for their often complex medical issues. Their ability to pursue certain employment opportunities and social or romantic relationships can be hindered by the so-called antifat bias (Lynagh, 2015; Miller, 2013). Additionally, psychiatric issues such as anxiety, depression, and poor or altered body image may arise as a consequence of this disorder and the previously discussed secondary effects so frequently associated with obesity.

Obesity and Sexual Dysfunction

Among the myriad of issues that plague those suffering from obesity, one that often goes overlooked is that of sexual dysfunction. Though there have been population studies examining the connection between obesity and sexual dysfunction, a tremendous amount remains unknown about this important topic. Given the high incidence of obesity worldwide, it may be the condition with the most profound impact on human sexuality. The problem of sexual dysfunction in the context of obesity is rather unique in that it can have etiologies that are organic, nonorganic, or mixed.

Obesity, through its effects on the body's vasculature and specific organs such as the lungs, liver, and heart, adversely impact the biology that underlies sexual function. Perhaps the organ system that is most directly affected by obesity is the endocrine system. The traditional view was that adipose tissue was simply a passive reservoir for energy storage; it is now considered to be a highly active metabolic and endocrine organ. Aberrations in the physiology of the hypothalamic-pituitary target organ may arise as a consequence of obesity. It also alters the metabolism of hormones such as testosterone and affects endothelial function via the release of adipokines by adipocytes, cytokines that essentially function as toxins that serve to augment the body's inflammatory response. Obesity may be the most common cause of testosterone deficiency in the developed world. More than half of men with a BMI greater than 30 have testosterone deficiency (Mulligan, 2006). However, our limited understanding of obesity's relation to sexuality is complicated by the fact that we do not yet know whether any impairments observed in obese individuals are directly related to their obesity or are instead a secondary consequence of comorbid conditions or even the medications used to treat them.

It is also possible that the psychology underlying the associations among obesity, sex, and sexuality plays a massive role in sexual performance. Many of the issues with self-esteem, body image, social interaction and integration, and perception of others and society at large can influence sexual behavior and performance. The aforementioned tension can complicate the relationship between obese patients and their healthcare providers. Furthermore, patients may be uncomfortable discussing topics such as body weight and sexual behavior, thereby contributing to the creation of a negative environment in which to address these issues and facilitate meaningful changes. Yet whether due to aberrations in physiology or psychology, it is clear that obesity dramatically impairs sexual function, desire, and satisfaction.

Obesity, Sleep Disorders, and Sexual Dysfunction

As mentioned, a number of sleep disorders are associated with obesity. These include sleep apnea, obesity hypoventilation syndrome, and insomnia. The relationship of each disorder to a state of adiposity is complex and perhaps not yet fully understood. Sleep apnea and obesity hypoventilation syndrome can be considered to exist along a continuum, as much of the pathophysiology of obesity hypoventilation syndrome is related to the adverse effects of sleep apnea occurring over an extended period of time. A 2013 study discussed a number of ways that obesity relates to and drives sleep apnea, by altering the anatomy of the neck so as to impair proper ventilation as well as affecting the concentrations of a variety of cytokines, including leptin and tumor necrosis factor-alpha, that play a role in maintaining normal respiratory drive and allowing the airways to function properly (Hargens, 2013). It is because of this that patients who suffer from sleep apnea often experience innumerable disturbances to their sleep during the course of a night, as the body's failure to breathe adequately interrupts an individual's state of rest. These repeated disruptions incline one to develop excessive daytime sleepiness because of an incapability to achieve sufficient sleep at night. Further, a long-term inability to achieve normal pulmonary function while sleeping can lead to the development of obesity hypoventilation syndrome, a disorder that manifests itself as chronic hypoxemia and hypercapnia and can eventually lead to *cor pulmonale* and right-sided heart failure.

The pathophysiology underlying insomnia, to the extent of current understanding, involves impairments in one's ability to sleep properly secondary to a state of increased arousal. With respect to how this relates to obesity, Hargens (2013) posits that the significantly higher prevalence of insomnia among obese patients compared to nonobese counterparts may arise as a consequence of some chronic psychological disturbance or stressor. The emotional aberrancies that underlie this discrepancy in the rates of insomnia may relate to the aforementioned psychological issues that are often specific to or more frequently observed in obese patients. What is particularly devastating about insomnia in the presence of pathologically elevated BMIs is that it may in fact reinforce obesity-promoting behaviors. The same 2013 study by Hargens noted that decreased leptin and increased ghrelin levels have been found to arise when the body does not achieve optimal sleep, thereby predisposing patients who suffer from insomnia to augment their caloric intake, a particular undesirable outcome in those who already struggle with a pathologically increased body mass.

As obese individuals strive to achieve an appropriate quantity and quality of sleep, their ability to maintain normal sexual function may be hindered. It is not difficult to conceive how the increased fatigue that can accompany insomnia and insufficient nocturnal sleep due to apneic disturbances may decrease sexual drive and activity simply because of a lack of energy. A 2015 study in *The Journal of Sexual Medicine* found that, among a group of female patients who were not selected on the basis of any particular body habitus, those who obtained less sleep experienced poorer sexual desire, genital arousal, and likelihood of engaging in sexual behavior with a partner the day after a suboptimal night of sleep (Kalmbach, 2015). In addition to this, a 2011 study discussed how in both men and women the awake state is associated with decreased concentrations of certain hormones that are critical in sexual physiology and behavior, such as testosterone and estrogen. Therefore, it is quite plausible that the sleep disorders such as sleep apnea and insomnia that are such prevalent comorbidities in obesity may predispose individuals to sexual dysfunction of a hormonal etiology (Andersen, 2011).

Obesity and Impairments in Male Sexual Function

A particularly strong relationship has been identified between obesity and erectile dysfunction (ED) in men. A 2008 review article in the *International Journal of Impotence Research* noted that men with a BMI of 28.7 or greater were roughly 30% more likely to suffer from ED than those with a BMI of 25 of below. Additionally, other studies report up to a 90% increased risk of ED associated with obesity (Esposito, 2009). Though this may arise through a number of mechanisms, it is easiest to consider the organic causes separately from those that are nonorganic.

Most often, when the etiology of ED comes about through some aberrancy in physiologic function, it is secondary to the impairment of blood flow. It has been proposed that its pathophysiology bears resemblance to that of atherosclerosis, with impairments in endothelial function and increased levels of inflammatory markers such as certain interleukins and C-reactive protein. Significantly higher levels of circulating endothelial microparticles (EMPs), which are extruded from cells upon their activation or programmed cell death and can contribute to dysfunction of the endothelium, have also been identified in obese individuals. Similarly elevated levels of EMPs have been seen in patients with such conditions as diabetes, acute coronary syndrome, or hypertension (Cheitlin, 2004). Further, many of the potential comorbidities of obese patients such as metabolic syndrome, diabetes, atherosclerosis, and peripheral artery disease also predispose one to the development of ED.

A number of psychological factors can influence an individual's ability to maintain normal erectile function as well. These include the previously discussed problems with self-esteem, such as feeling as though one is not desired, or having other concerns related to body image or perception by society and potential sexual partners. Additionally, ED may come about as a secondary consequence of an underlying psychiatric condition such as depression or anxiety. One or several of these underlying factors can be present in any given case, and it is therefore necessary for clinicians to tailor their workup to the individual patient when seeking the etiology or etiologies of ED.

Obesity commonly leads to a decrease in the concentration of testosterone. A 2013 paper found 40% to 50% reductions in levels of testosterone when obese pubertal and postpubertal males were compared to subjects with normal BMIs (Mogri, 2013). A study published in 1995 in *Metabolism* identified correlations between increased fatness and decreased levels of hormones such as testosterone, dehydroepiandrosterone, androstenedione, and sex hormone binding globulin (SHBG), which is responsible for binding and transporting sex hormones in the bloodstream. Researchers noted a positive correlation between higher amounts of body fat and increased levels of estrone (Tchernof, 1995).

The correlation between obesity and low testosterone remains a subject of considerable debate. It has historically been posited that low levels of testosterone in obese patients may be related to the aforementioned decrease of SHBG, which is associated with greater adiposity or the presence of increased amounts of adipose tissue, leading to greater peripheral conversion of testosterone to estrogens by the enzyme aromatase. However, Mogri (2013) identified no such increase in estrogen levels as testosterone concentrations fell. The authors suggest instead that the pathogenesis of low testosterone may be similar to the hypogonadotropic hypogonadism secondary to insulin resistance seen in type II diabetes mellitus, citing the findings from a 2006 study from *Endocrinology* that discussed insulin's role in stimulating gonadotropin-releasing hormone from the hypothalamus (Salvi, 2006). Though such studies underscore the shortcomings of our current understanding of the interplay between obesity and alterations to the endocrine system, they also highlight the need for further exploration of this topic.

There is also data within the medical literature suggesting that obese males may suffer from decreased fertility, likely due to lower sperm counts and/or impaired sperm motility. A 2013 review article highlighted a point initially raised in 1981 by

Shafik that the presence of increased amounts of surrounding adipose tissue could lead to higher temperatures in the testes and thereby exert adverse effects on the process of spermatogenesis (Sermondade, 2013; Shafik, 1981).

Obesity and Impairments in Female Sexual Function

Sexual dysfunction in the context of obesity is not exclusive to male patients. The 2006 study "Obesity and Sexual Quality of Life" (Kolotkin, 2006) found that women with class I obesity were between 2.31 and 2.89 times more likely to not enjoy sexual activity, have decreased sexual desire, struggle with sexual performance, and avoid sexual encounters. Further, those with class III obesity had a relative risk of at least 3.00 for all of the aforementioned parameters. When compared to the male patients included in the study, women actually reported greater adverse effects on sexual quality of life secondary to obesity (Kolotkin, 2006).

Any pathophysiologic mechanisms that may underlie problems with sexual function in obese women remain incompletely understood at this point in time. Nonetheless, we can extrapolate from the information that is presently available in order to better understand the physiology of the female sexual response. Proper endothelial function appears to be critical in women. Nitric oxide (NO) plays an essential role in the process of engorgement of the clitoris and labia, while both NO and vasoactive intestinal peptide are believed to regulate relaxation of and secretion from the vagina. It is therefore reasonable to raise the possibility that many of the same mechanisms through which obesity can bring about endothelial dysfunction in males, such as atherosclerosis, may also be at play in hindering proper sexual function in obese females (Berman, 2001).

Additionally, insulin resistance due to obesity may lead to hypogonadotropic hypogonadism. It is easy to conceive how aberrations in the hypothalamic-pituitary-adrenal and hypothalamic-pituitary-gonadal axes may lead to sexual dysfunction in women with increased adiposity. Alterations in normal concentrations of estrogens and androgens can impact sexual function through a variety of mechanisms, as low estrogen has been found to lead to impaired vaginal perfusion, fibrosis of the clitoris, and decreased thickness of the vaginal epithelium, while decreased androgen levels can manifest as impairments in sexual arousal, drive, and desire (Berman, 2001).

Many of the same psychological and emotional effects of obesity that often manifest as ED in males can also give rise to problems in females. Issues with body image and self-confidence, as well as major or subclinical depression, can decrease sexual desire and drive. Belief that societal bias may predispose obese women to be disregarded by many potential partners solely on the basis of their body habitus can further contribute to an unwillingness to pursue sexual or romantic relationships.

Furthermore, when obese female patients do engage in sexual behavior, their own performance and experience with respect to arousal, pleasure, and orgasm, as well as that of their partners, may be adversely affected by the aforementioned psychological hindrances.

Much like their male counterparts, the fertility of obese women may also suffer as a consequence of their disease. A 2015 paper in the journal *Hormone Molecular Biology and Clinical Investigation* noted an association between obesity and anovulation. The authors suggested that increased adiposity directly impacts oocytes and the endometrium, while also leading to problems with the optimal development of embryos (Best, 2015).

The data in this field remains limited as of this current time, and studies are warranted. The 2015 article, "Assessment of Female Sexual Function in a Group of Uncircumcised Obese Egyptian Women" attempts to shed light on this topic. In order to do so, the authors utilized the aforementioned FSFI and its diagnostic criteria for female sexual dysfunction and identified a relationship between an increased BMI and lower FSFI scores as well as impairments in arousal and orgasm (Elnashar, 2015).

Potential Reversibility of Obesity's Effects on Sexual Function

There is of course great hope that the negative effects of obesity on sexuality are reversible with attainment and maintenance of a normal or near-normal BMI. It is already known that 5% to 10% reductions in weight can have profound impacts on other areas of one's health, so it is conceivable that the same would be true with sexual function. However, because obesity is often a chronic and relapsing disorder, it has remained difficult to evaluate the true reversibility of the impact of obesity on sexual behavior.

Conclusion

Further elucidation of the exact nature of obesity's impact on sexual function is of considerable importance for the optimization of both health and quality of life in obese patients. The ability to achieve normal sexual function and the sequelae that so often come about as a result, such as romantic relationships, marriage, and the ability and opportunity to reproduce, are critical aspects of life to many people. Their absence may serve to further alienate certain individuals suffering from obesity and deprive them of a sense of fulfillment. Additionally, given some of the similarities between the proposed mechanisms of ED secondary to obesity and other obesity-related comorbidities such as atherosclerosis and diabetes, the possibility exists that the study of the pathophysiology underlying sexual dysfunction in obesity may yield valuable information about other, potentially more life-threatening disease processes. Last, and perhaps of greatest importance, society must abandon its inherent bias against and animosity toward those who struggle with obesity, as it is through acceptance and a concerted effort to strive toward achieving healthier lives for all those who may battle such issues that the greatest impacts on physical, psychological, and emotional health can be made.

REFERENCES

Andersen ML, Alvarenga TF, Mazaro-Costa R, Hachul HC, Tufik S (2011). The association of testosterone, sleep, and sexual function in men and women. *Brain Res.* 1416:80–104.

Berman JR, Goldstein I (2001). Female sexual dysfunction. *Urol Clin North Am.* 28(2):405–416.

Best D, Bhattacharya S (2015). Obesity and fertility. *Horm Mole Biol Clin Invest.* Oct;24(1):5–10.

Bray GA (2015). Obesity in adults: prevalence, screening, and evaluation. August 2015. https://www.uptodate.com/contents/obesity-in-adults-prevalence-screening-and-evaluation.

Cheitlin MD (2004). Erectile dysfunction: the earliest sign of generalized vascular disease? Editorial. *J Am Coll Cardiol.* 43(2):185–186.

Elnashar ARM, Ibrahim NH, Ahmed H-E, Hassanin AM, Elgawady MA (2015). Assessment of female sexual function in a group of uncircumcised obese Egyptian women. *Int J Impot Res.* 27(5):178–181.

Esposito K, Giugliano F, Ciotola M, De Sio M, D'Armiento M, Giugliano D (2009). Obesity and sexual dysfunction, male and female. *Obesit Metab.* 3:65.

Hargens TA, Kaleth AS, Edwards ES, Butner KL (2013). Association between sleep disorders, obesity, and exercise: a review. *Nat Sci Sleep.* 5:27–35.

Hinman NG, Burmeister JM, Kiefner AE, Borushok J, Carels RA (2015). Stereotypical portrayals of obesity and the expression of implicit weight bias. *Body Image.* 12:32–35.

Kalmbach DA, Arnedt JT, Pillai V, Ciesla JA (2015). The impact of sleep on female sexual response and behavior: a pilot study. *J Sex Med.* 12(5):1221–1232.

Klish WJ (2015). Comorbidities and complications of obesity in children and adolescents. September 10. https://www.uptodate.com/contents/comorbidities-and-complications-of-obesity-in-children-and-adolescents.

Kolotkin RL, Binks M, Crosby RD, Østbye T, Gress RE, Adams TD (2006). Obesity and sexual quality of life. *Obesity.* 14(3):472–479.

Lynagh M, Cliff K, Morgan PJ (2015). Attitudes and beliefs of nonspecialist and specialist trainee health and physical education teachers toward obese children: evidence for "anti-fat" bias. *J School Health.* 85(9):595–603.

Maseroli EE, Fanni M, Fambrini B, et al. (2016) Bringing the body of the iceberg to the surface: the Female Sexual Dysfunction Index-6 (FSDI-6) in the screening of female sexual dysfunction. *J Endocrinol Invest.* Apr;39(4):401–409.

Miller DP, Spangler JG, Vitolins MZ, Davis SW, Ip EH, Marion GS, Crandall SJ (2013). Are medical students aware of their anti-obesity bias? *Acad Med.* 88(7):978–982.

Mogri M, Dhindsa S, Quattrin T, Ghanim H, Dandona P (2013). Testosterone concentrations in young pubertal and post-pubertal obese males. *Clin Endocrinol.* 78(4):593–599.

Mulligan T, Frick MF, Zuraw QC, Stemhagen A, McWhirter C (2006). Prevalence of hypogonadism in males aged at least 45 years: the HIM study. *Int J Clin Pract.* 60:76–79.

Petersen JL, Hyde JS (2010). A meta-analytic review of research on gender differences in sexuality, 1993–2007. *Psychol Bull.* 136(1):21–38.

Rosen RC, Cappelleri JC, Smith MD, et al. (1999). Development and evaluation of an abridged, 5-item version of the International Index of Erectile Function (IIEF-5) as a diagnostic tool for erectile dysfunction. *Int J Impot Res.* 1999 Dec;11(6):319–326.

Rosen R, Brown C, Heiman J, et al. (2000). The Female Sexual Function Index (FSFI): a multidimensional self-report instrument for the assessment of female sexual function. *J Sex Marital Ther.* 26(2):191–208.

Salvi R, Castillo E, Voirol M-J, et al. (2006). Gonadotropin-releasing hormone-expressing neurons immortalized conditionally are activated by insulin: implication of the mitogen-activated protein kinase pathway. *Endocrinology.* 147(2):816–826.

Sermondade N, Faure C, Fezeu L, et al. (2013). BMI in relation to sperm count: an updated systematic review and collaborative meta-analysis. *Hum Reprod Update.* 19(3):221–231.

Shafik A, Olfat S (1981). Scrotal lipomatosis. *Br J Urol.* 53(1): 50–54.

Tchernof A, Després J-P, Bélanger A, et al. (1995). Reduced testosterone and adrenal C19 steroid levels in obese men. *Metabolism.* 44(4):513–519.

SECTION III

An Integrative Medicine Perspective on Psychiatric, Psychological, Physiological, and Behavioral Disturbances Relevant to Sexual Function

13

Integrative Treatment of Depression and Anxiety: An Effective Nonpharmacological Approach that Preserves Sexual Functioning

KATHERINE FALK

Depression, Anxiety, and Sexual Dysfunction

Sexual dysfunction is commonly associated with depression and anxiety; and many of the psychotropic medications used to treat depression or anxiety cause sexual dysfunction. Furthermore, sexual dysfunction, in and of itself, can be devastating and cause anxiety and/or depression, particularly in men with erectile dysfunction who have never before been symptomatic (Nurnberg, 2002, 2003; Seidman, 2002).

Depression is generally defined as a deep, unshakable sadness with diminished interest or pleasure in nearly all activities, difficulty concentrating, feelings of guilt, helplessness, hopelessness, despair, fatigue, low self-esteem, and worthlessness. It can be accompanied by thoughts of death and suicide. Individuals with depression may either sleep too little or too much; they may have a loss of appetite or over-eat. They may also have impaired ability to function in social situations or at work (American Psychiatric Association [APA], 2013). And they may also have impaired sexual functioning—decreased sexual desire and arousal in both men and women and problems having or maintaining erections in men (Kennedy, 1999).

Depression can also be part of bipolar disorder, characterized by cycling of mood. It is crucial to determine if the patient has a single episode of depression or has a cycling illness with recurrent episodes of depression, because the treatments are not the same.

Anxiety disorders are defined by excessive worry occurring more days than not for at least six months. Individuals have anxiety, worry, or physical symptoms that cause clinically significant distress or impairment in social, school, or occupational functioning. At least three of the following criteria are needed to make the diagnosis of anxiety disorder: restlessness, feeling keyed up or on edge, being easily fatigued, difficulty concentrating or mind going blank, irritability, muscle tension, and sleep disturbance (APA, 2013).

Men with mood or anxiety disorders have been found to have a higher incidence of premature ejaculation or erectile dysfunction than men who are not suffering from anxiety or depression. Women with anxiety disorder, panic disorder, or obsessive compulsive disorder (OCD) experience lower sexual desire, lower frequency of sexual contact, and higher frequency of sexual aversion disorder than women without mood or anxiety disorders. Those women with OCD reported more sexual dysfunction and less sexual satisfaction than individuals with panic disorder or controls (Van Minnen, 2000).

Avoidant behavior, which is characteristic of anxiety, puts both men and women at risk for sexual dysfunction (Rajkumar, 2015). Social phobia, marked by disabling anxiety in social gatherings, is associated with premature ejaculation. Furthermore, premature ejaculation can cause impaired self-esteem in men who are not suffering from either anxiety or depression, and anxiety about sexual performance can cause erectile dysfunction (Bradford, 2006; Figueira, 2001; Rajkumar, 2015).

Diagnosis and Treatment of Anxiety and Depression

While mood or anxiety disorders are treatable, they often go undiagnosed and untreated because of the difficulty in accessing care, caused, in part, by the stigma around mental illness and the shame that is often felt in getting help (Barney, 2006). Psychiatric conditions often are not covered by insurance as fully as medical conditions. As a result, many individuals with severe depression or anxiety never see a psychiatrist or other mental health specialist but may seek help from their primary care physician or do not seek help at all. Primary care physicians actually write more prescriptions for antidepressants than psychiatrists, and those suffering from severe depression will often see their primary care physician rather than a psychiatrist in the month before their death by suicide (McDowell, 2011).

The psychotropic medications used to treat depression and anxiety are often not effective; selective serotonin reuptake inhibitors (SSRIs), for example, are effective only 55% of the time (Fawcett, 1994; Moncrieff, 2007), a success rate that is not much better than that of placebos, which yield positive results 30% to 40% of the time (Sonawalla, 2002). Placebo results can actually be as high as 90% in some diseases and depend on positive beliefs and expectations of improved heath on the part of both the patient and the physician and a good relationship between them (Benson, 1996).

A meta-analysis published in the *New England Journal of Medicine* in 2009 examined data on 12 antidepressants, involving 12,564 patients, submitted to the Food and Drug Administration (FDA) over a 15-year period (Turner, 2008). The authors conducted a literature search to find the matching publications and discovered that studies yielding positive results were much more likely to appear in the medical literature than those yielding questionable or negative results. In fact, 31% of the 74 FDA-registered studies (3,449 study participants) were not published at all. As a result, 94% of the *published* studies supported the efficacy of antidepressants, even though only about half the *total* trials had found that these medications helped (Turner, 2008).

A similar reporting bias has been found in studies published on the effectiveness of second-generation antidepressants in the treatment of anxiety disorders. Studies involving seven SSRIs and two serotonin norepinephrine reuptake inhibitors (SNRIs) were evaluated. Trials considered positive by the FDA were five times more likely to be published than trials considered to be negative (Roest, 2015).

FIRST, DO NO HARM

Psychotropic medications, including antidepressants and anxiolytics, have serious sexual and other side effects. Problems in sexual function—diminished or loss of desire, arousal, or orgasm—can be the result of depression or, unfortunately, the result of its treatment (Baldwin, 2013). One study reports that 58% to 73% of individuals taking antidepressants have experienced sexual side effects (Dording, 2012). Integrative approaches, on the other hand, successfully treat anxiety and depression while preserving normal sexual function.

Prevalence of sexual dysfunction, both in the general population and in those prescribed antidepressants, has often been underestimated because most individuals do not disclose sexual problems unless specifically asked. Structured questionnaires or standardized instruments provide a more accurate response (Kronstein, 2015; Serretti, 2009; Waldinger, 2015).

Using a structured questionnaire, 72% to 80% of men who were prescribed SSRIs reported orgasm dysfunction (Serretti, 2009). On the other hand, a study (which used a Sexual Function Questionnaire) of men and women suffering from depression but not yet taking antidepressant medications found that the prevalence of orgasm dysfunction to be lower, with 50% of women and 40% of men reporting problems with orgasm. No correlation was found with severity of depression and specific symptoms of sexual dysfunction. Reduced arousal was more common in both men and women (40%–50%) than problems with ejaculation or orgasm

(15%–20%). In women (but not in men), problems with arousal and orgasm correlated with anxiety and personality problems (Kennedy, 1999).

Sexual side effects of antidepressants are dose related and affect all phases of sexual response, including decreased sexual desire; problems with arousal, such as diminished genital sensation; diminished vaginal lubrication; erectile dysfunction; delayed or premature ejaculation; and delayed female orgasm or complete anorgasmia (Clayton, 2009).

SSRIs and SNRIs (e.g., citalopram, fluoxetine, paroxetine, sertraline, and venlafaxine) have the highest rate of sexual side effects among antidepressants, ranging from 25% to 80% of patients (Waldinger, 2015). While usually reversible, there are case reports of sexual side effects not improving after discontinuation of medication (Csoka, 2008). There is some good news, however. Some antidepressants appear not to cause sexual side effects: these include bupropion, agomelatine, mirtazapine, vortioxetine, amineptine, and moclobemide (Waldinger, 2015).

Some side effects of antidepressant medications can be dangerous. For example, antidepressants can cause significant weight gain and type 2 diabetes (Andersohn, 2009). SSRIs have been shown to cause bone loss in older individuals with resulting increased incidence of fractures, debility, and death (Diem, 2007; Haney, 2007). While SSRI use can increase the risk of serious arrhythmias, including those that can cause sudden death (Narayan, 2009), no link was found between SSRI use and fatal or nonfatal coronary heart disease or myocardial infarction (Whang, 2009).

SSRIs can interfere with normal platelet function (Andrade, 2010), and older individuals taking SSRIs have 3.6 times the risk of gastrointestinal (GI) bleeding than individuals of similar age not taking antidepressants. Because nonsteroidal antiinflammatory drugs (NSAIDS) and aspirin also interfere with normal platelet function (Andrade, 2010), the risk of GI bleeding increases by a factor of 12 if NSAIDS or aspirin are taken along with SSRIs (Dalton, 2003). Concomitant use of SSRIs and NSAIDs also increases the risk of intracranial hemorrhage within 30 days of initial combination, and this risk is greater in men than in women. Taking either SSRIs or NSAIDs alone is not associated with cerebrovascular bleeding (Shin, 2015)

Many medications, including psychotropic medications, damage mitochondria (Neustadt, 2008; Souza, 1994). Mitochondria, located in every cell of the body, are responsible for generating energy through oxidative phosphorylation and the synthesis of adenosine triphosphate (ATP). Mitochondrial damage results in inhibition of the electron transport chain with consequent decreased ATP synthesis, decreased oxidative-phosphorylation, increased formation of harmful free radicals, and damaged mitochondrial DNA. Symptoms of mitochondrial damage can include symptoms such as fatigue, muscle pain, shortness of breath, or abdominal pain (Neustadt, 2007).

The medications that damage mitochondria include antidepressants (amitriptyline, amoxapine, citalopram, fluoxetine), antipsychotics (chlorpromazine, fluphenazine, haloperidol, risperidone, quetiapine, clozapine, olanzapine), mood stabilizers (valproic acid, lithium), and anti-anxiety agents (benzodiazepines such as diazepam and alprazolam). In addition, commonly used medications such as cholesterol-lowering drugs, aspirin, acetaminophen, indomethacin, naproxen, and barbiturates also damage mitochondria (Balijepalli, 2001; Chan, 2005; Chitturi, 2002; Lambert, 1999; Modica-Napolitano, 2003; Neustadt, 2007; Olszewska, 2013; Sarah, 1998; Souza, 1994; Xia, 1999).

Antidepressant use during pregnancy can also present problems to a fetus. Paroxetine use during the first trimester of pregnancy was found to be associated with an increased risk of any major congenital malformation and/or cardiac malformations (Bérard, 2016a, 2016b). Increased risk of autism spectrum disorder (ASD) is associated with SSRI use during the second and/or third trimester of pregnancy (Gentile, 2015). The highest risk of ASD was associated with exposure to more than one class of antidepressants during the second and/or third trimester. Use of antidepressants in the first trimester or the year before pregnancy was not associated with increased risk of ASD (Boukhris, 2016; Gentile, 2015).

IS THERE A BETTER WAY?

Integrative psychiatry offers a unique approach to patient care. The conventional/allopathic model of psychiatry focuses on the patient's symptoms and prescribes medications for those symptoms. Rather than merely identifying and treating symptoms, integrative psychiatry looks at the whole person and aims to address the root cause of the problem.

To fully understand an individual's illness and suffering, it is necessary to go deeper, to look beneath the symptoms—at issues of nutrition, biochemical individuality, metabolism, including thyroid and adrenal function, inflammation, GI health, detoxification, autoimmune illnesses, and lifestyle issues (Horrigan, 2012). Practitioners should inquire about sleep, relationships, community, work, leisure time, connection to the spiritual, health habits such as exercise and diet, and how the individual handles stress. Any or all of these issues may be involved in the cause of depression or anxiety and must be evaluated and addressed in an integrative approach to the treatment of depression or anxiety. Personalized treatments are developed to suit each person's lifestyle and to strengthen self-awareness and resources for self-care.

NUTRITION

What we eat is essential to the health of our brain and nervous system and therefore has a tremendous impact on mood. In fact, nutrition is as important a factor in the cause and treatment of mental illnesses as it is for the management of heart disease or diabetes (Mayer, 1997; Sarris, 2015). Studies have consistently shown that individuals who follow the Mediterranean diet—a mainly plant-based diet comprised largely of fruits, vegetables, whole grains, legumes, nuts, fish, and healthy fats such as olive oil—have a lower rate of mood, anxiety symptoms, and cognitive deficits than those who follow a diet high in processed foods, the Standard American Diet (Akbaraly, 2009; Bray, 2013; Dash, 2016; Lai, 2013; Psaltopoulou, 2013; Sánchez-Villegas, 2009).

Fish Oil

Adequate amounts of omega-3 fatty acids (docosahexaenoic acid [DHA] and eicosapentaenoic acid [EPA]) are essential for mood (Beydoun, 2013; Freeman, 2006b, 2011; Sublette, 2011). Sixty percent of the human brain is composed of fat, and approximately 30% of this must be obtained through the diet through consumption of fish or omega-3 supplementation (Hibbeln, 2008).

Countries with the highest consumption of fish have the lowest incidence of depression, suicidal behavior, aggression, and violence (Beydoun, 2013; Hibbeln, 2008, 2014). Rates of homicide are greater in countries with lower seafood consumption (Hibbeln, 2001). Higher omega-6/omega-3 ratio and low levels of DHA predict risk of suicidal behavior (Sublette, 2006). Iceland's unusually low rate of seasonal affective disorder has been attributed to its high consumption of fish (225 lb per person per year compared with 48 pounds in the United States; Cott, 2001). Eating less than 50 lb fish per year increases vulnerability for bipolar disorder (but not schizophrenia; Noaghiul, 2003).

The standard American diet, however, now has an excessive amount of omega-6 polyunsaturated fatty acids and a very high omega-6/omega-3 ratio, which promotes inflammation, cardiovascular disease, depression, and other psychiatric illnesses (Blasbalg, 2011).

Omega-3 fatty acids were found to be beneficial in the treatment of unipolar and bipolar depression, but only when the fish oil supplement had a higher proportion of EPA than DHA and the dose of EPA was not greater than 2,000 mg (Chetry, 2016; Freeman, 2006b; Liu, 2013; Martins, 2009; Ross, 2007; Sublette; 2011). The presence of severe anxiety along with depression was found to be associated with the lowest EPA and DHA levels (Liu, 2013). Adding omega-3 fatty acids to maintenance psychiatric medication improved the outcome of recurrent unipolar depression and also bipolar disorder (Nemets, 2002). Postpartum depression, which occurs in 10% to 20% of pregnancies, is 50 times more common in countries with low levels of seafood consumption and is inversely related to the consumption of fish (Hibbeln, 2002). DHA is depleted in women with frequent pregnancies, causing postpartum depression and perhaps even psychosis (Freeman, 2006, 2011).

Micronutrients

Our 21st-century soil and food supply is depleted of important nutrients, such as magnesium (Mg), iron, calcium, and zinc, compared to 50 years ago. This contributes to our nutritional deficiencies that affect our overall well-being, including mood, anxiety, brain function, and sexual health (Ramsey, 2013; Scheer, 2011; Thomas, 2003). In fact, one study has shown that those who follow the Standard American Diet were found to have a smaller hippocampus, the area of the brain responsible for learning and memory (Jacka, 2015).

In studies looking at the effect of single nutrients on mood, vitamins B1, B6, B12, D, E, and folate as well as iron, copper, zinc have been found to have the most evidence for affecting mood (Kaplan, 2007, 2015). However, a symphony of various nutrients promotes a healthy brain; therefore it is important to study the effects of multiple nutrients, rather than just one or a few.

Genetic mutations can give rise to individual variations in the need for certain vitamins and minerals; therefore, some individuals will have a greater need for particular micronutrients than other people. The conventional wisdom concerning micronutrients is that as long as Recommended Daily Allowances (RDAs) meet minimum standards and we are adequately nourished, a small deficiency of a vitamin or mineral will not cause medically significant problems. However, there is evidence that if just one element in the metabolic network is inadequate, repercussions can impact a large number of systems (Ames, 2002, 2004). Furthermore, inadequate micronutrient status can have implications for cognitive dysfunction and mental illness, as well as DNA damage, mitochondrial dysfunction, cancer, accelerated aging, and degenerative disease (Ames, 2002, 2004).

Micronutrient deficiencies that contribute to depression and anxiety may also contribute to sexual dysfunction. Furthermore, supplementation for actual nutritional deficiencies can alleviate symptoms of depression and anxiety, without causing sexual dysfunction, a hallmark of so many conventional drugs.

Iron

Because iron is a cofactor for the synthesis of the neurotransmitters serotonin, norepinephrine, epinephrine, and dopamine, it is an essential nutrient for a healthy brain and normal mood. However, iron deficiency is the most common nutritional deficiency in the United States and throughout the world (Auerbach, 2015; Stoltzfus, 2003) and is a frequent contributing cause of depression (Bodnar, 2005; Vahdat, 2007).

In childhood, the incidence of depression is equal among boys and girls. However, after puberty, there is a much higher incidence of depression in girls than in boys, most likely caused by iron deficiency due to menstruation, complicated by dieting and eating disorders among girls (Chen, 2013). Normal stores of iron are also important in thyroid function. The enzyme thyroid peroxidase (TPO), which is responsible for synthesizing thyroid hormone by adding iodine to thyroglobulin, is iron dependent (Zimmermann, 2006).

In evaluating iron deficiency, it is important to perform the following lab tests: serum iron, ferritin, total iron binding capacity, complete blood count, and mean corpuscular volume (MCV). Ferritin (the normal level is 100; Lord, 2008) is considered to be the best measure of both long-term iron stores and low iron (Herbert, 1992). If the low ferritin reflects iron-deficiency anemia, the hemoglobin and MCV also will be low. If there is no evidence of anemia, the low ferritin level is likely caused by leaky gut (Souroujon, 1982). High ferritin is suggestive of an inflammatory process, recent infection, autoimmune illness, or malnutrition (Coyne, 2006).

Magnesium

Mg is extremely important for overall health as well as for brain health, and its deficiency has been linked to depression and anxiety. It is required as a coenzyme in more than 300 biochemical reactions in the body, including DNA and RNA synthesis, ATP synthesis, and energy metabolism. Because ATP exists in cells primarily as MgATP, Mg is involved in virtually every metabolic process occurring in the body (de Baaij, 2015; Hsu, 1982). In fact, Mg is a cofactor for the synthesis of glutathione, s-adenocylmethionine, RNA, and DNA (Lord, 2008, 84).

Magnesium deficiency is common and is associated with aging (Costello, 1992); diarrhea or overuse of laxatives; chronic alcoholism; diabetes; long-term use of diuretics; and deficiency of selenium, vitamin B6, and vitamin D (Johnson, 2001). In addition, small intestine overgrowth of certain bacteria that produce tricarballylate will bind Mg and prevent its absorption, causing severe Mg deficiency, which will persist if not treated with appropriate antibiotics (Lord, 2008, 604). The typical American diet is deficient in Mg, which is found in nuts, leafy green vegetables, and unrefined grains. Increased Mg intake in food would be another benefit of a Mediterranean-style diet (Bahreini, 2013).

Common symptoms of Mg deficiency are muscle cramping, sleep disorders, anxiety, depression, and restless leg syndrome. Magnesium supplementation has been shown to alleviate depression and anxiety. Magnesium (450 mg) was found to be as effective as imipramine (50 mg) given daily in alleviating depression and anxiety of 23 elderly individuals with diabetes who had low Mg levels (Barragán-Rodríguez, 2008). In a study of premenstrual anxiety, women who were given Mg (200 mg) and B6 (50 mg), experienced significantly less anxiety and irritability than the control group (De Souza, 2000).

Because the majority of Mg exists inside the cells (only 1% of Mg is extracellular), it is important to test red blood cell (RBC) Mg levels, which better reflect the actual levels of Mg throughout the cells of the body. Because the majority of Mg is located inside the cells throughout the body, serum or plasma Mg are not accurate measures of functional Mg levels.

Zinc

Zinc, along with iron, is the most concentrated mineral in the human brain. More than 300 enzymes are zinc dependent. Zinc supplementation significantly reduced depression in a group of treatment-resistant depressed patients taking imipramine (Siwek, 2009). Low zinc level is a marker for treatment-resistant depression (Maes, 1997), while a high copper/zinc ratio has been used as a predictor of aggression and violence (Walsh, 1997). Zinc is released into the synapse of neurons along with glutamate, thereby modulating the N-methyl-D-aspartate (NMDA) receptors and creating a calming effect (Cichy, 2009).

Copper

Copper is essential for normal brain function and normal mood. It is a cofactor in the synthesis of norepinephrine, epinephrine, dopamine, and serotonin and also is involved in NMDA receptor modulation (Schlief, 2006). It is also required for the proper function of many other important enzyme systems, including those needed for cellular energy production and detoxification (Linder, 1996; Madsen, 2007). It is therefore crucial to diagnose and treat low levels of copper. Increased copper levels may be a factor in postpartum depression, as significantly higher copper levels have been detected in women with a history of postpartum depression, compared both with nondepressed women and with depressed women without a history of postpartum depression (Crayton, 2007).

Chromium

Chromium picolinate (400mcg) can successfully treat depression, even depression not responsive to antidepressant medication (McLeod 2000, 1999). In a study comparing chromium with placebo for treatment of atypical depression, the response rate was 70% and the remission rate was 60% in the group taking chromium compared with 0% response in the placebo group (Davidson, 2003). Chromium improves sensitivity of insulin receptors, which increases the transport of tryptophan across the blood-brain barrier and ultimately increases serotonin synthesis in the brain (McCarthy, 1994). In addition, it has been shown to increase peripheral and central tryptophan availability as well as increasing the content of 5-HT in the brains of rats (Attenburrow, 2002).

B Vitamins, SNPs, and Methylation

Folic acid, found in leafy green vegetables, enriched flour, or a multivitamin pill, is not active and requires an enzyme 5,10-methylenetetrahydrofolate reductase (MTHFR) to convert it into the active 5-methylenetetrahydrofolate (MTHF). MTHF then donates a methyl group to homocysteine, which becomes methionine and with 1 ATP becomes s-adenosylmethionine (SAM-e). MTHF, tetrahydrobiopterin (BH4), and SAM-e are examples of methyl donors. MTHF is also responsible for the recycling of BH4, a cofactor for the synthesis of neurotransmitters.

Deficiencies of folate, vitamin B12 or vitamin B6, as well as single nucleotide polymorphisms (SNPs) of the genes that involve methylation, such as MTHFR, methionine synthase reductase (MTRR), and cystathionine β synthase (CBS) have been linked to psychiatric disorders (Almeida, 2005; Mitchell, 2014; Sugden, 2006). Figure 13.1 depicts the methylation cycle.

There are two MTHFR mutations—C677T and A1298C. While an association has been found between MTHFR C677T variant and depression, schizophrenia, and bipolar disorder (Bjelland, 2003; Gilbody, 2006; Lewis, 2006), both mutations play an important role in health, and everyone should be tested for both mutations. In addition, decreased CBS and MTRR enzyme activity (caused by SNPs) will result in elevated homocysteine, which in turn will cause neurotoxic reactions in the brain and increased risk of cardiovascular disease (Bhatia, 2015).

There are two types of methylation disorders—those related to undermethylation and those related to overmethylation. However, having an MTHFR mutation does not guarantee undermethylation, because there are several other enzymes—and their SNPs—that also affect methylation.

SAM-e

SAM-e is an amino acid synthesized in our bodies by combining methionine with 1 ATP. As our body's most important methyl donor, SAM-e is essential for the synthesis of neurotransmitters; protection and repair of DNA; regulation of gene expression; synthesis of cell membranes; and phase 2 liver detoxification (Sugden, 2006). It is also a very effective antidepressant with a faster onset of action; it is

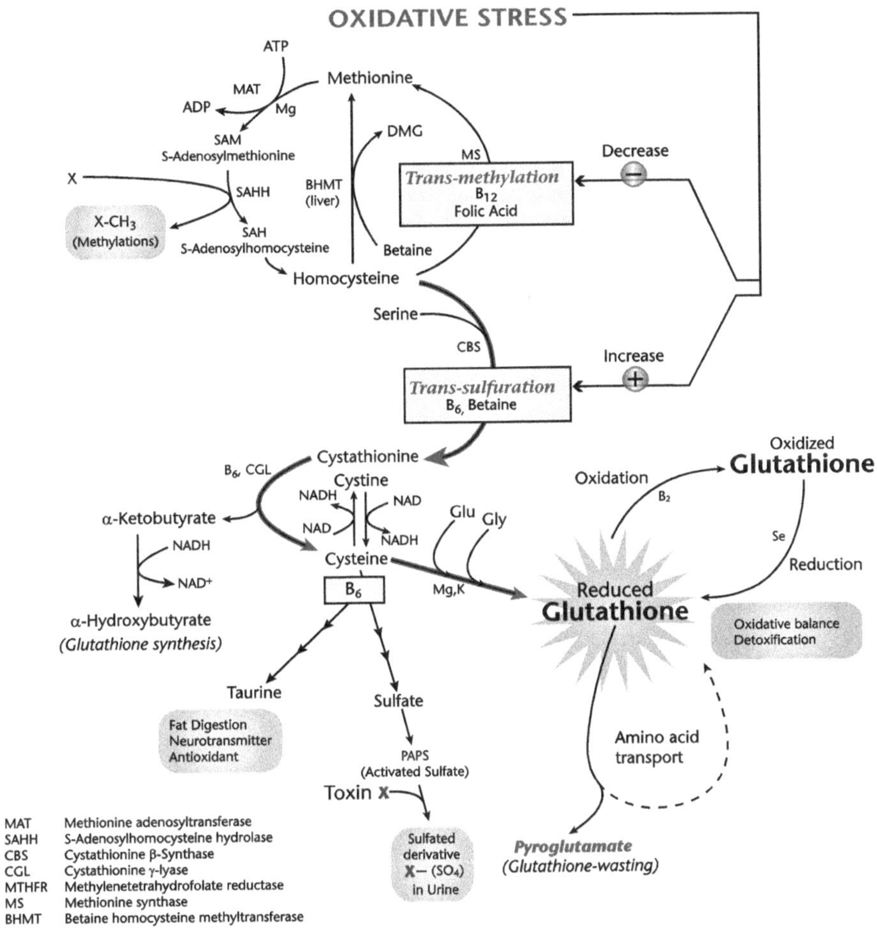

FIGURE 13.1. Methylation cycle.
Source: Lord RS, Braillley JA. (2008a). *Laboratory evaluations for integrative and functional medicine,* 2nd ed. Metametrix Institute, p. 22.

better tolerated than prescription antidepressants, with fewer side effects, no sexual side effects (Carney, 1987; Dording, 2012), and no weight gain (Bressa, 1994; Chiaie, 2002). SAM-e has been shown to have a positive effect on sexual function (erectile dysfunction and arousal in men), independent of improvement in their symptoms of depression (Dording, 2012). While pharmaceutical antidepressants can be toxic to the liver, SAM-e is safe for patients with liver disease. In a study of depressed patients with liver disease, not only did SAM-e help their depression, but it also improved their liver function, and a significant number of patients were removed from the liver transplant list (Mato, 1999).

SAM-e plays an important role in the regulation of mood. SAM-e potentiates the effect of pharmaceutical antidepressants (Berlanga, 1992; Papakostas, 2010, 2012), and higher plasma SAM-e levels are correlated with clinical improvement in depressed patients, regardless of the treatment (Bell, 1994). Furthermore, adequate blood levels of SAM-e are needed to achieve a treatment response using any antidepressant (Fava, 1994).

Although it is sold only over the counter in the United States, it is a prescription medication in some countries in Europe and used to treat depression, arthritis, and liver disease (Bottiglieri, 2002).

There have been over 40 clinical trials with SAM-e (Bottiglieri, 1994, 2002; Mischoulon, 2002). Studies have consistently found that SAM-e 1600 mg is superior to placebo and is as effective as imipramine 150 mg in treating major depressive disorder.

Tetrahydrobiopterin

BH4, which also is a methyl donor, has an important role in many biological processes associated with neurotransmitter formation, immune response, relaxation of blood vessels, and sexual response (Werner-Felmayer, 2002). It is a cofactor for the rate-limiting step in the synthesis of neurotransmitters, converting phenylalanine to tyrosine, tyrosine to dopamine, and tryptophan to serotonin. BH4 is also required for conversion of arginine into nitric oxide, which is essential for relaxation of the smooth muscles of blood vessels (Werner-Felmayer, 2002) and required for normal sexual response in men (Sommer, 2006). Lack of BH4 will result in low levels of neurotransmitters, leading to anxiety and depression, as well as low levels of nitric oxide, leading to hypertension and sexual dysfunction in men. Medications used to treat erectile dysfunction, such as sildernafil (Viagra), are effective because they increase levels of nitric oxide (Dishy, 2001; Sommer, 2006).

High levels of phenylalanine and low levels of tyrosine are a sign of possible BH4 deficiency. Because BH4 is recycled by MTHF, SNPs that interfere with synthesis of methylfolate can also cause low levels of BH4. BH4 can be given as a supplement.

Vitamin D

Vitamin D is crucial for mood (Berk, 2007; Hoogendijk, 2008) and cognition (Wilkins, 2006). Vitamin D3 is activated locally in the brain (Garcion, 2002), and vitamin D receptors have been found in the hippocampus, prefrontal cortex, hypothalamus, thalamus, and substantia nigra, regions of the brain linked to depression (Eyles, 2005). Furthermore, vitamin D is needed for the synthesis of brain-derived neurotrophic factor (Naveilhan, 1996; Pozzi, 2013) and at least one enzyme involved in neurotransmitter synthesis (Garcion, 2002; Puchacz, 1996). Vitamin D has also been shown to increase glutathione levels, suggesting that it plays an important role in brain detoxification (Garcion, 2002). In addition, it regulates gene expression (Hossein-Nezhad, 2013b) and is essential for general health (Cannell, 2008; Hossein-Nezhad, 2013a), reducing mortality from any cause (Autier, 2007), and even preventing cancer (Giovannucci, 2005, 2006, 2007, 2008; Holick, 2004; Lappe, 2007).

Normal levels of vitamin D3 are important for sexual functioning. Low 25-hydroxy vitamin D3 levels have been shown to be associated with low testosterone levels, and administering vitamin D3 has been shown to increase testosterone levels (Nimptsch, 2012; Tak, 2015). This is a critical finding since low testosterone levels not only cause sexual dysfunction but are also associated with premature death in men over 50 (Laughlin, 2008).

Vitamin D is made in the skin from cholesterol as a result of exposure to sunlight. Total-body sun exposure of 20 to 30 minutes, without sunscreen, provides the equivalent of 10,000 IU vitamin D a day (Vieth, 1999, 2006, 2007). This translates to approximately 70 to 100 ng per ml 25-hydroxy vitamin D3 in the serum, but only individuals who receive abundant sun exposure, such as surfers, lifeguards, and sun worshippers have this robust a level without supplementation; Hollis, 2007). Because melanin functions as a light filter in the skin, dark-skinned individuals are 90% less efficient in producing vitamin D in their skin compared with light-skinned individuals (Hossein-Nezhad, 2013a, 2013b) and are most at risk to be vitamin D deficient unless they take vitamin D supplementation (Holick, 2007).

Deficiency of vitamin D is epidemic (Holick, 2005, 2007; Hossein-Nezad, 2013). Everyone who does not get daily sun exposure (without sunscreen) should have daily supplementation. The recommended dose of vitamin D3 (cholecalciferol) is controversial. The RDA of vitamin D3 has been set at 600 or 800IU (Ross, 2011). However, much higher doses are needed for adequate physiological protection of the immune system and the brain (Cannell, 2008). Optimum levels of serum 25-hydroxy vitamin D3 are between 50 and 60 ng/mL and may be even as high as 80 ng/mL in cases of autoimmune illness (Holick, 2007).

Daily doses of 2,000 to 10,000 IU vitamin D3 are considered to be safe (Hossein-Nezhad, 2013a, 2013b). Vitamin D toxicity with hypercalcemia is observed when serum 25(OH)D levels are greater than 150 ng/ml (374 nmol/L) (Holick, 2007) and when intake of vitamin D3 was more than 40,000 IU daily (Vieth, 1999).

HERBS FOR DEPRESSION

St. John's Wort

St. John's wort (*Hypericum perforatum*) is an herb used to treat depression. There is abundant evidence of St. John's wort's effectiveness as an antidepressant for mild or moderate depression (Kalb, 2001; Linde, 2008) but not severe depression (Davidson, 2002). Because St. John's wort is inexpensive, is readily available without a prescription, and does not cause sexual side effects common to prescription antidepressants (Trautmann-Sponsel, 2004), it would seem to be the perfect herb for treating depression.

However, St Johns wort poses a real danger when used with other medications. Because St. John's wort results in selective induction of cytochrome P450 3A enzymes in the intestinal wall and in the liver, it needs to be used very carefully. The increased activity of these enzymes significantly speeds up the metabolism and detoxification of other medications and therefore routinely changes their effectiveness (Mansky, 2002). For example, in individuals taking St John's wort, digoxin levels were lowered 28% compared with the group that did not receive St John's wort (Brown, 2001; Jonte, 1999); cyclosporine levels decreased, causing rejection of heart transplants (Ruschitzka, 2000); and anticoagulation was impaired while taking Coumadin; efficacy of HIV medications decreased; and there was failure of birth control pills, resulting in unwanted pregnancies (Hammerness, 2003).

Saffron (Crocus sativus L)

At a cost of $11,000 per kg, saffron is the world's most expensive spice, made from the dried stigmas of the crocus flower. Four hundred fifty thousand hand-picked stigmas are needed to produce 1 kg of the spice (Lopresti, 2014; Melnyk, 2010). A fascinating paper favorably compares the current scientific research on saffron, culled from 76 articles and seven books in English and Persian, with the information contained in the Canon of Medicine, written by the Persian physician, Avicenna, in the 2nd century AD (Hosseinzadeh, 2012). The Canon of Medicine recommended saffron for treating depression, insomnia, seizures, and other ailments. It was also used as an aphrodisiac and to induce labor (Hosseinzadeh, 2013).

Saffron is effective for the treatment of mild to moderate depression (Moshiri, 2006). Studies have found saffron 30 mg/day to be safe, free of side effects, significantly better than placebo, and comparable to imipramine 100 mg/day (Akhondzadeh, 2004, 2005) and to fluoxetine 20 mg/day (Akhondzadeh, 2007; Hausenblas, 2013; Noorbala, 2005). Saffron has also been shown to be effective for treating the sexual side effects of fluoxetine in women, increasing arousal, lubrication, and improving dysparenuria (Kashani, 2013). Saffron in tablet form is available over the counter in France for the treatment of depression.

Saffron has been found to have anti-inflammatory, antioxidant, analgesic, anticancer, antiplatelet, and antispasmodic properties. Its antidepressant effects are thought to be due to affecting dopamine, norepinephrine, and serotonin, as well as its anti-inflammatory and antioxidant properties (Hosseinzadeh, 2012, 2013; Lopresti, 2014; Moshiri, 2015).

NEUROTRANSMITTERS AND AMINO ACIDS

Neurotransmitters are chemical messengers that transmit nerve impulses across a synapse. Serotonin, gaba-aminobutyric acid (GABA), glycine, and taurine are considered inhibitory neurotransmitters and, when levels are normal, are calming to the brain and body. Low levels of glycine or taurine will cause anxiety. Low levels of serotonin can cause depression, anger and aggression, anxiety, impulsivity, pain, overeating, insomnia, and diminished sexual responsiveness (Birdsall, 1998). Excitatory neurotransmitters—epinephrine, norepinephrine, dopamine, glutamate, phenylethylamine, and histamine—are stimulating to the brain and body. Low levels of these excitatory neurotransmitters may cause fatigue, lack of focus, and motivation problems. Dopamine is responsible for feelings of pleasure and satisfaction, and low levels of dopamine can cause cravings and addictions (Arias-Carrion, 2007; Di Chiara, 2007).

The amino acid tyrosine is responsible for the synthesis of epinephrine, norepinephrine, and dopamine, and the amino acid tryptophan is required for the synthesis of serotonin. Iron and BH4 are cofactors for the rate-limiting steps and a deficiency of these coenzymes will result in impaired synthesis of neurotransmitters. SSRI antidepressants help depression apparently by increasing serotonin levels in a neuron by blocking reuptake of serotonin into presynaptic storage vesicles.

Serotonin is unable to cross the blood-brain barrier and must be synthesized in the central nervous system from tryptophan or 5-hydroxytryptophan (5-HTP). Because L-tryptophan is an essential amino acid, it cannot be synthesized by the human body and must be acquired by taking a supplement or from food—from chocolate, peanuts, turkey, chicken, oats, bananas, dried prunes, milk, tuna fish, cheese, or bread (Richard, 2009). Furthermore, getting tryptophan into the brain is a challenge (Fernstrom, 2012). It is an active process, controlled by insulin receptors, and can be enhanced by eating a small carbohydrate snack along with a tryptophan supplement at bedtime (Bravo, 2013).

On the other hand, 5-HTP bypasses the rate-limiting step in the synthesis of serotonin (the conversion of L-tryptophan into 5-HTP) and easily crosses the blood-brain barrier. 5-HTP was found to be as effective as fluoxetine (Prozac; Jangid, 2013) and also successfully augmented antidepressant response when added to an antidepressant (Iovieno, 2011; Shaw, 2002). 5-HTP does not cause sexual side effects, but, while rare, side effects can include nausea, vomiting, diarrhea, headaches, and insomnia (Byerley, 1987).

Low levels of tryptophan and consequent compromised serotonergic activity is associated with aggressive behavior (Bjork, 2000), impulsivity (Dougherty, 1999; LeMarquand, 1999), depression (Maes, 1997; Richard, 2009), and long-term (but not short-term) memory problems (Riedel, 1999). Tryptophan supplementation has been shown to treat the symptoms of mild to moderate depression but not severe depression (Young, 1996) and can help treat seasonal affective disorder (Ghadirian, 1998, Lam, 1997). Premenstrual symptoms, particularly depression and aggression, are correlated with low levels of tryptophan and have been shown to improve with administering tryptophan (as well as SSRIs) in the late luteal phase of menstruation (Bond, 2001; Menkes, 1994; Steinberg 1999).

Decreased serotonin levels as well as deficiencies of Mg, B6, and BH4 can be caused by stress (Bottiglieri, 1992). Insulin resistance can decrease serotonin levels, both by inhibiting the tryptophan hydroxylase enzyme needed to convert tryptophan to 5-HTP and by increasing the conversion of tryptophan to kynurenine, making tryptophan less available for serotonin production, thus causing depression (Iovieno, 2011; Maes, 1990).

It is worth mentioning that this pathway of conversion of tryptophan to kynurenine is essential in pregnancy because it prevents rejection of the fetus as a foreign tissue (Maes, 2002). Proinflammatory cytokines stimulate the synthesis of the enzyme, indoleamine 2,3-dioxygenase (IDO), which is expressed at the maternal-fetal interface and causes a dramatic fall in tryptophan levels and inhibition of T cell activity. Tryptophan remains low throughout pregnancy but begins to increase during labor and for the next two days (Kohl, 2005; Kudo, 2004; Munn, 1998). However, tryptophan levels did not change in women with postpartum depression (Kohl, 2005). Postpartum depression and anxiety in the days following delivery may be caused by the effects of continued immune activation of IDO in the brain, leading to the preference to synthesize kynurenine, rather than serotonin from tryptophan (Bailara, 2006; Maes, 2002; Schröcksnadel, 2003). In a different study, women with postpartum depression were found to have inflammation with elevated levels of hsCRP and IL6 in the six months after delivery (Liu, 2016).

The obvious approach of supplementing L-tryptophan has been extensively studied, and both animal and human studies confirm effective antidepressant effects in mild to moderate depression (Young, 2007). A 3 gram load of L-tryptophan produced almost 100% rise in central nervous system serotonin synthesis (Young, 1981), and mixtures of free-form amino acids containing only 2% L-tryptophan along with the active form of vitamin B6 were effective in reducing the symptoms of major depression (Ille, 2007).

Because tryptophan is a precursor to melatonin, it has also been used successfully in the treatment of sleep disorders (Richard, 2009).

HORMONES

Thyroid

Adequate thyroid function is essential for normal brain function, sexuality, and mood (Bauer, 2003). Fifty percent of individuals with severe depression are hypothyroid (Sintzel, 2004), and the vast majority of individuals with severe hypothyroidism experience depression (Henley, 1997). Low-normal range thyroid functioning (or below normal range) has been found to be associated with greater severity of depression and increased numbers of episodes (Bauer 2003; Frye, 1999).

Levothyroxine (T4) is often prescribed as thyroid supplementation to treat a hypothyroid condition. However, many individuals cannot adequately convert T4, the storage form of the hormone, to T3 (triiodothyronine), the active form of the hormone (Ortiga-Carvalho, 2016). T3 potentiates the action of antidepressants (Coppen, 1972; Lifschytz, 2006; Sintzel, 2004) and is useful in treatment-resistant depression. Everyone with depression should receive a thorough thyroid workup, which includes TSH, free T3, T3, free T4, T4, reverse T3, and antithyroglobulin and TPO antibodies.

Adrenal

The initial response to a stressful situation is the release of corticotropinreleasing hormone (CRH) by the hypothalamus. CRH activates the hypothalamic-pituitary-adrenal (HPA) axis and the sympathetic nervous system, causing increased pituitary secretion of adrenocorticotropic hormone, which in turn causes increased secretion of epinephrine, norepinephrine, and cortisol by the adrenal glands. Epinephrine and norepinphrine stimulate the flight-or-flight response, which allows us to escape an impending threat. Norepinephrine makes us more aware, focused, and alert. Epinephrine stimulates an increase in heart rate, blood flow to our muscles, and the release of glucose from storage areas, thus allowing us to fight or flee—to stay and deal with the threat or to get away from what we perceive as a danger. Cortisol protects the body from the stress response by increasing levels of anti-inflammatory proteins and decreasing levels of proinflammatory proteins.

This anti-inflammatory effect of cortisol occurs only when stress is intermittent and time limited, such as the example of the zebra who has managed to get away from the lion and is now once again grazing peacefully on the savannah (Sapolsky, 1994). The worries and stresses of our modern world are constant and can create the pattern of a chronic stress response, with chronic elevations of cortisol, which in turn damage and destroy neurons in the region of the hypothalamus responsible for regulating CRH release (Sapolsky, 2000a, 2000b). This chronic stress response creates an erratic activation of the HPA axis, resulting in fatigue, depression, anxiety, a weakened immune system (Sapolsky, 2000c), and sexual dysfunction (Brotto, 2016; Kudielka, 2005). Elevated cortisol adversely affects sexual function by inhibiting gonadotropin releasing hormone, resulting in reduced estrogen and testosterone (Rivier, 1991; Tsigos, 2002).

Factors that increase inflammation—poor diet, obesity, physical inactivity, smoking, psychosocial stressors, altered GI permeability, allergies, gum disease, sleep deprivation, and vitamin D deficiency—also increase the risk for developing depression (Berk, 2013).

ANXIETY

Anxiety is a common reaction to ordinary life events and, at times, is experienced by everyone. However, anxiety is considered a psychiatric disorder when it disrupts one's daily life through excessive worry, occurring more days than not, for at least six months and causes clinically significant distress or impairment in social, school, or occupational functioning. It is associated with at least three of the following: restlessness, feeling keyed up or on edge, fatigue, difficulty concentrating or mind going blank, irritability, muscle tension, and sleep disturbance (APA, 2013). Individuals with an anxiety disorder often seek help from their primary care provider, concerned about physical symptoms and illnesses. Subtypes of anxiety disorder include generalized anxiety disorder, OCD, panic disorder, phobias, and posttraumatic stress disorder.

Antidepressant medication and benzodiazepines are often successful in treating anxiety as well as panic disorder. While one-third achieve remission, one out of five are not helped and follow an unremitting and chronic course (Katschnig, 1998; Marchesi, 2008). In addition, many patients stop these medications because of intolerable side effects, such as sedation, weight gain, addiction, sexual side effects, and relapse (Clayton, 2009). However, there are safe alternatives that relieve anxiety symptoms, without the possibility of side effects, addiction, or sexual dysfunction.

Inositol

Inositol in pharmacological doses of 12 to 18 g has been found to successfully treat depression, anxiety, panic disorders, and OCD (Einat, 2001; Fux, 1996). In double-blind studies, inositol, 18 g/day, was equal to fluvoxamine (Luvox) in the treatment of panic disorders (Palatnik, 2001) and was equal to SSRIs in reducing symptoms of OCD (Fux, 1996, 1999), without side effects and with better compliance than the SSRIs. While it is generally difficult to treat skin picking or trichotillomania, inositol was successful in reducing both of these symptoms (Seedat, 2001).

Lavender

Lavender, as a flower or oil, has been used for centuries as an aromatherapy for its calming properties. Recent studies have shown lavender oil to be effective in treating anxiety disorder as well as acute anxiety. For example, in a study of anxiety in preoperative patients, those who received lavender oil (Lavanin) had significantly less anxiety going into surgery than those who received either standard care and/or jojoba oil, which was used as placebo (Braden, 2009).

Silexan (also called Lavela WS 1265) is a patented product produced from Lavandula angustifolia flowers and authorized in Germany as a medicine for the treatment of anxiety. Silexan 80 mg was found to be equal to both lorazepam 0.5 mg (Woelk, 2010) and paroxetine 20 mg (Kasper, 2010, 2014) for the treatment of generalized anxiety disorder in double-blind placebo-controlled randomized studies. In addition, the lavender oil had no adverse sexual effects sedative effects or potential for abuse. Adverse effects were limited to GI symptoms—nausea, burping, and indigestion (Woelk, 2010). Both paroxetine and benzodiazepines are associated with adverse sexual effects. The mechanism of action of the anxiolytic effect of Silexan is thought to involve the serotonergic system, more specifically of the 5-HT1A receptor (Baldinger, 2014).

Matricaria Recutita (Chamomile)

The medicinal calming properties of chamomile are found in the flower heads of the plant, just prior to blooming. Chamomile is useful for conditions that require calming. It is anti-inflammatory, antispasmodic, antimicrobial, and sedative, and it is used for such ailments as upset stomach, colic in babies, ulcers, inflamed conditions of the skin, and of course anxiety (Ross, 2008, 2013). Randomized double-blind placebo-control studies demonstrated that chamomile was superior to placebo in treating mild to moderate generalized anxiety disorder (Amsterdam, 2009; Ross, 2013). While there are no known contraindications, those with allergy to daisies should be cautious.

Kava (Piper methysticum)

Kava was traditionally used for centuries as a calming nonalcoholic drink for ceremonies in the islands of the South Pacific. Kava targets the GABA pathway and has been found to be better than placebo (Sarris, 2011, 2013b) and as effective as tricyclic antidepressants and benzodiaepines in the treatment of anxiety disorders

(Boerner, 2003; Savage, 2015; Volz, 1997) and insomnia, (Lehrl, 2004) without the issues of tolerance, addiction (Lehrl, 2004), or sexual dysfunction. In fact, kava was found to increase sexual drive in women, compared to placebo, and showed no negative effects in men (Sarris, 2013a).

Kava was banned in Europe and Australia in 2002 because of liver toxicity associated with its use, first observed in 1999, but is now considered safe and is legal for importation and sale in Germany. Approximately 100 cases of liver toxicity worldwide have been connected to kava use, some so severe that they required liver transplantation (Stickel, 2003, 2015). Some individuals took doses of kava exceeding the 120 mg recommended daily dose and others already had impaired liver function. Furthermore, the batches of kava that made people ill were thought to be improperly manufactured. Instead of using roots and peeled stumps of noble kava varieties, which contain less than 1 mg/g of flavokavin B, the bark or peelings of non-noble varieties of kava were used, containing more than 5 mg/g flavokavin B. The peelings of the non-noble plant were exposed to sunlight, apparently making it even more toxic (Kuchta, 2015).

LIFESTYLE FACTORS AND MOOD AND ANXIETY

Lifestyle factors, including diet and nutrition, meditation, exercise, mind/body techniques, spirituality, and social support are effective in improving adrenal function and alleviating depression and anxiety, as well as other chronic illnesses including cardiovascular disease, diabetes, hypertension, and even cancer (Barbour, 2007, Dunn, 2005, Kirby, 2005, Sagner, 2014). But this is just part of a larger picture.

Epigenetics—environmental factors (what we eat, how much we sleep, how much we move, whether or not we meditate, love or feel loved, or are part of a community) turn on or off certain genes and are as important as the actual genes that we inherited. In fact, epigenetics influences the health of our chromosomes. Telomeres, located at the ends of chromosomes, protect the end of the chromosome from deterioration or from fusing with another chromosome. Telomeres become shorter in response to life stress (Epel, 2004), overall poor health, aging, and psychiatric illness (Lindqvist, 2015).

However, comprehensive changes in lifestyle, in only three months, can increase the length of telomeres in human immune cells, as demonstrated in a study of a group of patients with early prostate cancer (Ornish, 2008, 2013).

The following were the lifestyle interventions used in Ornish's study:

- a vegan diet high in whole foods, plant-based protein, fruits, vegetables, unrefined grains, and legumes and low in fat (approximately 10% of calories) and refined carbohydrates.

- moderate aerobic exercise (walking 30 minutes per day, six days per week).
- stress management (gentle yoga-based stretching, breathing, meditation, imagery, and progressive relaxation for 60 minutes daily).
- social support (60-minute support-group sessions once per week) (Ornish, 2013).

Stress Management

A causal relationship exists between stress and depression (Hammen, 2005), likely through the harmful effect of stress on the adrenal glands (Pariante, 2003). Interventions that help manage stress, including relaxation techniques (Manzoni, 2008), exercise (Carek, 2011, Ströhle, 2009), music (Lee, 2016; Pelletier, 2004), massage therapy (Diego, 2009), and meditation (Sharma, 2014), are nonpharmacologic approaches that are useful in the treatment of depression (Jacquart, 2014).

Meditation

Meditation is the act of focusing our attention on the present moment while observing or disregarding our thoughts, with the goal of attaining a sense of inner peace and calm. Different types of meditation, including transcendental meditation (Dillbeck, 1977; Orme-Johnson, 2014), mindfulness meditation (Marchand, 2012; Miller, 1995), compassion/lovingkindness meditation (Kearney, 2013; Salzberg, 2002), Zen meditation (Marchand, 2012), and movement-based meditations such as tai chi (Wang, 2013a) and qigong (Wang, 2013b) have been shown to improve mood and/or reduce anxiety. Proposed mechanisms include elicitation of the relaxation response, that is, mobilizing the parasympathetic nervous system and calming the sympathetic nervous system (Lazar, 2000) as well as changes in brain function (Davidson, 2003).

Social, Emotional, and Spiritual Support

Social isolation and loneliness are risk factors for depression, and of course those who are depressed often isolate themselves. A population-based study of more than 40,000 adults found a positive association between depression and self-reported lack of emotional and tangible support from others (Grav, 2012).

Spiritual dimensions, including a sense of meaning and purpose, also play a role in depression. Approaches designed to tap into the mind-body-spirit connection are being explored for understanding and treating depression. For example, two longitudinal studies found that feelings of gratitude appeared to result in higher levels of social support and lower levels of depression and stress. This suggests that helping patients cultivate gratitude might help prevent or reduce depression (Wood, 2008).

As these lifestyle practices improve mood and reduce anxiety and stress, they enhance a sense of well-being and a renewed interest and ability to enjoy life, including renewed interest in sex and improved sexual performance.

Exercise

Because physical exercise increases levels of endorphins and serotonin, it can be an effective treatment for depression and anxiety, as well as for certain aspects of sexual dysfunction (Cooney, 2013; Lorenz, 2012). A study of adults with major depressive disorder compared physical exercise alone, sertraline therapy alone, or a combination of exercise and sertraline. While there was significant improvement in all three groups, after 10 months those in the exercise-alone group had significantly lower relapse rate (Babyak, 2000).

Problems in genital arousal in women taking SSRIs were improved by exercise on a treadmill for 20 minutes at 80% of maximum heart rate, likely because of stimulation of the sympathetic nervous system. (Lorenz, 2012, 2014). Similarly, aerobic exercise significantly improved arterogenic erectile dysfunction (Lamina, 2011).

Light Therapy for Depression

Light therapy is a well-established effective nonpharmaceutical treatment for seasonal affective disorder (Magnusson, 2003, 2005; Tam, 2016), as well as premenstrual depression (Lam, 1999), depression during pregnancy, and postpartum depression (Epperson, 2004). Light therapy has recently been found to be effective for the treatment of nonseasonal major depressive disorder, alone and in combination with SSRIs (Lam, 2016).

Case Study

Mr B is a 55-year-old man, married with four children. He is a renowned sculptor; his work is in the permanent collections of prominent museums throughout the world. His wife, also 55, is a successful journalist. They have been married for 35 years. He is very involved with his children, who are all successful and thriving.

After his youngest daughter left for college three years ago, he began to have insomnia, felt depressed, and was not able to concentrate. He saw a psychiatrist, who prescribed fluoxetine (Prozac), 20 mg each day, which somewhat alleviated the depression, and clonazepam (Klonopin), 1 mg at bedtime, which helped the insomnia. However, he stated that he stopped feeling like himself and for the first time had problems with sexual functioning. He had diminished desire and erectile dysfunction with trouble getting and maintaining an erection. At a routine medical appointment, his total cholesterol was found to be high, at 250, and his internist prescribed atorvastin (Lipitor). Although he had been a marathon runner, he stopped exercising because he felt exhausted after exercise and developed pain in his calf muscles when he exercised. He gained 60 pounds.

He and his wife had always had an excellent sexual relationship, but since Mr B began having sexual problems, he felt helpless and ashamed and moved into his daughter's old bedroom, and, for the first time in their marriage, they slept in separate bedrooms. Their relationship deteriorated as Mr B began to drink alcohol excessively. His wife began an affair with her editor. Neither wanted a divorce.

In addition to severe depression and inability to maintain an erection, Mr B's symptoms included cold hands, migraines headaches, muscle cramps in his legs and feet, constipation, fatigue, and sensitivity to cold. When seen for the consultation, he was disheveled, depressed, suicidal, and having panic attacks. He had not bathed in two weeks and was drinking more than half a bottle of scotch almost every day. His symptoms of cold hands, migraine headaches, and erectile dysfunction indicate problems synthesizing nitric oxide from arginine (Cartledge, 2001), which could be due to low levels arginine, Mg, and/or tetrahydrobiopterin.

Laboratory testing revealed the following:

- Hypothyroid disorder with TSH 6, free T3 1.5, reverse T3 20 (normal values TSH 0.178-4.530, free T3 2-4.7, reverse T3 11–32);
- Stressed adrenal with high DHEA-S 531 ug/dL (normal 25–240)
- Low total testosterone 200 ng/ml (normal range 250–1100)
- Low vitamin D3 20 ng/mL (normal 50–60)
- Low B12 250pg/mL (while 200–900pg/mL is considered normal range by the labs, signs of B12 deficiency can be present even at levels of 500 [Lindenbaum 1994]. Therefore many physicians recommend vitamin B12 blood levels of 700 or even 750pg/mL).
- Low RBC Mg 3.5 ng/mil (normal 4–6)
- High homocysteine 20 umol/L (normal 5–12)
- Ferritin 40 ng/mL (normal 100)
- hsCRP mg/dL 12 (normal less than 3) indicating inflammation
- Total cholesterol 130mg/dL (normal 150–180mg/dL)
- MTHFR mutation—two copies of the C677T mutation

Significant medical issues:

- Hypothyroid disorder
- Low testosterone
- High homocysteine
- Leaky gut
- Inflammation with high hsCRP
- Low vitamin D3, B12
- Low RBC Mg
- Low cholesterol
- MTHFR DNA mutation
- Low ferritin—indicates leaky gut
- Mitochondrial damage
- Obesity
- Addictions

Mr B was treated with the following:

- SAM-e to treat depression and undermethylation. It was increased to a total of 1600 mg over the course of four weeks, with gradual tapering and discontinuation of Prozac.
- For leaky gut: probiotics, prebiotics; gluten-free diet; elimination of diet soda, sugar, most carbs; elimination of NSAIDS and acid blockers.
- Vitamins D3 10,000IU
- MTHF 5 mg (to address MTHFR mutation)
- Free form amino acid formula for anxiety (essential amino acids, amino acids that we are unable to synthesize but must get from food)
- BH4 (BH4 is a cofactor for the synthesis of all neurotransmitters and also cofactor for the synthesis of nitric oxide. Low levels can cause depression, anxiety, erectile dysfunction, cold hands and feet, and headaches).
- Mg 600 mg twice a day and bedtime
- Multivitamin
- B vitamins: B6, B12, niacin, biotin
- Fatty acids—fish oil, primrose oil, phosphatidyl choline
- Turmeric for inflammation
- Coenzyme Q10 300 mg twice a day
- Ashwagandha for stressed adrenal
- Natural desiccated thyroid for the hypothyroid condition
- For sleep: herbal sleep aide, combining L-theanine, 5-THP, and melatonin.

Both statins and antidepressants can damage mitochondria (Neustadt, 2008), causing extreme fatigue and pain with exercise. With the addition of coenzyme

Q10, Mr B had relief of his exhaustion and pain in his calf muscles and he began to exercise.

Within a few months, he resumed running 5 miles each day. He refused to go to Alcoholics Anonymous, but as his mood improved and sexual function normalized, he reduced his alcohol intake so that he was drinking one to two drinks a week. His wife broke off the affair with her editor. He and his wife went to couple's counseling, he moved back into their bedroom, and they resumed their sexual relationship. He reported that he felt normal again, energetic, productive, even happy. He and his wife took a series of cooking classes together, and they cook together most nights. He continued studying culinary arts and became an excellent chef.

Conclusion

Depression, anxiety, and sexual dysfunction are intertwined. Anxiety and depression—as well as the psychotropic medications used to treat them—often cause problems in sexual functioning. In addition to sexual side effects, the medications used to treat anxiety and depression also may have serious toxic effects.

Integrative psychiatry, on the other hand, provides a safe treatment for depression and anxiety, without the problematic side effects of the psychotropic medications. Instead of simply treating symptoms, integrative psychiatry aims to address the root causes of the illness and considers all factors that influence an individual's health including mind, body, spirit, social, and environmental factors. Whenever possible, it uses less toxic interventions including nutritional supplementation, herbs, homeopathy, mind/body techniques, and lifestyle change such as dietary modification, exercise, spiritual practice, and stress management.

An integrative psychiatry workup would therefore include a thorough history including lifestyle assessment and a comprehensive laboratory evaluation that addresses issues such as nutritional deficiencies, hormonal imbalances, and leaky gut, which all affect brain and mood. Recent scientific developments in the exploration of the microbiome and epigenetics suggest new directions for research, diagnosis, and treatment of medical and psychiatric conditions.

Finally, an integrative approach provides better outcomes, treating the underlying psychiatric condition as well as the sexual dysfunction, and promises improved quality of life and the relief of suffering.

REFERENCES

Akbaraly TN, Brunner EJ, Ferrie JE, Marmot MG, Kivimaki M, Singh-Manoux A (2009). Dietary pattern and depressive symptoms in middle age. *Br J Psychiatry.* 195(5):408–413.

Akhondzadeh S, Fallah-Pour H, Afkham K, Jamshidi AH, Khalighi-Cigaroudi F (2004). Comparison of Crocus sativus L. and imipramine in the treatment of mild to moderate depression: a pilot double-blind randomized trial. *BMC Complement Altern Med.* 4(12)1–5.

Akhondzadeh S, Tahmacebi-Pour N, Noorbala A, Amini H, Fallah-Pour H, Jamshidi A, Khani M (2005). Crocus sativus L. in the treatment of mild to moderate depression: a double-blind, randomized and placebo-controlled trial. *Phytother Res.* 19(2):148–151.

Akhondzadeh BA, Moshiri E, Noorbala AA, Jamshidi AH, Abbasi SH, Akhonzadeh, S (2007). Comparison of petal of Crocus sativus L. and fluoxetine in the treatment of depressed outpatients: a pilot double-blind randomized trial. *Prog Neuropsychopharmacol Biol Psych.* 31(2):439–442.

Almeida OP, Flicker L, Lautenschlager NT, Leedman P, Vasikaran S, Bockxmeer FM (2005). Contribution of the MTHFR gene to the causal pathway for depression, anxiety and cognitive impairment in later life. *Neurobiol Aging.* 26(2):251–257.

American Psychiatric Association (2013). *Diagnostic and statistical manual of mental disorders,* 5th ed. Washington, DC: American Psychiatric Publishing.

Ames BN, Elson-Schwab I, Silver EA (2002). High-dose vitamin therapy stimulates variant enzymes with decreased coenzyme binding affinity (increased K(m)): relevance to genetic disease and polymorphisms. *Am J Clin Nutr.* 75(4):616–658.

Ames BN (2004). A role for supplements in optimizing health: the metabolic tune-up. *Arch Biochem Biophys.* 423(1):27–234.

Amsterdam JD, Li Y, Soeller I, Rockwell K, Mao JJ, Shults J (2009). A randomized, double-blind, placebo-controlled trial of oral Matricaria recutita (chamomile) extract therapy for generalized anxiety disorder. *J Clin Psychopharmacol.* 29(4):378–382.

Andersohn F, Schade R, Suissa S, Garbe E (2009). Long-term use of antidepressants for depressive disorders and the risk of diabetes mellitus. *Am J Psychiatry.* May;166(5):591–598.

Andrade C, Sandarsh S, Chethan KB, Nagesh KS (2010). Serotonin reuptake inhibitor antidepressants and abnormal bleeding: a review for clinicians and a reconsideration of mechanisms. *J Clin Psychiatry. 71*(12):1565–1575.

Arias-Carrión O, Pŏppel E (2007). Dopamine, learning, and reward-seeking behavior. *Acta Neurobiol Exp (Wars).* 67(4):481–488.

Attenburrow M J, Odontiadis J, Murray B, Cowen P, Franklin M (2002). Chromium treatment decreases the sensitivity of 5-HT2A receptors. *Psychopharmacology.* 159(4):432–436.

Auerbach M, Adamson JW (2015). How we diagnose and treat iron deficiency anemia. *Am J Hematol.* 91(1):31–38.

Autier P, Gandini S (2007). Vitamin D supplementation and total mortality: a meta-analysis of randomized controlled trials. *Arch Intern Med.* 167(16):1730–1737.

Babyak M, Blumenthal JA, Herman S, Khatri P, Doraiswamy M, Moore K, Krishnan KR (2000). Exercise treatment for major depression: maintenance of therapeutic benefit at 10 months. *Psychosom Med.* 62(5):633–638.

Bahreini N, Gharipour M, Khosravi-Boroujeni H, Rouhi-Boroujeni H, Shiranian A, Salehi-Abargouei A, Sharifzadeh G (2013). Pivotal role of Mediterranean dietary regimen in the increase of serum magnesium concentration in patients with coronary artery disease. *J Nutr Metab.* 2013:1–5.

Baldinger P, Höflich AS, Mitterhauser M, Hahn A, Rami-Mark C, Spies M, Kasper S (2014). Effects of Silexan on the serotonin-1A receptor and microstructure of the human brain: a randomized, placebo-controlled, double-blind, cross-over study with molecular and structural neuroimaging. *Int J Neuropsychopharmacol.* 18(4):1–9.

Baldwin DS, Foong T (2013). Antidepressant drugs and sexual dysfunction. *Br J Psychiatry.* 202(6):396–397.

Balijepalli S, Kenchappa RS, Boyd MR, Ravindranath V (2001). Protein thiol oxidation by haloperidol results in inhibition of mitochondrial complex I in brain regions: comparison with atypical antipsychotics. *Neurochem Int.* 38(5):425–435.

Barbour KA, Edenfield TM, Blumenthal JA (2007). Exercise as a treatment for depression and other psychiatric disorders: a review. *J Cardiopulm Rehabil Prev.* Nov-Dec;27(6):359–367.

Barney LJ, Griffiths, KM Jorm AF, Christensen H (2006). Stigma about depression and its impact on help-seeking intentions. *Aust N Z J Psychiatry.* 40(1):51–54.

Barragán-Rodríguez L, Rodríguez-Moran M, Guerrero-Romero F (2008). Efficacy and safety of oral magnesium supplementation in the treatment of depression in the elderly with type 2 diabetes: a randomized, equivalent trial. *Magnes Res.* 21(4):218–223.

Bauer M, London ED, Silverman DH, Rasgon N, Kirchheiner J, Whybrow PC (2003). Thyroid, brain and mood modulation in affective disorder: insights from molecular research and functional brain imaging. *Pharmacopsychiatry.* 36(Suppl 3):S215–S221.

Bell KM, Potkin SG, Carreon D, Plon L (1994). S-adenosylmethionine blood levels in major depression: changes with drug treatment. *Acta Neurol Scand Suppl.* 89(154):15–18.

Benson MH, Friedman PR (1996). Harnessing the power of the placebo effect and renaming it "remembered wellness." *Ann Rev Med.* 47(1):193–199.

Bérard A, Chaabane S, Muanda FT, Boukhris T, Zhao J (2016a). Paroxetine use during pregnancy and the risk of cardiac defects. *Br J Clin Pharmacol.* Aug;82(2):566–567.

Bérard A, Iessa N, Chaabane S, Muanda FT, Boukhris T, Zhao J (2016b). The risk of major cardiac malformations associated with paroxetine use during the first trimester of pregnancy: a systematic review and meta-analysis. *Br J Clin Pharmacol.* 81(4):589–604.

Berk M, Sanders KM, Pasco JA, Jacka FN, Williams LJ, Hayles AL, Dodd S (2007). Vitamin D deficiency may play a role in depression. *Med Hypotheses.* 69(6):1316–1319.

Berk M, Williams LJ, Jacka FN, et al. (2013). So depression is an inflammatory disease, but where does the inflammation come from? *Bio Med Central.* 11(200):1–16.

Berlanga C, Ortega-Soto HA, Ontiveros M, Senties H (1992). Efficacy of S-adenosyl-L-methionine in speeding the onset of action of imipramine. *Psychiatry Res.* 44(3):257–262.

Beydoun MA. Kuczmarski MT, Beydoun, HA Hibbeln JR, Evans MK, Zonderman AB (2013). ω-3 fatty acid intakes are inversely related to elevated depressive symptoms among United States women. *J Nutr.* 143(11):1743–1752.

Bhatia P, Singh N (2015). Homocysteine excess: delineating the possible mechanism of neurotoxicity and depression. *Fund Clin Pharmacol.* 29(6):522–528.

Birdsall TC (1998). Therapeutic applications of taurine. *Altern Med Rev: J Clin Ther.* 3(2):128–136.

Bjelland I, Tell GS, Vollset SE, Refsum H, Ueland PM (2003). Folate, vitamin B12, homocysteine, and the MTHFR 677C→T polymorphism in anxiety and depression. *Arch Gen Psychiatry.* 60(6):618–626.

Bjork JM, Dougherty DM, Moeller FG, Swann AC (2000). Differential behavioral effects of plasma tryptophan depletion and loading in aggressive and nonaggressive men. *Neuropsychopharmacology.* 22(4):357–369.

Blasbalg TL, Hibbeln JR, Ramsden CE, Majchrzak SF, Rawlings RR (2011). Changes in consumption of omega-3 and omega-6 fatty acids in the United States during the 20th century. *Am J Clin Nutr.* 93(5):950–962.

Bodnar LM, Wisner KL (2005). Nutrition and depression: implications for improving mental health among childbearing-aged women. *Biol Psychiatry.* 58(9):679–685.

Boerner R, Sommer H, Berger W, Kuhn U, Schmidt U, Mannel M (2003). Kava-Kava extract LI 150 is as effective as opipramol and buspirone in generalised anxiety disorder—an 8-week randomized, double-blind multi-centre clinical trial in 129 out-patients. *Phytomedicine.* 10(4):38S–49S.

Bond A, Wingrove J, Critchlow G (2001). Tryptophan depletion increases aggression in women during the premenstrual phase. *Psychopharmacology.* 156(4):477–480.

Bottiglieri T (2002). S-Adenosyl-L-methionine (SAMe): from the bench to the bedside—molecular basis of a pleiotrophic molecule. *Am J Clin Nutr.* 76(5):1151S–1157S.

Bottiglieri T, Hyland K, Laundy M, Godfrey P, Carney MW, Toone BK, Reynolds EH (1992). Folate deficiency, biopterin and monoamine metabolism in depression. *Psych Med.* 22(4):871–876.

Bottiglieri T, Hyland K (1994). S-adenosylmethionine levels in psychiatric and neurological disorders: a review. *Acta Neurol Scand Suppl.* 89(154):19–26.

Boukhris T, Sheehy O, Mottron L, Bérard A (2016). Antidepressant use during pregnancy and risk of autism spectrum disorder in children. *JAMA Pediatr.* 170(2):117–124.

Braden R, Reichow S, Halm MA (2009). The use of the essential oil lavandin to reduce preoperative anxiety in surgical patients. *J Perianesthesia Nurs.* 24(6):348–355.

Bradford A, Meston CM (2006). The impact of anxiety on sexual arousal in women. *Beh Res Ther.* 44(8):1067–1077.

Bravo R, Matito S, Cubero J, et al. (2013). Tryptophan-enriched cereal intake improves nocturnal sleep, melatonin, serotonin, and total antioxidant capacity levels and mood in elderly humans. *Age (Dordr).* 35(4):1277–1285.

Bray GA (2013). Energy and fructose from beverages sweetened with sugar or high-fructose corn syrup pose a health risk for some people. *Adv Nutr.* 1;4(2):220–225.

Bressa GM (1994). S-adenosyl-l-methionine (SAMe) as antidepressant: meta-analysis of clinical studies. *Acta Neurol Scand Suppl.* 89(154):7–14.

Brotto L, Atallah S, Johnson-Agbakwu C, et al. (2016). Psychological and interpersonal dimensions of sexual function and dysfunction. *J Sex Med.* 13(4):538–571.

Brown RP, Gerbarg PL (2001). Herbs and nutrients in the treatment of depression, anxiety, insomnia, migraine, and obesity. *J Psychiatr Pract.* 7(2):75–91.

Byerley WF, Judd LL, Reimherr FW, Grosser BI (1987). 5-Hydroxytryptophan. *J Clin Psychopharmacol.* 7(3):127–137.

Cannell JJ, Hollis BW (2008). Use of vitamin D in clinical practice. *Alt Med Rev.* 13(1):6–20.

Carek PJ, Laibstain SE, Carek SM (2011). Exercise for the treatment of depression and anxiety. *Int J Psychiatry Med.* 41(1):15–28.

Carney MW, Toone BK, Reynolds EH (1987). S-adenosylmethionine and affective disorder. *Am J Med.* 83(5):104–106.

Cartledge J, Minhas S, Eardley I (2001). The role of nitric oxide in penile erection. *Expert Opin Pharmacother.* 2(1):95–107.

Chan K, Truong D, Shangari N, O'Brien PJ (2005). Drug-induced mitochondrial toxicity. *Expert Opinion Drug Metab Toxicol.* 1(4):655–669.

Chen M, Su T, Chen Y, et al. (2013). Association between psychiatric disorders and iron deficiency anemia among children and adolescents: a nationwide population-based study. *BMC Psychiatry.* 13(1):161.

Chetry BT, Hezghia A, Miller JM, et al. (2016). Omega-3 polyunsaturated fatty acid supplementation and white matter changes in major depression. *J Psychiatr Res.* 75:65–74.

Chiaie RD, Pancheri P, Scapicchio P (2002). Efficacy and tolerability of oral and intramuscular S-adenosyl-L-methionine 1,4-butanedisulfonate (SAMe) in the treatment of major depression: comparison with imipramine in 2 multicenter studies. *Am J Clin Nutr.* 76(5):1172S–1176S.

Chitturi S, George J (2002). Hepatotoxicity of commonly used drugs: nonsteroidal anti-inflammatory drugs, antihypertensives, antidiabetic agents, anticonvulsants, lipid-lowering agents, psychotropic drugs. *Sem Liver Dis.* 22(2):169–184.

Cichy A, Sowa-Kućma M, Legutko B, et al. (2009). Zinc-induced adaptive changes in NMDA/glutamatergic and serotonergic receptors. *Pharmacol Rep.* 61(6):1184–1191.

Clayton AH, Balon R (2009). The impact of mental illness and psychotropic medications on sexual functioning: the evidence and management. *J Sex Med.* 6(5):1200–1211.

Cooney G, Dwan K, Greig CA, Lawlor DA, Rimer J (2013). Exercise for depression. *Cochrane Database Syst Rev.* 311(23):2432.

Coppen A, Whybrow PC, Noguera R, Maggs R, Prange AJ Jr. (1972). The comparative antidepressant value of L-tryptophan and imipramine with and without attempted potentiation by liothyronine. *Arch Gen Psychiatry.* 26(3):234–241.

Costello RB, Moser-Veillon PB (1992). A review of magnesium intake in the elderly. A cause for concern? *Magnesium Res.* 5(1):61–67.

Cott J, Hibbeln JR (2001). Lack of seasonal mood change in Icelanders. *Am J Psychiatry.* 158(2):328–328.

Coyne, D. (2006). Iron indices: What do they really mean? *Kidney Int Suppl.* 69: S4–S8.

Crayton JW, Walsh, WJ (2007). Elevated serum copper levels in women with a history of post-partum depression. *J Trace Elements Med Biol.* 21(1):17–21.

Csoka AB, Bahrick A, Mehtonen OP (2008). Persistent sexual dysfunction after discontinuation of selective serotonin reuptake inhibitors. *J Sex Med.* 5(1):227–233.

Dalton SO, Johansen C, Mellemkjær L, Sørensen HT, Nørgård B, Olsen JH (2003). Use of selective serotonin reuptake inhibitors and risk of upper gastrointestinal tract bleeding: a population-based cohort study. *Arch Intern Med.* 163(1):59–64.

Dash SR, O'Neil A, Jacka FN (2016). Diet and common mental disorders: the imperative to translate evidence into action. *Front Public Health.* 29(4):81.

Davidson J (2002). Effect of Hypericum perforatum (St John's wort) in major depressive disorder: a randomized controlled trial. *JAMA.* 287(14):1807–1814.

Davidson JR, Abraham K, Connor KM, Mcleod, MN (2003). Effectiveness of chromium in atypical depression: a placebo-controlled trial. *Biol Psychiatry.* 53(3):261–264.

de Baaij JH, Hoenderop JG, Bindels RJ (2015). Magnesium in man: implications for health and disease. *Physiol Rev.* 95(1):1–46.

De Souza MC, Walker AF, Robinson PA, Bolland K (2000). A synergistic effect of a daily supplement for 1 month of 200 mg magnesium plus 50 mg vitamin B6 for the relief of anxiety-related premenstrual symptoms: a randomized, double-blind, crossover study. *J Women Health Gender-Based Med.* 9(2):131–139.

Di Chiara G, Bassareo V (2007). Reward system and addiction: what dopamine does and doesn't do. *Curr Opin Pharmacol.* 7(1):69–76.

Diego M, Field T (2009). Moderate pressure massage elicits a parasympathetic nervous system response. *Int J Neurosci.* 119:630–639.

Diem SJ, Blackwell TL, Stone KL, Yaffe K, Haney EM, Bliziotes MM (2007). Use of antidepressants and rates of hip bone loss in older women: the study of osteoporotic fractures. *Arch Intern Med.* 167(12):1240–1245.

Dillbeck MC (1977). The effect of the transcendental meditation technique on anxiety level. *J Clin Psychol.* 33(4):1076–1078.

Dishy V, Sofowora G, Harris PA, Kandcer M, Zhan F, Wood AJ, Stein CM (2001). The effect of sildenafil on nitric oxide-mediated vasodilation in healthy men. *Clin Pharmacol Ther.* 70(3):270–279.

Dording CM, Mischoulon D, Shyu I, Alpert JE, Papakostas GI (2012). SAMe and sexual functioning. *Eur Psychiatry.* 27(6):451–454.

Dougherty DM, Moeller FG, Bjork JM, Marsh DM (1999). Plasma L-tryptophan depletion and aggression. *Adv Exp Med Biol.* 467:57–65.

Dunn AL, Trivedi MH, Kampert JB, Clark CG, Chambliss HO (2005). Exercise treatment for depression: efficacy and dose response. *Am J Prevent Med.* 28(1):1–8.

Einat H, Belmaker R (2001). The effects of inositol treatment in animal models of psychiatric disorders. *J Affect Disorders.* 62(1–2):113–121.

Epel ES, Blackburn EH, Lin J, Dhabhar FS, Adler NE, Morrow JD, Cawthon RM (2004). Accelerated telomere shortening in response to life stress. *Proc Natl Acad Sci U S A*. 7;101(49):17312–17315.

Epperson CN, Terman M, Terman JS, et al. (2004). Randomized clinical trial of bright light therapy for antepartum depression. *J Clin Psychiatry*. 65(3):421–425.

Eyles DW, Smith S, Kinobe R, Hewison M, McGrath JJ (2005). Distribution of the vitamin D receptor and 1α-hydroxylase in human brain. *J Chem Neuroanatomy*. 29(1):21–30.

Fava M, Rosenbaum JF (1994). Preface transmethylation and the central nervous system. *Acta Neurol Scand Suppl*. 89(154):5–6.

Fawcett J (1994). Antidepressants: partial response in chronic depression. *Br J Psychiatr Suppl*. 165(26):37–41.

Fernstrom JD (2012). Large neutral amino acids: dietary effects on brain neurochemistry and function. *Amino Acids*. 45(3):419–430.

Figueira I, Possidente E, Marques C, Hayes K (2001). Sexual dysfunction: a neglected complication of panic disorder and social phobia. *Arch Sex Behav*. 30(4):369–377.

Freeman MP (2006). Omega-3 fatty acids and perinatal depression: a review of the literature and recommendations for future research. *Prostaglandins, Leukot Essent Fatty Acids*. 75(4–5):291–297.

Freeman MP, Hibbeln JR, Wisner KL, et al. (2006b). Omega-3 fatty acids: evidence basis for treatment and future research in psychiatry. *J Clin Psychiatry*. 67(12):1954–1967.

Freeman MP, Rapaport MH (2011). Omega-3 fatty acids and depression: from cellular mechanisms to clinical care. *J Clin Psychiatry*. 72(2):258–259.

Frye MA, Denicoff KD, Bryan AL, et al. (1999). Association between lower serum free T4 and greater mood instability and depression in lithium-maintained bipolar patients. *Am J Psychiatry*. 156(12):1909–1914.

Fux M, Levine J, Aviv A, Belmaker RH (1996). Inositol treatment of obsessive-compulsive disorder. *Am J Psychiatry*. 153(9):1219–1221.

Fux M, Benjamin J, Belmaker RH (1999). Inositol versus placebo augmentation of serotonin reuptake inhibitors in the treatment of obsessive–compulsive disorder: a double-blind cross-over study. *Int J Neuropsychopharmacol*. 2(3):193–195.

Garcion E, Wion-Barbot N, Montero-Menei CN, Berger F, Wion D (2002). New clues about vitamin D functions in the nervous system. *Trends Endocrinol Metab*. 13(3):100–105.

Gentile S (2015). Prenatal antidepressant exposure and the risk of autism spectrum disorders in children. Are we looking at the fall of Gods? *J Affect Disorders*. 182:132–137.

Ghadirian A, Murphy B, Gendron M (1998). Efficacy of light versus tryptophan therapy in seasonal affective disorder. *J Affect Disorders*. 50(1):23–27.

Gilbody S, Lewis S, Lightfoot T (2006). Methylenetetrahydrofolate reductase (MTHFR) genetic polymorphisms and psychiatric disorders: a huge review. *Am J Epidemiol*. 165(1):1–13.

Giovannucci E (2005). The epidemiology of vitamin D and cancer incidence and mortality: a review (United States). *Cancer Causes Control*. 16(2):83–95.

Giovannucci E, Liu Y, Rimm EB, et al. (2006). Prospective study of predictors of vitamin D status and cancer incidence and mortality in men. *J Natl Cancer Inst.* 98(7):451–459.

Giovannucci E (2007). Can vitamin D reduce total mortality? *Arch Intern Med.* 167(16):1709–1710.

Giovannucci E (2008). Vitamin D status and cancer incidence and mortality. *Adv Exp Med Biol.* 624:31–42.

Grav S, Hellzèn O, Romild U, Stordal E (2012). Association between social support and depression in the general population: the HUNT study, a cross-sectional survey. *J Clin Nurs.* 21(1–2):111–120.

Hammen C (2005). Stress and depression. *Ann Rev Clin Psychol.* 1:293–319.

Hammerness P, Basch E, Ulbricht C, et al. (2003). St John's wort: a systematic review of adverse effects and drug interactions for the consultation psychiatrist. *Psychosomatics.* 44(4):271–282.

Haney EM, Chan BK, Diem SJ, et al. (2007). Association of low bone mineral density with selective serotonin reuptake inhibitor use by older men. *Arch Intern Med.* 167(12):1246–1251.

Hausenblas HA, Saha D, Dubyak PJ, Anton SD (2013). Saffron (Crocus sativus L.) and major depressive disorder: a meta-analysis of randomized clinical trials. *J Integrat Med.* 11(6):377–383.

Henley WN, Koehnle TJ (1997). Thyroid hormones and the treatment of depression: an examination of basic hormonal actions in the mature mammalian brain. *Synapse.* 27(1):36–44.

Herbert V (1992). Iron disorders can mimic anything, so always test for them. *Blood Rev.* 6(3):125–132.

Hibbeln JR (2001). Seafood consumption and homicide mortality: a cross-national ecological analysis. *World Rev Nutrit Diet.* 88:41–46.

Hibbeln JR (2002). Seafood consumption, the DHA content of mothers' milk and prevalence rates of postpartum depression: a cross-national, ecological analysis. *J Affect Disorders.* 69(1–3):15–29.

Hibbeln JR (2008). Depression, suicide and deficiencies of omega-3 essential fatty acids in modern diets. *World Rev Nutrit Diet.* 99:17–30.

Hibbeln JR, Gow RV (2014). The potential for military diets to reduce depression, suicide, and impulsive aggression: a review of current evidence for omega-3 and omega-6 fatty acids. *Military Med.* 179(11):117–128.

Holick MF (2004). Sunlight and vitamin D for bone health and prevention of autoimmune diseases, cancers, and cardiovascular disease. *Am J Clin Nutrit.* 80(6):1678S–1688S.

Holick MF (2005). The vitamin D epidemic and its health consequences. *J Nutrit.* 135(11):2739S–2748S.

Holick MF (2007). Vitamin D deficiency. *N Engl J Med.* 357(3):266–281.

Hollis BW, Wagner CL, Drezner MK, Binkley NC (2007). Circulating vitamin D3 and 25-hydroxyvitamin D in humans: an important tool to define adequate nutritional vitamin D status. *J Steroid Biochem Mol Biol.* 103(3–5):631–634.

Hoogendijk WJ, Lips P, Dik MG, Deeg DJ, Beekman AT, Penninx BW (2008). Depression is associated with decreased 25-hydroxyvitamin D and increased parathyroid hormone levels in older adults. *Arch Gen Psychiat.* 65(5):508.

Horrigan B, Lewis S, Abrams DI, Pechura C (2012). Integrative medicine in America—how integrative medicine is being practiced in clinical centers across the United States. *Global Adv Health Med.* 1(3):18–94.

Hossein-Nezhad A, Holick MF (2013a). Vitamin D for health: a global perspective. *Mayo Clinic Proc.* 88(7):720–755.

Hossein-Nezhad A, Spira A, Holick MF (2013b). Influence of vitamin D status and vitamin D3 supplementation on genome wide expression of white blood cells: a randomized double-blind clinical trial. *PLoS One. 8*(3).

Hosseinzadeh H, Nassiri-Asl M (2012). Avicenna's (Ibn Sina) the canon of medicine and saffron (*Crocus sativus*): a review. *Phytotherapy Res.* 27(4):475–483.

Hosseinzadeh H, Nassiri-Asl M (2013). Avicenna's (Ibn Sina) the Canon of Medicine and saffron (Crocus sativus): a review. *Phytother Res.* 27(4):475–483.

Hsu JM, Rubenstein B, Paleker AG. (1982). Role of magnesium in glutathione metabolism of rat erythrocytes. *J Nutr.* 112(3):488–496.

Ille R, Spona J, Zickl M, et al. (2007). "Add-on" therapy with an individualized preparation consisting of free amino acids for patients with a major depression. *Eur Arch Psychiatry Clin Neurosci.* 257(4):222–229.

Iovieno N, Dalton ED, Fava M, Mischoulon D (2011). Second-tier natural antidepressants: review and critique. *J Affect Disorders.* 130(3):343–357.

Jacka FN, Cherbuin N, Anstey KJ, Sachdev P, Butterworth P (2015). Western diet is associated with a smaller hippocampus: a longitudinal investigation. *BMC Med.* 13(1):215.

Jacquart J, Miller KM, Radossi A, et al. (2014). The effectiveness of a community-based mind body group intervention for depression & anxiety: a pilot study. *Adv Mind-Body Med.* 20(5):6–13.

Jangid P, Malik P, Singh P, Sharma M, Gulia AK (2013). Comparative study of efficacy of l-5-hydroxytryptophan and fluoxetine in patients presenting with first depressive episode. *Asian J Psychiatr.* 6(1):29–34.

Johnson S (2001). The multifaceted and widespread pathology of magnesium deficiency. *Med Hypotheses.* 56(2):163–170.

Jonte A, Brockmuller J, Bauer S (1999). Pharmacokinetic interaction of digoxin with a herbal extract from St. John's Wort. *Clin Pharm Ther.* 66:338–345.

Kalb R, Trautmann-Sponsel RD, Kieser M (2001). Efficacy and tolerability of hypericum extract WS 5572 versus placebo in mildly to moderately depressed patients: a randomized double-blind multicenter clinical trial. *Pharmacopsychiatry.* 34(3):96–103.

Kaplan BJ, Crawford SG, Field CJ, Simpson JS. (2007). Vitamins, minerals and mood. Review. *Psychol Bull.* 133(5):747–760.

Kaplan BJ, Rucklidge JJ, Romijn A, McLeod K (2015). The emerging field of nutritional mental health: inflammation, the microbiome, oxidative stress, and mitochondrial function. Review. *Clin Psycholog Sci.* 3(6):964–980.

Kashani L, Raisi F, Saroukhani S, et al. (2013). Saffron for treatment of fluoxetine-induced sexual dysfunction in women: randomized double-blind placebo-controlled study. *Human Psychopharmacol.* 28(1):54–60.

Kasper S, Gastpar M, Müller WE, et al. (2010). Silexan, an orally administered Lavandula oil preparation, is effective in the treatment of "subsyndromal" anxiety disorder: a randomized, double-blind, placebo controlled trial. *Int Clin Psychopharmacol.* 25(5):277–287.

Kasper S, Gastpar M, Müller WE, Volz H, Möller H, Schläfke S, Dienel A (2014). Lavender oil preparation Silexan is effective in generalized anxiety disorder—a randomized, double-blind comparison to placebo and paroxetine. *Int J Neuropsychopharmacol.* 17(6):859–869.

Katschnig H, Amering M (1998). The long-term course of panic disorder and its predictors. Review. *J Clin Psychopharmacol.* 18(6 Suppl 2):6S–11S.

Kearney DJ, Malte CA, Mcmanus C, Martinez ME, Felleman B, Simpson TL (2013). Loving-kindness meditation for posttraumatic stress disorder: a pilot study. *J Traumatic Stress.* 26(4):426–434.

Kennedy SH, Dickens SE, Eisfeld BS, Bagby RM (1999). Sexual dysfunction before antidepressant therapy in major depression. *J Affect Disorders.* 56(2–3):201–208.

Kirby S (2005). The positive effect of exercise as a therapy for clinical depression. Review. *Nurs Times.* 2 4;101(13):28–29.

Kohl C, Walch T, Huber R, et al. (2005). Measurement of tryptophan, kynurenine and neopterin in women with and without postpartum blues. *J Affect Disorders.* 86(2–3):135–142.

Kronstein PD, Ishida E, Khin NA, et al. (2015). Summary of findings from the FDA regulatory science forum on measuring sexual dysfunction in depression trials. *J Clin Psychiatry.* 76(8):1050–1059.

Kuchta K, Schmidt M, Nahrstedt A (2015). German kava ban lifted by court: the alleged hepatotoxicity of kava (piper methysticum) as a case of ill-defined herbal drug identity, lacking quality control, and misguided regulatory politics. *Planta Medica.* 81(18):1647–1653.

Kudielka BM, Kirschbaum C (2005). Sex differences in HPA axis responses to stress: a review. *Biol Psychol.* 69(1):113–132.

Kudo Y, Boyd C, Spyropoulou I, et al. (2004). Indoleamine 2,3-dioxygenase: distribution and function in the developing human placenta. *J Reprod Immunology.* 61(2):87–98.

Lai JS, Hiles S, Bisquera A, Hure AJ, Mcevoy M, Attia J (2013). A systematic review and meta-analysis of dietary patterns and depression in community-dwelling adults. *Am J Clin Nutrit.* 99(1):181–197.

Lam RW, Levitan RD, Tam EM, Yatham LN, Lamoureux S, Zis AP (1997) Tryptophan augmentation of light therapy in patients with seasonal affective disorder. *Can J Psychiatry*. 42(4):303–306.

Lam RW, Carter D, Misri S, Kuan AJ, Yatham LN, Zis AP (1999). A controlled study of light therapy in women with late luteal phase dysphoric disorder. *Psychiatry Res*. 86(3):185–192.

Lam RW, Levitt AJ, Levitan RD, et al. (2016). Efficacy of bright light treatment, fluoxetine, and the combination in patients with nonseasonal major depressive disorder. *JAMA Psychiatry*. 73(1):56.

Lambert PD, McGirr KM, Ely TD, Kilts CD, Kuhar MJ (1999). Chronic lithium treatment decreases neuronal activity in the nucleus accumbens and cingulate cortex of the rat. *Neuropsychopharmacology*. 21(2):229–237.

Lamina S, Agabnusi E, Nwacha RC (2011). Effects of aerobic exercise in the management of erectile dysfunction: a meta analysis study on randomized controlled trials. *Ethiop J Health Sci*. 21(3):195–201.

Lappe JM (2007). Vitamin D and calcium supplementation reduces cancer risk: results of a randomized trial. *Am J Clin Nutrit*. 85(6):1586–1591.

Laughlin GA, Barrett-Connor E, Bergstrom J (2008). Low serum testosterone and mortality in older men. *J Clin Endocrinol Metab*. 93(1):68–75.

Lazar SW, Bush G, Gollub RL, Fricchione GL, Khalsa G, Benson H (2000). Functional brain mapping of the relaxation response and meditation. *NeuroReport*. 11(7):1581–1585.

Lee KS, Jeong HC, Yim JE, Jeon MY (2016). Effects of music therapy on the cardiovascular and autonomic nervous system in stress-induced university students: a randomized controlled trial. *J Altern Complement Med*. 22(1):59–65.

Lehrl S (2004). Clinical efficacy of kava extract WS 1490 in sleep disturbances associated with anxiety disorders: results of a multicenter, randomized, placebo-controlled, double-blind clinical trial. *J Affect Disorders*. 78(2):101–110.

Lemarquand D, Benkelfat C, Pihl RO, Palmour RM, Young SN (1999). Behavioral disinhibition induced by tryptophan depletion in nonalcoholic young men with multigenerational family histories of paternal alcoholism. *Am J Psychiatry*. 156:1771–1779.

Lewis SJ, Lawlor DA, Smith GD, Araya R, Timpson N, Day IN, Ebrahim S (2006). The thermolabile variant of MTHFR is associated with depression in the British Women's Heart and Health Study and a meta-analysis. *Mol Psychiatry*. 11(4):352–360.

Lifschytz T, Segman R, Shalom G, et al. (2006). Basic mechanisms of augmentation of antidepressant effects with thyroid hormone. *Curr Drug Targets*. 7(2):203–210.

Linde K, Berner MM, Kriston L (2008). St John's wort for major depression. *Cochrane Database Syst Rev*. 8(4):CD000448.

Lindenbaum J, Rosenberg IH, Wilson PW, Stabler SP, Allen RH (1994). Prevalence of cobalamin deficiency in the Framingham elderly population. *Am J Clin Nutr*. Jul;60(1):2–11.

Linder M C (1996). Copper biochemistry and molecular biology. *Am J Clin Nutrit*. 63(5):797S–811S.

Lindqvist D, Epel ES, Mellon SH, et al. (2015). Psychiatric disorders and leukocyte telomere length: underlying mechanisms linking mental illness with cellular aging. *Neurosci Biobehav Rev*. 55:333–364.

Liu H, Zhang Y, Gao Y, Zhang Z (2016). Elevated levels of Hs-CRP and IL-6 after delivery are associated with depression during the 6 months post partum. *Psychiatry Res*. 243:43–48.

Liu JJ, Galfalvy HC, Cooper TB, Oquendo MA, Grunebaum MF, Mann JJ, Sublette ME (2013). Omega-3 polyunsaturated fatty acid (PUFA) status in major depressive disorder with comorbid anxiety disorders. *J Clin Psychiatry*. 74(7):732–738.

Lopresti AL Drummond PD (2014). Saffron (*Crocus sativus*) for depression: a systematic review of clinical studies and examination of underlying antidepressant mechanisms of action. *Hum Psychopharmacol: Clin Exp*. 29(6):517–527.

Lord RS, Bralley JA (2008). *Laboratory evaluations for integrative and functional medicine*. Duluth, GA: Metametrix Institute.

Lorenz TA, Meston CM (2012). Acute exercise improves physical sexual arousal in women taking antidepressants. *Ann Behav Med*. 43(3):352–361.

Lorenz TA, Meston CM (2014). Exercise improves sexual function in women taking antidepressants: results from a randomized crossover trial. *Depress Anxiety*. 31(3):188–195.

Madsen E, Gitlin JD (2007). Copper and iron disorders of the brain. *Ann Rev Neurosci*. 30(1):317–337.

Maes M, Jacobs M, Suy E, Vandewoude M, Minner B, Raus J (1990). Effects of dexamethasone on the availability of l-tryptophan and on the insulin and FFA concentrations in unipolar depressed patients. *Biol Psychiatry*. 27(8):854–862.

Maes M, Vandoolaeghe E, Neels H, et al. (1997). Lower serum zinc in major depression is a sensitive marker of treatment resistance and of the immune/inflammatory response in that illness. *Biol Psychiatry*. 42(5):349–358.

Maes M, Verker R, Bonaccorso S, Ombelet W, Bosmans E, Scharpé S (2002). Depressive and anxiety symptoms in the early puerperium are related to increased degradation of tryptophan into kynurenine, a phenomenon which is related to immune activation. *Life Sciences*. 71(16):1837–1848.

Magnusson A, Boivin D (2003). Seasonal affective disorder: an overview. *Chronobiol Int*. 20(2):189–207.

Magnusson A, Partonen T (2005). The diagnosis, symptomatology, and epidemiology of seasonal affective disorder. *CNS Spectrums*. 10(8):625–634.

Mansky PJ, Straus SE (2002). St. John's Wort: more implications for cancer patients. *J Natl Cancer Inst*. 94(16):1187–1188.

Manzoni GM, Pagnini F, Castelnuovo G, Molinari E (2008). Relaxation training for anxiety: a ten-years systematic review with meta-analysis. *BMC Psychiatry*. 8:41.

Marchand WR (2012). Mindfulness-based stress reduction, mindfulness-based cognitive therapy, and zen meditation for depression, anxiety, pain, and psychological distress. *J Psychiatr Pract*. 18(4):233–252.

Marchesi C (2008). Pharmacological management of panic disorder. *Neuropsychiatr Disease Treat.* 4(1):9–106.

Martins JG (2009). EPA but not DHA appears to be responsible for the efficacy of omega-3 long chain polyunsaturated fatty acid supplementation in depression: evidence from a meta-analysis of randomized controlled trials. *J Am Coll Nutrit.* 28(5):525–542.

Mato JM, Cámara J, Paz JF, et al. (1999). S-Adenosylmethionine in alcoholic liver cirrhosis: a randomized, placebo-controlled, double-blind, multicenter clinical trial. *J Hepatol.* 30(6):1081–1089.

Mayer A (1997). Historical changes in the mineral content of fruits and vegetables. *Br Food J.* 99(6):207–211.

McCarthy M (1994). Enhancing central and peripheral insulin activity as a strategy for the treatment of endogenous depression—an adjuvant role for chromium picolinate? *Med Hypotheses.* 43(4):247–252.

McDowell AK, Lineberry TW, Bostwick JM (2011). Practical suicide-risk management for the busy primary care physician. *Mayo Clinic Proc.* 86(8):792–800.

McLeod MN, Gaynes BN, Golden RN (1999). Chromium potentiation of antidepressant pharmacotherapy for dysthymic disorder in 5 patients. *J Clin Psychiatry.* 60(4):237–240.

McLeod MN, Golden RN (2000). Chromium treatment of depression. *Int J Neuropsychopharmacol.* 3(4):311–314.

Melnyk JP, Wang S, Marcone MF (2010). Chemical and biological properties of the world's most expensive spice: Saffron. *Food Res Int.* 43(8):1981–1989.

Menkes DB, Coates DC, Fawcett J (1994). Acute tryptophan depletion aggravates premenstrual syndrome. *J Affect Disorders.* 32(1):37–44.

Miller JJ, Fletcher K, Kabat-Zinn J (1995). Three-year follow-up and clinical implications of a mindfulness meditation-based stress reduction intervention in the treatment of anxiety disorders. *Gen Hosp Psychiatry.* 17(3): 192–200.

Mischoulon D, Fava M (2002). Role of S-adenosyl-L-methionine in the treatment of depression: a review of the evidence. *Am J Clin Nutrit.* 76(5):1158S–1161S.

Mitchell ES, Conus N, Kaput J (2014). B vitamin polymorphisms and behavior: evidence of associations with neurodevelopment, depression, schizophrenia, bipolar disorder and cognitive decline. *Neurosci Biobehav Rev.* 47:307–320.

Modica-Napolitano JS, Lagace CJ, Brennan WA, Aprille JR (2003). Differential effects of typical and atypical neuroleptics on mitochondrial function in vitro. *Arch Pharmacol Res.* 26(11):951–959.

Moncrieff J (2007). Are antidepressants as effective as claimed? No, they are not effective at all. *Can J Psychiatry.* 52(2):96–97; discussion 102.

Moshiri E, Basti AA, Noorbala A, Jamshidi A., Abbasi SH, Akhondzadeh S (2006). Crocus sativus L. (petal) in the treatment of mild-to-moderate depression: a double-blind, randomized and placebo-controlled trial. *Phytomedicine.* 13(9–10):607–611.

Moshiri M, Vahabzadeh M, Hosseinzadeh H (2015). Clinical applications of saffron (Crocus sativus) and its constituents: a review. *Drug Res (Stuttg).* 65(6):287–295.

Munn DH, Zhou M, Attwood JT, et al. (1998). Prevention of allogeneic fetal rejection by tryptophan catabolism. *Science.* 21;281(5380):1191–1193.

Narayan SM, Stein MB (2009). Do depression or antidepressants increase cardiovascular mortality? *J Am Coll Cardiol.* 53(11):959–961.

Naveilhan P, Neveu I, Wion D, Brachet P (1996). 1,25-Dihydroxyvitamin D3, an inducer of glial cell line-derived neurotrophic factor. *NeuroReport.* 7(13):2171–2175.

Nemets B, Stahl Z, Belmaker RH (2002). Addition of omega-3 fatty acid to maintenance medication treatment for recurrent unipolar depressive disorder. *Am J Psychiatry.* 159(3):477–479.

Neustadt J, Pieczenik SR (2008). Medication-induced mitochondrial damage and disease. *Mol Nutr Food Res.* 52(7):780–788.

Nimptsch K, Platz EA, Willett WC, Giovannucci E (2012). Association between plasma 25-OH vitamin D and testosterone levels in men. *Clin Endocrinology.* 77(1):106–112.

Noaghiul S, Hibbeln JR (2003). Cross-national comparisons of seafood consumption and rates of bipolar disorders. *Am J Psychiatry.* 160(12):2222–2227.

Noorbala AA, Akhondzadeh S, Tahmacebi-Pour N, Jamshidi A (2005). Hydro-alcoholic extract of Crocus sativus L. versus fluoxetine in the treatment of mild to moderate depression: a double-blind, randomized pilot trial. *J Ethnopharmacol.* 97(2):281–284.

Nurnberg HG, Seidman SN, Gelenberg AJ, Fava M, Rosen R, Shabsigh R (2002). Depression, antidepressant therapies, and erectile dysfunction: clinical trials of sildenafil citrate (Viagra®) in treated and untreated patients with depression. *Urology.* 60(2):58–66.

Nurnberg HG, Hensley PL, Gelenberg AJ, Fava M, LaurielloJ, Paine S (2003). Treatment of antidepressant-associated sexual dysfunction with sildenafil: a randomized controlled trial. *JAMA.* 289(1):56–64.

Olszewska A, Szewczyk A (2013). Mitochondria as a pharmacological target: magnum overview. *IUBMB Life.* 65(3):273–281.

Orme-Johnson DW, Barnes VA (2014). Effects of the transcendental meditation technique on trait anxiety: a meta-analysis of randomized controlled trials. *J Alt Complement Med.* 20(5):330–341.

Ornish D, Lin J, Daubenmier J, Weidner G, Epel E (2008). Increased telomerase activity and comprehensive lifestyle changes: a pilot study. *The Lancet Oncology.* 9(11), 1048–1057.

Ornish D, Lin J, Chan JM, Epel E, Kemp C et al. (2013). Effect of comprehensive lifestyle changes on telomerase activity and telomere length in men with biopsy-proven low-risk prostate cancer: 5-year follow-up of a descriptive pilot study. *Lancet Oncol.* 14(11):1112–1120.

Ortiga-Carvalho TM, Chiamolera MI, Pazos-Moura CC, Wondisford FE (2016). Hypothalamus-pituitary-thyroid axis. *Compr Physiol.* 13;6(3):1387–1428.

Palatnik A, Frolov K, Fux M, Benjamin J (2001). Double-blind, controlled, crossover trial of inositol versus fluvoxamine for the treatment of panic disorder. *J Clin Psychopharmacol.* 21(3):335–339.

Papakostas GI, Mischoulon D, Shyu I, Alpert JE, Fava M (2010). S-adenosyl methionine (SAMe) augmentation of serotonin reuptake inhibitors for antidepressant nonresponders with major depressive disorder: a double-blind, randomized clinical trial. *Am J Psychiatry*. 167(8):942–948.

Papakostas GI, Shelton RC, Zajecka JM, et al. (2012). L-methylfolate as adjunctive therapy for SSRI-resistant major depression: results of two randomized, double-blind, parallel-sequential trials. *Am J Psychiatry*. 169(12):1267–1274.

Pariante CM (2003). Depression, stress, and the adrenal axis. *J Neuroendocrinol.* 15(8):811–812.

Pelletier CL (2004). The effect of music on decreasing arousal due to stress: a meta-analysis. *J Music Ther.* 41(3):192–214.

Pozzi F, Aloe L, Frajese G, Frajese G (2013). Vitamin D (calcifediol) supplementation modulates NGF and BDNF and improves memory function in postmenopausal women: a pilot study. *Res Endocrinol.* 2013:1–11.

Psaltopoulou T, Sergentanis TN, Panagiotakos DB, Sergentanis IN, Kosti R, Scarmeas N (2013). Mediterranean diet, stroke, cognitive impairment, and depression: a meta-analysis. *Ann Neurol*. 74(4):580–591.

Puchacz E, Stumpf WE, Stachowiak EK, Stachowiak MK (1996). Vitamin D increases expression of the tyrosine hydroxylase gene in adrenal medullary cells. *Mol Brain Res*. 36(1):193–196.

Rajkumar RP, Kumaran AK (2015). Depression and anxiety in men with sexual dysfunction: a retrospective study. *Compr Psychiatry.* 60:114–118.

Ramsey D, Muskin PR (2013). Vitamin deficiencies and mental health: how are they linked? *Curr Psychiatry*. 12(1):37–44.

Richard DM, Dawes MA, Mathias CW, Acheson A, Hill-Kapturczak N, Dougherty DM (2009). L-tryptophan: basic metabolic functions, behavioral research, and therapeutic indications. *Int J Tryptophan Res*. 2:45–60.

Riedel WJ, Klaassen T, Deutz NE, Someren AV, Praag HM (1999). Tryptophan depletion in normal volunteers produces selective impairment in memory consolidation. *Psychopharmacology*. 141(4):362–369.

Rivier C, Rivest S (1991). Effect of stress on the activity of the hypothalamic-pituitary-gonadal axis: peripheral and central mechanisms. *Biol Reprod*. 45(4):523–532.

Roest AM, Jonge PD, Williams CD, Vries YA, Schoevers RA, Turner EH (2015). Reporting bias in clinical trials investigating the efficacy of second-generation antidepressants in the treatment of anxiety disorders: a report of 2 meta-analyses. *JAMA Psychiatry*. 72(5):500–510.

Ross AC, Manson JE, Abrams SA, Aloia JF, Brannon PM, Clinton SK, et al. (2011). The 2011 report on dietary reference intakes for calcium and vitamin D from the Institute of Medicine: what clinicians need to know. *J Clin Endocrinol Metab*. 96(1):53–58.

Ross BM, Seguin J, Sieswerda LE (2007). Omega-3 fatty acids as treatments for mental illness: which disorder and which fatty acid? Review. *Lipids Health Dis*. 18;6:21.

Ross SM (2008). Chamomile: a spoonful of medicine. *Compr Psychiatry*. 22(1):56–57.

Ross SM (2013). Psychophytomedicine: an overview of clinical efficacy and phytopharmacology for treatment of depression, anxiety, and insomnia. *Holistic Nurs Pract.* 28(4):275–280.

Ruschitzka F, Meier PJ, Turina M, Lüscher TF, Noll G (2000). Acute heart transplant rejection due to Saint John's wort. *Lancet.* 12;355(9203):548–549.

Sagner M, Katz D, Egger G, et al. (2014). Lifestyle medicine potential for reversing a world of chronic disease epidemics: from cell to community. *Int J Clin Pract.* 68(11):1289–1292.

Salzberg S (2002). *Lovingkindness: the revolutionary art of happiness.* Boston: Shambhala.

Sánchez-Villegas A, Delgado-Rodríguez M, Alonso A, et al. (2009). Association of the Mediterranean dietary pattern with the incidence of depression. *Arch Gen Psychiatry.* 66(10):1090–1098.

Sapolsky RM (1994). *Why zebras don't get ulcers.* New York: W.H. Freeman

Sapolsky RM (2000a). Glucocorticoids and hippocampal atrophy in neuropsychiatric disorders.Review. *Arch Gen Psychiatry.* 57(10):925–935.

Sapolsky RM (2000b). The possibility of neurotoxicity in the hippocampus in major depression: a primer on neuron death. *Biol Psychiatry.* 15;48(8):755–765.

Sapolsky RM (2000c). Stress hormones: good and bad. *Neurobiol Dis.* 7(5):540–542.

Sarah M, Poonam K (1998). Diazepam induced early oxidative changes at the subcellular level in rat brain. *Mol Cell Biochem.* 178, 41–46.

Sarris J, Laporte E, Schweitzer I (2011). Kava: a comprehensive review of efficacy, safety, and psychopharmacology. *Aust N Z J Psychiatry.* 45(1):27–35.

Sarris J, Stough C, Teschke R, et al. (2013a). Kava for the treatment of generalized anxiety disorder RCT: analysis of adverse reactions, liver function, addiction, and sexual effects. *Phytother Res.* 2013 Nov;27(11):1723–1728.

Sarris J, Stough C, Teschke R, et al. (2013b). Kava in the treatment of generalized anxiety disorder: a double-blind, randomized, placebo-controlled study. *J Clin Psychopharmacol.* 33(5):643–648.

Sarris J, Logan AC, Akbaraly TN, et al. (2015). Nutritional medicine as mainstream in psychiatry. *Lancet Psychiatry.* 2(3):271–274.

Savage KM, Stough CK, Byrne GJ, et al. (2015). Kava for the treatment of generalised anxiety disorder (K-GAD): study protocol for a randomised controlled trial. *Trials.* 16(1):493.

Scheer R, Moss D (2011). Dirt poor: have fruits and vegetables become less nutritious? *Scientific American.* http://www.scientificamerican.com/article/soil-depletion-and-nutrition-loss/

Schlief ML, Gitlin JD (2006). Copper homeostasis in the CNS: a novel link between the NMDA receptor and copper homeostasis in the hippocampus. *Mol Neurobiol.* 33(2):81–90.

Schröcksnadel K, Widner B, Bergant A, Neurauter G, Schröcksnadel H, Fuchs D (2003). Tryptophan degradation during and after gestation. *Adv Exp Med Biol.* 527:77–83.

Seedat S, Stein DJ, Harvey BH (2001). Inositol in the treatment of trichotillomania and compulsive skin picking. *J Clin Psychiatry.* 62(1):60–61.

Seidman SN (2002) Exploring the relationship between depression and erectile dysfunction in aging men. Review. *J Clin Psychiatry.* 63(Suppl 5):5–12; discussion 23–25.

Serretti A, Chiesa A (2009). Treatment-emergent sexual dysfunction related to antidepressants: a meta-analysis. *J Clin Psychopharmacol.* 3:259–266.

Sharma M, Rush SE (2014). Mindfulness-based stress reduction as a stress management intervention for healthy individuals: a systematic review. *J Evid Based Complement Altern Med.* 19(4):271–286.

Shaw K, Turner J, Del Mar C (2002). Are tryptophan and 5-hydroxytryptophan effective treatments for depression? A meta-analysis. *Aust N Z J Psychiatry.* 36(4):488–491.

Shin JY, Park MJ, Lee SH, et al. (2015). Risk of intracranial haemorrhage in antidepressant users with concurrent use of non-steroidal anti-inflammatory drugs: nationwide propensity score matched study. *BMJ.* 14;351:h3517.

Sintzel F, Mallaret M, Bougerol T (2004). [Potentializing of tricyclics and serotoninergics by thyroid hormones in resistant depressive disorders]. *Encephale.* 30(3):267–275. [Article in French]

Siwek M, Dudek D, Paul IA, et al. (2009). Zinc supplementation augments efficacy of imipramine in treatment resistant patients: a double blind, placebo-controlled study. *J Affect Disord.* 118(1–3):187–195.

Sommer F, Klotz T, Steinritz D, Bloch W (2006). Evaluation of tetrahydrobiopterin (BH4) as a potential therapeutic agent to treat erectile dysfunction. *Asian J Androl.* 8(2):159–167.

Sonawalla SB, Rosenbaum JF (2002). Placebo response in depression. *Dialogues Clin Neurosci.* 4(1):105–113.

Souroujon M, Ashkenazi A, Lupo M, Levin S, Hegesh E (1982). Serum ferritin levels in celiac disease. *Am J Clin Pathol.* 77(1):82–86.

Souza ME, Polizello AC, Uyemura SA, Castro-Silva O, Curti C (1994). Effect of fluoxetine on rat liver mitochondria. *Biochem Pharmacol.* 48(3):535–541.

Steinberg S, Annable L, Young SN, Liyanage N (1999). A placebo-controlled clinical trial of L-tryptophan in premenstrual dysphoria. *Biol Psychiatry.* 45(3):313–320.

Stickel F, Baumüller HM, Seitz K, Vasilakis D, Seitz G, Seitz HK, Schuppan D (2003). Hepatitis induced by Kava (Piper methysticum rhizoma). *J Hepatol.* 39(1):62–67.

Stickel F, Shouval D (2015). Hepatotoxicity of herbal and dietary supplements: an update. *Arch Toxicology.* 89(6):851–865.

Stoltzfus RJ (2003). Iron deficiency: global prevalence and consequences. *Food Nutr Bull.* 24(4):S99–S103.

Ströhle A (2009). Physical activity, exercise, depression and anxiety disorders. *J Neural Transm.* 116(6).

Sublette ME, Ellis SP, Geant AL, Mann JJ (2011). Meta-analysis of the effects of eicosapentaenoic acid (EPA) in clinical trials in depression. *J Clin Psychiatry.* 72(12):1577–1584.

Sublette ME, Hibbeln JR, Galfalvy H, Oquendo MA, Mann JJ (2006). Omega-3 polyunsaturated essential fatty acid status as a predictor of future suicide risk. *Am J Psychiatry.* 163(6):1100–1102.

Sugden C (2006). One-carbon metabolism in psychiatric illness. *Nutr Res Rev.* 19(1):117–136.

Tak YJ, Lee LG, Kim YJ, et al. (2015). Serum 25-hydroxyvitamin D levels and testosterone deficiency in middle-aged Korean men: a cross-sectional study. *Asian J Androl.* 17(2):324–328.

Tam EM (2016). Efficacy of bright light treatment, fluoxetine, and the combination in patients with nonseasonal major depressive disorder. *JAMA Psychiatry.* 73(1):56.

Thomas D (2003). A study on the mineral depletion of the foods available to us as a nation over the period 1940 to 1991. *Nutr Health.* 17(2):85–115.

Trautmann-Sponsel RD, Dienel A (2004). Safety of Hypericum extract in mildly to moderately depressed outpatients: a review based on data from three randomized, placebo-controlled trials. *J Affect Disord.* 15;82(2):303–307.

Tsigos C, Chrousos GP (2002). Hypothalamic-pituitary-adrenal axis, neuroendocrine factors and stress. *J Psychosom Res.* 53(4):865–871.

Turner EH, Matthews AM, Linardatos E, Tell RA, Rosenthal R (2008). Selective publication of antidepressant trials and its influence on apparent efficacy. *N Engl J Med.* 358(3):252–260.

Vahdat Shariatpanaahi M, Vahdat Shariatpanaahi Z, Moshtaaghi M, Shahbaazi SH, Abadi A (2007). The relationship between depression and serum ferritin level. *Eur J Clin Nutr.* 61(4):532–535.

Van Minnen A, Kampman M (2000). The interaction between anxiety and sexual functioning: a controlled study of sexual functioning in women with anxiety disorders. *Sex Relation Ther.* 15(1):47–57.

Vieth R (1999). Vitamin D supplementation, 25-hydroxyvitamin D concentrations, and safety. *Am J Clin Nutr.* 69(5):842–856.

Vieth R (2006). What is the optimal vitamin D status for health? *Prog Biophys Mol Biol.* 92(1):26–32.

Vieth R (2007). Vitamin D toxicity, policy, and science. *J Bone Miner Res.* 22(Suppl 2):V64–V68.

Volz H, Kieser M (1997). Kava-kava extract WS 1490 versus placebo in anxiety disorders—a randomized placebo-controlled 25-week outpatient trial. *Pharmacopsychiatry.* 30(1):1–5.

Waldinger MD (2015). Psychiatric disorders and sexual dysfunction. *Handb Clin Neurol.* 130:469–489.

Walsh WJ, Isaacson HR, Rehman F, Hall A (1997). Elevated blood copper/zinc ratios in assaultive young males. *Physiol Behav.* 62(2):327–329.

Wang F, Lee EO, Wu T, Benson H, Fricchione G, Wang W, Yeung A (2013a). The effects of tai chi on depression, anxiety, and psychological well-being: a systematic review and meta-analysis. *Int J Behav Med.* 21(4):605–617.

Wang F, Man JK, Lee EO, et al. (2013b). The effects of Qigong on anxiety, depression, and psychological well-being: a systematic review and meta-analysis. *Evid Based Complement Alt Med.* 1–16.

Werner-Felmayer G, Golderer G, Werner E (2002). Tetrahydrobiopterin biosynthesis, utilization and pharmacological effects. *Curr Drug Metab.* 3(2):159–173.

Whang W, Kubzansky LD, Kawachi I, et al. (2009). Depression and risk of sudden cardiac death and coronary heart disease in women. *J Am Coll Cardiol.* 53(11):950–958.

Wilkins CH, Sheline YI, Roe CM, Birge SJ, Morris JC (2006). Vitamin D deficiency is associated with low mood and worse cognitive performance in older adults. *Am J Geriatr Psychiatry.* 14(12):1032–1040.

Woelk H, Schläfke S (2010). A multi-center, double-blind, randomised study of the Lavender oil preparation Silexan in comparison to Lorazepam for generalized anxiety disorder. *Phytomedicine.* 17(2):94–99.

Wood AM, Maltby J, Gillett R, Linley PA, Joseph S (2008). The role of gratitude in the development of social support, stress, and depression: two longitudinal studies. *J Res Personality.* 42(4):854–871.

Xia Z, Lundgren B, Bergstrand A, Depierre JW, Nässberger L (1999). Changes in the generation of reactive oxygen species and in mitochondrial membrane potential during apoptosis induced by the antidepressants imipramine, clomipramine, and citalopram and the effects on these changes by Bcl-2 and Bcl-X(L). *Biochem Pharmacol.* 57(10):1199–1208.

Young SN (1996). Behavioral effects of dietary neurotransmitter precursors: basic and clinical aspects. *Neurosci Biobehav Rev.* 20(2):313–323.

Young SN, Gauthier S (1981). Effect of tryptophan administration on tryptophan, 5-hydroxyindoleacetic acid and indoleacetic acid in human lumbar and cisternal cerebrospinal fluid. *J Neurol Neurosurg Psychiatry.* 44(4):323–328.

Young SN (2007). How to increase serotonin in the human brain without drugs. *J Psychiatry Neurosci.* 32(6):394–399.

Zimmermann MB (2006). The influence of iron status on iodine utilization and thyroid function. *Ann Rev Nutr.* 26(1):367–389.

14

Inflammation and the Gut–Brain–Mood Connection: A New Perspective in Psychiatry and Sexual Dysfunction

KATHERINE FALK, BARBARA BARTLIK, AND JANET MINDES

This is an exciting time in psychiatry. Factors such as inflammation, oxidative stress, and the dysfunction of the gastrointestinal (GI) tract and the gut microbiome are now understood to be among the causative factors in depression, other psychiatric illnesses (Maes, 2012), and sexual dysfunction (Falsetta, 2015; Urkmez, 2016). This is a new appreciation of psychiatric illness as a systemic illness, occurring in the body as well as in the brain, and it is a significant departure from the conventional neurotransmitter-based approach to the treatment of depression and anxiety (Kaplan, 2015).

Depression and Inflammation

Inflammation itself can cause the symptoms of depression, and the stress of a psychiatric illness can trigger an inflammatory response (Baganz, 2013; Berk, 2013a, 2013b; Raison, 2006, 2013). Elevated levels of proinflammatory markers—cytokines such as interleukin-6 (IL-6), interleukin-1-B (IL-1-B), tumor necrosis factor (TNF), lymphokines, and interferon—are seen in some individuals suffering from depression (Furtado, 2015a, b; Raison, 2006; Schmidt, 2016b) and other psychiatric illnesses (Borovcanin, 2012; Hamdani, 2012).

Normalization of inflammatory markers has been documented in patients who have experienced successful treatment of their depression (Eller, 2008; Hannestad, 2011), while treatment failure has been associated with persistence of elevated inflammatory markers (Raison, 2006). In addition, anti-inflammatory medications, such as nonsteroidal anti-inflammatory drugs (NSAIDS), may help to alleviate depression (Na, 2014).

Mutations in the genes for IL-1B and TNF-alpha, the primary inflammatory cytokines, have been shown to be associated with an increased risk for depression and a decreased responsiveness to antidepressant medications (Raison, 2006). Finally, antidepressant medications have been shown to have anti-inflammatory properties, which inhibit proinflammatory cytokines and protect neurons from inflammatory assault (Daniele, 2015; Noto, 2014).

Proinflammatory cytokines can directly cause depression through the activation of the enzyme indoleamine 2,3-dioxygenase, diverting tryptophan into kynurenic acid (rather than serotonin) in the microglial cells of the brain, thus lowering the levels of tryptophan and serotonin (Davis, 2015; Wichers, 2004). Figure 14.1 shows the tryptophan pathways under the conditions of B6 deficiency or inflammation.

Although kynurenic acid is neuroprotective (Moroni, 1999; Nozaki, 1992), the kynurenine metabolites quinolinic acid and 3-hydroxy-kynurenine are neurotoxic and stimulate N-methyl-D-aspartate receptors in the hippocampus, leading to nerve cell death and hippocampal atrophy (Hashimoto, 2007; Müller, 2010; Wichers, 2004). The hippocampus is essential for learning and memory, but as much as a 20% loss of hippocampal volume can occur with depression and may explain the cognitive deficits that can accompany, and exacerbate, major depression (Sapolsky, 2001; Shirazi, 2015).

Because of the brain's particular vulnerability to oxidative damage, lowering oxidative stress with antioxidant interventions is an important strategy for treating psychiatric illness (Betteridge, 2000; Berk, 2007, 2008a, 2013a, 2013b, 2014; Lafleur, 2006; Oliver, 2015; Ng, 2008).

Of note, the mood stabilizers lithium and valproate, but not antidepressants, are thought to work by targeting glutathione (GSH) activity within the brain cells, thus decreasing inflammation (Cui, 2007).

GLUTATHIONE AND N-ACETYLCYSTEINE

GSH, the body's most important intracellular antioxidant and its precursor n-acetylcysteine (NAC), also a powerful antioxidant (Dean, 2011), have been found to be effective for the treatment of depression (Berk, 2013a, 2014), cocaine dependence, and other addictions (Deepmala, 2015; LaRowe, 2013), bipolar disorder (Berk, 2007; Berk, 2008b), obsessive-compulsive disorder (Lafleur, 2006), and trichotillomania (Grant, 2009).

Trichotillomania, a severe form of obsessive-compulsive disorder, marked by compulsive pulling out of hair, is generally unresponsive to conventional treatment

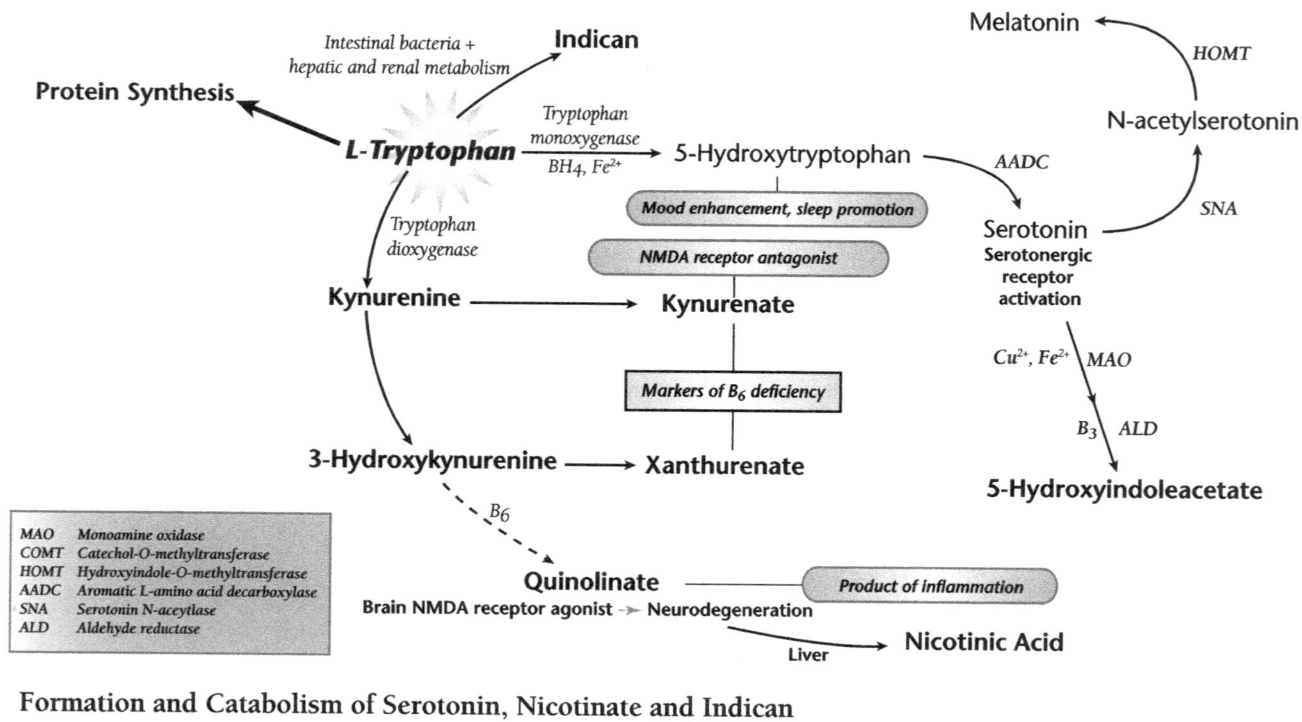

FIGURE 14.1. Tryptophan pathways.
Source: Lord RS, Braillev JA (2008). *Laboratory evaluations for integrative and functional medicine*, 2nd ed. Metametrix Institute, p. 220.

with psychiatric medications or psychotherapy. However, treatment with NAC has been shown to successfully treat this condition better than placebo or conventional treatment (Grant, 2009; Oliver, 2015).

CURCUMIN

Curcumin, the most active ingredient of the Indian spice turmeric, stimulates glutathione synthesis (Biswas, 2005) and is a powerful anti-inflammatory agent and antioxidant, acting as a COX-2 inhibitor (Lee, 2011) and a free radical scavenger (Pari, 2008), effectively lowering markers of inflammation and oxidative stress.

Curcumin has been extensively studied for over 50 years and has been shown to affect cytokines, enzymes, growth factors, growth factor receptors, transcription factors, and genes regulating apoptosis (Aggarwal, 2007; Pari, 2008). Curcumin inhibits platelet aggregation and thrombosis; protects against myocardial infarction; protects the liver; reduces cholesterol; enhances wound healing; has antibacterial and antiviral properties; inhibits HIV replication; reduces the symptoms associated with rheumatoid arthritis, multiple sclerosis (MS), type 2 diabetes, and Alzheimer's disease; and it has been shown to be effective in treating cancer (Pari, 2008). Finally, studies have shown curcumin to be effective in treating major depressive disorder (MDD; Al-Karawi, 2016; Lopresti, 2014, 2015; Sanmukhani, 2014; Schmidt, 2016a; Yu, 2015).

Curcumin's antidepressant activity is related not only to its anti-inflammatory properties but also to its ability to act as a monoamine oxidase inhibitor, resulting in increased availability of serotonin and dopamine at synapses, regulation of the hypothalamic-pituitary-adrenal (HPA) axis, modulation of neurotrophic factors, and hippocampal neuroplasticity (Kaufmann, 2016).

In one study using 500mg curcumin twice daily, improvement in MDD was seen after four to eight weeks (Singh, 2006). Patients with atypical depression (depression accompanied by oversleeping and overeating; Singh, 2006) had a better outcome than those diagnosed with MDD (Lopresti, 2014), suggesting that there is greater inflammation in atypical depression. Another study of patients with MDD, comparing curcurmin, cucurmin and fluoxetine, and placebo, found that the group that received a combination of curcumin plus fluoxetine responded the best (Sanmukhani, 2014).

FISH OIL

Inflammation is now considered to be the underlying issue in chronic illnesses, including mood and sexual disorders (Falsetta, 2015; Urkmez, 2016). The resolution of inflammation and the return to the previous baseline is now understood to be an active process, controlled by fatty acids found in docosahexaenoic acid (DHA) and eicosapentaenoic acid (EPA) of fish oil, called lipoxins, resolvins, protectins, and maresins (Serhan, 2005, 2015; Moro, 2016). These specialized proresolving mediators (SPMs), derived from essential fatty acids, have the ability to coordinate the resolution of the inflammatory response (Basil, 2016).

For example, resolvin brings about the release of anti-inflammatory cell signaling molecules called cytokines, which then direct the cleanup of the inflammatory reaction, bringing the inflammatory reaction to an end. Once the inflammation has stopped, and the debris is removed by white blood cells, individuals are able to have an improved state of health (Basil, 2016).

Many people with chronic inflammatory disease already eat fish or take fish oil supplements but are unable to stop the inflammatory process. However, SPMs, derived from essential fatty acids, have the ability to coordinate the resolution of the inflammatory response (Basil, 2016).

"LET FOOD BE YOUR MEDICINE"—HIPPOCRATES

What we eat is critical to our health. The ideal diet for improving mood, gut, and sexual health, is an anti-inflammatory diet, such as the Mediterranean diet, which has an abundance of fish protein, unsaturated fats, fruits, vegetables, fiber, and whole grains; it is high in omega-3s compared to omega-6s, and it is low in saturated fatty acids (Galland, 2010). A diet low in fiber but high in meat and fat is associated with inflammation, obesity, insulin resistance, dyslipidemia, and problems with mood and sexual function (Albenberg, 2014; Esposito, 2006).

Sexual dysfunction is associated with elevated levels of triglycerides, caused by eating excessive amounts of sugar and other carbohydrates. In mice experiments, hypertriglyceridemia is associated with significant impairment of the copulatory pattern, increased ejaculatory latencies, and reduced intracavernous pressure (Srilatha, 2006). In humans, epidemiological evidence links hypertriglyceridemia with erectile dysfunction (ED; Ahn, 2007; Roumeguère, 2003). On the other hand, lifestyle modifications that improve antioxidant status, such as exercise, reduced consumption of unhealthy fats and sugar, weight loss, supplementation with omega-3 fatty acids, and improved antioxidant status through diet and/or supplements, all reduce inflammation and increase nitric oxide (NO) production, resulting in better erectile function (Meldrum, 2012).

Dr. Andrew Weil's anti-inflammatory food pyramid makes it easy to know what to eat and how much (Weil, 2016) (see Figure 14.2).

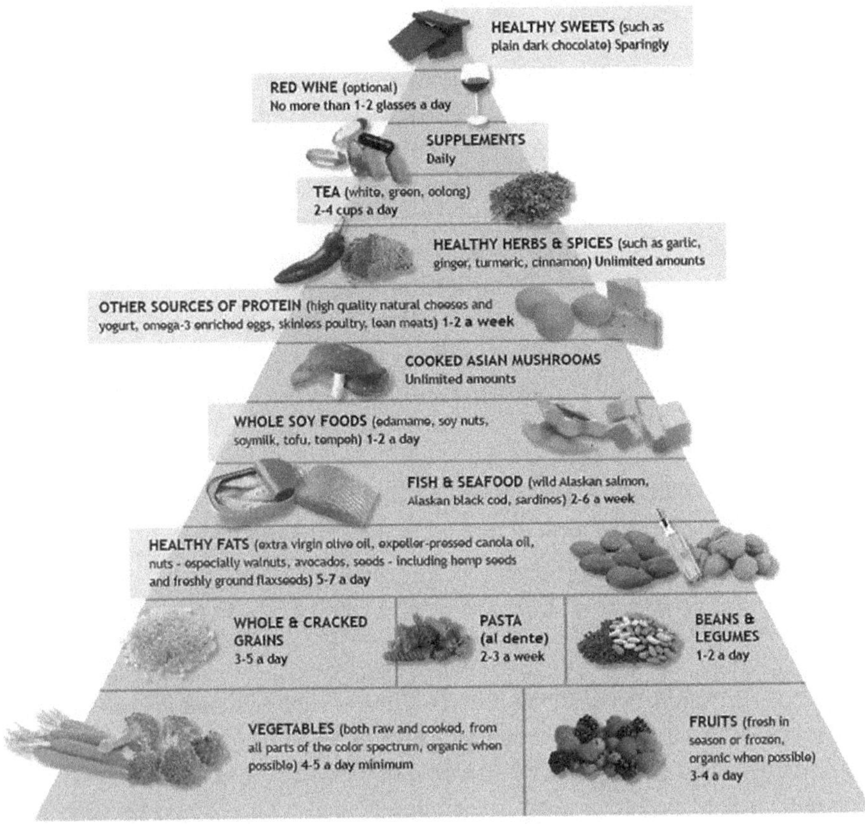

FIGURE 14.2. Dr Andrew Weil's food pyramid.
Source: https://www.drweil.com/diet-nutrition/anti-inflammatory-diet-pyramid

The Gut–Brain–Mood Connection

Inflammation of the GI tract influences mood because the gut and the brain are in direct bidirectional communication (Collins, 2012; Slyepchenko, 2014, 2016, 2017). The vagus nerve, which travels from the brainstem to the gut, plays an important role in communicating from the brain to the gut and from the gut to the brain (Bravo, 2011). The microbiome in the GI tract communicates with the brain, influencing our mood, our behavior, cognition, and pain through its effect on hormones, neurotransmitters, fatty acids, and cytokines and other immune factors (Cryan, 2012; Maes, 2008a, 2012; O'Mahony, 2015; Slyepchenko, 2017).

THE MICROBIOME

We are in the midst of an astounding revolution in medicine—the recognition of the role of the gut microbiome in illness and health. There is growing awareness of the extensive and complex role of gut health and communication within the gut-brain axis, in regulating mood and behavior, and in modulating pathophysiological processes linking depression and noncommunicable medical comorbidities such as diabetes and obesity (Luppino, 2010; Slyepchenko, 2016, 2017). Ultimately, sexual behavior and function are impacted as well.

The microbiome consists of communities of bacteria, parasites, yeast, viruses, and their genes, found throughout the body—on the mucous membranes of our GI tract as well as on the skin and in the nose, vagina, and anus (Lloyd-Price, 2016). It is estimated that approximately100 trillion organisms of more than 10,000 different species live in and on our bodies. They outnumber the cells of our own body by a factor of 10:1. In fact, 99% of the DNA in our body is bacterial DNA; only 1% of our DNA is the DNA inherited from our parents (Lepage, 2013). The sheer numbers of these micro-organisms attest to their significance in our health.

These micro-organisms are not just tourists taking a ride. They are constantly working, making us healthy or making us sick, depending on which organisms are living in the GI tract. The microbiome increases or decreases inflammation, helps make us thin or obese (Tilg, 2011; Tsai, 2009), and influences neurotransmitters in the brain, initiation and maintenance of REM sleep, mood, reactivity of the HPA axis, and how we handle stress and pain (Cryan, 2011a; Forsythe, 2010; Galland, 2014; Mayer, 2014, 2015), depending on which bacteria are present. They help us digest food, absorb nutrients and convert them to energy, regulate metabolism, and synthesize vitamins and neurotransmitters; they also aid detoxification (Cryan, 2012). Figure 14.3 demonstrates the various pathways involved in the bidirectional communication between the gut microbiota and the brain, which include cortisol, cytokines, and the vagus nerve and enteric nervous system.

The majority of our serotonin is produced in the gut by the bacteria Streptococcus, Enterococcus and E. coli (Galland, 2014) and binds to our receptor sites, resulting in increased gut motility, platelet aggregation, bone development, and regulation of cardiac function and immune responses (Mawe, 2013; Ridaura, 2015; Yano, 2015). Serotonin does not cross the blood-brain barrier but must be synthesized in the brain from tryptophan or 5-hydroxytryptophan (Birdsall, 1998).

Because the gut of the newborn is colonized during the birth process through the vaginal canal, children born by Cesarean section do not develop the same microbiome as those born through the vaginal canal (Dominguez-Bello, 2010; Funkhouser, 2013) and have a significant increased risk of autism (Curran, 2015a), attention deficit hyperactivity disorder (Curran, 2015b), celiac disease (Decker, 2010), allergies (Almqvist, 2014), and obesity (Goldani, 2013). Some obstetricians now place a sponge in the mother's vagina and immediately after the baby

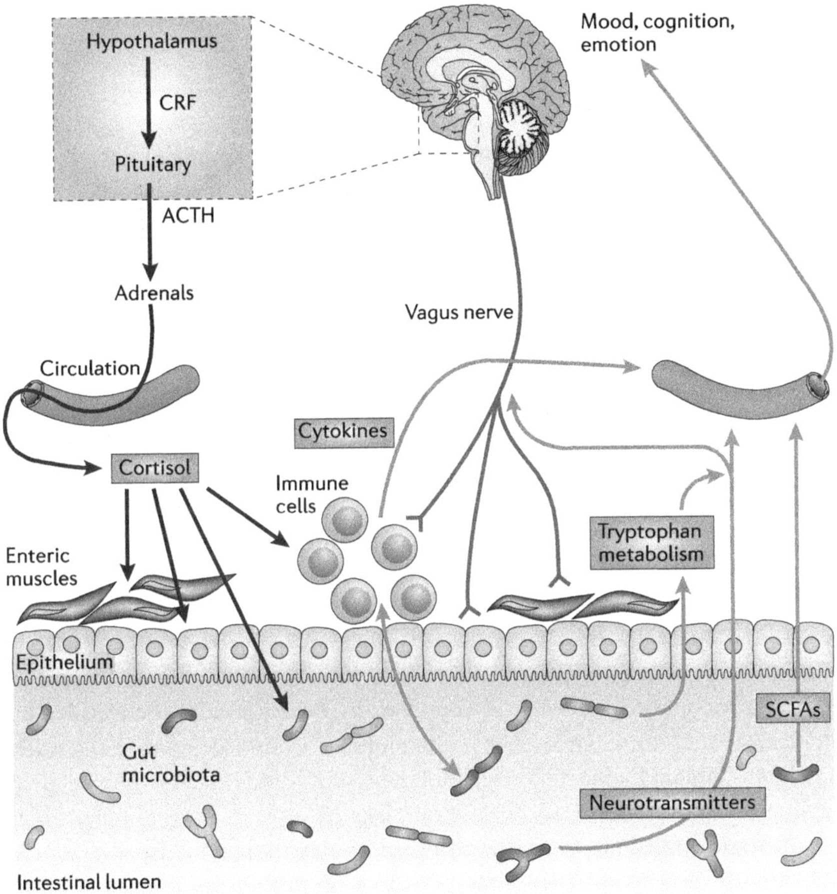

FIGURE 14.3. Bidirectional communication between the gut microbiota and the brain. *Source*: Cryan JF, Dinan TG. (2012) Mind-altering microorganisms: the impact of the gut microbiota on brain and behavior. *Nat Rev Neurosci.* 13:701–712.

is born by C-section swab the baby's face with the fluids from the mother's vagina (Mueller, 2015). By the age of one year, children have developed a complex microbiome similar to that of an adult (Grenham, 2011).

Animal studies attest to the influence of the microbiome on mood and behavior. Germ-free mice—mice born surgically and then raised in a sterile environment—had leaky gut, a poorly developed blood-brain barrier (Braniste, 2014; Cryan, 2011b, 2012), anxiety-type behavior (Neufeld, 2011), and lower levels of brain-derived neurotrophic factor (BDNF) in the cortex and hippocampus (Desbonnet, 2015) compared with mice exposed to their mother's bacteria at birth (Neufeld, 2011). BDNF is a protein that protects the nervous system, enabling survival, growth, and differentiation of new neurons and synapses in the brain and nervous system (Huang, 2001).

Stress has been found to suppress beneficial bacteria growth in studies in both animals (Dinan, 2013) and humans (Liu, 2015). One study found fewer lactobacilli in the feces of college students during exam week, compared with their first days of the semester (Knowles, 2008). Exercise, long seen as an antidote to stress, has been shown to improve the gut microbiome in exercised versus sedentary obesity-prone rats (Welly, 2016). Exercise also caused decreased insulin resistance, increased fat oxidation, improved mitochondrial function in brown adipose tissue, and increased spontaneous physical activity (Welly, 2016)

In experiments using mice, the intact vagus nerve was found to be essential for communication between the gut bacteria and the brain (Bravo, 2011; Cryan, 2011a). For example, *L. rhamnosus* had a direct effect on central nervous system GABA receptors in mice, causing increased GABA receptor expression in the cingulate and prelimbic areas of the cerebral cortex and reduced GABA receptor expression in the hippocampus, amygdala, and locus coeruleus, compared with the control mice. These effects were not found in mice whose vagus nerve was severed, indicating that the intact vagus nerve is essential for communication between the gut bacteria and the brain (Bravo, 2011).

Finally, a poor gut microbial milieu may have a very direct and immediate impact on mood and behavior. A recent study demonstrated that fecal microbiota transplantation from humans who suffered from a depressive disorder to microbiota-depleted adult rats triggered depressive- and anxiety- like behaviors, as measured by standardized behavioral tasks of rodent affective behavior and neurobiological markers typically observed in depression in the recipient rats (Kelly, 2016). Fecal samples from the humans with depression showed fewer bacterial species with lower diversity of microbial populations, compared with those collected from controls.

In summary, it is clear that certain bacteria in the gut have beneficial effects on neurotransmission and the stress response, as well as general health. Probiotics should be part of the treatment plan for treating anxiety, mood disorders, stress, and sexual dysfunction (Bravo, 2012; Slyepchenko, 2014).

LEAKY GUT (DYSBIOSIS): IMPLICATIONS FOR SEXUAL HEALTH AND MOOD

The lining of the GI tract of a healthy person forms a protective barrier consisting of a single layer of epithelial cells connected to each other by tight junctions. Proteins of the tight junction (e.g., occludin) interact with scaffold proteins (zonulin), which are responsible for the structural and functional integrity of the intestine. Leaky gut occurs when the tight junctions are damaged, creating increased permeability, which allows undigested or partially digested food, toxins, proteins, bacteria, and other micro-organisms to cross the lining of the gut and enter the bloodstream (Camilleri, 2012; Rao, 2012; Turner, 2009; Wang, 2000).

Stress and elevated cortisol, often caused by depression (Maes, 2012), can injure tight junctions and alter gut permeability, changing the composition of the gut microbiome, thus creating even more inflammation (Demaude, 2006; Dinan, 2012), which in turn will worsen the depression and also sexual functioning. Gut permeability and alterations in the composition of the microbiome are also caused by infections, food allergies, and other illnesses (Farhadi, 2003; Rao, 2012); commonly used medications such as antibiotics (Blaser, 2011), aspirin (Farhadi, 2003), and NSAIDS (Davies, 1998; Farhadi, 2003; Montenegro, 2014); drinking alcohol (Elamin, 2014; Farhadi, 2003); and the Standard American Diet, which is high in saturated fat and carbohydrates (Greer, 2011; Grenham, 2011).

Glutamine

Glutamine alone or in combination with N-acetyl cysteine and zinc restores the integrity of the tight junctions, reduces intestinal permeability, and helps maintain the normal intestinal barrier function (Galland, 2014; Rao, 2012; Rapin, 2010).

Zinc

Zinc is an important trace mineral, needed for 300 enzymatic reactions throughout the body. It is an antioxidant and anti-inflammatory, critical for the function of the immune system, sexual function, growth, and cognition. It improves intestinal permeability by repairing the disruption of the barrier function of the intestinal lining caused by damage to the terminal junctions (Hendricks, 1988; Mahmood, 2007; Sturniolo, 2001). Zinc carnosine is particularly effective for repairing the gut because carnosine prevents oxidative modification of proteins (Boldyrev, 2005; Skrovanek, 2014).

Zinc also is crucial for sexual functioning: low levels of zinc are correlated with hypogonadism, low serum tetosterone levels, low sperm count in the ejaculate, low thyroid function, as well as poor immune functioning, alopecia, and skin and nail problems (Lord, 2008, 94–97; Prasad, 2014).

Copper is often given along with zinc to avoid copper depletion. The ratio of zinc/copper is 9:1 (Lord, 2008, 98).

Gluten Sensitivity

Sensitivity to gluten, a protein found in wheat, barley, rye, oats, spelt, kamut, and triticale, causes leaky gut and leads to a systemic inflammatory response. Wheat is one of the world's primary sources of food, providing up to 50% of the caloric intake in both industrialized and developing countries (Tovoli, 2015). Compared with wheat grown for thousands of years, our modern wheat contains much larger amounts of gliadin, the toxic part of gluten (Molberg, 2005).

The prevalence of gluten sensitivity has increased throughout the world, perhaps as more people are eating wheat, often in the form of Western-style foods and fast foods. In fact, in traditionally rice-eating countries, such as India, consumption of wheat has increased (Catassi, 2007), and celiac disease, as well as non-celiac gluten sensitivity (NCGS), are on the rise (Tovoli, 2015).

Celiac disease is an autoimmune disorder, triggered by the ingestion of gluten, the protein component of wheat, in genetically susceptible individuals (Briani, 2008). This is a serious illness complicated by infertility, sexual dysfunction (Farthing, 1982; Sher, 1994), and an increased risk of cancer (Green, 2007). Treatment involves a gluten-free diet, which can quickly eliminate the symptoms, but the response is poor because of difficulty adhering to such a diet (Green, 2007). NCGS causes an increase in the production of inflammatory cytokines, leading to activation and release of zonulin, which then causes increased intestinal permeability (Lammers, 2008; Wang, 2000) and the cascade of the problems of leaky gut, as well as psychiatric and/or neurological symptoms.

Common GI symptoms that accompany NCGS are abdominal pain, bloating, diarrhea, or constipation. NCGS can also present without GI symptoms (Hadjivassiliou, 2010), as a systemic autoimmune disorder with symptoms such as headache, fatigue, brain fog, pain in muscles or joints, anemia, skin rash, and psychiatric and/or neurological symptoms including seizures, cerebellar ataxia, peripheral neuropathy, autism, depression, anxiety, and psychosis (Catassi, 2015). Gluten sensitivity can also masquerade as neurological illnesses such as MS (Volta, 2010), and, because the presenting symptoms often do not involve GI symptoms, the sensitivity to gluten can go unrecognized (Jackson, 2012).

A study of individuals with NCGS found objective markers of systemic immune activation, antibody reactivity to microbial antigens, epithelial cell damage, compromised intestinal epithelial barrier integrity, and microbial translocation. When the study participants ate a gluten-free diet, they showed improvement of symptoms, along with a significant decline of markers of immune activation and gut epithelial cell damage (Uhde, 2016).

Many individuals with NCGS do not test positive for the various immune markers for gluten sensitivity. However, a strict gluten-free diet will often result in improvement (or disappearance) of symptoms in individuals who are, in fact, sensitive to gluten, and reintroducing the gluten into the diet will result in reappearance of symptoms (Catassi, 2015; Volta, 2010).

SIGNS OF LEAKY GUT

Symptoms of leaky gut include bloating after meals, excessive flatulence, and diarrhea. Other indications of having leaky gut may include skin problems, such as rosacea, acne, eczema, and psoriasis (Bowe, 2014; Humbert, 1991; Pike, 1986); gluten sensitivity; food allergies; chemical sensitivities; seasonal allergies; asthma (Perlmutter, 2015, 36); hormonal imbalances such as premenstrual syndrome or polycystic ovary syndrome (PCOS; Zhang, 2015); mood and anxiety disorders (Dettore, 2013; Dinan, 2012; Steele, 1976); and/or diagnosis of an autoimmune disease such as Hashimoto's thyroiditis (Pasquali, 2013), celiac disease (Rostami, 2001), psoriasis, systemic lupus erythematosis (Tristano, 2009), or rheumatoid arthritis (Fasano, 2012; Hill, 2003). Leaky gut can also be associated with chronic fatigue syndrome (Blazquez, 2009), MS (Hulter, 1995), and sexual dysfunction (Dashti, 2016).

Leaky gut can be diagnosed by clinical presentation and also by measuring serum lipopolysaccharide (LPS) or antibodies to LPS (Maes, 2008b, 2012). High levels of LPS have been found in individuals with depression (Maes, 2012), Alzheimer's disease, Parkinson's disease, amyotrophic lateral sclerosis, and autism (Maes, 2008b).

RISK FACTORS FOR DEVELOPING LEAKY GUT

The risk for developing leaky gut begins prior to birth. Antibiotics, steroids, and C-section will increase the risk of the baby suffering from leaky gut and its many complications. In addition, being breastfed for less than one month, having frequent ear or throat infections and antibiotics as a child, removal of tonsils, needing steroid medication for more than one week, taking antibiotics at least once every two or three years, and taking acid-blocking drugs all increase the risk of developing leaky gut (Perlmutter, 2015, 16–17). Traumatic brain injury has been shown to damage tight junctions, leading to leaky gut, within 72 hours of the head trauma (Bansal, 2009; Feighery, 2008).

HOW TO HEAL LEAKY GUT

The following are recommendations for treating leaky gut:

- Eliminate gluten completely (Fasano, 2012).
- Identify specific food allergies and eliminate those foods (Karakuła-Juchnowicz, 2017).
- Reduce (or, if possible, eliminate) consumption of refined carbohydrates, sugars, and artificial sweeteners (Suez, 2015).
- Decrease intake or stop drinking alcohol (Elamin, 2014).

- Avoid NSAIDS, aspirin, acid blockers, and antibiotics, which damage to the microbiome (Blaser, 2011; Davies, 1998; Elamin, 2014; Montenegro, 2014).
- Supplement with zinc carnosine (Hendricks, 1988; Sturniolo, 2001); glutamine (Rao, 2012; Rapin, 2010); aloe, a versatile medicinal plant, which is antibacterial and antioxidant, treats digestive problems, and lowers blood sugar in diabetics (Nandal, 2012); quercetin, which reduces intestinal inflammation by inhibiting histamine release from mast cells in the intestine (Shaik, 2006; Suzuki, 2009); probiotic supplements (Farhadi, 2003; Logan, 2005); and vitamin D, which plays an important role in maintaining intestinal barrier function, stimulating the expression of several tight junction proteins and reducing epithelial permeability and mucosal inflammation (Pastorelli, 2013).
- Eat an anti-inflammatory diet that includes prebiotics—fermented foods containing healthy bacteria such as kimchi, sauerkraut, kombucha, and kefir.
- Manage stress with regular exercise (Hawrelak, 2004), meditation, and breathing exercises.

INFLAMMATION, LEAKY GUT, AND SEXUAL FUNCTION

Healthy sexuality is inseparable from a person's overall state of health, which can be compromised by stress; insufficient physical activity; inadequate restorative sleep; a suboptimal diet of refined carbohydrates, caffeine, meat, or dairy products from animals treated with hormones and antibiotics and including xenoestrogens from foods sprayed with pesticides or stored in certain plastic containers (Hyman, 2007). This will likely result in a myriad of problems including inflammation, microbiome imbalance, leaky gut, obesity, and sexual dysfunction. The conventional medical approach for addressing sexual dysfunction is to use medications, procedures, and devices that are for the most part focused on ED, with varied results (Hatzimouratidis, 2005). On the other hand, sexual problems can be helped by an integrative approach that addresses and attempts to correct the underlying issues.

Obesity

Our waistline is said to be a measure of our health. Abdominal (central) obesity, as well as the visceral fat covering major organs in the belly, such as the liver, intestines, and kidneys, secrete hormones and inflammatory cytokines. In fact, the fat itself actually becomes inflamed, causing a cascade of inflammatory reactions that will cause obesity, metabolic syndrome, autoimmune diseases, and mood and cognitive problems and also will disrupt normal hormone function, causing sexual dysfunction (Perlmutter, 2015, 102–104).

Lack of restorative sleep is now seen as an important cause of obesity, because it results in higher levels of cortisol and ghrelin, a hormone that stimulates hunger, and lower levels of leptin, a hormone that controls feeling satiated (Broussard, 2016; Spiegel, 2004a, 2004b).

Men with central obesity, metabolic syndrome, and insulin resistance have an increased incidence of premature ejaculation (Kradag, 2013), reduced testosterone levels (Corona, 2014), and more ED problems compared to normal weight men (Bajos, 2010; Corona, 2014; Esposito, 2008, Kradag, 2013; Saad, 2015). Metabolic syndrome and ED both were improved on the Mediterranean diet (Esposito, 2006).

Studies of overweight women are contradictory concerning the impact of weight on aspects of sexual function. Some showed negative consequences for sexual desire, arousal, lubrication, orgasm, and satisfaction (Ambler, 2012; Esposito, 2007; Kradag, 2013; Mozafari, 2015); others showed no difference between heavier and normal weight women (Jarzabek-Bielecka, 2015; Nackers, 2015). One study of obese premenopausal women who lost weight in a structured 16-week program of diet and exercise revealed significant improvements in arousal and lubrication, frequency of sex, and sexual satisfaction, perhaps because of improved endothelial function and decreased insulin resistance (Aversa, 2013).

Obesity in both men and women is associated with alteration in gut flora, with a reduced ratio of healthy (bifidobacteria and lactobacillus) to unhealthy (enterococcus, clostridia, E. coli, proteus, and pseudomonas) gut flora. This is likely caused by a high-fat, high-sugar, low-fiber diet (D'Aversa, 2013; Zhang, 2015), which also causes increased gut permeability (Cani, 2007a, 2007b; Turner, 2009). Most importantly, a low-fiber diet does not provide the prebiotics needed by bifidobacter, one of the healthy gut flora (Zhang, 2015). Bifidobacter processes insoluble dietary fiber into short chain fatty acids that feed the cells of the intestinal mucosa, increasing the production of mucus and maintaining the integrity of the tight junctions of cells of the gut lining (Tremellen, 2016; Willemsen, 2003).

Obesity in men is related to hypogonadism, low testosterone levels, diminished libido, erectile disorder, depression, functional musculoskeltal limitations and muscle weakness, and lethargy (Fui, 2014; Erskine, 2017; Luppino, 2010; Saboor Aftab, 2013; Vgontzas, 2006). Excess adipose tissue leads to a decrease in testosterone in two ways: first by the conversion of testosterone to estrogen through the action of aromatase and second by the negative feedback that increased estrogen has on the pituitary. This results in decreased luteinizing hormone (LH), and hence, decreased testosterone production. With obesity, leptin levels rise and inhibit hypothalamic pituitary (HP) function and the production of testosterone by the Leydig cells, which are adjacent to the seminiferous tubules in the testicle and produce testosterone in the presence of LH. Proinflammatory cytokines associated with obesity (Cani, 2009; Wu, 2011) also impair HP function and contribute to hypogonadism (Saboor Aftab, 2013). Low testosterone levels reduce sex drive and functioning in men (Bolona, 2007) and also in women (Shifren, 2000). As mentioned, hypogonadism in men, related to obesity, metabolic syndrome, and insulin resistance can result in reduced testosterone levels and ED (Bajos, 2010; Corona, 2014; Esposito, 2008, Kradag, 2013). Finally, poor diet, often a major feature of obesity, metabolic syndrome, and their sequelae, can weaken the integrity of the intestinal mucosal barrier (D'Aversa, 2013; Tremellen, 2012), causing material from the bowel lumen to pass into systemic circulation. This can contribute to chronic inflammation, which also impairs both testosterone production and sexual function (Tremellen, 2016).

PCOS and Leaky Gut

Increased gut permeability is considered to play an important role in the development of PCOS (Tremellen, 2012; Zhang, 2015). Serum zonulin levels are increased in women with PCOS, which correlate with insulin resistance and history of anovulation (Zhang, 2015). PCOS is also associated with an increase in white blood cell count (Herlihy, 2011; Ruan, 2009), C-reactive protein (Ruan, 2009), and increased levels of proinflammatory cytokines and chemokines (Duleba, 2012). Seventy percent of women with PCOS live in a chronic state of inflammation and insulin resistance (Traub, 2011).

While the majority of women with PCOS are obese, not all are, and normal weight women with PCOS usually have insulin resistance (Morales, 1996). Zhang (2015) hypothesized that dysbiosis and increased gut permeability allow for the passage of lipopolysaccharides from cell walls of gram-negative bacteria (E. coli) into the systemic circulation, creating a state of chronic inflammation. This interferes with insulin receptor function, creating insulin resistance, which then increases production of androgens in the ovaries, impairing normal follicle formation. Therefore, it is possible that dysbiosis leads to the symptoms of PCOS, with abnormal ovarian function, menstrual irregularity, infertility, hyperandrogenism, seborrhea, acne, hirsutism, and multiple ovarian cysts (Layegh, 2016; Zhang, 2015).

Many women with PCOS experience sexual dysfunction despite having elevated testosterone levels (Dashti, 2016). For example, PCOS is associated with impaired orgasm (Stovall, 2012). Compared to controls, women with PCOS, matched to partner status and frequency of sexual intercourse, were significantly less satisfied with their sex lives and found themselves less sexually attractive (Hahn, 2005). They did, however, have sexual thoughts and fantasies just as often as controls. PCOS is associated with an increase in anxiety and depression (Dèttore, 2013; Steele, 1976), known risk factors for impaired sexual functioning (Benson, 2009).

Inflammation and Sexuality

Inflammation adversely affects other issues of female sexuality. For example, vulvodynia, a condition in which intense vulvar pain interferes with the ability to experience sexual pleasure, has been found to have inflammatory origins (Falsetta, 2015). Tissue from vulvodynia patients demonstrated a fibroblast-mediated proinflammatory response to the dimorphic fungus Candida albicans, when compared to controls.

The prevalence of ED is increasing worldwide. Modifiable risk factors for ED include excessive alcohol consumption, obesity, a poor diet high in sugar and other carbohydrates (Maiorino, 2015), high triglycerides (Ahn, 2007), chronic insufficient sleep (Broussard, 2016; Spiegel, 2004a, 2004b), sedentary lifestyle, and smoking cigarettes (Meldrum, 2012; Roumeguère, 2003).

All of these risk factors are associated with a proinflammatory state that decreases the availability of NO (Traish, 2007). NO is required to achieve an erection (Cartledge, 2001). NO also is needed for engorgement of female genital tissue (Musicki, 2009; Ottesen, 1987), and factors that diminish NO also may negatively affect women's sexual function. However, genital engorgement is not as critical for female sexual performance as it is in the male. Female sexual dysfunction is more complex and less well understood (Laan, 2008; Traish, 2010; Vaccaro, 2015).

Inflammatory conditions have been shown to negatively affect sexual function, even when the patient is otherwise asymptomatic. Men with elevated prostate-specific antigen levels were given biopsies of their prostate. Those men with asymptomatic prostatitis had problems in sexual function and maintaining an erection (Urkmez, 2016), suggesting that even asymptomatic inflammation can cause sexual dysfunction.

The ideal approach to treating both PCOS in women and hypogonadism in men is improved diet, lifestyle changes, and weight loss. However, many people find this difficult to achieve, so another approach would be to give the prescription diabetes medication Metformin. Metformin improves insulin sensitivity and leads to an improvement in ovarian function (Thessaloniki, 2008; Teede, 2011). In fact, some studies also demonstrate improvement in ED with Metformin (Seftel, 2012). In men, aromatase inhibitors may be helpful because they reduce estrogen. Testosterone supplementation may be helpful to both men and to women with low testosterone levels.

Stress and Sexual Dysfunction

Finally, stress has an influence on sexual function, from the psychological to the biological, via several pathways. Stress causes the release of cortisol, which makes the body less sensitive to leptin. Leptin is the satiety hormone that signals the brain that enough food has been consumed, so the person is no longer hungry. Therefore, when cortisol is high, as in times of stress, a person is hungrier and will tend to eat more. Cortisol also is upregulated when individuals have had too little sleep, which promotes excessive release of ghrelin, the hunger hormone, leading to increased food consumption. If these factors become chronic, insulin resistance and obesity can develop. When a person is consuming a healthy diet and living an optimal lifestyle, blood sugar and insulin levels are more likely to be naturally well controlled and hormonal function is more likely to be intact. This optimum state of general health underlies and can improve sexual function.

It is worth underscoring that stressors, diet, mood status, and other factors interact in complex ways. A recent double-blind, randomized study found that stress can override the anti-inflammatory benefits of healthy food choices (Kiecolt-Glaser, 2017). Standardized assessment of prior-day stressors combined with assessment of inflammatory biomarkers revealed that, in the presence of prior-day stress, the inflammatory response to a healthy, high oleic sunflower oil meal looked similar to an unhealthy, high saturated fat meal. In the absence of prior-day stress, the healthy meal had the expected anti-inflammatory impact. In addition, women in the study who had a history of major depression had higher postmeal blood pressure responses than those without that history. These data show how recent stressors and a history of major depression can affect metabolism and promote inflammatory and atherogenic responses on a daily basis. Therefore, reducing sources of inflammation and improving health should ideally involve a coordinated approach.

INFLAMMATION AS "SICKNESS SYNDROME" AND ITS BEHAVIORAL SEQUELAE

When animals and human beings are ill, a long-recognized set of adaptive behavioral changes called sickness syndrome occurs spontaneously, as a response to infection and inflammation (Hart, 1988; Kelley, 2003; Takenaka, 2005). These usually include lethargy, low mood, anxiety, loss of appetite, sleepiness, increased sensitivity to pain, reduction in grooming, and poor concentration. This motivational state focuses the body's energies on combating the illness and also perturbs our normal behaviors. Shared human experience and direct observation suggest that this behavioral syndrome might well include decreased interest in sexual behavior. There is some evidence that this is the case. In animals, peripheral inflammation is accompanied by neurobehavioral alterations such as depression of feeding, exploratory, and sexual behaviors (Takenaka, 2005). Inflammatory processes can have a large impact on social behavior, causing reduced motivation to engage in social interaction (Arakawa, 2011), of which sexual behavior is a subtype.

There appears to be little human research on this biobehavioral question with respect to sexuality, but we are animals too. It is reasonable to think that most people have more interest in sex when in good health and possessing, to use the old-fashioned phrase, "animal spirits," compared to when they are ill. Conditions in which chronic rather than acute inflammation plays a role, such as depression, might reasonably be assumed to have a similar, if weaker, effect on inhibiting sexual interest. Depression is known for promoting withdrawal from others.

As established in this chapter, gut dysbiosis is a major driver of chronic inflammation and low mood. In an animal model, research has shown that mice whose suboptimal gut flora predisposed them to inflammation showed behaviors similar to those of sickness syndrome, including anxiety and depressive-like behaviors. When the mice were fed anti-inflammatory DHA, the gut microbial populations improved, and these changes significantly correlated with the behavioral changes such as reduced anxiety and depressive-like behaviors (Davis, 2016). (In this study, this prosocial improvement was observed in the male mice only; there is evidence that in females concentrations of n-3 polyunsaturated fatty acids (PUFA) are higher due to interactions with oestradiol and progesterone, while testosterone reduces n-3 PUFA levels (Childs, 2008).

Chronic inflammation may have a negative impact on human behavioral motivation to engage in sexual behavior at any age. As we age physiologically, hormones wane, and there is an increased burden of low-grade inflammation present in most older adults that affects sexual functioning in various ways (Graziottin, 2015). This may well dampen sexual interest in later years—when objectively many still might have fulfilling sexual lives—whether for reason of compromised sexual mechanics and emotional inhibition caused by that reality or even possibly from a biologically driven reduction in behavioral motivation for sex, due to chronic inflammation.

Conclusion

Hippocrates was right. Health starts in the gut. In our 21st-century world, creating and maintaining excellent gut health is a challenge. Stress, antibiotics, poor nutrition caused by eating a diet that promotes inflammation—all cause problems in the gut, resulting in a body and brain that are inflamed and out of balance. Inflammation, also caused by obesity, is often accompanied by a decrease in sex hormones and low sex drive and ED. Improving the health of the GI tract and reducing inflammation with diet modifications and stress reduction improves mood, anxiety, and overall physical health and sexual health.

REFERENCES

Aggarwal BB, Sundaram C, Malani N, Ichikawa H (2007). Curcumin: the Indian solid gold. *Adv Exp Med Biol.* 595:1–75.

Ahn TY, Park JK, Lee SW, et al. (2007). Prevalence and risk factors for erectile dysfunction in Korean men: results of an epidemiological study. *J Sex Med.* Sep;4(5): 1269–1276.

Albenberg LG, Wu GD (2014). Diet and the intestinal microbiome: associations, functions, and implications for health and disease. *Gastroenterology.* 146(6):1564–1572.

Al-Karawi D, Al Mammori DA, Tayyar Y (2016). The role of curcumin administration in patients with major depressive disorder: mini meta-analysis of clinical trials. *Phytother Res.* 30:175–183.

Almqvist C, Öberg AS (2014). The association between caesarean section and asthma or allergic disease continues to challenge. *Acta Paediatrica.* 103(4):349–351.

Ambler DR, Bieber EJ, Diamond MP (2012). Sexual function in elderly women: a review of current literature. *Rev Obstet Gynecol.* 5:16–27.

Arakawa H, Cruz S, Deak T (2011). From models to mechanisms: odorant communication as a key determinant of social behavior in rodents during illness-associated states. *Neurosci Biobehav Rev.* Oct;35(9):1916–1928.

Aversa A, Bruzziches R, Francomano D, et al. (2013). Weight loss by multidisciplinary intervention improves endothelial and sexual function in obese fertile women. *J Sex Med.* Apr;10(4):1024–1033.

Baganz NL, Blakely RD (2013). A dialogue between the immune system and brain, spoken in the language of serotonin. *ACS Chem Neurosci.* 4(1):48–63.

Bajos N, Wellings K, Laborde C, Moreau C (2010). Sexuality and obesity, a gender perspective: results from French national random probability survey of sexual behaviors. *BMJ.* 340:c2573.

Bansal V, Costantini T, Kroll L, et al. (2009). Traumatic brain injury and intestinal dysfunction: uncovering the neuro-enteric axis. *J Neurotrauma.* 26(8):1353–1359.

Basil MC, Levy BD (2016). Specialized pro-resolving mediators: endogenous regulators of infection and inflammation. *Nat Rev Immunol.* 16(1):51–67.

Benson S, Hahn S, Tan S, Mann K, Janssen OE (2009). Prevalence and implications of anxiety in polycystic ovary syndrome: results of an Internet-based survey in Germany. *Hum Reprod.* 24(6):1446–1451.

Berk M (2007). The glutathione precursor N-acetyl cysteine as a treatment for oxidative stress in bipolar disorder: a double-blind randomised placebo-controlled trial. *Bipolar Disord.* 9(2):3.

Berk M, Ng F, Dean O, Dodd S, Bush AI (2008a). Glutathione: a novel treatment target in psychiatry. *Trends Pharmacol Sci.* 29(7):346–351.

Berk M, Copolov DL, Dean O, et al. (2008b). N-acetyl cysteine for depressive symptoms in bipolar disorder—a double-blind randomized placebo-controlled trial. *Biol Psychiatry.* 64(6):468–475.

Berk M, Malhi GS, Gray LJ, Dean OM (2013a). The promise of N-acetylcysteine in neuropsychiatry. *Trends Pharmacol Sci.* 34(3):167–177.

Berk M, Williams LJ, Jacka FN, et al. (2013b). So depression is an inflammatory disease, but where does the inflammation come from? *BMC Med.* 11:200.

Berk M, Dean OM, Cotton SM, et al. (2014). The efficacy of adjunctive N-acetylcysteine in major depressive disorder: a double-blind, randomized, placebo-controlled trial. *J Clin Psychiatry.* 75(6):628–636.

Betteridge DJ (2000). What is oxidative stress? *Metabolism.* 49(2):3–8.

Birdsall TC (1998). 5-hydroxytryptophan: a clinically effective serotonin precursor. *Altern Med Rev.* 3(4):271–280.

Biswas SK, McClure D, Jimenez LA, Megson IL, Rahman I (2005). Curcumin induces glutathione biosynthesis and inhibits NF-kappaB activation and interleukin-8 release in alveolar epithelial cells: mechanism of free radical scavenging activity. *Antioxid Redox Signal.* 7(1–2):32–41.

Blaser M (2011). Antibiotic overuse: stop the killing of beneficial bacteria. *Nature.* 24;476(7361):393–394.

Blazquez A, Alegre J, Ruiz E (2009). Women with chronic fatigue syndrome and sexual dysfunction: past, present, and future. *J Sex Marital Ther.* 35(5):347–359.

Boldyrev AA (2005). Protection of proteins from oxidative stress: a new illusion or a novel strategy? *Ann N Y Acad Sci.* 1057:193–205.

Bolona E, Uraga MV, Haddad RM, et al. (2007). Testosterone use in men with sexual dysfunction: a systematic review and meta-analysis of randomized, placebo-controlled trials. *Mayo Clin Proc.* Jan; 82(1):20–28.

Borovcanin M, Jovanovic I, Radosavljevic G, et al. (2012). Elevated serum level of type-2 cytokine and low IL-17 in first episode psychosis and schizophrenia in relapse. *J Psychiatr Res.* 46(11):1421–1426.

Bowe W, Patel NB, Logan AC (2014). Acne vulgaris, probiotics and the gut-brain-skin axis: from anecdote to translational medicine. *Benef Microbes.* 5(2):185–199.

Braniste V, Al-Asmakh M, Kowal C, et al. (2014). The gut microbiota influences blood-brain barrier permeability in mice. *Sci Transl Med.* 6(263):1–11.

Bravo JA, Forsythe P, Chew MV, et al. (2011). Ingestion of Lactobacillus strain regulates emotional behavior and central GABA receptor expression in a mouse via the vagus nerve. *Proc Natl Acad Sci U S A.* 108(38):16050–16055.

Bravo JA, Julio-Pieper M, Forsythe P, et al. (2012). Communication between gastrointestinal bacteria and the nervous system. Review. *Curr Opin Pharmacol.* 12(6):667–672.

Briani C, Samaroo D, Alaedini A (2008). Celiac disease: from gluten to autoimmunity. *Autoimmun Rev.* 7(8):644–650.

Broussard JL, Van Cauter E (2016). Disturbances of sleep and circadian rhythms: novel risk factors for obesity. *Curr Opin Endocrinol Diabetes Obes.* 23(5):353–359.

Camilleri M, Madsen K, Spiller R, Greenwood-Van Meerveld B, Verne GN (2012). Intestinal barrier function in health and gastrointestinal disease. *Neurogastroenterol Motil.* 24(6):503–512.

Cani PD, Amar J, Iglesias MA (2007a). Metabolic endotoxemia initiates obesity and insulin resistance. *Diabetes.* 56:1761–1772.

Cani PD, Neyrinck AM, Fava F, et al. (2007b). Selective increases of bifidobacteria in gut microflora improve high-fat-diet-induced diabetes in mice through a mechanism associated with endotoxaemia. *Diabetologia.* 50:2374–2383.

Cani PD, Delzenne NM (2009). Interplay between obesity and associated metabolic disorders: new insights into the gut microbiota. *Curr Opin Pharmacol.* 9:737–743.

Cartledge J, Minhas S, Eardley I (2001). The role of nitric oxide in penile erection. *Expert Opin Pharmacother.* 2(1):95–107.

Catassi C, Cobellis G (2007). Coeliac disease epidemiology is alive and kicking, especially in the developing world. *Dig Liver Dis.* 39:908–910.

Catassi C (2015). Gluten sensitivity. *Ann Nutr Metab.* 67(Suppl 2):16–26.

Childs CE, Romeu-Nadal M, Burdge GC, Calder PC (2008). Gender differences in the n-3 fatty acid content of tissues. *Proc Nutr Soc.* 67(1):19–27.

Collins SM, Surette M, Bercik P (2012). The interplay between the intestinal microbiota and the brain. *Nat Rev Microbiol.* 10(11):735–742.

Corona G, Rastrelli G, Filippi S, Vignozzi L, Mannucci E, Maggi M (2014). Erectile dysfunction and central obesity: an Italian perspective. *Asian J Androl.* 16(4):581–591.

Cryan JF (2011a). Ingestion of Lactobacillus strain regulates emotional behavior and central GABA receptor expression in a mouse via the vagus nerve. *Proc Natl Acad Sci U S A.* 108(38):16050–16055.

Cryan JF, O'Mahony SM (2011b). The microbiome-gut-brain axis: from bowel to behavior. *Neurogastroenterol Motility.* 23(3):187–192.

Cryan JF, Dinan TG (2012). Mind-altering microorganisms: the impact of the gut microbiota on brain and behavior. *Nat Rev Neurosci.* 13:701–712.

Cui J, Shao L, Young L, Wang J (2007). Role of glutathione in neuroprotective effects of mood stabilizing drugs lithium and valproate. *Neuroscience.* 144(4):1447–1453.

Curran EA, Dalman C, Kearney PM, et al. (2015a). Association between obstetric mode of delivery and autism spectrum disorder: a population-based sibling design study. *JAMA Psychiatry.* 72(9):935–942.

Curran EA, O'Neill SM, Cryan JF, Kenny LC, Dinan TG, Khashan AS, Kearney PM (2015b). Research review: birth by caesarean section and development of autism spectrum disorder and attention-deficit/hyperactivity disorder: a systematic review and meta-analysis. *J Child Psychol Psychiatry.* 56(5):500–508.

Daniele S, Zappelli E, Martini C (2015). Trazodone regulates neurotrophic/growth factors, mitogen-activated protein kinases and lactate release in human primary astrocytes. *J Neuroinflammation.* 12:225.

Dashti S, Latiff LA, Hamid HA, et al. (2016). Sexual dysfunction in patients with polycystic ovary syndrome in Malaysia. *Asian Pac J Cancer Prev.* 17(8):3747–3751.

D'Aversa F, Tortora A, Ianiro G, Ponziani FR, Annicchiarico BE, Gasbarrini A (2013). Gut microbiota and metabolic syndrome. *Intern Emerg Med.* 8(Suppl 1):S11–S115.

Davies NM (1998). Review article: non-steroidal anti-inflammatory drug-induced gastrointestinal permeability. *Aliment Pharmacol Ther.* 12(4):303–320.

Davis DJ, Hecht PM, Jasarevic E, et al. (2016). Sex-specific effects of docosahexaenoic acid (DHA) on the microbiome and behavior of socially isolated mice. *Brain Behav Immun.* Sep 9;pii: S0889-1591(16)30412-3.

Davis I, Liu A (2015). What is the tryptophan kynurenine pathway and why is it important to neurotherapeutics? *Expert Rev Neurotherap.* 15(7):719–721.

Dean O, Giorlando F, Merk, M (2011). N-acetylcysteine in psychiatry: current therapeutic evidence and potential mechanisms of action. *J Psychiatry Neurosci.* 36(2):78–86.

Decker E, Engelmann G, Findeisen A, et al. (2010). Cesarean delivery is associated with celiac disease but not inflammatory bowel disease in children. *Pediatrics.* 125(6):e1433–e1440.

Deepmala J, Slattery J, Kumar N, et al. (2015). Clinical trials of N-acetylcysteine in psychiatry and neurology: a systematic review. *Neurosci Biobehav Rev.* 55:294–321.

Demaude J, Salvador-Cartier C, Fioramont J, Ferrier L, Bueno L (2006) Phenotypic changes in colonocytes following acute stress or activation of mast cells in mice: implications for delayed epithelial barrier dysfunction. *Gut.* 55:655–661.

Desbonnet L, Clarke G, Traplin A, et al. (2015). Gut microbiota depletion from early adolescence in mice: implications for brain and behaviour. *Brain Behav Immun.* 48:165–173.

Dèttore D, Pucciarelli M, Santarnecchi E (2013). Anxiety and female sexual functioning: an empirical study. *J Sex Marital Ther.* 39(3):216–240.

Dinan TG, Cryan JF (2012). Regulation of the stress response by the gut microbiota: implications for psychoneuroendocrinology. *Psychoneuroendocrinology.* 37(9):1369–1378.

Dinan TG, Stanton C, Cryan JF (2013). Psychobiotics: a novel class of psychotropic. *Biol Psychiatry.* 74(10):720–726.

Dominguez-Bello MG, Costello EK, et al. (2010). Delivery mode shapes the acquisition and structure of the initial microbiota across multiple body habitats in newborns. *Proc Natl Acad Sci U S A.* 107(26):11971–11975.

Duleba AJ, Dokras A (2012). Is PCOS an inflammatory process? *Fertil Steril.* 97(1):7–12.

Elamin E, Masclee A, Dekker J, Jonkers D (2014). Ethanol disrupts intestinal epithelial tight junction integrity through intracellular calcium-mediated Rho/ROCK activation. *Am J Physiol Gastrointest Liver Physiol.* 15;306(8):G677–G685.

Eller T, Vasar V, Shlik J, Maron E (2008). Pro-inflammatory cytokines and treatment response to escitalopram in major depressive disorder. *Prog Neuropsychopharmacol Biol Psychiatry.* 15;32(2):445–450.

Erskine RM, Tomlinson DJ, Morse CI, et al. (2017). The individual and combined effects of obesity- and ageing-induced systemic inflammation on human skeletal muscle properties. *Int J Obes. (Lond).* Jan;41(1):102–111.

Esposito K, Ciotola M, Giugliano F, et al. (2006). Mediterranean diet improves erectile function in subjects with the metabolic syndrome. *Int J Impot Res.* 18(4):405–410.

Esposito K, Ciotola M, Giugliano F, Bisogni C, Schisano B (2007). Association of body weight with sexual function in women. *Int J Impot Res.* 19(4):353–357.

Esposito K, Giugliano F, Ciotola M, De Sio M (2008). Obesity and sexual dysfunction, male and female. *Int J Impot Res.* 20:358–365.

Farthing MJ, Edwards CR, Rees LH, Dawson AM (1982). Male gonadal function in coeliac disease: 1. Sexual dysfunction, infertility, and semen quality. *Gut.* 23(7):608–614.

Falsetta ML, Foster DC, Woeller CF, et al. (2015). Identification of novel new mechanisms involved in generating localized vulvodynia pain. *Am J Obstet Gynecol.* 213(1):38. e1–e12.

Farhadi A, Banan A, Fields J, Keshavarzian A (2003). Intestinal barrier: an interface between health and disease. *J Gastroenterol Hepatol.* 18(5):479–497.

Fasano A (2012). Leaky gut and autoimmune diseases. *Clin Rev Allerg Immun.* 42(1):71–78.

Feighery L, Smyth A, Keely S, Baird AW, O'Connor WT, Callanan JJ, Brayden DJ (2008). Increased intestinal permeability in rats subjected to traumatic frontal lobe percussion brain injury. *J Trauma.* 64(1):131–137; discussion 137–138.

Forsythe P, Sudo N, Dinan T, Taylor VH, Bienenstock J (2010). Mood and gut feelings. *Brain Behav Immun.* 24(1):9–16.

Fui MN, Dupuis P, Grossmann M (2014). Lowered testosterone in male obesity: mechanisms, morbidity and management. *Asian J Androl.* Mar-Apr;16(2):223–231.

Funkhouser LJ, Bordenstein SR (2013). Mom knows best: the universality of maternal microbial transmission. *PLoS Biol.* 11(8):1–9.

Furtado M, Katzman MA (2015a). Examining the role of neuroinflammation in major depression. *Psychiatr Res.* 229(1– 2):27–36.

Furtado M, Katzman MA (2015b). Neuroinflammatory pathways in anxiety, posttraumatic stress, and obsessive compulsive disorders. *Psychiatr Res.* 229(1–2):37–48.

Galland, L (2010). Diet and inflammation. *Nutr Clin Pract.* 25(6):634–640.

Galland, L (2014). The gut microbiome and the brain. *J Med Food.* (12):1261–1272.

Goldani M, Barbieri M, Silva AA, Gutierrez MR, Bettiol H, Goldani HA (2013). Cesarean section and increased body mass index in school children: two cohort studies from distinct socioeconomic background areas in Brazil. *Nutr J.* 12(1):104.

Grant JE, Odlaug BL, Kim SW (2009). N-acetylcysteine, a glutamate modulator, in the treatment of trichotillomania: a double-blind, placebo-controlled study. *Arch Gen Psychiatry.* 66(7):756–763.

Graziottin A, Gambini D (2015). Anatomy and physiology of genital organs—women. *Handb Clin Neurol.* 130:39–60.

Green PHR, Cellier C (2007). Celiac disease. *N Engl J Med.* 357:1731–1743.

Greer JB, O'Keefe SJ (2011). Microbial induction of immunity, inflammation, and cancer. *Front Physiol.* 26;1:168.

Grenham S, Clarke G, Cryan J, Dinan TG (2011). Brain–gut–microbe communication in health and disease. *Front Physiol.* 2:94.

Hadjivassiliou M, Sanders DS, Grünewald RA, Woodroofe N, Boscolo S, Aeschlimann D (2010). Gluten sensitivity: from gut to brain. *Lancet Neurol.* 9(3):318–330.

Hahn S, Janssen OE, Tan S, et al. (2005). Clinical and psychological correlates of quality-of-life in polycystic ovary syndrome. *Eur J Endocrinol.* Dec;153(6): 853–860.

Hamdani N, Tamouza R, Leboyer M (2012). Immuno- inflammatory markers of bipolar disorder: a review of evidence. *Front Biosci.* 4(1):2170–2182.

Hannestad J, Dellagioia N, Bloch M (2011). The effect of antidepressant medication treatment on serum levels of inflammatory cytokines: a meta-analysis. *Neuropsychopharmacology.* 36(12):2452–2459.

Hart BL (1988). Biological basis of the behavior of sick animals. *Neurosci Biobehav Rev.* 12(2):123–37.

Hashimoto K, Sawa A, Iyo M (2007). Increased levels of glutamate in brains from patients with mood disorders. *Biol Psychiatry.* 62(11):1310–1316.

Hatzimouratidis K, Hatzichristou DG (2005). A comparative review of options for treatment of erectile dysfunction: Which treatment for which patient? *Drugs.* 60(12):1621–1650.

Hawrelak JA, Myers SP (2004). The causes of intestinal dysbiosis: a review. *Altern Med Rev.* 9(2):180–197.

Hendricks KM, Walker WA (1988). Zinc deficiency in inflammatory bowel disease. *Nutr Rev.* 46(12):401–408.

Herlihy AC, Kelly RE, Hogan JL, O'Connor N, Farah N, Turner MJ (2011). Polycystic ovary syndrome and the peripheral blood white cell count. *J Obstet Gynaecol.* 31(3):242–244.

Hill J, Bird H, Thorpe R (2003). Effects of rheumatoid arthritis on sexual activity and relationships. *Rheumatology.* Feb; 42(2):280–286.

Hulter BM, Lundberg PO (1995). Sexual function in women with advanced multiple sclerosis. *J Neurol Neurosurg Psychiatry.* Jul;59(1):83–86.

Huang EJ, Reichardt LF (2001). Neurotrophins: roles in neuronal development and function. *Ann Rev Neurosci.* 24:677–736.

Humbert P, Bidet A, Treffel P, Drobacheff C, Agache P (1991). Intestinal permeability in patients with psoriasis. *J Dermatol Sci.* 2(4):324–326.

Hyman M (2007). The life cycles of women: restoring balance. *Altern Ther Health Med.* 13(3):10–16.

Jackson JR, Eaton WW, Cascella NG, Fasano A, Kelly DL (2012). Neurologic and psychiatric manifestations of celiac disease and gluten sensitivity. *Psychiatr Q.* 83(1):91–102.

Jarzabek-Bielecka G, Wilczak M, Potasinska-Sobkowska A, et al. (2015). Overweight, obesity and female sexuality in perimenopause: a preliminary report. *Prz Menopauzalny.* 14, 97–104.

Kaplan BJ, Rucklidge JJ, Romijn A, McLeod K (2015). The emerging field of nutritional mental health: inflammation, the microbiome, oxidative stress, and mitochondrial function. *Clin Psychol Sci.* 2015:1–17.

Karakuła-Juchnowicz H, Szachta P, Opolska A, et al. (2017). The role of IgG hypersensitivity in the pathogenesis and therapy of depressive disorders. *Nutr Neurosci.* 20:110–118.

Kaufmann FN, Gazal M, Bastos CR., Kaster MP, Ghisleni G (2016). Curcumin in depressive disorders: an overview of potential mechanisms, preclinical and clinical findings. *Eur J Pharmacol.* 5;784192–784198.

Kelley KW, Bluthé RM, Dantzer R, et al. (2003). Cytokine-induced sickness behavior. *Brain Behav Immun.* 17(Suppl 1):S112–S118.

Kelly JR, Borre Y, O'Brien C, et al. (2016). Transferring the blues: depression-associated gut microbiota induces neurobehavioral changes in the rat. *J Psychiatr Res.* 82:109–118.

Kiecolt-Glaser JK, Fagundes CP, Andridge R, et al. (2017). Depression, daily stressors and inflammatory responses to high-fat meals: when stress overrides healthier food choices. *Mol Psychiatry.* 22:476–482.

Knowles SR, Nelson EA, Palombo EA (2008). Investigating the role of perceived stress on bacterial flora activity and salivary cortisol secretion: a possible mechanism underlying susceptibility to illness. *Biol Psychology.* 77(2):132–137.

Kradag H, Oner O, Karaoglan A, Itil O (2013). Body mass index and sexual dysfunction in males and females in a population study. *Bull Clin Psychopharmacol.* 24:76–83.

Laan E, van Driel EM, van Lunsen RH (2008). Genital responsiveness in healthy women with and without sexual arousal disorder. *J Sex Med.* Jun;5(6):1424–1435.

Lafleur DL, Pittenger C, Kelmendi B, et al. (2006). N-acetylcysteine augmentation in serotonin reuptake inhibitor refractory obsessive-compulsive disorder. *Psychopharmacology.* 184(2):254–256.

Lammers KM, Lu R, Brownley J, et al. (2008). Gliadin induces an increase in intestinal permeability and zonulin release by binding to the chemokine receptor CXCR3. *Gastroenterology.* Jul;135(1):194–204.

LaRowe SD, Kalivas PW, Nicholas JS, et al. (2013). A double-blind placebo-controlled trial of N-acetylcysteine in the treatment of cocaine dependence. *Am J Addict.* 22(5):443–452.

Layegh P, Mousavi Z, Farrokh Tehrani D, Parizadeh SM, Khajedaluee M (2016). Insulin resistance and endocrine-metabolic abnormalities in polycystic ovarian syndrome: Comparison between obese and non-obese PCOS patients. *Int J Reprod Biomed (Yazd).* Apr;14(4):263–270.

Lee KH, Abas F, Alitheen NB, Shaari K, Lajis NH, Ahmad S (2011). A curcumin derivative, 2,6-bis(2,5-dimethoxybenzylidene)-cyclohexanone (BDMC33) attenuates prostaglandin E2 synthesis via selective suppression of cyclooxygenase-2 in IFN-γ/LPS-stimulated macrophages. *Molecules.* 16(11):9728–9738.

Lepage P, Leclerc MC, Joossens M, Mondot S, Blottière HM (2013). A metagenomic insight into our gut's microbiome. *Gut.* 62(1):146–158.

Liu X, Cao S, Zhang X (2015). Modulation of gut microbiota–brain axis by probiotics, prebiotics, and diet. *J Ag Food Chem.* 63(36):7885–7895.

Lloyd-Price J, Abu-Ali G, Huttenhower C (2016). The healthy human microbiome *Genome Med.* 8:51.

Logan AC, Katzman M (2005). Major depressive disorder: probiotics may be an adjuvant therapy. *Med Hypotheses.* 64(3):533–538.

Lopresti AL, Maes M, Maker GL, Hood SD, Drummond PD (2014). Curcumin for the treatment of major depression: a randomised, double-blind, placebo controlled study. *J Affect Dis.* 167:368–375.

Lopresti AL, Maes M, Meddens MJ, Maker GL, Arnoldussen E, Drummond PD (2015). Curcumin and major depression: a randomised, double-blind, placebo-controlled trial investigating the potential of peripheral biomarkers to predict treatment response and antidepressant mechanisms of change. *Eur Neuropsychopharmacol.* 25(1):38–50.

Lord RS, Brailley JA (2008). *Laboratory evaluations for integrative and functional medicine*, 2nd ed. Duluth, GA: Metametrix Institute.

Luppino FS, de Wit LM, Bouvy PF, et al. (2010). Overweight, obesity, and depression: a systematic review and meta-analysis of longitudinal studies. *Arch Gen Psychiatry.* Mar;67(3):220–229.

Maes M (2008a). The cytokine hypothesis of depression: inflammation, oxidative & nitrosative stress (IO&NS) and leaky gut as new targets for adjunctive treatments in depression. *Neuroendocrinol Lett.* 29(3):287–291.

Maes M, Kubera M, Leunis JC (2008b). The gut-brain barrier in major depression: intestinal mucosal dysfunction with an increased translocation of LPS from gram negative enterobacteria (leaky gut) plays a role in the inflammatory pathophysiology of depression. *Neuroendocrinol Lett.* 29(1):117–124.

Maes M, Kubera M, Leunis JC, Berk M (2012). Increased IgA and IgM responses against gut commensals in chronic depression: further evidence for increased bacterial translocation or leaky gut. *J Affect Disord.* 1;141(1):55–62.

Mahmood A, FitzGerald AJ, Marchbank T, et al. (2007). Zinc carnosine, a health food supplement that stabilizes small bowel integrity and stimulates gut repair processes. *Gut.* 56(2):168–175.

Maiorino MI, Bellastella G, Esposito K (2015). Lifestyle modification and erectile dysfunction: what can be expected. *Asian J Androl.* Jan-Feb;17(1):5–10.

Mawe GM, Hoffman JM (2013). Serotonin signalling in the gut—functions, dysfunctions and therapeutic targets. *Nat Rev Gastroenterol Hepatol.* 10(8):473–486.

Mayer EA, Knight R, Mazmanian SK, Cryan JF, Tillisch K (2014). Gut microbes and the brain: paradigm shift in neuroscience. *J Neurosci.* 12;34(46):15490–15496.

Mayer EA, Tillisch K, Gupta A (2015). Gut/brain axis and the microbiota. *J Clin Invest.* 125(3):926–938.

Meldrum DR, Gambone JC, Morris MA, Esposito K, Giugliano D, Ignarro LJ (2012). Lifestyle and metabolic approaches to maximizing erectile and vascular health. *Int J Impot Res.* 24(2):61–68.

Molberg O, Uhlen AK, Jensen T, et al. (2005). Mapping of gluten T-cell epitopes in the bread wheat ancestors: implications for celiac disease. *Gastroenterology.* Feb;128(2):393–401.

Montenegro L, Losurdo G, Licinio R, et al. (2014). Non steroidal anti-inflammatory drug induced damage on lower gastro-intestinal tract: Is there an involvement of microbiota? *Curr Drug Saf.* 9(3):196–204.

Morales AJ, Laughlin GA, Bützow T, Maheshwari H, Baumann G, Yen SS (1996). Insulin, somatotropic, and luteinizing hormone axes in lean and obese women with polycystic ovary syndrome: common and distinct features. *J Clin Endocrinol Metab.* Aug;81(8):2854–2864.

Moro MK, Nagahashi M, Ramanathan R, Takabe K, Wakai T (2016). Resolvins and omega three polyunsaturated fatty acids: clinical implications in inflammatory diseases and cancer. *World J Clin Cases.* 16;4(7):155–164.

Moroni, F (1999). Tryptophan metabolism and brain function: focus on kynurenine and other indole metabolites. *Eur J Pharmacol.* 375(1–3):87–100.

Mozafari M, Khajavikhan J, Jaafarpour M, et al. (2015). Association of body weight and female sexual dysfunction: a case control study. *Iran Red Crescent Med. J.* 17:e24685.

Mueller NT, Bakacs E, Combellick J, Grigoryan Z, Dominguez-Bello MG (2015). The infant microbiome development: mom matters. *Trends Mol Med.* 21(2):109–117.

Müller N, Schwarz MJ (2010). The immunological basis of glutamatergic disturbance in schizophrenia: towards an integrated view. *J Neural Transm Suppl.* 72:269–280.

Musicki B, Liu T, Burnett AL (2009). Endothelial nitric oxide synthase regulation in female genital tract structures. *J Sex Med.* Mar;6(Suppl 3):247–253.

Na KS, Lee KJ, Lee JS, Cho YS, Jung H (2014). Efficacy of adjunctive celecoxib treatment for patients with major depressive disorder: a meta-analysis. *Prog Neuropsychopharmacol Biol Psychiatry.* 48:79–85.

Nackers LM, Appelhans BM, Segawa E, et al. (2015). Associations between body mass index and sexual functioning in midlife women. *Menopause.* 22:1175–1181.

Nandal U, Bhardwaj RL (2012). Aloe vera: a valuable wonder plant for food, medicine and cosmetic use. *Int J Pharm.* 13(1):59–67.

Neufeld KM, Kang N, Bienenstock J, Foster JA (2011). Reduced anxiety-like behavior and central neurochemical change in germ-free mice. *Neurogastroenterol Motil.* 23(3):255–264.

Ng F, Berk M, Dean O, Bush AI (2008). Oxidative stress in psychiatric disorders: evidence base and therapeutic implications. *Int J Neuropsychopharmacol.* 11(6):851–876.

Noto C, Rizzo LB, Mansur RB, McIntyre RS, Maes M, Brietzke E (2014). Targeting the inflammatory pathway as a therapeutic tool for major depression. *Neuroimmunomodulation.* 21(2–3):131–139.

Nozaki K, Beal MF (1992). Neuroprotective effects of L-kynurenine on hypoxia-ischemia and NMDA lesions in neonatal rats. *J Cereb Blood Flow Metab.* 12(3):400–407.

Oliver G, Dean O, Camfield D, Blair-West S, Ng C, Berk M, Sarris J (2015). N-acetyl cysteine in the treatment of obsessive compulsive and related disorders: a systematic review. *Clin Psychopharmacol Neurosci.* 13(1):12–24.

O'Mahony S, Clarke G, Borre Y, Dinan T, Cryan J (2015). Serotonin, tryptophan metabolism and the brain-gut-microbiome axis. *Behav Brain Res.* 277:32–48.

Ottesen B, Pedersen B, Nielsen J, Dalgaard D, Wagner G, Fahrenkrug J (1987). Vasoactive intestinal polypeptide (VIP) provokes vaginal lubrication in normal women. *Peptides.* Sep-Oct;8(5):797–800.

Pari L, Tewas D, Eckel J (2008). Role of curcumin in health and disease. *Arch Physio Biochem.* 114(2):127–149.

Pasquali D, Maiorino MI, Renzullo A, et al. (2013). Female sexual dysfunction in women with thyroid disorders. *J Endocrinol Invest.* Oct; 36(9):729–733.

Pastorelli L, De Salvo C, Mercado JR, Vecchi M, Pizarro TT (2013). Central role of the gut epithelial barrier in the pathogenesis of chronic intestinal inflammation: lessons learned from animal models and human genetics. *Front Immunol.* 17;4:280.

Perlmutter D (2015). *Brain maker: the power of gut microbes to heal and protect your brain—for life.* New York: Little, Brown.

Pike, M., G, Heddle RJ, Boulton P, Turner MW, Atherton DJ (1986). Increased intestinal permeability in atopic eczema. *J Invest Dermatol.* Feb;86(2):101–104.

Prasad AS (2014). Zinc is an antioxidant and anti-inflammatory agent: its role in human health. *Front Nutr.* 1;1:14.

Raison CL, Capuron L, Miller AH (2006). Cytokines sing the blues: inflammation and the pathogenesis of depression. *Trends Immunol.* 27(1):24–31.

Raison CL, Miller AH (2013). Do cytokines really sing the blues? *Cerebrum.* 10:1–16.

Rao R, Samak G (2012). Role of glutamine in protection of intestinal epithelial tight junctions. *J Epithel Biol Pharmacol.* 5(Suppl 1-M7):47–54.

Rapin JR, Wiernsperger N (2010). Possible links between intestinal permeability and food processing: a potential therapeutic niche for glutamine. *Clinics (Sao Paulo).* 6:635–643.

Ridaura V, Belkaid Y (2015). Gut microbiota: the link to your second brain. *Cell.* 161(2):193–194.

Rostami K, Steegers EA, Wong WY, Braat DD, Steegers-Theunissen RP (2001). Coeliac disease and reproductive disorders: a neglected association. *Eur J Obstet Gynecol Reprod Biol.* 96(2):146–149.

Roumeguère T, Wespes E, Carpentier Y, Hoffmann P, Schulman CC (2003).Erectile dysfunction is associated with a high prevalence of hyperlipidemia and coronary heart disease risk. *Eur Urol.* Sep;44(3):355–359.

Ruan X, Dai Y (2009). Study on chronic low-grade inflammation and influential factors of polycystic ovary syndrome. *Med Princ Pract.* 18(2):118–122.

Saad F (2015). The emancipation of testosterone from niche hormone to multi-system player. *Asian J Androl.* Jan-Feb;17(1):58–60.

Saboor Aftab SA, Kumar S, Barber TM (2013). The role of obesity and type 2 diabetes mellitus in the development of male obesity-associated secondary hypogonadism. *Clin Endocrinol. (Oxf).* Mar;78(3):330–337.

Sanmukhani J, Satodia V, Trivedi J, et al. (2014). Efficacy and safety of curcumin in major depressive disorder: a randomized controlled trial. *Phytother Res.* 28(4):579–585.

Sapolsky RM (2001). Depression, antidepressants, and the shrinking hippocampus. *Proc Natl Acad Sci U S A.* 98(22):12320–12322.

Schmidt FM, Kirkby KC, Lichtblau N (2016a). Inflammation and immune regulation as potential drug targets in antidepressant treatment. *Curr Neuropharmacol.* 14:674–687.

Schmidt FM, Schröder T, Kirkby K, Sander C, Suslow T et al. (2016b). Pro- and anti-inflammatory cytokines, but not CRP, are inversely correlated with severity and symptoms of major depression. *Psychiatry Res.* 30;239:85–91.

Seftel AD (2012). Re: addition of metformin to sildenafil treatment for erectile dysfunction in eugonadal non-diabetic men with insulin resistance: a prospective, randomized, double blind pilot study. *J Urol.* 187(5):1787–1788.

Serhan CN (2005). Novel omega—3-derived local mediators in antiinflammation and resolution. *Pharmacol Ther.* 105(1):7–21.

Serhan CN, Chiang N, Dalli J (2015). The resolution code of acute inflammation: novel pro-resolvig lipid mediators in resolution. *Semin Immunol.* May;27(3):200–215.

Shaik B, Castellani ML, Perrella A, et al. (2006). Role of quercetin (a natural herbal compound) in allergy and inflammation. *J Biol Regul Homeost Agents.* 20(3–4):47–52.

Sher KS, Jayanthi V, Probert CSJ, Stewart CR, Mayberry JF (1994). Infertility, obstetric and gynaecological problems in coeliac sprue. *Dig Dis.* 12(3):186–190.

Shifren J, Braunstein G, Simon J, Cass P (2000). Transdermal testosterone treatment in women with impaired sexual functioning after oophorectomy. *N Eng J Med.* 343:682–688.

Shirazi SN, Friedman AR, Kaufer D, Sakhai SA (2015). Glucocorticoids and the brain: neural mechanisms regulating the stress response. *Adv Exp Med Biol.* 872:235–252.

Singh T, Williams K (2006). Atypical depression. *Psychiatry (Edgmont).* 3(4):33–39.

Skrovanek S, DiGuilio K, Bailey R, Huntington W, Urbas R, Mayilvaganan B, Mercogliano G, Mullin JM (2014). Zinc and gastrointestinal disease. *World J Gastrointest Pathophysiol.* 15;5(4):496–513.

Slyepchenko A, Carvalho AF, Cha DS, Kasper S, McIntyre RS (2014). Gut emotions—mechanisms of action of probiotics as novel therapeutic targets for depression and anxiety disorders. *CNS Neurol Disord Drug Targets.* 13(10):1770–1786.

Slyepchenko A, Maes M, Machado-Veira R, et al. (2016). Intestinal dysbiosis, gut hyperpermeability and bacterial translocation: missing links between depression, obesity and type 2 diabetes? *Curr Pharm Des.* 22:6087–6106.

Slyepchenko A, Maes M, Jacka FN, et al. (2017). Gut microbiota, bacterial translocation, and interactions with diet: pathophysiological links between major depressive disorder and non-communicable medical comorbidities. *Psychother Psychosom.* 86(1):31–46.

Spiegel K, Tasali E, Penev P, Van Cauter E (2004a). Brief communication: sleep curtailment in healthy young men is associated with decreased leptin levels, elevated ghrelin levels, and increased hunger and appetite. *Ann Intern Med.* 141(11):846–850.

Spiegel K, Leproult, R, L'hermite-Balériaux M, et al. (2004b). Leptin levels are dependent on sleep duration: relationships with sympathovagal balance, carbohydrate regulation, cortisol, and thyrotropin. *J Clin Endocrinol Metab.* 89(11):5762–5771.

Srilatha B, Adaikan PG (2006). Characterization of hypertriglyceridemia-induced erectile dysfunction. *Urology.* 67(3):642–646.

Steele TE, Finkelstein SH, Finkelstein FO (1976). Marital discord, sexual problems, and depression. *J Nerv Ment Dis.* 162(4):225–237.

Stovall DW, Scriver JL, Clayton AH, Williams CD, Pastore LM (2012). Sexual function in women with polycystic ovary syndrome. *J Sex Med.* 9(1):224–230.

Sturniolo GC, Di Leo V, Ferronato A, D'Odorico A, D'Incà R (2001). Zinc supplementation tightens "leaky gut" in Crohn's disease. *Inflamm Bowel Dis.* 7(2):94–98.

Suez J, Korem T, Zilberman-Schapira G, Segal E, Elinav E (2015). Non-caloric artificial sweeteners and the microbiome: findings and challenges. *Gut Microbes.* 6(2):149–155.

Suzuki T, Hara H (2009). Quercetin enhances intestinal barrier function through the assembly of zonula [corrected] occludens-2, occluding, and claudin-1 and the expression of claudin-4 in Caco-2 cells. *J Nutr.* 139(5):965–974.

Takenaka M, Kanada S, Hamazaki T, Watanabe S (2005). Dietary supplementation with n-3 polyunsaturated fatty acids attenuates the depression of food-motivated behavior during zymosan-induced peritonitis. *Biol Pharm Bull.* Jul;28(7):1291–1293.

Teede HJ, Misso ML, Deeks AA, et al. (2011). Guideline development groups. Assessment and management of polycystic ovary syndrome: summary of an evidence-based guideline. *Med J Aust.* 195, S65–S112.

Thessaloniki ESHRE/ASRM-Sponsored PCOS Consensus Workshop Group (2008). Consensus on infertility treatment related to polycystic ovary syndrome. *Hum Reprod.* 23:462–477.

Tilg H, Kaser A (2011). Gut microbiome, obesity, and metabolic dysfunction. *J Clin Invest.* 121(6):2126–2132.

Tovoli F, Masi C, Guidetti E, Negrini G, Paterini P, Bolondi L (2015). Clinical and diagnostic aspects of gluten related disorders. *World J Clin Cases.* 16;3(3):275–284.

Traish AM, Goldstein I, Kim NN (2007). Testosterone and erectile function: from basic research to a new clinical paradigm for managing men with androgen insufficiency and erectile dysfunction. *Eur Urol.* 52(1):54–70.

Traish AM, Botchevar E, Kim NN (2010). Biochemical factors modulating female genital sexual arousal physiology. *J Sex Med.* Sep;7(9):2925–2946.

Traub ML (2011). Assessing and treating insulin resistance in women with polycystic ovarian syndrome. *World J Diabetes.* Mar 15; 2(3):33–40.

Tremellen K, Pearce K (2012). Dysbiosis of Gut Microbiota (DOGMA)—a novel theory for the development of Polycystic Ovarian Syndrome. *Med Hypotheses.* Jul;79(1):104–112.

Tremellen K (2016). Gut Endotoxin Leading to a Decline In Gonadal function (GELDING)—a novel theory for the development of late onset hypogonadism in obese men. *Basic Clin Androl.* Jun 22;26:7.

Tristano AG (2009). The impact of rheumatic diseases on sexual function. *Rheumatol Int.* 29(8):853–860.

Tsai F, Coyle WJ (2009). The microbiome and obesity: Is obesity linked to our gut flora? *Curr Gastroenterol Rep.* 11(4):307–313.

Turner JR (2009). Intestinal mucosal barrier function in health and disease. *Nat Rev Immunol.* 9:799–809.

Uhde M, Ajamian M, Caio G, et al. (2106). Intestinal cell damage and systemic immune activation in individuals reporting sensitivity to wheat in the absence of coeliac disease. *Gut.* 65:1930–1937.

Urkmez A, Yuksel OH, Uruc F, et al. (2016). The effect of asymptomatic histological prostatitis on sexual function and lower urinary tract symptoms. *Arch Esp Urol.* May; 69(4): 185–191.

Vaccaro C (2015). The use of magnetic resonance imaging for studying female sexual function: a review. *Clin Anat.* 28(3):324–330.

Volta U, De Giorgio R (2010). Gluten sensitivity: An emerging issue behind neurological impairment? *Lancet Neurol.* 9(3):233–235.

Vgontzas AN, Bixler EO, Chrousos GP (2006). Obesity-related sleepiness and fatigue: the role of the stress system and cytokines. *Ann N Y Acad Sci.* Nov;1083:329–344.

Wang W, Uzzau S, Goldblum SE, Fasano A (2000). Human zonulin, a potential modulator of intestinal tight junctions. *J Cell Sci.* Dec;113(Pt 24):4435–4440.

Weil A (2016). Dr Andrew Weil's Anti-Inflammatory Food Pyramid. Weil Lifestyle, https://www.drweil.com/diet-nutrition/anti-inflammatory-diet-pyramid.

Welly RJ, Liu TW, Zidon TM, et al. (2016). Comparison of diet versus exercise on metabolic function and gut microbiota in obese rats. *Med Sci Sports Exerc.* 48(9):1688–1698.

Wichers MD, Maes M (2004). The role of indoleamine 2,3-dioxygenase (IDO) in the pathophysiology of interferon-alpha-induced depression. *J Psychiat Neurosci.* 29(1):11–17.

Willemsen LE, Koetsier MA, van Deventer SJ, van Tol EA (2003). Short chain fatty acids stimulate epithelial mucin 2 expression through differential effects on prostaglandin E(1) and E(2) production by intestinal myofibroblasts. *Gut.* 52:1442–1447.

Wu GD, Chen J, Hoffmann C, et al. (2011). Linking long-term dietary patterns with gut microbial enterotypes. *Science.* 334:105–108.

Yano J, Yu K, Donaldson G, et al. (2015). Indigenous bacteria from the gut microbiota regulate host serotonin biosynthesis. *Cell.* 161(2):264–276.

Yu JJ, Pei LB, Zhang Y, Wen ZY, Yang JL (2015). Chronic supplementation of curcumin enhances the efficacy of antidepressants in major depressive disorder: a randomized, double-blind, placebo-controlled pilot study. *J Clin Psychopharmacol.* 35(4):406–410.

Zhang D, Zhang L, Yue F, Zheng Y, Russell R (2015). Serum zonulin is elevated in women with polycystic ovary syndrome and correlates with insulin resistance and severity of anovulation. *Eur J Endocrinol.* 172(1):29–36.

15

Understanding Sexual Addiction and Hypersexuality: An Integrative Approach to Treatment

MONICA MEYER

Introduction

Many different labels have been proposed to describe a clinically observed pattern of problematic and destructive sexual behavior, most commonly characterized by engagement in excessive or compulsive sexual behaviors despite adverse consequences. Some of the proposed terms for this syndrome are based on conceptual models for understanding the disorder that provide implications for etiology and treatment. Most notably, these tend to cluster into the general categories of an addiction model (sexual addiction; Carnes, 1983), an obsessive-compulsive model (sexual compulsivity, compulsive sexual behavior; Coleman, 1987, 1990, 1992; Raymond, 2003), and an impulse-control model (sexual impulsivity; Barth, 1987; Kafka, 2007; Rinehart, 1997; Schwartz, 2003, 2005). Still, other authors propose more atheoretical nomenclature such as paraphilia-related disorders (Kafka, 1993, 2007) and hypersexual disorder (Kafka, 2010) to describe this pattern of problematic sexual behavior.

An exhaustive theoretical review of this ongoing nosological debate is beyond the scope of this chapter (see Carnes, 2014; Garcia, 2010; Goodman, 2001; Kafka, 2010; Kingston, 2008; Kor, 2013; Ley, 2014; Steele, 2013). Each conceptual model contributes meaningfully to this discussion by offering unique implications for etiology and treatment. Given the controversy surrounding this debate, along with the relatively small, albeit growing, body of research in this area, this chapter uses an inclusive, multitheoretical perspective when presenting recent research findings illuminating how several different models may inform effective intervention strategies. This approach mirrors the collaborative and integrative spirit of healing of this book. Throughout this chapter, different terms such as "sexual addiction," "compulsive sexual behaviors," and "hypersexuality" are included to represent research from their respective theoretical viewpoints. At the same time, this author's theoretical bias is toward an addiction model, given training and experience in the treatment of sexual addiction; therefore the overarching structure of this chapter relies more heavily on sexual addiction theory and terminology.

Symptoms and Consequences of Sexual Addiction

DIAGNOSTIC CRITERIA

Diagnostic criteria for hypersexual disorder were recently proposed for consideration in the Sexual Disorders section of the *Diagnostic and Statistical Manual of Mental Disorders* (fifth edition [DSM-5]; Reid, 2012). These broad, behaviorally based and theoretically neutral criteria were derived from the *Diagnostic and Statistical Manual of Mental Disorders* (fourth edition, text revision) diagnostic criteria for hypoactive sexual desire disorder and paraphilias (American Psychiatric Association [APA], 2000; Kafka, 2010) and were designed to allow for multiple conceptual and etiological frameworks. According to the proposed criteria, diagnosis for hypersexual disorder requires four or more of five criteria related to repetitively engaging in sexual fantasies, urges, and behavior over an excessive amount of time, in response to dysphoric mood states, in response to stressful life events, following unsuccessful efforts to control, and while disregarding the risk of physical or emotional harm to self or others (Reid, 2012).

While hypersexual disorder was not included in the final publication of the DSM-5 (APA, 2013), the section previously labeled Substance-Use Disorders was significantly restructured, such that substance use and pathological gambling disorders are now subsumed under the new heading of Addiction and Related Disorders. Additionally, Internet gaming disorder was included in the DSM-5 "Conditions for Further Study" chapter. These structural changes in the DSM-5 are viewed by some researchers as a recognition of the many parallels that have emerged in recent literature between chemical addictions and behavioral or process addictions, though additional research is needed to establish which behaviors of excess (such as gambling, sex, or spending) fit in this conceptualization (Carnes, 2014; Petry, 2015).

Over the past several decades, a number of other researchers have attempted to identify a consistent set of diagnostic criteria for problematic hypersexual behavior (Carnes, 1983, 1991, 2005; Coleman, 2003; Goodman, 1998; Orford, 1978; Stein, 2001). Interestingly, despite controversy over etiology and nomenclature, several key consistencies have been observed across these different sets of diagnostic indicators. The most common consistencies relate to continuation of problematic sexual behavior despite adverse consequences, engagement in sexual behaviors during time allotted for other obligations, and ineffective attempts to limit or control certain sexual behaviors (Carnes, 2014). While the hypersexuality label provides a valuable overarching construct for varying presentations of these behaviors, Carnes (2014) proposed a set of 10 diagnostic criteria for sexual addiction that are more consistent with the diagnostic criteria for a substance use disorder found in the DSM-5 Addiction and Related Disorders section. Several of these criteria overlap significantly with those found in the proposed hypersexual disorder diagnosis (excessive time spent engaging in sexual behavior; unsuccessful efforts to control or reduce sexual behavior; continuation of sexual behavior despite social or physical harm), while others describe more traditional addiction-related aspects of symptom presentation such as tolerance and withdrawal (increased intensity, frequency, or risk of sexual behavior to achieve the desired effect; becoming upset, anxious, restless, or irritable if unable to engage in sexual behavior). Authors suggest cutoff scores for diagnosis of sexual addiction ranging from mild (two to three criteria) to severe (six or more criteria; Carnes, 2014). Without universally accepted diagnostic criteria, estimating prevalence is challenging; several authors have estimated prevalence rates between 3% and 6% (Black, 2000; Carnes, 1991; Coleman, 2003; Kuzma, 2008; Sussman, 2011a).

There are a number of measures designed to assess for problematic sexual behaviors, and the content differs based on the various conceptualizations of the phenomenon. Complete assessment for sexual addiction may include items related to the diagnostic criteria for the disorder as well as additional qualitative information about the specific types of sexual behaviors engaged in and the consequences experienced. While complete sexual addiction/hypersexuality assessments should be administered only by clinicians trained to interpret these measures, healthcare providers can utilize a screening tool to determine which patients would benefit from further evaluation. The PATHOS brief screening instrument for sexual addiction assessment was created by Carnes (2012) and queries via six items (Table 15.1).

Table 15.1. PATHOS Screening Tool

P—Preoccupied	Do you often find yourself preoccupied with sexual thoughts?
A—Ashamed	Do you hide some of your sexual behavior from others?
T—Treatment	Have you ever sought help for sexual behavior you did not like?
H—Hurt Others	Has anyone been hurt emotionally because of your sexual behavior?
O—Out of Control	Do you feel controlled by your sexual desire?
S—Sad	When you have sex, do you feel depressed afterwards?

The PATHOS screening tool ratings have shown good sensitivity and specificity, using a cutoff score of 3 when distinguishing between clinical and nonclinical subject samples in identifying a sex addiction problem. The authors point out that far more men currently seek treatment for sex addiction than women and, therefore, more men were sampled in this study, which included outpatients receiving treatment for sex addiction (n = 646, 86.8% male), individuals receiving residential treatment for sex addiction (n = 64, 100% male), and undergraduate college students (n = 203, 23.2% male). The authors point out that additional data from samples with more gender, age, and racial diversity are needed to further validate the potential usefulness of this screening tool.

TYPES OF BEHAVIORS

Categories of sexual behaviors that make up sexual compulsivity and hypersexuality are generally divided into paraphilic and nonparaphilic. Paraphilic behaviors are outlined in the DSM-V (APA, 2014) and differ from hypersexual or compulsive sexual behaviors in that they are characterized as socially unconventional or "deviant" forms of sexual arousal that cause significant distress or impairment in functioning. These include pedophilia, exhibitionism, voyeurism, sexual masochism, sexual sadism, transvestic fetishism, fetishism, and frotteurism.

Nonparaphilic hypersexuality is generally characterized by an excessive or disinhibited pattern of conventional or socially accepted sexual behavior (Coleman, 2003; Kafka, 2007, 2010, 2014). The overarching behavioral categories that appear consistently in the literature include compulsive masturbation, multiple partners/promiscuity/sexual conquest, anonymous sex, compulsive use of pornography/cybersex, phone sex, multiple relationships/relationship addiction, compulsive/coercive sex within a relationship, paying for sex, intrusive/boundary-violating sex, exploitation of power/position, public/group sex, object sex, and sex with drugs.

There is a high degree of comorbidity between paraphilias and nonparaphilic hypersexual behaviors. In one survey, 86% of outpatient males with a paraphilia met criteria for nonparaphilic hypersexuality at some point in their lifetime (Kafka, 1999). This pattern of overlap has led some researchers to include several paraphilic behaviors in the typology of sexually addictive/compulsive behavior (Coleman, 2003; Green, 2015). Because several of the paraphilic typologies may involve illegal or sexual-offending behaviors (voyeurism, exhibitionism, frotteurism, and pedophilia), there is a need for an expanded conceptualization of paraphilic sexual offenses that can incorporate the comorbidity of sexual addiction/compulsivity. While there is some degree of overlap between recommended treatment methods for sexual addiction/compulsivity and sexual offending, the divergent treatment needs of these populations call for research examining ways to effectively blend or integrate sexual addiction and sexual-offending treatment methodologies that preserves their effectiveness and reconciles theoretical incongruities (e.g., the 12-step concept of powerlessness in sexual addiction treatment vs. the concept of personal responsibility in sexual offender treatment).

THE CONTROVERSIAL AND CRITICAL NATURE OF DIFFERENTIAL DIAGNOSIS

The inclusion of any specific sexual behavior in the DSM is highly controversial. Sexual addiction/compulsivity as a construct has been criticized as sex-negative, pathologizing of sexual minorities (Levine, 1988), overly pathologizing of normal sexual behavior (Ley, 2014), or simply a reflection of sexual conservatism (Reay, 2013) and socially enforced religiosity and moralism (Grubbs, 2015a). Such criticisms are prudent, considering that the DSM historically has been used to inappropriately pathologize certain healthy variations of human sexuality simply because they were considered "deviant" or "abnormal." The most notable example was homosexuality, which was removed from the DSM list of mental illnesses in 1973 (APA, 1973).

When conceptualizing sexual behaviors through the lens of sexual addiction/compulsivity, it is essential to understand that diagnosis and treatment of this behavior pattern is not an attempt to pathologize or change a specific sexual behavior in and of itself. Rather, the purpose is to acknowledge and clinically address the problematic or maladaptive *manner in which these behaviors are engaged*, as well as the destructive impact of these behaviors on the client's ability to function and sustain intimate relationships. For this reason, thorough assessment and differential diagnosis are essential to take into account the multitude of other diagnoses and biopsychosocial factors that may present in a manner similar to sexual addiction. In addition to scrupulous differential diagnosis, clinicians can help safeguard against the risk of overpathologizing based on cultural norms by working from a sex-positive stance in which sexual sobriety or recovery is defined in a way that is fluid enough to promote active sexual health; incorporate gender, sexual orientation, and erotic diversity; and fit within the client's value system.

One important thing to consider when working with sex addicts is that their addictive/compulsive sexual behaviors may not be an accurate representation of how they experience their sexual arousal outside of the addictive behavior pattern; sometimes, aspects of the addictive arousal patterns are not fully integrated or accepted by the client. This may be especially true for compulsive pornography users or cybersex addicts. Accurate differential diagnosis of sexual addiction is complex, and a complete discussion is beyond our scope; however, several common differential diagnosis dilemmas are presented throughout the chapter in order to illustrate the potentially harmful effects of misdiagnosis. See Box 15.1 for a brief discussion of the nuances of sex addiction and sexual orientation conflict, and see Box 15.2 for an exploration of the intersection among sex addiction, sadomasochistic arousal, and bondage, domination, sadism, and masochism (BDSM) subculture.

CONSEQUENCES

The consequences of sexual addiction/compulsivity are vast and provide possibly the most compelling evidence of the disorder's legitimacy and the need for effective treatment. As Kafka (2010b) wrote, "This is a significant sexual disorder associated with public health consequences." Sexual addiction is associated with significant physical and medical consequences, including sexually transmitted diseases (STDs), unplanned pregnancies, abortions, rape, domestic violence, overuse of Viagra and other sexual aids (Gutman, 2006), and injury to the body or genitals as a result of sexual behaviors. Some individuals report exhaustion and lack of sleep from long nights of viewing pornography or seeking potential sex partners (Carnes, 1991; Coleman, 1992; Kafka, 2010). In addition to the direct

Box 15.1 Differential Diagnosis: Sexual Orientation Conflict and Sexual Addiction

Given the controversy around pathologizing sexually marginalized groups, a notable and relatively common differential diagnosis dilemma arises when a self-identified heterosexual male presents with out of control or "addictive" sexual behaviors or fantasies that all revolve around themes of sexual contact with men. The inaccurate assumption that this is a gay man experiencing sexual orientation conflict may prompt the clinician to utilize gay-affirming therapy. On the other hand, the inaccurate assumption that this man is a sex addict may prompt the clinician to pathologize and treat these same-sex-themed behaviors as "an addiction," an approach that can closely resemble reparative or conversion therapy (a widely criticized therapeutic practice of attempting to change a person's sexual orientation or associated behaviors and urges).

Appropriate diagnosis and selection of treatment approach rely upon an understanding of how normal experiences of sexual orientation conflict (confusion about sexual orientation, incongruities between sexual urges/behaviors and personal values or sense of identity, sexual promiscuity) can appear similar to the experiences of a heterosexual sex addict who gravitates toward gay pornography or sex with men due to other factors that may be unrelated to his sexual orientation (see Kort [2014] for an in-depth exploration of this issue).

Box 15.2 Differential Diagnosis: Kink, BDSM, and Sadomasochism in Sex Addiction

Another example of a presenting issue with the potential for inaccurate labeling is a sex addict whose pornographic or behavioral sexual interests have shifted to BDSM, pain exchange, violence, or even rape-related content. Assess for a pattern of escalation over time, interest or engagement in these behaviors with nonconsenting persons, and clinically significant distress or functional impairment to identify appropriate diagnostic considerations and applicable clinical approaches while bearing in mind the potential harmful consequences of assigning pathology where there is none:

- Diagnostic consideration of sexual sadism/masochism disorder
- Diagnostic consideration of sexual addiction with sadomasochistic interest/arousal
- Assessing history and risk of sexual contact with a nonconsenting person
- Assessing safety precautions and risk of harm to self and/or consenting participants
- Consideration of a sex positive psychoeducation approach to normalize healthy BDSM practices and orient the client to reputable resources and cultural connection if desired.

Recovering sex addicts are not forbidden from integrating BDSM, kink, polyamorous, or other alternative sexual practices and identities into their healthy sex plan (Katehakis, 2010). However, in the early phases of treatment and recovery, it is essential to understand that the problematic sexual behaviors of a sex addict are often, by definition, out of control, boundary violating, and destructive, and they may significantly conflict with the mutual and informed negotiation of boundaries and consent that is considered central in BDSM culture. Also, while this is not the case for nonclinical samples of BDSM practitioners, early research findings demonstrate a significant association between trauma symptomology and sadomasochistic behaviors in a large sample of treatment-seeking male and female sex addicts (Hopkins, 2016). Therefore, in some sexually addicted clients, sadomasochistic sexual interests may be part of a more complex clinical presentation in which these sexual behaviors represent attempts to regulate negative affect, repetition of past trauma, inability to form secure attachments, or escalation of intensity, rage, and shame.

costs of commodified sexual services and materials (e.g., prostitution, pornography, strip clubs), sexual addiction may also result in significant career and financial consequences, such as reduced productivity, loss of job or promotion, and costs or work loss associated with treatment or legal proceedings (Carnes, 1991; Coleman, 1992).

Subjective psychological distress is not a universal feature of sexual addiction (Rosenberg, 2014b), especially prior to and during the early phases of treatment, when denial and distorted thinking often serve to justify the behaviors and conceal their true impact. However, for many, sexual addiction has been associated with a litany of negative emotional consequences, including depression and anxiety (Kaplan, 2010), loneliness (Yoder, 2005), guilt, shame, loss of integrity, loss of self-esteem, and loss of spiritual connection (Carnes, 1991). Given the intimate nature of sexual addiction, it is not surprising that it has a destructive impact on relationships through marital discord, reduced interest in sex with the primary partner, severe pair-bond dysfunction, loss of primary relationship, and divorce (Carnes, 1991; Kafka, 2010). Sexual addiction/compulsivity may also initially present as sexual dysfunction, including discrepant sexual desire, lack of desire within the primary relationship, or erectile dysfunction (Kafka, 2014). While these sexual dysfunctions may be comorbid with sexual addiction, sexual dysfunction may also develop as a consequence of sexual addiction. For example, what started as a slight discrepancy in sexual desire may become more substantially polarized as the addict's persistent pressuring for sex is repeatedly met with increased rejection by the partner.

When a sex addict is in a romantic relationship, perhaps the most devastating consequence of the addiction is the traumatic impact it has on the addict's partner. Steffens (2006) found that most partners discover evidence of compulsive sexual behavior themselves without voluntary or planned disclosure from the addict, causing extreme shock and surprise. However, even structured therapeutic disclosure of compulsive sexual behavior is a significant distressing event that adversely impacts the partner's mental health and ability to function. Some authors found that the majority of partners in their study met criteria for posttraumatic stress disorder (PTSD) following discovery or disclosure of the addicts' sexual behaviors (Steffens, 2006). Partners of sex addicts also report decreased marital satisfaction, decreased marital intimacy, and increased marital distress (Reid, 2010; Schneider, 2000a, 2000b). The next section expands on the specific relational impact of cybersex addiction. These findings emphasize the importance of a systemic treatment model that not only addresses the compulsive sexual behaviors but also prioritizes consideration of the partner's trauma and the impact on the couple's relationship.

Pornography and Cybersex

DISTINGUISHING BETWEEN HEALTHY AND PROBLEMATIC CYBERSEX USE

While extensive research has been conducted on the negative impact of online sexual media, several researchers note that the Internet can offer outlets for healthy sexual expression by providing a safe, private space for sexual exploration and education, free access to sexual information and virtual communities in which lesbian, gay, bisexual, and transsexual, erotic minority, or other disenfranchised individuals may connect (Cooper, 1999; Dhuffar, 2015; Döring, 2009). In one community sample of individuals who use the Internet for sexual purposes, most respondents indicated that online sexual activity neither interfered with nor jeopardized their lives. However, 8.3% of respondents reported using the Internet for sex more than 11 hours per week, and, unlike recreational cybersex users, they were more likely to experience financial, legal, occupational, emotional, or relational problems related to cybersex use (Cooper, 2000).

Addiction researchers suggest that convenience features such as accessibility, affordability, and anonymity (Cooper, 1998) may facilitate engagement in sexual behaviors online that are associated with increased risk and potential for Internet sex addiction (Griffiths, 2012). In addition to the ease and convenience of access, Internet pornography has been referred to as a "supranormal stimulus" because of its potential ability to overstimulate the brain's natural reward circuitry in comparison to the far more limited visual sexual stimuli that was available during the evolution of this neural circuitry (Hilton, 2013). Compulsive cybersex use shares many of the features of more general sexual addiction, including preoccupation with cybersex, inability to stop or limit use of cybersex, and experiencing negative consequences including danger, distress, or reduced ability to function. Cybersex, as a construct, continues to morph in response to advances in technology and is broadly defined here as any use of the Internet for sexual gratification.

Carnes (2001) describes how lengthy exposure to the availability of virtually endless sexual options on the Internet can quickly alter a person's arousal template, the sexual belief system or map that shapes the various types of stimuli that an individual experiences as erotic. According to Carnes, this shift tends to happen in predictable phases, first with a rapid expansion in the variety of sexual behaviors experienced as stimulating. In the sex addict, this initial expansion eventually narrows in focus to an obsessional fixation on new, highly specific sexual behaviors. As compulsive pornography use escalates over time, an addict's arousal template may begin to shift toward a preference for heightened risk, intensity, and reward through increasingly violent, illicit, forbidden, or bizarre pornographic content. Compulsive cybersex users often report lengthy online sessions searching for the "perfect" image or scenario, becoming oblivious to the hours passing as well as their bodies' cues for hunger, thirst, and fatigue (Delmonico, 2001). When they finally identify, and typically achieve orgasm while viewing, the "perfect" cybersex image, scenario, or fantasy, compulsive cybersex users regularly report disappointment at the fleeting nature of their satisfaction, and they are often left with feelings of guilt, shame, emptiness, and exhaustion (Philaretou, 2005). Alteration of the arousal template related to pornography addiction can complicate the process of differential diagnosis. While it is valuable to explore a client's preferences for specific themes or genres of pornographic content, these themes may not be representative of the client's primary sexual interest. Box 15.3 explores the escalation

Box 15.3 Differential Diagnosis: Illegal Pornography, Pedophilia, and Cybersex Addiction

A notable example of how the arousal template can be influenced by exposure to Internet pornography is the cybersex addict who recently began viewing illegal child sexual abuse images (CSAI; also known as child pornography) as part of a long-standing pattern of daily compulsive Internet pornography use. The client reports having watched every type of pornography available and has observed his porn preferences getting "darker" in recent years as "normal" or nonboundary-violating pornographic content no longer keeps his interest. This client has no previous legal record and denies current or historical attraction to children, and results from his psychosexual evaluation and risk assessment determine that he represents a low risk for future offending.

Given the data, this client's interest in CSAI appears to be an escalation of his cybersex addiction, so diagnosis and treatment of pedophilia would not be indicated, as this sexual interest may be adequately addressed by sex addiction treatment. Research has begun to examine what proportion of CSAI viewers are also at risk for exploiting children through the involvement in the production of these images or engaging in sexual contact with minors (Berlin, 2012). While potential and actual harm to children should always be evaluated thoroughly by a professional who specializes in such assessments of risk, it is beneficial for clinicians to be aware that escalation of a cybersex addiction can potentiate an interest in CSAI in clients with no other signs of pedophilic sexual interest.

of cybersex into illegal sexual images and related implications for differential diagnosis.

IMPACT OF CYBERSEX ADDICTION ON INTIMATE RELATIONSHIPS

One major feature of sexual addiction is an increasing tendency to engage in sexual objectification of others, defined by Fredrickson (1997, 174) as treating someone "*as a body* (or collection of body parts) valued predominantly for its use to (or consumption by) others." Research has shown that some women consider their male partner's pornography use to be a form of sexual objectification, and for these women there is an associated increase in their experience of interpersonal sexual objectification, internalization, eating disorder symptomatology, body surveillance, body shame, relationship anxiety, and negative affect (Tylka, 2015). Excessive sexual objectification can promote insecurities in body image and sexual performance in both the viewer and the partner, regardless of gender.

Among men, difficulty finding an acceptable partner and forming intimate relationships has been associated with cybersex frequency, cybersex craving (Weinstein, 2015), and exposure to narrow ideals of female sexual attractiveness (Schooler, 2006). In children and adolescents, cybersex can provide a kind of short-cut to sexual satisfaction that allows the user to bypass the important developmental work of moving through the stages of courtship (Carnes, 2001). This can produce a significant developmental gap between the sexual templates formed by pornographic images residing within an adolescent or young adult who has profound limitations in relational sexual maturity. This impact on sexual development and intimacy may have broader implications for the development of healthy social adjustment in general (Owens, 2012).

Within a committed relationship, compulsive cybersex use has been associated with reduced intimacy and relationship satisfaction (Schneider, 2000a). A Dutch longitudinal study revealed that, at least for men, the association between use of online sexually explicit material and relationship quality appears to be bidirectional: husbands in poor relationships used more cybersex materials, and use of more cybersex materials lead husbands to experience their relationships as poorer. However, cybersex use did not appear to affect long-term sexual satisfaction in this nonclinical sample (Muusses, 2014). One interpretation of the mixed results in the literature is that cybersex use, per se, may not negatively impact relationship and sexual satisfaction for most people, but for compulsive or addicted cybersex users and their partners, these out-of-control online sexual behaviors can have a devastating impact on intimacy, safety, and trust.

In addition to the negative effects of objectification, cybersex addiction within the context of a romantic relationship also carries with it the impact of intimate betrayal. When confronted about their online sexual behaviors, cybersex addicts may provide reasons why the behavior is not problematic or should not negatively impact their partner. These justifications typical revolve around the idea that use of pornography or online communication with cybersex partners is normative, only fantasy, not real, or somehow less significant than physical sexual contact with an extramarital partner. The resulting assumption is that cybersex use is not technically "cheating" (Weiss, 2015). However, numerous studies have shown that partners of cybersex addicts often do consider cybersex to be a form of infidelity and experience it as just as emotionally painful as live, offline affairs (Schneider, 2000, 2012). Even when sexual behaviors were limited to online only, the majority of partners still reported loss of trust, negative relationship consequences, and identifying themselves as the victims of trauma. Schneider (2012) uses the term "betrayal trauma" to normalize a partner's emotional response to secretive cybersex use, which can include feeling unattractive, doubting self-worth, searching for evidence of betrayal, attempting to monitor or control the user's access to cybersex, and even doubting their own reality after repeatedly being told lies (Schneider, 2012; Weiss, 2015).

IMPACT OF CYBERSEX ADDICTION ON SEXUAL FUNCTIONING

There is very limited research on the effect of cybersex addiction on the addict's sexual functioning, and that research has produced mixed results. Of course, sex addicts may present with comorbid sexual dysfunction, but more research is needed to establish whether certain sexual dysfunctions may be associated with, or even exacerbated by, cybersex addiction. One illuminating study found that, when compared to matched healthy volunteers, a small sample of men with compulsive sexual behaviors reported significantly more difficulty with arousal and erectile functioning in the context of an intimate sexual relationship but not while using pornography (Voon, 2014). A larger study of Italian adolescents found that 10% of participants who admitted to using pornography reported a reduced sexual interest in potential real-life partners. They also found that 25% of those who viewed pornography regularly reported experiencing abnormalities in their sexual functioning (Capogrosso, 2013). However, another large-scale epidemiological study found no evidence of an association between pornography use and male sexual dysfunctions (Landripet, 2015).

Much of this research was prompted by a substantial grassroots movement of young men sharing personal anecdotal accounts of erectile dysfunction, low sexual desire, and delayed ejaculation associated with excessive pornography use. These web-based communities (including YourBrainOnPorn.com, NoFap.com, and FightTheNewDrug.org) are designed to raise awareness about the potential for online pornography to lead to addiction and what is referred to as "porn-induced sexual dysfunctions" or "porn-induced erectile dysfunction." Though additional research is needed to establish the exact nature of the association between pornography and sexual dysfunction, a substantial portion of young people seeking treatment for sexual addiction/compulsivity are familiar with these accounts and endorse similar struggles with sexual functioning, which they attribute to excessive pornography use. Anecdotal observations of this phenomenon by the author are included in Box 15.4.

Within a sexual relationship, the contrast between a heightened interest in pornography and low arousal/sexual dysfunction during partnered sex can exacerbate feelings of rejection and insecurity in the partner. Additionally, as cybersex addiction disrupts intimacy and safety within a relationship, this can also deeply affect a partner's sexual functioning, with sexual responses varying drastically, sometimes even within the same person. Some partners attempt to increase sexual frequency, attempt to improve their appearance or sexual performance, or try to incorporate pornography into partnered sex. These attempts typically do not change the cybersex addict's behavior and often leave the partner with feelings of hopelessness, rejection, shame, and anger. Some partners begin to seek sexual satisfaction alone or outside of the relationship, and others become completely shut down sexually

Box 15.4 Anecdotal Observations of Sexual Dysfunction in Cybersex Addicts

In working with this population, this author has observed male clients report several fairly common presentations of sexual dysfunction that they often experience as secondary to excessive cybersex use and/or masturbation. The most notable of these is the experience of situational erectile dysfunction in male cybersex addicts. For these men, lack of desire or inability to achieve an erection for partnered sex is experienced as a function of excessive pornography use and masturbation. Sometimes this is attributed to the frequency of masturbation; however, many clients report that the escalating pornographic content they view has become so explicit, intense, and shocking that "normal," partnered sex is no longer experienced as sufficiently novel or arousing. Another commonly reported issue is delayed ejaculation, in which excessive masturbation, often to pornography, has led to an inability to achieve orgasm except by manual self-stimulation. In addition to shaping subjective sexual arousal around highly explicit pornographic content, these clients may also have essentially conditioned their bodies to only respond to the specific rhythm and pressure of masturbation, which, over time, may inhibit their ability to achieve orgasm through other forms of partnered sexual contact.

(Schneider, 2012; Weiss, 2015). General sexual arousal or specific sexual behaviors can take on new meaning and become painful reminders of the trauma of betrayal (Carnes, 2011).

Sexual Addiction: A Disorder of the Brain, Mind, and Heart

While no large-scale epidemiological studies of sexual addiction currently exist, some theoretical framework for identifying the possible etiology and function of these behaviors is necessary to inform effective interventions. In light of the substantial disagreement in the scientific community about whether addiction, obsessive-compulsive, impulse control, or some other theory best explains this pattern of out-of-control sexual behavior, this chapter utilizes interpersonal neurobiology as a broader unifying framework for understanding sex addiction and discussing treatment strategies. This framework allows for integration of research from competing models by conceptualizing sexual addiction from the perspectives of three separate yet interconnected dimensions of the human experience: the brain, the mind, and relationships (Siegel, 2010). Specifically, sexual addiction can be understood from the theoretical perspectives of dysfunctional neural reward circuitry, maladaptive affect regulation, and disordered intimacy (or insecure attachment). Each of these lenses is presented with supporting theory and research along with their respective explanations for the presence of commonly observed psychiatric comorbidities and implications for effective treatment strategies.

THE BRAIN: SEXUAL ADDICTION AS DYSFUNCTIONAL NEURAL REWARD CIRCUITRY

Recent observations of similarities between the neurobiological presentation of chemical addictions and that of other rewarding behavioral processes has led many researchers to conceptualize problematic, excessive sexual behaviors as an addictive disorder with a predictable set of pathological neuroplastic changes in the brain (Blum, 2012; Goodman, 2008; Hilton, 2013; Olsen, 2011). Behavioral or process addictions, more generally, have gained growing acceptance in the addiction field, most notably since 2011, when the American Society for Addiction Medicine changed its definition of addiction to "a primary, chronic disease of brain reward, motivation, memory and related circuitry. Dysfunction in these circuits leads to characteristic biological, psychological, social and spiritual manifestations. This is reflected in an individual pathologically pursuing reward and/or relief by substance use *and other behaviors*" (emphasis added). The neuroscience of chemical addictions as well as early research pointing toward commonalities observed in sexual and other behavioral addictions are reviewed to illuminate the possible etiological influence of neurochemical reward abnormalities and resulting treatment implications.

The neurobiological presentation of addiction was initially established in the chemical dependence literature. The rewarding effects of addictive substances are generally attributed to their ability to increase dopamine activity in the nucleus accumbens (Budygin, 2012; Di Chiara, 1988), sometimes at levels much higher than natural rewards. The intensity of the reward motivates some people to use drugs habitually, so the brain adjusts to this repeated intense flood of dopamine with a set of neuroplastic changes that downregulate the dopamine reward response by producing less dopamine or reducing the number of available dopamine receptors (Blum, 2012). This downregulation can take place independent of genetic makeup, causing the low dopamine function associated with compulsive drug intake (Volkow, 2005), heightened craving and motivation to use drugs, and an overall reduction in the ability to experience pleasure or reward (Blum, 2012; Chen, 2011). Put another way, the addicted individual "wants" or craves the rewarding stimulus but then does not "like" or enjoy it as much as expected. This discrepancy between wanting and liking, or anticipatory and consummatory reward, is thought to produce compulsive reward-seeking behavior (Finlayson, King, 2007; Stice, 2010; Volkow, 2012; Voon, 2014).

Recent research has demonstrated that addiction does not just affect dopamine reward circuitry but also a host of other neurotransmitters that impact complex behavioral processes. One example is the effect of chronic drug use on the balance between synaptic and nonsynaptic glutamate. This glutamate imbalance produces pathological neuroplastic changes that impair the corticostriatal communication that normally allows the prefrontal cortex to regulate reward-seeking behaviors (Kalivas, 2009). Research studies using functional magnetic resonance imaging and positron emission tomography have shown that addiction also affects memory, motivation, executive function, mood, and interoception (sensitivity to stimuli originating inside of the body), which helps explain the constellation of behavioral changes observed in a chemically addicted person (see Volkow [2012] for review). Not everyone who is exposed to substances of abuse will become addicted; genetic deficits in reward processing may be activated epigenetically (Sussman, 2011b) by interacting with environmental and developmental factors to produce vulnerabilities to addiction and influence gravitation toward a specific drug or behavior of choice (Blum, 2012; Volkow, 2005).

In addition to drugs of abuse, natural rewards produced by behaviors like eating, shopping, gambling, video gaming, Internet use, and sexual activity may also develop into compulsive behaviors due to similar neuroplastic changes in addiction-related brain circuitry (Leeman, 2013; Olsen, 2011, Tao, 2010). Similarities in neurobiology and clinical presentation have led some authors to hypothesize that chemical addiction and compulsive engagement in other naturally rewarding behaviors share an underlying biopsychological addictive process (Blum, 2012; Goodman, 2008; Olsen, 2011). Because neurobiological research on sexual addiction is in its early stages, findings from research on process addictions more generally may hold important implications for sexual addiction theory and treatment. Specifically, the concepts of food addiction and compulsive overeating are of particular relevance to future sexual addiction research and intervention strategies because eating and sex are both experienced as hedonic and they are essential for survival (Olsen, 2011). A growing body of research has demonstrated an overlap between the neurochemical reward properties of hedonic foods and drugs and the implications for our understanding of compulsive overeating and obesity (Finlayson, 2007; Luppino, 2010; Stice, 2010; Volkow, 2012).

Several studies examining the neural responses to pornography suggest that excessive or compulsive use of pornography may be associated with neuroplastic changes that are consistent with those commonly observed in chemical addiction (Hilton, 2013; Kuhn, 2014; Voon, 2014). Similar to the amplified reward value of certain hedonic foods in the modern era, the increased novelty, variety, and accessibility of Internet pornography may produce a "supranormal stimulus" that increases the addictive potential of technologically mediated sexual behaviors (Hilton, 2013). Compared to healthy controls, individuals with compulsive sexual behavior demonstrated greater engagement of corticostriatal limbic circuitry following exposure to sexually explicit cues as well as dissociation between liking and wanting responses to sexual cues, both of which are hallmarks of addiction (Voon, 2014). Also, self-reported hours of pornography use have been correlated with reductions in gray matter volume and functional connectivity to the prefrontal cortical areas, which may suggest neuroplastic change from an overstimulation of the brain's reward center similar to that found in chemical addiction research (Kühn, 2014).

Reflecting on the growing body of evidence that chemical and process addictions share a set of neuroplastic changes in the brain's reward circuitry, it is not surprising that these various addictive and compulsive disorders are highly comorbid. A large sample of sex addicts demonstrated high rates of lifetime prevalence of at least one other addictive behavior (69%–80%; Carnes, 2005). Comorbidity rates are significant between substance abuse and gambling addiction (20%–30%), Internet addiction (10%), exercise addiction (15%), and food addiction (25%; Sussman, 2011a). Substance use disorders and sexual addiction/hypersexuality are estimated to have comorbidity rates ranging from 18% to 40% (Deneke, 2015; Kafka, 2002; Stavro, 2013; Sussman, 2011a). Specific substances of abuse (marijuana, methamphetamine, cocaine) seem to be more frequently associated with sexual addiction than others (Deneke, 2015; Guss, 2000), and these specific substance use disorders may be considered risk factors that substantiate additional screening for sexual addiction. Comorbid addictions may produce unique treatment needs as these addictive processes do not simply coexist but often interact with each other creating complex, interconnected behavioral patterns that, if not addressed holistically, may promote relapse (Carnes, 2009).

THE MIND: SEXUAL ADDICTION AS MALADAPTIVE AFFECT REGULATION

Hypersexual behavior has been widely associated with problems related to affect, including neuroticism, alexithymia, emotional dysregulation, depression, anxiety, and vulnerability to stress (Kafka, 2010; Reid, 2008, 2011, 2014). Hypersexual clients have been found to experience significantly less positive emotion and significantly greater negative emotion, including self-hostility and shame (Reid, 2010). While these observations are simply correlational, they have led some researchers to theorize that hypersexual individuals use sex as an impulsive or compulsive form of pleasure-seeking to cope with or self-medicate their negative affective experiences (Adams, 2001; Reid, 2010, 2011).

The obsessive-compulsive and impulse-control conceptualizations of this syndrome are both predicated in affect regulation but differ primarily in their implied motivational states. Compulsivity functions to reduce anxiety or negative emotional states, while impulsivity functions to increase pleasure or positive emotional states (Rinehart, 1997). Due to the high degree of overlap in these constructs, a complete conceptual framework must allow for the contribution of both compulsivity and impulsivity to accurately describe the affect regulation function of hypersexual behavior (Kingston, 2008). Addiction is one model that accounts for the dual function of engaging in problematic behaviors both to experience pleasure, typically in the early stages (impulsivity), and to avoid or escape uncomfortable emotional states like cravings or withdrawal symptoms (compulsivity), more commonly in the later stages (Goodman, 1993; Koob, 2009). The function of addictive behavior as a maladaptive coping strategy for affect dysregulation is a widely recognized concept in the behavioral addictions literature (Adams, 2001; Goodman, 2008; Rosenberg, 2014a).

Support for this conceptualization can be theoretically inferred from the extensive comorbidity of sexual addiction with psychiatric disorders involving dysregulations in mood, anxiety, and impulse control. In a review of the literature on psychiatric comorbidity, Kafka (2010) reported a consistent finding of multiple lifetime comorbid mood, anxiety, substance use, and impulse disorder diagnoses among the majority of hypersexual individuals. Raymond (2003) examined the psychiatric comorbidity of a small sample of individuals with compulsive sexual behavior and found that 88% had a current Axis I disorder and 100% met criteria at some point in their lifetime. Specifically, the authors found a 71% lifetime prevalence of mood disorders, 96% lifetime prevalence of anxiety disorders, 71% lifetime prevalence of substance use disorders, and 46% prevalence of personality disorders. In a recent study, Deneke (2015) screened inpatients in chemical dependency treatment for sexual addiction to examine prevalence rates and comorbidities. Individuals who were at risk for sexual addiction were more likely than the other chemical dependency inpatients to be diagnosed with a mood disorder, impulse-control disorder, eating disorder, or PTSD. In another study among Indian outpatients diagnosed with mood and/or anxiety disorders, 25% percent screened positive for hypersexual disorder (Nair, 2013).

The affect regulation function of sexual addiction appears to be closely related to the construct of experiential avoidance: engaging in strategies to reduce the impact of uncomfortable internal experiences, even when doing so causes harm (Hayes, 1996). An example could be excessive viewing of online pornography to cope with loneliness and depression. The behavior is reinforced due to the immediate relief it provides; however, over the long term, excessive pornography viewing may result in even greater isolation, paradoxically increasing the loneliness and depression that the behavior was originally intended to reduce. Levin (2012) found that viewing pornography was associated with anxiety and problems related to viewing only among those with clinically significant levels of experiential avoidance. Their findings suggest that experiential avoidance may play a role in determining whether viewing pornography becomes problematic. The role of experiential avoidance and affect regulation in sex addiction has major implications for treatment by placing an emphasis on addressing psychiatric comorbidities and guiding interventions to provide alternative healthy coping and mindfulness- and acceptance-based strategies for regulation.

Affect dysregulation is one of the mechanisms by which trauma is implicated in the etiology and function of sexual addiction. Traumatic experiences in childhood or adulthood may produce or exacerbate negative affect states, and compulsive sexual behaviors may function as an attempt to self-sooth (Howard, 2007). This theory is supported by the finding that individuals exposed to adverse childhood experiences are at increased risk for alcoholism, drug abuse, depression, suicide attempts, high number of sexual partners, and STDs (Felitti, 1998). Adams (2001) proposed that affect dysregulation among sex addicts may originate in early developmental attachment failure. The authors note that anxious and disorganized attachment patterns render a child incapable of tolerating distress. They are resistant to being comforted or soothed by a caregiver and instead rely on their own external action to self-sooth. These authors draw a link between affective dysregulation and limitations in the sex addict's ability to achieve and maintain trusting, intimate, and coregulatory relationships. The relationship between sex addiction and intimacy, and attachment and trauma, is explored further in the next section.

THE HEART: SEXUAL ADDICTION AS DISORDERED INTIMACY AND ATTACHMENT

Understanding sexual addiction through the theoretical lenses of dysfunctional neural reward circuitry and maladaptive affect regulation creates an accurate, yet incomplete, depiction of its possible etiology, function, and impact. Intimacy and attachment make up an essential interpersonal component of sexual addiction conceptualization by integrating some of the most deeply meaningful, and often profoundly painful, features of this syndrome. Disordered attachment and developmental trauma have been implicated in the development of chemical and behavioral addictions by numerous authors (Adams, 2001; Carnes, 1991; Flores, 2004; van der Kolk, 1989).

Early research from a small survey of recovering sex addicts found that, when all types of abuse were accounted for, 97% of respondents endorsed some history of trauma (Carnes, 1991). A recent large-scale study of sex addicts established more contemporary prevalence rates of abuse by gender: sexual abuse of males, 25.6%; sexual abuse of females, 60.3%; emotional/physical abuse of males, 40.7%; emotional/physical abuse of females, 74.0% (Jore, 2015). These authors noted that experiencing some type of abuse was associated with slightly more symptoms of sexually compulsive behavior, though the effect size was small. In contrast, a smaller study found that sexual compulsivity in adulthood is associated with childhood emotional abuse, childhood exposure to pornography, and parental sex addiction but not childhood sexual or physical abuse (McPherson, 2013).

Reactions to childhood sexual abuse (CSA) are varied and complex, which may help to contextualize the inconsistent findings in the literature. Individuals who have experienced CSA demonstrate a spectrum of sexual responses ranging from withdrawal and dysfunction to hypersexuality and sexual compulsion. Additionally, the gender and age of the child both influence which type of behavior is more likely to develop (Aaron, 2012). Even as children, victims of CSA may demonstrate a range of sexual responses including hypersexual, exposing, and victimizing sexual behaviors, and rates of these behaviors are significantly elevated for children who experienced onset of abuse prior to age seven (McClellan, 1996). Individuals who experienced more severe CSA (involving the use of penetration and/or force) report more adult sexual risk behavior, including a higher number of lifetime partners, higher number of lifetime STD diagnoses, and trading sex for money or drugs (Senn, 2007). While the hypersexual behaviors associated with CSA do not necessarily constitute sexual addiction, they are common among sex addicts, demonstrating one way in which childhood trauma may be linked to compulsive sexual behavior.

The association between childhood trauma and the development of sex addiction can be partly explained by the affective dysregulation that results from early trauma. In theory, victims of early trauma and neglect experience high levels of hyperarousal and affect dysregulation due to the absence or inconsistency of effective coregulation by a parent or caregiver (Adams, 2001; van der Kolk, 1989). As discussed in the previous section, adverse or traumatic experiences in childhood place adults at increased risk for addictions of all types, as well as a myriad of other mental health problems (Felitti, 1998). Traumatized individuals with dysregulated affect may resort to external sources of regulation such as those found in various addictions (Adams, 2001; van der Kolk, 1989). As many professionals who work with this population have observed, trauma survivors also consistently tend to re-enact their trauma through patterns of experiencing abusive relationships, dangerous situations, aggression toward others, and harm to self. Additionally, engaging in risky sex increases the odds of becoming the victim of coerced sex in adulthood (Griffee, 2012). This process plays a major role in sexual addiction as the specific sexual acting out behavior of many sex addicts often involves clear repetition of traumatic experiences that result in additional adult trauma. This compulsion to repeat or re-create trauma is explained by van der Kolk (1989) as "addiction to the trauma," a process involving the endogenous opioid response, autonomic arousal, and state dependent learning.

For many sex addicts, developmental trauma may have prevented the formation of a secure attachment bond. Attachment theory originated to describe the relationship between a mother and her infant (Bowlby, 1969) and was later applied to adult romantic love by Hazan and Shaver (1987). Fraley's (2000a,

2000b) Experiences in Close Relationships–Revised measure has been used in various studies to examine adult attachment styles among sex addicts. Using a college sample, Bartholomew (1990) established normative rates of each of the four attachment styles (college age: Secure—57%; Preoccupied—10%; Fearful-Avoidant—15%; Dismissing—18%). A Secure attachment denotes comfort with intimacy and autonomy, a sense of lovability, and an expectation that others will be accepting and responsive to one's needs. This contrasts with three insecure styles, which include Preoccupied (anxious in relationships, feels unlovable and strives for the acceptance of others), Fearful-Avoidant (fearful of intimacy and socially avoidant, sense of unworthiness, and expects untrustworthiness and rejection from others), and Dismissing (avoidance of intimacy, sense of worthiness, strives for independence from others) attachment (Zapf, 2008).

In a small sample of male sex addicts, only 8% were found to have Secure attachment, while the remainder of the sample were found to have Preoccupied (28%), Fearful-Avoidant (44%), and Dismissing (20%) attachment styles (Zapf, 2008). A recent large-scale study of individuals seeking treatment for sexual addiction supported previous research demonstrating that sex addicts tended to have less secure attachment styles than the normative population (Jore, 2015); however, the observed discrepancies were less pronounced in this larger sample (male sex addicts: Secure—36%; Preoccupied—26%; Fearful-Avoidant—25%; Dismissing—13%). Authors note that individuals with a Dismissing attachment style may be underrepresented in clinical populations as they may be less motivated by relationship distress to seek treatment. The consistent finding that individuals with sexual addiction are less securely attached than their healthy counterparts (Faisandler, 2012; Jore, 2015; Zapf, 2008) supports the contribution of conceptualizing sex addiction as an intimacy disorder.

Additional support for this conceptualization can be inferred from the high prevalence of disordered personality traits and personality-based difficulties experienced by sex addicts (Jore, 2013, 2015) and hypersexuals (Carpenter, 2013). A recent study examining the relationships between specific types of sexual acting out behaviors and Minnesota Multiphasic Personality Inventory–2 personality profiles identified five classes of sex addicts using latent profile analysis (De Guzmán, 2016). This classification system identified how certain types of sexual acting out behaviors appear to correspond with specific personality profiles in a pattern of escalating clinical acuity. Class 1 sex addicts were generally lacking in comorbid psychopathology, which offers support for sex addiction as a clinically distinct mental health construct. Class 2 sex addicts tended to have avoidant, dependent, depressive, and self-defeating personality traits. These individuals were more apt to seek out isolated and self-stimulating sexual behaviors like pornography and masturbation. Those in Class 3 were high on narcissistic and, to a lesser extent, antisocial personality traits, and they tended to engage in paying for sex and use of seduction. Class 4 sex addicts demonstrated more antisocial, sadistic, and depressive personality traits and were more likely to endorse intrusive or exploitive sexual behaviors as well as substance-related problems. Finally, Class 5 sex addicts were the most clinically acute group. This class had more borderline and trauma-related personality traits and tended to engage in more fantasy and covertly intrusive sexual behaviors such as voyeurism and exhibitionism.

Recent literature illuminating more modest relationships between childhood trauma, attachment, and sex addiction than those found in early studies have challenged researchers and clinicians to broaden their conceptualization of sex addiction to include other etiological factors besides trauma. Riemersma (2013) helped to clarify these discrepancies by differentiating between the classic sexual addiction presentation described in early literature (defined by a history of developmental trauma, insecure attachment, comorbid addictions, and mood disorders) and a contemporary form of sexual addiction more common among current youth (rapid onset, less evidence of traumatic origins, often organized around exposure to Internet sexual media). It is important to acknowledge that, while all sexual addiction impacts intimacy, many sex addicts may be securely attached and lacking a history of trauma. Simple exposure to a highly rewarding stimulus (such as Internet pornography) as well as other psychosocial factors in adulthood can produce a sexual addiction in those with no history of trauma. However, even in the absence of overt developmental trauma or insecure attachment, incorporating an exploration of both early developmental and adult attachment experiences into therapy provides a rich and emotionally laden context in which to examine the unique etiology and function of addictive sexual behavior for each specific client.

While no causal links can be made between trauma, attachment styles, personality traits, and the development of sex addiction, these data begin to paint a more detailed picture of the complex set of interpersonal factors that must be addressed in order to achieve effective treatment and recovery. Though only in the early stages of being validated by research, the integration of trauma, attachment, and personality traits in conceptualizing various sexual addiction profiles can help to illuminate why certain sex addicts gravitate toward specific types of sexual behaviors. These conceptual explanations, while hypothetical and somewhat analytic in nature, are often extremely helpful in therapy as mechanism for deriving meaning from a pattern of behaviors that are often experienced as highly illogical, values incongruent and confusing to the client.

Finally, a conceptual model that includes an interpersonal component is essential to describe the most devastating impact of sexual addiction—the impact on the individual's intimate relationships. Whether the sex addiction manifests as compulsive use of pornography, serial infidelity, or sexual boundary violations, sex addicts often end up isolated and alone. Even for those with secure attachment and no history of childhood trauma, the impact of deception and betrayal actively undermines safety, trust, and intimacy in clients' relationships. If trauma, disordered attachment, and lack of intimacy are major factors in the development, function, and impact of sexual addiction, then effective treatment must certainly look beyond the intrapersonal emotional functioning of the individual and include interpersonal, relationship-centered, and trauma-informed interventions.

Integrative Treatment of Sexual Addiction

Use of interpersonal neurobiology as a unifying conceptual framework offers a particularly relevant way to organize this discussion for several reasons. The field of interpersonal neurobiology is not rooted in illness or dysfunction but instead focuses on three universal and interacting dimensions of the human experience: the brain (and, by extension, the body and nervous system), the mind, and relationships. Through therapy and intentional action, clients can promote wellness through physical and neuroplastic healing, enhanced emotional well-being, and improved relational connection. Importantly, this conceptual framework leaves room for the addiction model, the obsessive-compulsive model, the impulse-control model, and the hypersexual disorder model while also adding in the essential systemic component of interpersonal relationships. Lastly, Siegel (2010) discusses the implications of recent advances in our understanding of neuroplasticity that show that the intentional actions of the mind can alter the structure of the brain. These actions of the mind can include a number of therapeutic techniques aimed at changing thought patterns and behaviors; however, arguably one of the most broadly applicable and health-promoting intentional actions of the mind is the practice of mindfulness. Siegel goes on to describe integration as a fundamental sign of health, and he discusses nine specific domains of integration. This approach implicitly embodies the spirit of integrative medicine through the promotion of neural, intrapersonal, and interpersonal connection as a means of cultivating not only an absence of illness but holistic health. It also explicitly structures and guides treatment to include integrative interventions, with mindfulness at the core.

To date, there are no methodologically rigorous, large-scale randomized control trials assessing the efficacy of any therapeutic or pharmacological interventions for sexual addiction or hypersexuality (Hook, 2014), leaving researchers and clinicians to rely upon uncontrolled studies and case reports or extrapolate applicable treatment methodologies from existing literature on chemical and other behavioral addictions and related psychiatric disorders (Miles, 2016; Rosenberg, 2014b). Despite varying theoretical approaches to diagnosing and conceptualizing problematic hypersexual behaviors, several general principles consistently emerge from the literature describing treatment recommendations. Taken together, these general principles combine into a set of standard multimodal, multidisciplinary treatment approaches that draw upon cognitive-behavioral, insight-oriented, group, couple/family, psychoeducational, peer support, physical health, and pharmacological interventions that can be integrated into a unique, client-specific treatment plan (Kafka, 2014; Rosenberg, 2014a, 2014b). These standard treatment recommendations is presented in the context of the three dimensions of interpersonal neurobiology with a focus on how complimentary, alternative, and integrative methodologies can be used to augment traditional interventions to form a more comprehensive approach to treatment.

INTEGRATIVE TREATMENT OF THE SEXUALLY ADDICTED BRAIN: NEUROCHEMICAL, MICRONUTRIENT, AND LIFESTYLE BALANCE

One of the most obvious first-order goals in sexual addiction and hypersexuality treatment is to assist the client in reducing or completely abstaining from engagement in specific problematic sexual behaviors. Treatment models can generally be divided into one of two categories: abstinence-based and harm reduction. Unlike chemical addiction recovery, abstinence-based recovery from sexual addiction does not require complete abstinence from all sexual behavior. Rather, abstinence or sobriety is defined as abstaining from specific dysfunctional, problematic, or "bottom line" sexual behaviors that are considered by the client to be addictive/compulsive, boundary violating, values-incongruent, and/or associated with negative consequences (Carnes, 2015; Kafka, 2014). Harm reduction models, in contrast, seek to moderate or limit the frequency or manner of engagement in the problematic sexual behaviors to a level that minimizes risk or negative consequences. Harm reduction models are typically viewed as a remote second choice by addiction professionals, if they are considered at all, and these are often only negotiated after a client has experience repeated detrimental relapses (Rosenberg, 2014a).

Of course, because even abstinence-based models of sexual addiction recovery do not recommend complete abstinence from sexual behavior and occasional relapse is considered a typical part of the recovery process, it can become difficult, at times, to make meaningful distinctions between abstinence-based and harm reduction models in the treatment of sexual addiction and behavioral addictions more generally. One rationale for abstinence proposed by sexual addiction researchers is based on the assumption that the neuroplastic dopamine reward abnormalities associated with chemical and process addictions may be reversible, to an extent, with a period of abstinence. Various researchers recommend negotiating a three- to six-month period of sexual celibacy early in the treatment process, offering rationales that include neuroplastic regeneration of dopamine reward function, spiritual or personal growth, increased safety in the romantic partnership, improved ability to navigate reintegration of healthy sexual practices, and mitigation of additional negative consequences (Carnes, 2010; Kafka, 2014; Kaplan, 2010; Rosenberg, 2014a).

Psychopharmacology

Research on the efficacy of psychopharmacological interventions for behavioral addictions (Grant, 2016) and sexual addiction more specifically (Naficy, 2013) is in its very early stages, with outcomes showing some promising results that point out directions for future study. Psychopharmacological agents aimed at reducing the frequency of problematic sexual behaviors typically do so directly by targeting impulsivity/compulsivity or indirectly by lowering overall sex drive or regulating dysphoric affect (Kafka, 2014). A number of single case reports have been published supporting the use of naltrexone (Bostwick, 2008; Grant, 2001; Raymond, 2010) and naltrexone with serotonin reuptake inhibitors (Raymond, 2002), clomipramine, and valproic acid (Gulsun, 2007) in the treatment of compulsive sexual behaviors (Kaplan, 2010). Naltrexone acts as an opioid receptor antagonist and is thought to attenuate addictive or compulsive behavior through modulation of the dopaminergic mesolimbic pathway. Serotonin reuptake inhibitors have been used in the treatment of compulsive sexual behaviors both to address common comorbid mood and anxiety disorders as well as to enhance central serotonin levels to produce a reduction in sexual arousal and disinhibited sexual appetitive behaviors (Coleman, 2000; Kafka, 2014; Malladi, 2005; Wainberg, 2006).

In addition to depression, sex addiction is highly comorbid with attention deficit hyperactivity disorder (ADHD), and some researchers suggest bupropion and psychostimulants such as methylphenidate and dextroamphetamine because they are more activating and may help with depressed mood, inattention, and distractibility (Kafka, 2014). The dopamine-enhancing effect of psychostimulants can produce a counterproductive increase in sexual motivation and compulsivity in some patients, so non-habit-forming medication such as atomoxetine may be considered a safer option for this population. Benzodiazepines, opiates, and other habit-forming medications should also be avoided when possible or prescribed with caution due to the potential for this population to become multiply addicted (Rosenberg, 2014b). Lastly, dopamine agonist treatment in patients with Parkinson's disease has been associated with an increased risk of impulse-control disorders, including compulsive gambling, spending, binge-eating, and sexual behavior (Garcia-Ruiz, 2014; Moore, 2014; Weintraub, 2010). These findings illuminate possible neurochemical etiological factors related to sexual compulsivity and encourage careful screening for compulsive behaviors in patients being treated with dopamine agonists.

Nutrition and Supplements

In addition to psychotropic medications, nutritional interventions may be helpful in the treatment of sexual addiction and its common comorbidities by supporting the overall health and functioning of the body and the brain. Nutritional interventions may be helpful in addressing addiction and mental health issues through a variety of mechanisms including direct improvement of brain function, supplementation of micronutrients, management of food sensitivities, and reduction of inflammation through a healthy balance of intestinal microbiota.

Dopamine reward abnormalities are implicated in addiction, impulse control, and compulsivity conceptualizations of hypersexuality. Blum (2012) conceptualized these and other mental health issues to be associated with a genetically predisposed hypodopaminergic state termed "reward deficiency syndrome" (RDS). In addition to genetic testing to determine risk for RDS, these authors recommend nutrigenomic and neuromodulation solutions aimed at activating dopaminergic pathways to augment the more traditional recommended treatment areas of psychotherapy, self-help organizations, spirituality, and holistic treatments that promote well-being (Blum, 2012). Specifically, authors cite early evidence that nutrigenomic-based dopamine agonist therapy with natural, nonaddictive, customizable neuroadaptogen-Namino-acid therapy variants enhances dopaminergic functionality in a manner that is safer and less intense than typical pharmaceutical agents (Blum, 2012, 2015). Authors also recommend cranial electrical stimulation (Mindes, 2015), meditation, relaxation, yoga, exercise, music therapy, and acupuncture in addition to traditional psychotherapy and self-help groups (Blum, 2011, 2012).

Dietary supplements have been found by some researchers to be as effective as a psychostimulants in the treatment of ADHD. The supplements in this study were intended to increase dopamine and norepinephrine synthesis by loading with tyrosine, the precursor to these neurotransmitters, and using B-vitamins, iron, and copper as cofactors. Phospholipids and essential fatty acids were used in this study for cell membrane repair, improved synaptic physiology and neurotransmitter efficiency. Essential fatty acids and probiotics were also used to promote gut enterocyte repair (Harding, 2003). This study illustrates some of the primary nutritional interventions for increasing dopamine function. Dopamine is synthesized from the amino acid tyrosine, which can be taken as a supplement or found in foods like apples, bananas, beets, fava beans, watermelon, wheat germ, and protein-rich animal products like meat, cheese, and eggs. Other foods and supplements that can increase dopamine include turmeric, ginkgo biloba, L-theanine (green tea), and fermented foods high in probiotics such as sauerkraut, yogurt, and kefir.

Improvements in dopamine function achieved through the nutrient supplementation suggested here would likely impart positive changes in mood and motivation. However, given that sexual addiction is commonly treated in tandem with a myriad of comorbid psychiatric conditions, it is also important to consider nutritional therapy and micronutrient supplementation approaches that specifically target the unique presenting comorbidities of each client. The use of nutrition and dietary supplements to treat various psychiatric conditions has been reviewed in the literature (Fugh-Berman, 1999; Lakhan, 2008; Rucklidge, 2013), with notable findings for L-methylfolate (Roman, 2011), St. John's wort, kava, folate, tryptophan, phenylalanine, *S*-adenosylmethionine, and omega-3 polyunsaturated fatty acids (Sarris, 2009). One more recent area of focus in the nutritional therapy literature is the gut microbiome. There is growing evidence that increased gastrointestinal permeability (leaky gut) may play a role in the inflammatory pathophysiology of depression, alcoholism, and other psychiatric disorders (Glaus, 2014; Forsythe, 2010; Foster, 2013; Leclercq, 2014; Maes, 2008; Perlmutter, 2015). This research has implications for probiotic nutritional interventions (via fermented foods or supplements) for improved mental health as well as consideration of additional dietary changes that may mitigate inflammation and oxidative stress (e.g., removal of gluten, assessment for other food sensitivities, increased antioxidant consumption; Bested, 2013; Perlmutter, 2015; Selhub, 2014).

Neurofeedback and Cranial Electrotherapy Stimulation

Other nonpharmaceutical approaches that function by directly targeting brain activity include entrainment and feedback devices such as cranial electrotherapy stimulation (CES) and neurofeedback. While no specific research has been conducted on the use of either of these treatment approaches with sex addicts or hypersexuals specifically, there is an established body of literature supporting their use for treatment of commonly comorbid conditions of anxiety, depression, insomnia, and ADHD (for reviews, see Mindes [2015]; US Food and Drug Administration [2012]; Yucha [2008]). Given their utility in supporting mood regulation, attention, and sleep, augmentation of traditional psychotherapy approaches with the use of CES and neurofeedback devices may help clients manage negative mood states, cope with urges and cravings, and reduce impulsivity in a manner that facilitates and reinforces use of the skills being developed in therapy.

In addition to some of the specific treatment recommendations mentioned here, many sexual addiction treatment protocols include mention of more general principles of physical health and wellness to support the recovery process. Recommendations aimed at sexual addiction as well as other comorbid mood disorders and psychiatric conditions include balanced nutrition, maintaining normal weight, regular aerobic exercise, seven to eight hours of sleep nightly, and regular medical checkups with a physician (including testing for STDs or infections; Amen, 2015; Blum, 2012; Carnes, 2015; Lofgreen, 2012; Luppino, 2010).

INTEGRATIVE TREATMENT OF THE SEXUALLY ADDICTED MIND: HEALTHY COGNITION, REGULATED AFFECT, AND TRAUMA RESOLUTION

Much of the published literature reviewing treatment of sexual addiction and hypersexuality has focused on multimodal psychotherapeutic approaches aimed at attenuating problematic sexual behavior through cognitive restructuring, behavioral activation, affect regulation, insight development, trauma resolution, and relapse prevention planning (Adams, 2001; Carnes, 2015; Kafka, 2014; Miles, 2016; Rosenberg, 2014a, 2014b). These recommendations tend to draw from several different theoretical approaches, including cognitive-behavioral, psychodynamic, mindfulness-based, and psychoeducational, though research assessing the effectiveness of these various approaches tends to favor cognitive-behavioral and mindfulness-based approaches. Based on the needs of the client, a combination of individual, group, couple, and family therapy modalities are often recommended and may be delivered at an outpatient, intensive outpatient, or residential/inpatient level of care depending on level of acuity, psychiatric and addiction-related comorbidities, and relapse history. Authors specifically emphasize the value of group therapy as a mechanism for combating shame and isolation (Adams, 2001; Coleman, 2003) and promoting intimacy and connection. While this section focuses on the intrapsychic aspects of psychotherapy, the interpersonal aspects of therapy, including group, couple, and family modalities that promote intimacy and interpersonal connection, are explored in greater detail in the next section.

Cognitive-Behavioral Therapy Approaches

Because ambivalence toward behavior change is a common feature of addictive disorders, treatment for sex addiction typically begins with an assessment of the client's motivation to change. Clients enter treatment with varying levels of willingness to change, and some may consider mitigation of consequences or acquiescence to a partner or family member's request as the primary motivation for treatment. For clients who enter therapy with some ambivalence about giving up certain sexual behaviors or other aspects of the recovery process, therapeutic approaches like motivational interviewing may be warranted to support the client in identifying intrinsic motivation and fostering a more stable willingness to prepare for, initiate, and sustain behavior change (Carnes, 2015; Del Guidice,

2007; Rosenberg, 2014b). Throughout the early phases of treatment, cognitive-behavioral therapy (CBT) interventions tend to target denial and other types of distorted thinking that maintain the problematic pattern of behavior.

In adults with proposed hypersexual disorder, engaging in sexual behavior often involves disregarding the risk for physical or emotional harm to self or others. While this population typically reports clinically significant distress providing intrinsic motivation for change, impairment in multiple areas of life that cannot be attributed to drugs of abuse, manic episodes, or other medical conditions may also provide therapeutic leverage for treatment when hypersexuals are in the early stages motivation to change.

CBT may also be used to assist the client in identifying a predictable cycle or pattern of problematic behavior so that treatment techniques can target client-specific behavioral rituals, environmental cues/stressors, and negative affect states that may precede a relapse (Table 15.2) (Behun, 2012). Relapse prevention focused therapy typically relies heavily on establishing behavioral changes that support sexual sobriety. Diagnostic criteria for hypersexual disorder considered by the APA for inclusion in the DSM-V (APA, 2013; Reid, 2012) implicitly suggest specific behavioral targets to consider in sex addiction treatment and relapse prevention. For example, if excessive time is occupied by sexual fantasies, urges, and behaviors, treatment interventions may target regulation of the time available for these activities. If low mood (e.g., anxiety, depression, boredom, irritability) or stressors consistently precipitate episodes of problematic sexual behavior, mood stabilizers, stress reduction practices, exercise, and increased social contact may help support behavior change.

More substantial life changes may be considered for those facing significant environmental stressors that will likely not be adequately attenuated by affect-regulation and behavioral coping skills. For example, a client with a current career or living situation that is directly interfering with sexual sobriety may weigh the costs and benefits of changing jobs or moving into a sober living environment, especially early in the recovery process (Carnes, 2015; Kafka, 2014; Weiss, 2015).

In addition to attenuating thoughts and behaviors directly linked to problematic sexual behaviors, psychotherapy should also target comorbid psychiatric disorders, substance use disorders, and other compulsive or addictive behavior patterns, as these conditions may interact with and predispose a client to sexual addiction relapse (Adams, 2001; Carnes, 2009). While the motivational interviewing, CBT, and relapse prevention techniques reviewed previously may be implemented to address comorbid conditions as appropriate, therapeutic interventions with a broader focus on improving affect regulation more generally may reinforce versatile skills relevant to coping with negative mood states, tolerating distress, and managing urges and cravings. In addition to mood and anxiety disorders, sex addicts have been found to struggle with other emotional issues, including alexithymia, stress proneness (Reid, 2008), neuroticism (Reid, 2011),

Table 15.2. Sexual Addiction Relapse Prevention Behavioral Targets[a]

	Relapse Prevention Goal	*Specific Examples of Behavioral Targets*
Access to Addictive Stimulus	Limit or remove access to former partners, secret forms of communication, pornographic stimuli, and other "stash."	Email accounts and addresses, secret cell phones, contact information of former partners, text messages, photos, videos, adult website memberships, digital and hard copy pornography, other sexual paraphernalia, smartphone applications for sex/dating.
Secure Technology	Secure points of access to Internet and technology including home/work computer, smartphone, work phone, laptops, tablets, and e-readers	Limit Internet access to several known devices equipped with monitoring/filtering software with third-party accountability. Consider removing Internet browser and application store from smartphone and identify possible loopholes in Internet access via applications with internal browsers.
Environmental Changes	Reduce or remove triggers for acting out and create an environment conducive to recovery.	Keep computer/tablets in a central public location, re-arrange furniture in work or home office, remove curtains or blinds from office window, redecorate home/office, replace mattress.
Habit or Routine Changes	Create daily routine that minimizes unstructured time and circumvents typical opportunities to act out.	Adjust route to work to avoid strip clubs, keep a structured daily schedule, schedule free time with others to avoid isolation, avoid triggering social situations, former partners, and acting out friends.
Additional Accountability	Build in additional sources of accountability for staying on track or handling high-risk scenarios.	Random periodic urine drug testing, repeat polygraph examinations (approximately six months apart), third-party financial accountability, limit/remove access to cash, "bookend" risky situations with supportive check-in phone calls, utilize smartphone's geolocation, FaceTime, and social media technology for accountability, travel with family, friends, or sober companion.

[a] Monica Meyer, PhD, 2017.

self-hostility, and shame (Adams, 2001; Gilliland, 2011; Reid, 2010). Several treatment modalities that promote affect regulation are reviewed with an emphasis on approaches that integrate some form of mindfulness practice as a core component of treatment.

Mindfulness and Acceptance-Based Approaches

Traditional CBT approaches have been augmented and adapted in recent decades to create various new approaches that utilize mindfulness and acceptance as core mechanisms for regulating affect, restructuring cognition, and promoting behavior change. These include dialectical behavior therapy, acceptance and commitment therapy (ACT), mindfulness-based stress reduction, and many others. Mindfulness has been applied to addiction treatment and relapse prevention with notable success, and researchers are beginning to hypothesize about the potential neurobiological mechanisms that may account for the observed reduced relapse rates, including attenuation of neural responses to the experiences of craving and negative affect. Authors theorize that this may signify a degree of reversal or repair of the neuroadaptive changes associated with addiction and relapse (Brewer, 2013; Garland, 2014; Witkiewitz, 2013). While the neurobiological mechanisms of mindfulness have yet to be established by research, this theory is consistent with the main principles of interpersonal neurobiology, which include the assumption that intentional actions of the mind (specifically mindfulness) can produce neural plastic change in the brain (Siegel, 2010).

Mindfulness has been operationally defined in the literature by two essential components: maintaining attention on present-moment experiences and doing so with an attitude of nonjudgmental acceptance (Bishop, 2004). Mindfulness has been implicated as a potential target for future study in the treatment of sex addiction (Grubbs, 2015b; Katehakis, 2009; Reid, 2010, 2014) and behavioral addictions more generally (Shonin, 2014) by multiple authors due to its potential to promote modulation of dysphoric and self-critical affect (shame) through kindness and self-compassion, nonreactive tolerance of distress and urges to act out, reduced autonomic arousal, and improved identification of judgments, criticisms, and thought distortions (Reid, 2010, 2014; Shonin, 2014). In a male sample that included hypersexual patients and nonhypersexual controls, mindfulness was shown to be a robust negative predictor of hypersexuality, accounting for unique variance beyond that explained by emotional dysregulation, impulsivity, and stress proneness (Reid, 2014). While there is a clear interest in the literature for exploring mindfulness-based sex addiction treatment approaches, minimal research has been conducted to date evaluating the effectiveness of this modality in attenuating hypersexual behavior in this population.

One small multiple-baseline study of treatment-seeking problematic porn viewers found that an eight-session ACT protocol resulted in an 85% reduction in pornography viewing, and results remained stable at three-month follow-up (Twohig, 2010). A primary mechanism of action for ACT is a reduction in experiential avoidance, a construct reviewed previously in this chapter as a moderating factor whether or not viewing pornography was related to problems such as anxiety and negative consequences (Levin, 2012). While more research is needed, levels of experiential avoidance may be valuable to consider as a potential risk factor, viable treatment target, or as one indicator for determining which clients may respond positively to mindfulness-based interventions for problematic hypersexuality. Box 15.5 presents anecdotal reflections from the author regarding the use of mindfulness techniques to support sexual sobriety.

Box 15.5 Anecdotal Reflections on the Compatibility between Mindfulness and Abstinence-Based Sexual Sobriety Definitions

One interesting consideration is the compatibility of mindfulness-based treatment protocols like dialectical behavior therapy or ACT with an abstinence-based model for sexual recovery. A prominent feature of mindfulness-based treatment approaches includes redirecting the tendency to struggle against, become consumed by, or attempt to change one's uncomfortable internal experiences. While the behavioral goal of abstinence from certain sexual activities would appear to be compatible with this approach, some clients self-impose a more rigid sobriety definition that may deem certain internal mental processes like sexual fantasy to constitute a relapse. This typically presents as an issue for clients whose notion of morality, purity, or relational fidelity requires a mind free from engagement in sexual objectification and fantasy.

Based on the mechanisms of action in mindfulness, clients who react to these internal experiences as relapse risks will likely judge them negatively and make active attempts to avoid or prevent them, leading to the paradoxical increase in sexual thoughts and emotional distress that often results from these avoidance strategies. This author typically observes this phenomenon in clients with highly anxious, obsessive-compulsive tendencies and very rigid or moralistic values related to sexuality. In therapy with this population, it may be helpful to introduce mindfulness techniques with more benign thought patterns and internal experiences to promote comfort with the approach before moving on to more salient internal experiences such as sexual thoughts, urges, and fantasies. When these clients begin to rely upon and integrate the mindfulness-based skill of allowing thoughts to flow through the mind unencumbered by avoidance and judgmental labels, many describe the result as a profound sense of freedom and relief along with notable reductions in the frequency and intensity of sexual thoughts and cravings.

Psychodynamic and Insight-Oriented Approaches

In contrast with the affect regulation and behavior modification goals of CBT and mindfulness-based interventions, psychodynamic, insight-oriented, and trauma-focused theoretical approaches are also noted frequently in the sex addiction treatment literature, though they have notably less empirical support (Carnes, 2015; Kafka, 2014; Rosenberg, 2014b). The application of these theoretical approaches to sexual addiction treatment tends to focus on psychosexual development, family of origin relationships, attachment insecurities, traumatic experiences, and defense mechanisms (Adams, 2001; Kafka, 2014; Katehakis, 2009; Rosenberg, 2014b). While psychodynamic, insight-oriented, and trauma resolution approaches to treatment target symptomatology that overlaps considerably with the comorbid mood and anxiety disorders and affect dysregulation patterns reviewed previously in this section, the major contribution of psychodynamic theory to sex addiction treatment is presented in the context of developmental trauma and attachment in the next section. These therapeutic approaches typically inform case conceptualization and clinical interventions aimed at deepening the client's insight into how the pattern of sexually addictive behaviors may serve the intrapsychic function of defending the ego or resolving unconscious internal conflicts. Box 15.6 presents a sample psychodynamic case conceptualization for a sex addict with exhibitionistic behavior.

Somatic and Body-Based Approaches

The discussion of integrative treatments for affect regulation and trauma resolution concludes with a review of several body-based trauma treatment modalities aimed at reprocessing or renegotiating the emotional, cognitive, and somatic experiences that often accompany traumatic memories. Eye movement desensitization and reprocessing (EMDR) therapy has been shown by research to be effective in relieving distress associated with traumatic memories by targeting dysfunctional associations using bilateral stimulation accompanied by reprocessing and resourcing techniques (Regan, 2015). EMDR treatment involves some focus on tracking of physical sensations and body scanning, though the proposed mechanism of action revolves largely around restructuring negative cognitions and desensitizing the negative emotions from traumatic memories.

More comprehensive integration of the body in trauma treatment can be found in modalities like somatic experiencing (SE; Levine, 1997) and brainspotting (Grand, 2011). While there is no current efficacy research on brainspotting, the approach combines aspects of EMDR, SE, mindfulness, and self-observation to resolve distressing internal processes with a primary focus on somatic sensations.

Box 15.6 Sample Psychoanalytic Case Conceptualization

Consider, briefly, the case example of Adam, a 45-year old, never-married, male sex addict with significant insecurities about his physical appearance, genital size, sexual performance, and overall sense of inadequacy that appeared most prominently in his relationships with women. A psychodynamic theoretical approach may offer a valuable insight into the underlying function of Adam's counterintuitive behavior of exposing his genitals to women in public when he found them to be attractive and intimidating. Most immediately, his acting out defense can be understood as a covert form of aggression intended to express his displaced rage through shocking and offensive sexual behavior. The bold shamelessness of Adam's exhibitionism could be conceptualized as a reaction formation defense or an opposing action intended to protect his true sense of shame and fear of rejection over his genital size.

A closer examination of the interpersonal impact of his behavior reveals the more primitive projective identification defense in which projection of his own sense of inadequacy, self-criticism, and rage over his anticipation of interpersonal rejection leads him to behave in a manner that elicits these exact anticipated reactions of judgment, anger, and fear in his victims. While this behavior ultimately still results in the rejection Adam fears most, the emotional impact may be more superficial due to the brief, anonymous, and covert nature of his intrusions. By essentially demanding that random attractive women notice and respond to his sexuality without asking for consent, Adam has managed to avoid the stages of courtship that would normally require him to identify and emotionally invest in a specific woman of interest and allow her to get to know him more intimately. This avoidance may function to protect Adam from the deeper, more painful rejection he might experience if he took the risk of being truly "exposed" and vulnerable by expressing his desire for a specific woman and allowing her an authentic opportunity to respond.

SE is a somatic trauma therapy aimed at discharge and regulation of excessive autonomic arousal through the process of guiding the client's attention to interoceptive, kinesthetic, and proprioceptive experiences (Levine, 1997, 2010; Payne, 2015). SE researchers draw from established literature on the relationship between interoceptive function and psychiatric disorders along with the utility of interventions focused on mindfulness meditation, meditative movement, and attending to interoception to support their proposed neurophysiological rationale for the mechanisms involved in the use of SE in the treatment of trauma and chronic stress (Payne, 2015).

SE offers several unique benefits in its application to trauma treatment for sexually addicted clients. Clinicians utilizing nonsomatic trauma treatments that are highly rooted in the trauma narrative or family of origin dynamics may find it challenging to work toward validating experiences of victimization in a manner that does not undermine current responsibility for their client's behaviors that have negatively impacted others. Also, clients in early recovery from any addiction may be at increased risk of relapse during difficult or activating trauma work, especially if some of their sexual behaviors show a trend toward trauma repetition in some way. Lastly, most sex addicts could benefit from interventions focused on development of empathy for those impacted by their behaviors. Somatic experiencing fits very naturally in the treatment of trauma in sexually addicted clients because it is a gentle, slow modality that focuses more on awareness of the body than the content of the trauma narrative, making it a safer trauma treatment option than most for clients with a high potential to become triggered. Also, shifting the focus off of the narrative and into the body helps the client regulate and renegotiate dysregulated somatic and emotional trauma symptomology and provide relief with less risk of accountability being undermined by clients who may be at risk of overidentifying with the victim role. Framing SE as a mindfulness and interoceptive awareness modality reinforces its potential efficacy in addressing addictions and other psychiatric disorders based on reports from the mindfulness literature discussed previously. Specific implications for improving a sex addict's empathy and attunement through SE include its potential for use to augment victim empathy and emotional restitution work with partners, family members, and any victims of sexual boundary crossing.

Finally, some sex addicts have a complex and contentious relationship with their bodies. Sex addiction has the potential to impact the relationship with the body in a variety of ways, potentially resulting in an overreliance on sexual touch at the expense of nonsexual touch; physical harm or injury to the body (resulting from excessive, aggressive, or dangerous sexual activity, contraction of STDs, neglected self-care); comorbid compulsive issues around food, exercise, and weight control; as well as body image issues (impact of pornography exposure, objectification of self, obsessions over appearance, physique, or genital size). Integrating SE and other body-based treatment modalities may help sexually addicted clients foster a more caring, self-compassionate relationship with their bodies through mindful awareness, improved interoception, and nonjudgmental acceptance.

In addition to clinical psychotherapy approaches, several other body-based and somatic treatment modalities have been implicated for improving mindful awareness and promoting affect regulation in populations with addictions and mental illness. A review of current literature demonstrated that body-oriented psychotherapies are found to be generally effective at improving subjective symptoms of depression and anxiety, somatization, social insecurity, activity levels, motor tension, and general well-being (Röhricht, 2009). This review focused on a variety of body-oriented psychotherapies with mechanisms of action ranging from self-awareness, perceptual functional body-mind work, energetic and expressive body-mind work, grounding and catharsis, and creative dance movement therapies. Additional research focusing on mindfulness, body-based, and somatic modalities in the treatment of affect dysregulation and addiction supports the general efficacy of integrating mindfulness meditation, intentional movement practices such as yoga and tai chi (Khanna, 2013; Shapiro, 2007; van der Kolk, 2014), auricular and full body acupuncture (Behere, 2009; Blum, 2011), and regulation of breath, heart rate variability, and autonomic arousal through physiological measurement devices such as electroencephalograph biofeedback (Yucha, 2008).

The efficacy of some complementary and alternative treatment modalities have been called into question due to lack of methodologically rigorously supporting research (Ernst, 2002). In a review of complementary and alternative therapies for the treatment of substance use disorders, Behere (2009) reported significant promising evidence for biofeedback, acupuncture, and herbal medicines (specifically kudzu for reducing alcohol consumption, ibogaine for reducing withdrawal and craving) only, citing lack of research supporting various other complementary and alternative modalities. Even when scientific evidence of treatment efficacy is absent from the literature, additional exploration of mind-body interventions such as yoga, meditation, rhythmic auditory stimulation, and individual and communal spiritual practices are worth pursuing as these activities are often low in cost, lacking in contraindications or side effects, and may be valuable if they are experienced by the client as promoting healing, connection, and a subjective sense of well-being. This attitude may be especially applicable to behavioral health, a field of medicine in which even the most efficacious interventions are only functionally effective to the extent that clients are willing to participate in them. Therefore, when considering the therapeutic contribution of a benign, low-cost alternative treatment modality, research describing its efficacy for attenuating the discrete symptoms being targeted should be considered alongside other factors such the impact of participation in the modality on a client's active engagement, willing participation, and subjective satisfaction with treatment as a whole.

INTEGRATIVE TREATMENT OF SEXUALLY ADDICTED RELATIONSHIPS: ATTACHMENT, INTIMACY REPAIR, AND SPIRITUAL CONNECTION

Re-establishing Secure Attachment

While treatments that target arousal levels, neurochemistry, and affect regulation are likely to promote symptom reduction and behavioral change, these improvements may be most meaningful when they support the client in establishing or rebuilding secure interpersonal relationships. Social support and connection is necessary for all humans to flourish. A focus on relational healing is especially important for treating sex addiction because problems with relationships are pervasively implicated in the development, manifestation, and impact of the disorder.

Treatment approaches target relational healing using a variety of therapeutic approaches; however, the most important mechanism for supporting a client's relational healing is the therapeutic relationship itself. The therapeutic relationship is most effective as a mechanism of action in treatment when therapists are able to regulate their own affect and demonstrate attunement and empathy so the client has the experience of feeling safe, known, and valued.

Many sexually addicted clients present with symptom constellations that include insecure attachment and histories of developmental trauma, so sex addiction treatment must effectively integrate trauma-focused and relational interventions (Creeden, 2004). In addition to improving affect regulation with the mindfulness and body-focused modalities previously discussed, clients may benefit from exploration of early attachment wounds through experiential therapy and inner-child work. Based on the assumption that effective attachment figures were not available for modeling healthy nurturing, one goal of this approach is to support the client in developing an adult ego state than can effectively re-parent the self through self-nurturing and setting compassionate limits (Mellody, 2011). Other approaches, like internal family systems (Schwartz, 1994), which utilizes similar concepts around the function of various parts of self, have been implicated in the treatment of developmental trauma (van der Kolk, 2014).

Intimacy and Acceptance in Group Therapy

Group therapy provides a unique environment in which to explore relational issues, and it is often recommended as a preferred modality for sex addiction treatment (Adams, 2001; Carnes, 2015; Kafka, 2014; Rosenberg, 2014a, 2014b). One study found no reduction in problematic sexual behavior in response a targeted group therapy treatment; however, results showed that the modality was effective in decreasing depressive symptoms and improving quality of life (Orzack, 2006). As a behavioral health condition in which lapses into previous sexual behaviors are common, group therapy and other relational interventions may provide supportive connections that offer ancillary benefits, even as clients struggle to maintain behavior change.

When considering group therapy targets other than attenuation of problematic sexual behavior, intimacy and shame are among the most prevalent. This modality can be very effective in reducing shame through several interacting mechanisms. Being part of a group of people who can all relate based on their common struggle with sex addiction can help to combat the stigma associated with sex addiction. Some of the shame experienced by sex addicts may be result of repeatedly acting out against their values and hurting the people they love. This makes it threatening to be known by others, and sex addicts often deal with this by isolating from others, maintaining only superficial relationships, and lying. To address these tendencies most effectively, group therapy should challenge clients to practice accountability and rigorous honesty by sharing the most shameful aspects of their story with other group members present, as this allows clients to experience being intimately known and accepted (Adams, 2001).

Group therapy may also benefit clients in becoming more interpersonally effective by creating a semistructured social environment that provides opportunities to practice newly acquired skills in peer relationships. This unique setting supports the client in moving toward more functional independence by offering support around effectively transferring relational therapeutic progress to peer interactions with increased safety from the group structure and regulating influence of the therapist. Groups that prioritize processing relational dynamics between members and provide a structure for in-the-moment, specific interpersonal feedback can be very powerful for addressing disordered personality features, patterns of deception, interpersonal manipulation or grooming, and socially or sexually inappropriate behavior, even in resistant clients.

Integrating Treatment of the Partner and the Couple

While most of the literature and research on sex addiction treatment has focused on the addict, there is a growing body of literature exploring the application of various trauma-focused modalities in treating partners of sex addicts (Hentsch-Cowles, 2013; Steffens, 2010; Zitzman, 2005). While studies of efficacy are not currently available, typical treatment recommendations for working with partners may include assessing for suicide/self-harm risk, providing tools for affect regulation, identifying additional sources of social support, working toward a coherent narrative of the betrayal trauma, identifying trauma triggers, re-establishing a sense of safety through boundary setting, describing the impact of the betrayal trauma, assessing for issues related to sexual functioning, and providing sexual health resources as needed (Carnes, 2011; Fisher, 2009; Steffens, 2010).

Partner treatment methods that were more prevalent in the past tended to underemphasize the significantly traumatic impact of not only the betrayal surrounding infidelity but also the discovery of the betrayal, the disclosure process, and the pattern of repeated and systematic deception by the addict (Carnes, 2011; Steffens, 2010). Some treatment approaches are organized around conceptualizing the partner as codependent and primarily focus on identifying ways in which the partner's own trauma reactivity or family of origin dynamics may have contributed to a dyadic pattern in the relationship that provided the necessary context for sex addiction to develop (Mellody, 2011). While partners of sex addicts are just as likely to suffer from psychopathology or developmental trauma as any other client, proponents of a more trauma-focused conceptualization for treating partners point out the risk of diagnosing a partner as mentally ill or personality disordered when the maladaptive behaviors and functional impairments used to justify those diagnoses may be more accurately described as a context-dependent post-traumatic stress response to intimate betrayal trauma. Proponents of this trauma-focused partner conceptualization draw from attachment and PTSD literature to support treatment of partners that is focused on validating the trauma, regulating affect and re-establishing a sense of safety and trust (Steffens, 2010).

Couples therapy addressing the impact of sexual addiction on the relationship is typically recommended once the addict has made a behavioral commitment to engage in the treatment process. Couples therapy in early recovery is typically focused on negotiating immediate boundaries to support emotional safety, preparation and processing of a therapeutically supported full disclosure, and basic tools for communication, regulation, and structure (Schneider, 2013). For couples seeking to repair the relationship, modalities like emotionally focused couple therapy may help with treatment goals related to rebuilding trust, deepening intimacy, and promoting secure attachment (Butler, 2006; Reid, 2006; Seedall, 2008). Once basic levels of trust and intimacy are established, couples therapy may focus on reintegrating healthy sexuality back into the relationship (Katehakis, 2010), which may involve treatment for sexual dysfunction (Manley, 1999). Integrative interventions such as yoga, tantra, breath work, and sensate focus may be helpful in guiding couples through sexual reintegration and exploration (Katehakis, 2010).

Spirituality and Social Support

There is a great deal of agreement in the literature around the value of spirituality, community, and social support in promoting sex addiction recovery (Adams, 2001; Carnes, 2015; Kafka, 2014; Rosenberg, 2014b). Whether or not addiction conceptualization is the best fit for a particular case, involvement in the 12-step recovery community is often recommended as a relational, widely accessible, free resource for social support and connection with others who have similar struggles (Kafka, 2014; Wright, 2010).

The 12-step tradition reinforces the value of belief in God or a Higher Power for sex addicts working toward recovery. However, most authors note that those in treatment for sexual addiction may benefit from some form of spirituality, if perceived as a supportive influence, and spirituality as defined by the client may or may not include a personified god or specific religious affiliation. Interestingly, many religious traditions involve practices that are the basis for certain integrative healing modalities such as mindfulness, meditation, gratitude/compassion, fasting and participation in group rituals involving synchronized, rhythmic movement, dance, music, drumming, and chanting (Blum 2012; Galanter, 2006).

Conclusion and Case Study

The clinical presentation of those struggling with excessive, problematic sexual behavior is highly variable and complex and warrants consideration not only of current models for conceptualizing the problematic sexual behavior itself (addiction, compulsivity, impulse control, hypersexuality) but also of comorbid mood, anxiety and personality disorders, co-occurring chemical and process addictions, history of trauma and attachment dysfunction, and psychosocial and relational stressors as well as careful differential diagnosis that explores various alternative factors (psychiatric, medical, psychosocial, behavioral, and cultural) that may be contributing to the client's current presentation and subjective distress.

The addiction model for conceptualizing excessive, problematic sexual behavior allows for integration of several prominent theories for conceptualizing this syndrome (dopamine reward abnormalities, affect dysregulation, and attachment dysfunction), while interpersonal neurobiology provides a comprehensive theoretical framework for organizing various integrative treatment approaches for sex addiction into a holistic model that addresses mind, body, and relational wellness. Treatment-related improvements in any one of these three interconnected systems have the potential to effect change in the other two, further reinforcing behaviors and connections that support recovery and health.

Human sexuality is a complex and multifaceted phenomenon, so treatment of problematic sexual behaviors should typically include multitheoretical, multimodal interventions aimed at addressed the unique constellation of comorbid and contributing factors with a client-specific approach. Applicable integrative treatment approaches from the sexual health, addictions, psychiatric disorders, and trauma specialties have produced a wide variety of treatment options, but mindfulness appears so consistently and promotes mutually reinforcing positive changes so universally that it seems to warrant consideration as a key foundational element in effective sex addiction and trauma treatment.

INTEGRATIVE CASE STUDY

See Box 15.7 for an integrative case presentation.

Box 15.7 Integrative Treatment of Masochistic Sexual Addiction, Alcoholism, Depression, and PTSD

Paul was a 42-year-old Caucasian male seeking inpatient treatment for the third time after relapsing on alcohol following approximately one year of sustained sobriety outside of a controlled environment. Paul presented with PTSD, profound treatment-resistant depression, severe alcoholism with chronic relapse pattern, and sex addiction with masochistic, humiliating, and degrading sexual behaviors involving direct replication of scenarios and sensations from his history of severe childhood sexual and physical abuse. His multiple previous attempts at achieving chemical sobriety were each interrupted by a predictable pattern of heightened trauma reactivity, typically in the form of anxiety, panic attacks, intrusive memories, and night terrors that escalated in intensity over time. Paul would begin engaging in masturbation to masochistic fantasies largely organized around memories of his childhood physical and sexual trauma as a mechanism for self-soothing and coping with his escalating traumatic reactivity when he was not drinking alcohol. Masturbation and fantasy would eventually give way to viewing masochistic-themed pornography, which would provide temporary relief followed by intense shame and a strong compulsion to act out various scenes of his childhood abuse with anonymous partners and paid sex workers.

Paul had a history of involvement in the BDSM and fetish communities and recognized these expressions of sexuality as healthy options for others but described his masochistic sexual behaviors as highly compulsive, ego dystonic, and retraumatizing, indicating that afterward he felt like he was less than trash. Paul would eventually attempt to attenuate the immense shame from his sexual behaviors with the only reliable coping mechanism he had left: drinking alcohol to the point of blacking out. This cycle of trauma reactivity, sexual acting out, shame, and drinking had precipitated his last three posttreatment alcohol relapses. This is the first time Paul has sought treatment for his sexual addiction.

Treatment was delivered over seven weeks at an inpatient level of care followed by 12 weeks at an intensive outpatient level of care with a supported sober living environment. This treatment trajectory allowed for the use of intensive trauma and addiction interventions within the safety of an inpatient setting to establish initial sobriety. Transition to the intensive outpatient level of care allowed for treatment to continue in a setting that fostered increasing levels of independent utilization of regulation and support resources while Paul contended with increased contact with relational and environmental stressors outside of a controlled setting. A combination of Patrick Carnes' 30 Task Model for treatment of sexual addiction and Pia Mellody's Model of Developmental Immaturity for treatment of developmental trauma provided the foundation for Paul's treatment. These models draw upon cognitive-behavioral, relapse-prevention, psychodynamic, attachment-focused, and experiential treatment modalities aimed at addressing the specific constellation of presenting sexual addiction and trauma symptomology found in clients like Paul. Some specific integrative interventions that were particularly effective in Paul's treatment included

- Micronutrient Supplementation: Vitamin D, fish oil, red yeast rice (in lieu of Lipitor).
- Pharmacological Treatment of Depression: changed from Lexapro to Effexor and added Deplin. Showed rapid and robust improvement in mood after four weeks.
- Sleep: CES and Minipress supported improvement in sleep onset and maintenance after about two weeks.
- Trauma Treatment and Nervous System Regulation:
    - Utilized both Somatic Experiencing and EMDR with positive response and fairly rapid relief of night terrors.
    - Neurofeedback twice weekly and HRV biofeedback twice weekly
    - Daily formal mindfulness meditation and weekly mindful meal
- Additional Body-Focused Modalities: yoga three times a week, tai chi once a week, acupuncture full body and auricular once a week, dance and movement therapy (four sessions), daily exercise, and weekly consultation with a registered dietician.

One major factor influencing the effectiveness of Paul's clinical work was the establishment of interpersonal trust and safety that supported his decision to continue treatment with a long-term step-down plan that included a safe transition to the intensive outpatient level of care. Interpersonally, Paul's high intrinsic motivation for treatment elicited positive countertransferential responses and authentic positive regard from several of the clinicians working on his team, and the mutual and reciprocal bonding experienced in these relationships began the process of healing his major attachment wounding while also supporting his ongoing participation in treatment with safe and trusted individuals.

REFERENCES

Aaron M (2012). The pathways of problematic sexual behavior: a literature review of factors affecting adult sexual behavior in survivors of childhood sexual abuse. *Sex Addict Compulsivity.* 19(3):199–218.

Adams KM, Robinson DW (2001). Shame reduction, affect regulation, and sexual boundary development: essential building blocks of sexual addiction treatment. *Sex Addict Compulsivity.* 8(1):23–44.

Amen DG (2015). *Change your brain, change your life: the breakthrough program for conquering anxiety, depression, obsessiveness, lack of focus, anger, and memory problems.* New York: Crown.

American Psychiatric Association (1973). Homosexuality and sexual orientation disturbance: Proposed change in *Diagnostic and Statistical Manual*, 2nd ed., 6th printing, p. 44: Position statement (retired).

American Psychiatric Association (2000). Task force on DSM-IV. *Diagnostic and Statistical Manual of Mental Disorders*: DSM-IV-TR. Washington, DC: American Psychiatric Association, 4th ed.

American Psychiatric Association (2013). Proposed for consideration in Sexual Disorders section of *Diagnostic and Statistical Manual of Mental Disorders*, 5th ed.

Barth RJ, Kinder BN (1987). The mislabeling of sexual impulsivity. *J Sex Marital Ther.* 13(1):15–23.

Bartholomew K (1990). Avoidance of intimacy: an attachment perspective. *J Soc Pers Relat.* 7(2):147–178.

Behere RV, Muralidharan K, Benegal V (2009). Complementary and alternative medicine in the treatment of substance use disorders—a review of the evidence. *Drug Alcohol Rev.* 28(3):292–300.

Behun RJ, Sweeney V, Delmonico DL, Griffin EJ (2012). Filtering and monitoring Internet content: a primer for helping professionals. *Sex Addict Compulsivity.* 19(1–2):140–155.

Berlin FS, Sawyer D (2012). Potential consequences of accessing child pornography over the Internet and who is accessing it. *Sex Addict Compulsivity.* 19(1–2):30–40.

Bested AC, Logan AC, Selhub EM (2013). Intestinal microbiota, probiotics and mental health: from Metchnikoff to modern advances: Part II—contemporary contextual research. *Gut Pathogens.* 5(1):1.

Bishop SR, Lau M, Shapiro S, et al. (2004). Mindfulness: a proposed operational definition. *Clin Psychol Sci Pract.* 11(3):230–241.

Black DW (2000). The epidemiology and phenomenology of compulsive sexual behavior. *CNS Spectrums.* 5(1):26–72.

Blum K, Chen AL, Giordano J, et al. (2012). The addictive brain: all roads lead to dopamine. *J Psychoactive Drugs.* 44(2):134–143.

Blum K, Giordano J, Morse S, et al. (2011). Hypothesizing synergy between acupuncture/auriculotherapy and natural activation of mesolimbic dopaminergic pathways: putative natural treatment modalities for the reduction of drug hunger and relapse. *IIOAB Lett.* 1(1).

Blum K, Simpatico T, Badgaiyan RD, Demetrovics Z, Fratantonio J (2015). Coupling Neurogenetics (GARS™) and a nutrigenomic based dopaminergic agonist to treat reward deficiency syndrome (RDS): targeting polymorphic reward genes for carbohydrate addiction algorithms. *J Reward Defic Syndr.* 1(2):75–80.

Bostwick JM, Bucci JA (2008). Internet sex addiction treated with naltrexone. *Mayo Clinic Proc.* 83:226–230.

Bowlby J (1969, 1982). *Attachment and loss.* New York: Basic Books. http://abebe.org.br/wp-content/uploads/John-Bowlby-Attachment-Second-Edition-Attachment-and-Loss-Series-Vol-1-1983.pdf.

Brewer JA, Elwafi HM, Davis JH (2013). Craving to quit: psychological models and neurobiological mechanisms of mindfulness training as treatment for addictions. *Psychol Addict Behav.* 27(2):366.

Budygin EA, Park J, Bass CE, Grinevich VP, Bonin KD, Wightman RM (2012). Aversive stimulus differentially triggers subsecond dopamine release in reward regions. *Neuroscience.* 201:331–37.

Butler MH, Seedall RB (2006). The attachment relationship in recovery from addiction. Part 1: relationship mediation. *Sex Addict Compulsivity.* 13(1):289–315.

Capogrosso P, Colicchia M, Ventimiglia E, et al. (2013). One patient out of four with newly diagnosed erectile dysfunction is a young man—worrisome picture from the everyday clinical practice. *J Sex Med.* 10:1833–1841.

Carnes P (1983). *Out of the shadows.* Minneapolis: CompCare.

Carnes P (1991). *Don't call it love: recovery from sexual addiction.* New York: Bantam.

Carnes PJ (2001). Cybersex, courtship, and escalating arousal: factors in addictive sexual desire. *Sex Addict Compulsivity.* 8(1):45–78.

Carnes P (2009). *Recovery zone, Vol. 1: Making changes that last: the internal tasks.* Carefree, AZ: Gentle Path Press.

Carnes S (2011). *Mending a shattered heart: a guide for partners of sex addicts.* Carefree, AZ: Gentle Path Press.

Carnes P (2015). *Facing the shadow: starting sexual and relationship recovery: a gentle path to beginning recovery from sex addiction.* Carefree, AZ: Gentle Path Press.

Carnes PJ, Green BA, Merlo LJ, Polles A, Carnes S, Gold MS (2012). PATHOS: a brief screening application for assessing sexual addiction. *J Addict Med.* Mar;6(1):29–34.

Carnes PJ, Hopkins TA, Green BA (2014). Clinical relevance of the proposed sexual addiction diagnostic criteria: relation to the Sexual Addiction Screening Test–Revised. *J Addict Med.* Nov-Dec;8(6):450–461.

Carnes PJ, Murray RE, Charpentier L (2005). Bargains with chaos: sex addicts and addiction interaction disorder. *Sex Addict Compulsivity.* 12(2–3):79–120.

Carpenter BN, Reid RC, Garos S, Najavits LM (2013). Personality disorder comorbidity in treatment-seeking men with hypersexual disorder. *Sex Addict Compulsivity.* 20(1–2):79–90.

Chen TJ, Blum K, Chen AL, et al. (2011). Neurogenetics and clinical evidence for the putative activation of the brain reward circuitry by a neuroadaptagen: proposing an addiction candidate gene panel map. *J Psychoactive Drugs.* 43(2):108–127.

Coleman E (1987). Sexual compulsivity: definition, etiology, and treatment considerations. *J Chem Depend Treat.* 1(1):189–204.

Coleman E (1990). The obsessive-compulsive model for describing compulsive sexual behavior. *Am J Prevent Psychiatry Neurology.* 2(3):9–14.

Coleman E (1992). Is your patient suffering from compulsive sexual behavior? *Psychiatr Ann.* 22:320–325.

Coleman E, Gratzer T, Nesvacil L, Raymond NC (2000). Nefazodone and the treatment of nonparaphilic compulsive sexual behavior: a retrospective study. *J Clin Psychiatry.* 61:282–284.

Coleman E, Raymond N, McBean A (2003). Assessment and treatment of compulsive sexual behavior. *Minn Med.* 86(7):42–47.

Cooper A (1998). Sexuality and the Internet: surfing into the new millennium. *Cyber Psychol Behav.* 1(2):187–193.

Cooper A, Delmonico DL, Burg R (2000). Cybersex users, abusers, and compulsives: new findings and implications. *Sex Addict Compulsivity.* 7(1–2):5–29.

Cooper A, Putnam DE, Planchon LA, Boies SC (1999). Online sexual compulsivity: getting tangled in the net. *Sex Addict Compulsivity.* 6(2):79–104.

Creeden K (2004). The neurodevelopmental impact of early trauma and insecure attachment: re-thinking our understanding and treatment of sexual behavior problems. *Sex Addict Compulsivity.* 11(4):223–247.

De Guzmán IN, Arnau RC, Green BA, et al. (2016). Empirical identification of psychological symptom subgroups of sex addicts: an application of latent profile analysis. *Sex Addict Compulsivity.* 23(1):34–55.

Del Giudice MJ, Kutinsky J (2007). Applying motivational interviewing to the treatment of sexual compulsivity and addiction. *Sex Addict Compulsivity.* 14(4):303–319.

Delmonico DL, Griffin EJ, Moriarty J (2001). *Cybersex unhooked: a workbook for breaking free from compulsive online sexual behavior.* Center City, MN: Hazelden Educational Press.

Deneke E, Knepper C, Green BA, Carnes PJ (2015). Comparative study of three levels of care in a substance use disorder inpatient facility on risk for sexual addiction. *Sex Addict Compulsivity.* 22(2):109–125.

Dhuffar MK, Griffiths MD (2015). A systematic review of online sex addiction and clinical treatments using CONSORT evaluation. *Curr Addict Rep.* 2(2):163–174.

Di Chiara G, Imperato A (1988). Drugs abused by humans preferentially increase synaptic dopamine concentrations in the mesolimbic system of freely moving rats. *Proc Natl Acad Sci U S A.* 85:5274–5278.

Döring NM (2009). The Internet's impact on sexuality: a critical review of 15 years of research. *Comput Hum Behav.* 25(5):1089–1101.

Ernst E (2002). Complementary therapies for addictions. *Addiction.* 97(12): 1491–1492.

Faisandler K, Taylor J, Salisbury R (2012). What does attachment have to do with out-of-control sexual behavior? *N Z J Psychol.* 41(1):19–29.

Felitti VJ, Anda RF, Nordenberg D, Williamson DF, Spitz AM, Edwards V, Koss MP, Marks JS (1998). Relationship of childhood abuse and household dysfunction to many of the leading causes of death in adults: the Adverse Childhood Experiences (ACE) Study. *Am J Prevent Med.* 14(4):245–258.

Finlayson G, King N, Blundell JE (2007). Liking vs. wanting food: importance for human appetite control and weight regulation. *Neurosci Biobehav Rev.* 31:987–1002.

Fisher AD, Corona G, Bandini E, Mannucci E, Lotti F, Boddi V, Forti G, Maggi M (2009). Psychobiological correlates of extramarital affairs and differences between stable and occasional infidelity among men with sexual dysfunctions. *J Sex Med.* 6(3):866–875.

Flores PJ (2004). *Addiction as an attachment disorder.* Lanham, MD: Rowman & Littlefield.

Forsythe P, Sudo N, Dinan T, Taylor VH, Bienenstock J (2010). Mood and gut feelings. *Brain Behav Immun.* 24(1):9–16.

Foster JA, Neufeld KAM (2013). Gut–brain axis: how the microbiome influences anxiety and depression. *Trends Neurosci.* 36(5):305–312.

Fraley RC, Brennan KA, Waller NG (2000a). The Experiences in Close Relationships–Revised (ECR-R) Questionnaire. http://internal.psychology.illinois.edu/~rcfraley/measures/ecrritems.htm.

Fraley RC, Waller NG, Brennan KA (2000b). An item response theory analysis of self-report measures of adult attachment. *J Pers Soc Psychol.* Feb;78(2):350–365.

Fugh-Berman A, Cott JM (1999). Dietary supplements and natural products as psychotherapeutic agents. *Psychosom Med.* 61(5):712–728.

Galanter M (2006). Spirituality and addiction: a research and clinical perspective. *Am J Addict.* 15(4):286–292.

Garcia FD, Thibaut F (2010). Sexual addictions. *Am J Drug Alcohol Abuse.* 36(5):254–260.

Garcia-Ruiz PJ, Castrillo JCM, Alonso-Canovas A, et al. (2014). Impulse control disorder in patients with Parkinson's disease under dopamine agonist therapy: a multicentre study. *J Neurol Neurosurg Psychiatry.* 85(8):840–844.

Garland E, Froeliger B, Howard M (2014). Mindfulness training targets neurocognitive mechanisms of addiction at the attention-appraisal-emotion interface. *Front Psychiatry.* 10(4):173.

Gilliland R, South M, Carpenter BN, Hardy SA (2011). The roles of shame and guilt in hypersexual behavior. *Sex Addict Compulsivity.* 18(1):12–29.

Glaus J, Vandeleur CL, von Känel R, et al. (2014). Associations between mood, anxiety or substance use disorders and inflammatory markers after adjustment for multiple covariates in a population-based study. *J Psychiatr Res.* 58:36–45.

Goodman A (1993). Diagnosis and treatment of sexual addiction. *J Sex Marital Ther.* 19(3):225–251.

Goodman A (1998). Sexual addiction: diagnosis and treatment. *Psychiatric Times.* 15(10):1–7.

Goodman A (2001). What's in a name? Terminology for designating a syndrome of driven sexual behavior. *Sex Addict Compulsivity.* 8(3–4):191–213.

Goodman A (2008). Neurobiology of addiction: an integrative review. *Biochem Pharmacol.* 75(1):266–322.

Grand D (2011). Brainspotting a new brain-based psychotherapy approach. *Trauma Gewalt.* 3:276–285.

Grant JE, Chamberlain SR (2016). Pharmacotherapy for behavioral addictions. *Curr Behav Neurosci Rep.* 3(1):67–72.

Grant JE, Kim SW (2001). A case of kleptomania and compulsive sexual behavior treated with naltrexone. *Ann Clin Psychiatry.*13:229–231.

Green BA, Arnau RC, Carnes PJ, Carnes S, Hopkins TA (2015). Structural Congruence of the Sexual Dependency Inventory, 4th edition. *Sex Addict Compulsivity.* 22(2):126–153.

Griffee K, O'Keefe SL, Stroebel SS, Beard KW, Swindell S, Young DH (2012). On the brink of paradigm change? Evidence for unexpected predictive relationships among sexual addiction, masturbation, sexual experimentation, and revictimization, child sexual abuse, and adult sexual risk. *Sex Addict Compulsivity.* 19(4):225–264.

Griffiths, MD (2012). Internet sex addiction: a review of empirical research. *Addict Res Theory.* 20(2):111–124.

Grubbs JB, Exline JJ, Pargament KI, Hook JN, Carlisle RD (2015a). Transgression as addiction: religiosity and moral disapproval as predictors of perceived addiction to pornography. *Arch Sex Behav.* 44(1):125–136.

Grubbs JB, Hook JN, Griffin BJ, Davis DE (2015b). Evaluating outcome research for hypersexual behavior. *Curr Addict Rep.* 2(3):207–213.

Gulsun M, Gulcat Z, Aydin H (2007). Treatment of compulsive sexual behaviour with clomipramine and valproic acid. *Clin Drug Invest.* 27:219–223.

Guss JR (2000). Sex like you can't even imagine: "crystal," crack and gay men. *J Gay Lesbian Psychother.* 3(3–4):105–122.

Gutman SA (2006). Why addiction has a chronic, relapsing course: the neurobiology of addiction: implications for occupational therapy practice. *Occup Therap Mental Health.* 22(2):1–29.

Harding KL, Judah RD, Gant CE (2003). Outcome-based comparison of Ritalin® versus food-supplement treated children with AD/HD. *Alt Med Rev.* 8(3):319–330.

Hayes SC, Wilson KG, Gifford EV, Follette VM, Strosahl K (1996). Experiential avoidance and behavioral disorders: a functional dimensional approach to diagnosis and treatment. *J Consult Clin. Psychol.* 64(6):1152.

Hazan C, Shaver P (1987). Romantic love conceptualized as an attachment process. *J Personality Social Psychol.* 52(3):511.

Hentsch-Cowles G, Brock LJ (2013). A systemic review of the literature on the role of the partner of the sex addict, treatment models, and a call for research for systems theory model in treating the partner. *Sexual Addiction &Sex Addict Compulsivity.* 20(4):323–335.

Hilton DL (2013). Pornography addiction—a supranormal stimulus considered in the context of neuroplasticity. *Socioaffect Neurosci Psychol.* 19;3.

Hook JN, Reid RC, Penberthy JK, Davis DE, Jennings DJ (2014). Methodological review of treatments for nonparaphilic hypersexual behavior. *J Sex Marital Ther.* 40(4):294–308.

Hopkins TA, Brawner CA, Meyer M, Zawilinski L, Carnes PJ, Green BA (2016). MMPI-2 correlates of sadomasochism in a sexual addiction sample: contrasting and men and women. *Sex Addict Compulsivity.* 23(1):114–140.

Howard MD (2007). Escaping the pain: examining the use of sexually compulsive behavior to avoid the traumatic memories of combat. *Sex Addict Compulsivity.* 14(2):77–94.

Jore JK (2013). Maladaptive personality traits and sex addiction profiles. Unpublished master's thesis. Hattisburg, MS: University of Southern Mississippi.

Jore JK, James JS, Arnau RC, Green BA, Schultz DM, Carnes PJ (2015). The multivariate relationships between self-reported history of abuse and sexually compulsive behaviors and cognitions. Poster presented at the 27th Annual Convention of the Association for Psychological Science, New York.

Kafka MP (1993). Update on paraphilias and paraphilia-related disorders. *Curr Affect Illness*. 12(6):5–13.

Kafka MP (2007). Paraphilia-related disorders: the evaluation and treatment of nonparaphilic hypersexuality. *Principles Pract Sex Ther*. 4:442–476.

Kafka MP (2010). Hypersexual disorder: a proposed diagnosis for DSM-V. *Arch Sex Behav*. 39(2):377–400.

Kafka MP (2014). Nonparaphilic hypersexuality disorder. *Principles Pract Sex Ther*. 280–304.

Kafka MP, Hennen J (1999). The paraphilia-related disorders: an empirical investigation of nonparaphilic hypersexuality disorders in outpatient males. *J Sex Marital Ther*. 25(4):305–319.

Kafka MP, Hennen J (2002). A DSM-IV Axis I comorbidity study of males (n = 120) with paraphilias and paraphilia-related disorders. *Sexual Abuse*. 14(4):349–366.

Kalivas PW (2009). The glutamate homeostasis hypothesis of addiction. *Nat Rev Neurosci*. 10(8):561–572.

Kaplan MS, Krueger RB (2010). Diagnosis, assessment, and treatment of hypersexuality. *J Sex Res*. 47(2–3):181–198.

Katehakis A (2009). Affective neuroscience and the treatment of sexual addiction. *Sex Addict Compulsivity*. 16(1):1–31.

Katehakis A (2010). Erotic intelligence: igniting hot, healthy sex while in recovery from sex addiction. Deerfield Beach, FL: Health Communications, Inc.

Katehakis A (2016). Sex addiction as affect dysregulation: a neurobiologically informed holistic treatment (norton series on interpersonal neurobiology). New York, NY: WW Norton & Company.

Khanna S, Greeson JM (2013). A narrative review of yoga and mindfulness as complementary therapies for addiction. *Complement Therap Med*. 21(3):244–252.

Kingston DA, Firestone P (2008). Problematic hypersexuality: a review of conceptualization and diagnosis. *Sex Addict Compulsivity*. 15(4):284–310.

Koob GF (2009). Neurobiological substrates for the dark side of compulsivity in addiction. *Neuropharmacology*. 56:18–31.

Kor A, Fogel YA, Reid RC, Potenza MN (2013). Should hypersexual disorder be classified as an addiction? *Sex Addict Compulsivity*. 20(1–2):27–47.

Kort J (2014). *Is my husband gay, straight, or bi? A guide for women concerned about their men*. New York: Rowman & Littlefield.

Kühn S, Gallinat J (2014). Brain structure and functional connectivity associated with pornography consumption: the brain on porn. *JAMA Psychiatry*. 71(7):827–834.

Kuzma JM, Black DW (2008). Epidemiology, prevalence, and natural history of compulsive sexual behavior. *Psychiatr Clin North Am*. 31:603–611.

Lakhan SE, Vieira KF (2008). Nutritional therapies for mental disorders. *Nutr J.* 7(1):2.

Landripet I, Štulhofer A (2015). Is pornography use associated with sexual difficulties and dysfunctions among younger heterosexual men?. *J Sex Med.* 12(5):1136–1139.

Leclercq S, Matamoros S, Cani PD, Francois J, Starkel P, Windey K, Delzenne NM (2014). SY39-1 increased intestinal permeability is related to gut bacterial dysbiosis and behavioral markers of addiction severity in alcohol-dependent subjects. *Alcohol Alcoholism.* 49(Suppl 1):i33–i33.

Leeman RF, Potenza MN (2013). A targeted review of the neurobiology and genetics of behavioural addictions: an emerging area of research. *Can J Psychiatry.* 58(5):260–273.

Levine MP, Troiden RR (1988). The myth of sexual compulsivity. *J Sex Res.* 25(3):347–363.

Levine P (1997). *Waking the tiger.* Berkeley CA: North Atlantic Books.

Levine PA (2010). *In an unspoken voice: how the body releases trauma and restores goodness.* Berkeley CA: North Atlantic Books.

Levin ME, Lillis J, Hayes SC (2012). When is online pornography viewing problematic among college males? Examining the moderating role of experiential avoidance. *Sex Addict Compulsivity.* 19(3):168–180.

Ley D, Prause N, Finn P (2014). The emperor has no clothes: a review of the "pornography addiction" model. *Curr Sex Health Rep.* 6(2):94–105.

Lofgreen CA (2012). *The storm of sex addiction: rescue and recovery.* Minneapolis: Hillcrest Publishing Group.

Luppino FS, de Wit LM, Bouvy PF, Stijnen T, Cuijpers P, Penninx BW, Zitman, FG (2010). Overweight, obesity, and depression: a systematic review and meta-analysis of longitudinal studies. *Arch Gen Psychiatry.* 67(3):220–229.

Maes M, Kubera M, Leunis JC (2008). The gut-brain barrier in major depression: intestinal mucosal dysfunction with an increased translocation of LPS from gram negative enterobacteria (leaky gut) plays a role in the inflammatory pathophysiology of depression. *Neuroendocrinol Lett.* 29(1):117–124.

Malladi SS, Singh AN (2005). Hypersexuality and its response to citalopram in a patient with hypothalamic hamartoma and precocious puberty. *Int J Neuropsychopharmacol.* 8:1–2.

Manley G (1999). Treating chronic sexual dysfunction in couples recovering from sex addiction and sex coaddiction. *Sex Addict Compulsivity.* 6(2): 111–124.

McClellan J, McCurry C, Ronnei M, Adams J, Eisner A, Storck M (1996). Age of onset of sexual abuse: relationship to sexually inappropriate behaviors. *J Am Acad Child Adolesc Psychiatry.* 35:1375–1383.

McPherson S, Clayton S, Wood H, Hiskey S, Andrews L (2013). The role of childhood experiences in the development of sexual compulsivity. *Sex Addict Compulsivity.* 20(4):259–278.

Mellody P, Miller AW, Miller JK (2011). *Facing codependence: what it is, where it comes from, how it sabotages our lives.* New York: HarperCollins.

Miles LA, Cooper RL, Nugent WR, Ellis RA (2016). Sexual addiction: a literature review of treatment interventions. *J Hum Behav Social Environ.* 26(1):89–99.

Mindes J, Dubin MJ, Altemus M (2015). Cranial electric stimulation. In H Knotkova, D Rasche (Eds.), *Textbook of neuromodulation: principles, methods and clinical applications* (2015 ed.). New York: Springer.

Moore TJ, Glenmullen J, Mattison DR (2014). Reports of pathological gambling, hypersexuality, and compulsive shopping associated with dopamine receptor agonist drugs. *JAMA Intern Med.* 174(12):1930–1933.

Muusses LD, Finkenauer C, Kerkhof P, Billedo CJ (2014). A longitudinal study of the association between compulsive internet use and wellbeing. *Comput Human Behav.* 36:21–28.

Naficy H, Samenow CP, Fong TW (2013). A review of pharmacological treatments for hypersexual disorder. *Sex Addict Compulsivity.* 20(1–2):139–153.

Nair D, Pawar A, Kalra G, Shah N (2013). An Indian study of hypersexual disorder in patients with anxiety and mood disorders. *Sex Addict Compulsivity.* 20(4):292–305.

Olsen CM (2011). Natural rewards, neuroplasticity, and non-drug addictions. *Neuropharmacology.* 61(7):1109–1122.

Orford J (1978). Hypersexuality: implications for a theory of dependence. *Addiction.* 73(3):299–310.

Orzack MH, Voluse AC, Wolf D, Hennen J (2006). An ongoing study of group treatment for men involved in problematic Internet-enabled sexual behavior. *Cyberpsychol Behav.* 9(3):348–360.

Owens EW, Behun RJ, Manning JC, Reid RC (2012). The impact of Internet pornography on adolescents: a review of the research. *Sex Addict Compulsivity.* 19(1–2):99–122.

Payne P, Levine PA, Crane-Godreau MA (2015). Corrigendum: Somatic Experiencing: using interoception and proprioception as core elements of trauma therapy. *Front Psychol.* 6.

Perlmutter D (2015). *Brain maker: the power of gut microbes to heal and protect your brain-for life.* New York: Hachette.

Petry NM, Rehbein F, Ko CH, O'Brien CP (2015). Internet gaming disorder in the DSM-5. *Curr Psychiatr Rep.* 17(9):72.

Philaretou AG, Mahfouz AY, Allen KR (2005). Use of Internet pornography and men's well-being. *Int J Mens Health.* 4(2):149.

Raymond NC, Coleman E, Miner MH (2003). Psychiatric comorbidity and compulsive/impulsive traits in compulsive sexual behavior. *Compr Psychiatry.* Sep-Oct;44(5):370–380.

Raymond NC, Grant JE, Coleman E (2010). Augmentation with naltrexone to treat compulsive sexual behavior: a case series. *Ann Clin Psychiatry.* 22(1):56–62.

Raymond NC, Grant JE, Kim SW, Coleman E (2002). Treatment of compulsive sexual behaviour with naltrexone and serotonin reuptake inhibitors: two case studies. *Int Clin Psychopharmacol.* 17:201–205.

Reay B, Attwood N, Gooder C (2013). Inventing sex: the short history of sex addiction. *Sex Culture*. 17(1):1–9.

Regan J (2015). Efficacy of Eye Movement Desensitization and Reprocessing Therapy in the treatment of trauma: a systematic review (Master of Social Work Clinical Research Paper 513). http://sophia.stkate.edu/msw_papers/513.

Reid RC (2010). Differentiating emotions in a sample of men in treatment for hypersexual behavior. *J Soc Work Pract Addict*. 10(2):197–213.

Reid RC, Bramen JE, Anderson A, Cohen MS (2014). Mindfulness, emotional dysregulation, impulsivity, and stress proneness among hypersexual patients. *J Clin Psychol*. 70(4):313–321.

Reid RC, Carpenter BN, Hook JN, et al. (2012). Report of findings in a DSM-5 field trial for hypersexual disorder. *J Sex Med*. 9(11):2868–2877.

Reid RC, Carpenter BN, Spackman M, Willes DL (2008). Alexithymia, emotional instability, and vulnerability to stress proneness in patients seeking help for hypersexual behavior. *J Sex Marital Ther*. 34(2):133–349.

Reid RC, Stein JA, Carpenter BN (2011). Understanding the roles of shame and neuroticism in a patient sample of hypersexual men. *J Nerv Mental Dis*. 199(4):263–267.

Reid RC, Woolley SR (2006). Using emotionally focused therapy for couples to resolve attachment ruptures created by hypersexual behavior. *Sex Addict Compulsivity*. 13(2–3):219–239.

Riemersma J, Sytsma M (2013). A new generation of sexual addiction. *Sex Addict Compulsivity*. 20(4):306–322.

Rinehart NJ, McCabe MP (1997). Hypersexuality: Psychopathology or normal variant of sexuality? *Sex Marital Ther*. 12(1):45–60.

Röhricht F (2009). Body oriented psychotherapy. The state of the art in empirical research and evidence-based practice: a clinical perspective. *Body Move Dance Psychother*. 4(2):135–156.

Roman MW, Bembry FH (2011). L-Methylfolate (Deplin®): a new medical food therapy as adjunctive treatment for depression. *Issues Mental Health Nurs*. 32(2):142–143.

Rosenberg K, Feder L (2014a). An introduction to behavioral addictions: behavioral addictions, criteria, evidence, and treatment. In KP Rosenberg, LC Feder (Eds.), *Behavioral addictions: criteria, evidence, and treatment* (pp. 1–17). New York: Academic Press

Rosenberg KP, Carnes P, O'Connor S (2014b). Evaluation and treatment of sex addiction. *J Sex Marital Ther*. 40(2):77–91.

Rucklidge JJ, Kaplan BJ (2013). Broad-spectrum micronutrient formulas for the treatment of psychiatric symptoms: a systematic review. *Expert Rev Neurotherap*. 13(1):49–73.

Sarris J, Kavanagh DJ (2009). Kava and St. John's wort: current evidence for use in mood and anxiety disorders. *J Alt Complement Med*. 15(8):827–836.

Schneider JP (2000a). Effects of cybersex addiction on the family: results of a survey. *Sex Addict Compulsivity*. 7(1–2):31–58.

Schneider JP (2000b). A qualitative study of cybersex participants: gender differences, recovery issues, and implications for therapists. *Sex Addict Compulsivity*. 7(4):249–278.

Schneider JP, Corley MD (2013). Disclosure of extramarital sexual activities by persons with addictive or compulsive. *Clin Manag Sex Addict.* 2:137–161.

Schneider JP, Weiss R, Samenow C (2012). Is it really cheating? Understanding the emotional reactions and clinical treatment of spouses and partners affected by cybersex infidelity. *Sex Addict Compulsivity.* 19(1–2):123–139.

Schooler D, Ward LM (2006). Average Joes: men's relationships with media, real bodies, and sexuality. *Psycholog Men Masc.* 7(1):27.

Schwartz BK, ed. (2003). *Correctional psychology: practice, programming, and administration.* Kingston NJ: Civic Research Institute.

Schwartz BK (2005). *The sex offender: current treatment modalities and systems issues.* Kingston NJ: Civic Research Institute.

Schwartz, R. (1994). *The internal family systems model.* New York: Guilford Press.

Seedall RB, Butler MH (2008). The attachment relationship in recovery from addiction. Part 2: substantive enactment interventions. *Sex Addict Compulsivity.* 15(1):77–96.

Siegel D (2010). *Mindsight.* New York: Bantam Books.

Selhub EM, Logan AC, Bested AC (2014). Fermented foods, microbiota, and mental health: ancient practice meets nutritional psychiatry. *J Physiolog Anthrop.* 33(1):1.

Senn TE, Carey MP, Vanable PA, Coury-Doniger P, Urban M (2007). Characteristics of sexual abuse in childhood and adolescence influence sexual risk behavior in adulthood. *Arch Sex Behav.* 36(5):637–645.

Shapiro SL, Brown KW, Biegel GM (2007). Teaching self-care to caregivers: effects of mindfulness-based stress reduction on the mental health of therapists in training. *Train Educ Prof Psychology.* 1(2):105.

Shonin E, Van Gordon W, Griffiths MD (2014). Mindfulness as a treatment for behavioural addiction. *J Addict Res Ther.* 5:e122.

Stavro K, Rizkallah E, Dinh-Williams L, Chiasson JP, Potvin S (2013). Hypersexuality among a substance use disorder population. *Sex Addict Compulsivity.* 20(3): 210–216.

Steele VR, Staley C, Fong T, Prause N (2013). Sexual desire, not hypersexuality, is related to neurophysiological responses elicited by sexual images. *Socioaffect Neurosci Psychol.* 16:3.

Steffens B, Means M (2010). *Your sexually addicted spouse: how partners can cope and heal.* New York: New Horizon Press.

Steffens BA, Rennie RL (2006). The traumatic nature of disclosure for wives of sexual addicts. *Sex Addict Compulsivity.* 13(2–3):247–267.

Stein DJ, Black DW, Shapira NA, Spitzer RL (2001). Hypersexual disorder and preoccupation with Internet pornography. *Am J Psychiatry.* 158(10):1590–1594.

Stice E, Yokum S, Blum K, Bohon C (2010). Weight gain is associated with reduced striatal response to palatable food. *J Neurosci.* 30:13105e13109.

Sussman S, Lisha N, Griffiths M (2011a). Prevalence of the addictions: a problem of the majority or the minority? *Eval Health Profess.* 34(1):3–56.

Sussman S, Sussman AN (2011b). Considering the definition of addiction. *Int J Environ Res Pub Health.* 8(10):4025–38.

Tao R, Huang X, Wang J, Zhang H, Zhang Y, Li M (2010). Proposed diagnostic criteria for Internet addiction. *Addiction.* 105:556e564.

Twohig MP, Crosby JM (2010). Acceptance and commitment therapy as a treatment for problematic Internet pornography viewing. *Behav Ther.* 41(3):285–295.

Tylka TL, Kroon Van Diest AM (2015). You looking at her "hot" body may not be "cool" for me: integrating male partners' pornography use into objectification theory for women. *Psychol Women Q.* 39(1):67–84.

US Food and Drug Administration (2012). Petitions to request change in classification for cranial electrotherapy stimulators. http://google2.fda.gov/search?q=cranial+electrotherapy+stimulator&client=FDAgov&site=FDAgov&lr=&proxystylesheet=FDAgov&requiredfields=-archive%3AYes&output=xml_no_dtd&getfields=*

Van der Kolk BA (1989). The compulsion to repeat the trauma. *Psychiatr Clin North Am.* 12(2):389–411.

Van Der Kolk B (2014). *The body keeps the score.* New York: Viking.

Volkow N, Li TK (2005). The neuroscience of addiction. *Nat Neurosci.* 8(11):1429–1430.

Volkow ND, Wang GJ, Fowler JS, Tomasi D (2012). Food and drug reward: overlapping circuits in human obesity and addiction. *Behav Neurosci.* 11:1e24.

Voon V, Mole TB, Banca P, et al. (2014). Neural correlates of sexual cue reactivity in individuals with and without compulsive sexual behaviours. *PloS One.* 11;9(7):e102419.

Wainberg ML, Muench F, Morgenstern J, et al. (2006). A double-blind study of citalopram versus placebo in the treatment of compulsive sexual behaviors in gay and bisexual men. *J Clin Psychiatry.* 67:1968–1973.

Weinstein AM, Zolek R, Babkin A, Cohen K, Lejoyeux M (2015). Factors predicting cybersex use and difficulties in forming intimate relationships among male and female users of cybersex. *Front Psychiatry.* 6:54.

Weintraub D, Koester J, Potenza MN, et al. (2010). Impulse control disorders in Parkinson disease: a cross-sectional study of 3090 patients. *Arch Neurol.* 67(5):589–595.

Weiss R, Schneider JP (2015). *Always turned on: facing sex addiction in the digital age.* Carefree AZ: Gentle Path Press.

Witkiewitz K, Lustyk MKB, Bowen S (2013). Retraining the addicted brain: a review of hypothesized neurobiological mechanisms of mindfulness-based relapse prevention. *Psychol Addict Behav.* 27(2):351.

Wright PJ (2010). Sexual compulsivity and 12-step peer and sponsor supportive communication: a cross-lagged panel analysis. *Sex Addict Compulsivity.* 17(2):154–169.

Yoder VC, Virden TB III, Amin K (2005). Internet pornography and loneliness: An association? *Sex Addict Compulsivity.* 12(1):19–44.

Yucha C, Montgomery D (2008). *Evidence-based practice in biofeedback and neurofeedback.* Wheat Ridge, CO: AAPB.

Zapf JL, Greiner J, Carroll J (2008). Attachment styles and male sex addiction. *Sex Addict Compulsivity.* 15(2):158–175.

Zitzman ST, Butler MH (2005). Attachment, addiction, and recovery: conjoint marital therapy for recovery from a sexual addiction. *Sex Addict Compulsivity.* 12(4):311–337.

16

Complementary and Alternative Treatments for Female Sexual Pain

SUSAN KELLOGG SPADT AND JENNIFER YONAITIS FARIELLO

Introduction

Despite the widespread use of the traditional medical multimodal treatment model used in the treatment of chronic sexual pain (including topical anaesthetics, topical hormonal or corticosteroids, analgesics, tricyclic antidepressants, muscle relaxants, neurologics, and anti-inflammatory medication), alternative treatment modalities are often overlooked and underutilized, although they can be very effective as adjuvant therapies for women with sexual pain or as sole treatment strategies. The alternative and complementary treatments that are the focus of this chapter include psychobehavioral therapy, physical therapy (PT), alternative therapy, and lifestyle strategies.

Psychobehavioral therapy focuses on reducing sexual pain, restoring sexual function, and strengthening the couple relationship by targeting the thoughts, emotions, behaviors, and couple interactions associated with sexual pain. Interventions can be done with an individual, with a couple, or in a group format. A key treatment strategy is cognitive behavioral therapy (CBT), which assists the woman with sexual pain and her partner to identify and differentiate between thoughts, behaviors, and reactions that are facilitative and pain-reducing versus those that are maladaptive and maintain pain and distress.

Cognitive-Behavioral Therapy and Mindfulness

CBT is one of the most commonly used and studied psychological intervention for sexual pain. In this treatment, the woman/partner are educated about the multidimensional impact of pain on each individual and on the relationship. The role of psychological and relationship factors in coping with pain and in the maintenance of pain are explored. Women are taught how to use pain journals and mindfulness to raise awareness about factors that positively or negatively influence pain and to track progress toward recovery. Focusing on the pain experience in this way can facilitate attention to and correction of personal maladaptive coping strategies that increase pain and sexual dysfunction, such as catastrophizing and hypervigilance to pain and anxiety, and can identify behaviors in the partner that might intensify distress, such as overly solicitous or hostile interactions and/or ambivalence in terms of relationship intimacy (LeClerc, 2014; Rosen, 2014). Another focus of CBT is enabling women and their partners to make optimum use of approach behaviors, optimize their emotional regulation, learn better communication and self-assertiveness, find ways to reconnect through nonsexual physical and emotional intimacy, and expand the sexual repertoire away from intercourse and share experiences of desire, arousal, and sexual intimacy. Exercises associated with CBT include breathing, discussing the impact of the pain on the romantic and sexual relationship, identifying distressing thoughts, identifying facilitators of arousal and desire, communicating sexual preferences and needs, and using cognitive techniques to introduce a space between the individual/couple and their thoughts and emotions. Throughout the course of therapy, the therapist continuously focuses on skill consolidation and maintenance of gains. Issues such as childhood trauma and significant relationship conflict are also addressed if considered integral to the onset or maintenance of sexual pain (Pazmany, 2014).

EVIDENCE SUPPORTING CBT

Bergeron (2008) investigated the efficacy of a combination of group CBT in two different randomized trials of women with provoked vestibulodynia (PVD). In the first study, which compared vestibulectomy surgery, biofeedback, and CBT, participants who received CBT reported significant improvements in pain at a six-month follow-up. At a 2.5-year follow-up, their ratings of pain during intercourse were equivalent to those of women who underwent surgery. In another study, participants were randomly assigned to either a topically applied corticosteroid cream or to group CBT for a 13-week treatment period. Intent-to-treat multilevel analyses showed that participants of both groups reported statistically significant reductions on pain measures from baseline to posttreatment and six-month follow-up, although the CBT group reported significantly more pain reduction at six-month follow-up. At posttreatment, women in the CBT condition were significantly more satisfied with their treatment, displayed significantly less pain catastrophizing, and reported significantly better global improvements in sexual functioning than women assigned to the topical application. Findings suggest that CBT may yield a positive impact on more dimensions of sexual pain than does a topical treatment (Bergeron, 2016).

In a randomized trial involving a mixed group of 50 women with sexual pain Masheb (2009) also found that CBT yielded significantly greater improvements in pain and sexual function than supportive psychotherapy. More recently, Corsini-Munt (2014) prospectively examined the preliminary efficacy of a 12-session couple therapy intervention in nine couples coping with sexual pain. Participants reported significant improvements in women's pain, as well as in sexuality outcomes, pain-related cognitions, anxiety, and depression for both members of the couple, in addition to high treatment satisfaction. Brotto (2015) also prospectively evaluated a four-session mindfulness-based, group psychoeducational intervention in 85 women with sexual pain. Pain self-efficacy, catastrophizing, hypervigilance, as well as sexual distress and pain during gynecological examination all significantly improved from pre- to posttreatment. In summary, individual and group psychological treatments represent empirically validated, noninvasive therapeutic options for sexual pain, while couple interventions remain to be rigorously assessed. Several authors suggest that an additive effect can be achieved with concomitant psychobehavioral and PT treatment for sexual pain, and therapy has been proposed as a first-line intervention in an interdisciplinary approach to sexual pain treatment (Bergeron, 2010).

Physical Therapy

Women with sexual pain commonly exhibit increased muscle tension and myofascial trigger points in the pelvic, abdominal, back, and pelvic floor muscles and have increased restrictions and decreased mobility in these areas. This widespread muscle tension can contribute to a self-perpetuating sexual pain cycle, producing symptoms such stabbing, shooting, or referred pain; vulvovaginal sensitivity; burning with penetration; pain with intravaginal thrusting; pain with orgasm; and/or postcoital pain (Fitzgerald, 2003; Prendergast, 2003; Travell, 1992).

Treatment of sexual pain associated with pelvic muscle dysfunction can be exceedingly complex because the pelvic floor muscles surround the urethral opening, the anorectal opening, and the vaginal opening. As a result, dysfunction of the muscles can contribute to pain during or after sexual activity; altered arousal and/or orgasm patterns; urinary frequency, urgency, retention, burning, and slow stream; abdominal pain; bloating; and constipation, diarrhea, and pain with defecation. Due to this widespread and variable symptomatology, women with these pain disorders often see multiple specialists (in conjunction with PT) in their efforts to become pain-free.

Women with sexual pain exhibit tenderness of the pelvic floor muscles and surrounding genital tissues as a result of muscle overactivity and shortening, inflammation, altered perfusion, and neural patterns. Pelvic floor PT utilizes a variety of modalities, including pelvic and core mobilization and stabilization; connective tissue, visceral, and neural mobilization; internal and external myofascial trigger point release, biofeedback, and electrical stimulation (Gentilcore-Saulnier, 2010; Hartmann, 2008). PT restores the proper length of the pelvic floor muscles and tissues, reduces neural tension, and decreases pain, which facilitates sexual comfort (Bergeron, 2001; Fitzgerald, 2003; Hartmann, 2001, 2010; Hartmann, 2008; Prendergast, 2003).

TECHNIQUES

After a comprehensive assessment, the pelvic floor physical therapist begins therapy with manual connective tissue manipulation to decrease restrictions in pelvic muscles and nerves and improve circulation to ischemic areas and those with pelvic congestion. A crucial aspect of pelvic floor and sexual rehabilitation is PT-assisted stretching of the muscles of the hips, back, gluteus, and abdomen. The pelvic floor physical therapist augments stretching exercises with strengthening techniques that address muscle weakness, allowing for balance and stability. Education regarding the progression and rate of strengthening is very important because restricted neural and connective tissue can become more irritated if these techniques are initiated prematurely (Travell, 1992). In addition to external tissue release and core stabilization, direct internal transvaginal/transrectal release of muscles is a central aspect of pelvic floor PT and sexual rehabilitation (Hartmann, 2001, 2010; Travell, 1992). Internal manual therapy techniques reduce tissue adherence and restore skeletal alignment. These interventions allow women to sit more comfortably, participate in work/home activities, and enjoy improved bladder, bowel, and sexual function. During internal PT, therapists educate patients through proprioceptive neuromuscular facilitation, verbal cueing, biofeedback, and electrical stimulation techniques that coordinate mind–body control of the pelvic floor and downtrain overactive muscles (Fitzgerald, 2003; Weiss, 2000, 2001)

PT HOME PROGRAM

As part of pelvic floor rehabilitation, the physical therapist develops a "home program" that may include abdominal, hip, pelvic, lower extremity, and back stretching; daily relaxation and breathing exercises; patient-performed manual techniques on external and internal tissue; and bladder and bowel retraining as indicated

(Fitzgerald, 2003). Vaginal dilators may be recommended to normalize the tone of the muscles, desensitize previously hypersensitive areas of the vulva/vagina, and restore sexual function (Bergeron, 2003; Hartmann, 2001, 2010; Kellogg-Spadt, 2012). When needed, techniques such as guided imagery and mindfulness meditation can be incorporated into the patient's home program to facilitate neural downregulation. Learning relaxation, self-release, and gentle strengthening techniques can empower patients and instill confidence, so that sexual rehabiliation can continue at home after discharge from therapy (Bo, 2007; Hilton, 2011).

EVIDENCE SUPPORTING PT

Pelvic floor PT (inclusive of manual techniques and mobilization and stabilization exercises) is recommeded as a primary or secondary treatment for sexual pain in women by several professional associations, including the American College of Obstetricians and Gynecologists (2011) and the American Urologic Association (2012). In addition, several research studies document the efficacy of pelvic floor PT for women with sexual pain.

Oyama (2004) found that manual PT released pelvic floor muscle spasm and decreased sexual pain in 21 pre- and postmenopausal women with high tone pelvic floor muscle dysfunction and dyspareunia. Weiss (2001) evaluated the effectiveness of manual pelvic floor PT in subjects with sexual pain, of whom 70% had moderate to marked improvement of their symptoms after PT. Hartmann (2001) reported significant improvement in 87% of women suffering from sexual pain disorders after they underwent pelvic floor PT. In this study, a 10-year follow-up survey reported that 92% of subjects had mild or no current symptoms (Hartmann, 2008). Fitzgerald (2009) conducted a multicenter feasibility trial of 48 men and women with pelvic pain and dyspareunia of urologic origin. The subjects were randomized to either specialized myofascial physical therapy (MPT) or global therapeutic massage (GTM) for 10 treatments. The improvements in the MPT group were significantly higher than in the GTM group (57% for MPT vs. 21% for GTM; $p = 0.03$). The MPT group demonstrated significant improvement in sexual function. In a follow-up trial, Fitzgerald (2012) found similar results in a multicenter randomized controlled trial of 81 pre- and postmenopausal women with sexual pain and pelvic floor tenderness related to bladder disorders. Subjects were divided into MPT or GTM treatment groups. After 10 treatments, 59% of the MPT group reported moderate or marked improvement compared to 26% of the GTM group ($p = 0.0012$).

Several authors have demonstrated the utility of pelvic floor muscle therapy for improving sexual function in cohorts of women with vestibulodynia, a well-known cause of sexual pain (Glazer, 1995; Reissing, 2004). In 2002, Bergeron reported on the usefulness of six to eight sessions of manual pelvic floor PT and adjunctive techniques including education, biofeedback, electrical stimulation, and home programs in the treatment of vestibulodynia. Approximately 71% of the patients in the investigation reported complete to moderate improvement in pain. Dionisi (2008) observed improvement in sexual pain in 75% of participants with vestibulodynia after 10 sessions of PT, electrical stimulation, biofeedback, and home exercises. A success rate of 52% was reported in a retrospective study by Glazer (1995); a similar study by McKay (2001) treated women with vestibulodynia with PT and biofeedback. A randomized trial by Bergeron (2001) compared psychosexual therapy, PT, and vestibulectomy, finding significant reduction in sexual pain from all three therapies. Danielsson (2006) compared PT and biofeedback to topical lidocaine in the treatment of vestibulodynia-related pain. The study concluded that PT resulted in improvement in 66% of patients, which was maintained at 12-month follow-up. Goetsch (2007) noted that PT was effective in facilitating sexual function after vestibulectomy. Finally, in a prospective study involving 11 participants with vestibulodynia, Gentilcore-Saulnier (2010) demonstrated that PT led to normalization of pelvic floor muscles, significant reduction in pain during vaginal palpation, and decreased pain during gynecologic examination and sexual intercourse.

Although more research is warranted, data suggests that pelvic floor muscle PT treatment can be a key element toward alleviating discomfort and restoring sexual function (Fitzpatrick, 2012; Hartmann, 2001; Prendergast, 2003).

Interdisciplinary Treatments: Combining Approaches

An interdisciplinary model of care is espoused by many sexual pain experts and has been the recommendation of the Third International Consultation on Sexual Medicine for women's sexual pain disorders (van Lankveld, 2010). A crucial step involved in success of this model is to provide patient education about the interconnectivity of biomedical and behavioral factors involved in the onset and maintenance of sexual pain. When used by a group of collaborating health professionals who are knowledgeable about one another's approaches, advantages of this model include more engaged and hopeful women and partners, increased coherence among the professionals on the treatment team, and multiple dimensions of sexual pain being targeted simultaneously rather than sequentially.

There are two published, uncontrolled quantitative studies evaluating a multimodal approach to the treatment of sexual pain, with both reporting major improvements in women's sexual function (Backman, 2008; Spoelstra, 2011). A retrospective qualitative study evaluated 29 women who participated in a multidisciplinary treatment program. Twenty-seven subjects reported a significant benefit, with nine reporting complete absence of pain (Munday, 2007). Another study conducted among 19 women indicated that participants noted gaining increased knowledge and tools to manage the pain, improved psychological well-being, validation and support, and an enhanced sense of empowerment (Sadownik, 2012). These promising findings emphasize the need for randomized trials aimed at evaluating the efficacy of an integrated approach to care, over and above the efficacy of single modalities.

Alternative and Lifestyle Treatments for Sexual Pain

Although less studied that psychobehavioral and PT, alternative treatment approaches can produce promising results in women with sexual pain. One prospective pilot study showed that eight participants reported improvements in sexual pain after taking part in 10 one-hour acupuncture sessions (Curran, 2010). Another pilot trial evaluated the effect of six hypnotherapy sessions conducted with eight women with vestibulodynia and sexual pain. Results indicated significant decreases in gynecologic examination pain, intercourse pain, and noncoital vulvar pain. Overall sexual function and sexual satisfaction increased at posttreatment. (Pukall, 2007). Since alternative therapy options such as acupuncture and hypnotherapy are devoid of adverse effects, future studies are warranted.

Topical Treatments

Four small studies have evaluated topical application of capsaicin 0.025% to 0.05% cream for sexual pain associated with hypersensitivity in the vulvar vestibule in a total of 92 subjects. Capsaicin is a vanillyl amide derived from the pepper plant that activates A-delta sensory neurons and unmyelinated C fibers. Repeated applications of the compound facilitate degeneration and depolarization of C- afferent neurons and depletion of neuropeptides, resulting in desensitization. In various studies, female subjects reported between 59% and 77% mean reduction in vulvar pain after applying cream one to five times daily for at least four weeks. Results ranged from a modest reduction in tenderness on physical examination to complete relief of symptoms with return to pain-free sexual intercourse. Tolerability of capsaicin was increased if lidocaine gel was applied to the skin or if subjects took oral analgesics prior to application of the study cream (Friedrich, 1988; Murina, 2004; Steinberg, 2005; Zyczynski, 1997).

Lifestyle Strategies

HYGIENE AND EXERCISE

Lifestyle changes for women who have sexual pain have been proposed by the American College of Obstetricians and Gynecologists (2011) and by the authors of the Vulvodynia Guideline (Haefner, 2005) who conclude that gentle care of the vulva can contribute significantly to sexual pain reduction. Specific strategies that they suggest for women include wearing nonconstrictive cotton underwear during the day and none during sleep; avoiding pantyhose and tight pants; using mild soap for bathing and water only for cleansing the vulva; avoiding douches, commercial vaginal wipes, deodorants, bubble bath, deodorant tampons and pads; using watery based nonscented, nonwarming, nonallergenic lubricants for intercourse; applying cool gel packs to the vulvar area before and/or after sex to reduce pain and swelling; avoiding physical activities that place pressure directly on the vulva such as bicycling; and changing from moist or constrictive clothing after physical exercise.

DIET AND NUTRITION

The management of sexual pain becomes more holistic when clinicians consider the comorbid conditions and associated pain generators. It has been proposed that diet modification can help alleviate sexual pain in patients with certain chronic pain conditions.

Sexual Pain Associated with Vulvar Inflammation

The first and most well-known report on diet's influence on vulvar pain was a case study on the use of calcium citrate for a woman with high levels of urinary oxalate. The authors reported that the patient's symptoms were decreased after three months of dietary modification, and she was pain free after a year (Solomons, 1991). This report became the basis for the postulate that high urinary and tissue oxalate levels could exacerbate vulvar pain and that a low-oxalate diet with calcium citrate supplemetation would ameliorate vulvar pain. A subsequent study compared urinary oxalate excretion in women with vulvar and sexual pain to controls and evaluated the role of dietary intervention in 59 women with pain. Results failed to show a difference in the 24-hour urinary excretion of oxalate between affected women and controls; however, 24% of affected women treated with the low oxaltae diet reported a decrease in genital redness, 19% were able to tolerate sexual intercourse with mild pain, and 10% were able to have intercourse without pain. Authors of the study concluded that the role of urinary oxalates as a trigger for vulvar pain was questionable and did not endorse the low oxalate diet as the standard of care for sexual pain (Baggish, 1997). In two case-controlled follow-up studies, no difference was found between between patients with sexual pain versus controls with respect to dietary intake of oxalate nor increased risk of developing sexual pain as a result of dietary oxalate ingestion (Harlow, 2008; Sarma, 1999).

Sexual Pain Associated with Interstitial Cystitis/Bladder Pain Syndrome

Up to 87% of women with chronic bladder conditions report sexual pain (Nickel, 2004). Koziol evaluated 565 women with interstitial cystitis/bladder pain syndrome (IC/BPS) and found that 60.8% of women had pain in the genital area (Koziol, 1994). Management of sexual pain associated with IC/BPS initially includes behavioral modification and dietary changes (Fariello, 2013). Shorter (2007) used a validated questionnaire to determine which foodstuffs either exacerbated or improved bladder symptoms. Results found that bladder symptoms in 90% of respondents were negatively affected by artificial sweeteners; caffeinated, carbonated, and alcoholic beverages; spicy foods; and acidic fruits and fruit juices. A subsequent survey of 598 IC/BPS patients found similar results (Bassaly, 2011). Several studies suggest that ingestion of the dietary supplements calcium glycerophosphate or sodium bicarbonate improve painful symptoms in women who have sexual pain related to bladder inflammation (Friedlander, 2012).

For women with sexual pain related to IC/BPS, individual products that exacerbate pain can be identified through the use of food and pain journals. It is equally beneficial for women to note the onset of bladder symptoms with relation to sexual activity. Instructions for a "bladder and sexual pain elimination diet" are as follows. The first week, women eat "normally," without making changes to their diet. The following two weeks, women observe a strict, "low acid IC diet" and eliminate all bladder-irritating condiments such as artificial sweeteners and spicy sauces. At the start of the fourth week, women can begin to introduce one previously eliminated food or condiment every two to three days. In this manner, women can compile an individualized list of edibles that are bladder friendly while not restricting their diet more than necessary and compromising nutrition (Friedlander, 2012).

Conclusion

Female sexual pain represents a complex clinical challenge that affects women, couples, and relationships. Although there are a variety of treatment options available, there is no one treatment that is effective for all women, thus treatment must be individualized. Complementary and alternative medicine can be an excellent option for first-line treatment or be employed as adjuvant therapy to compliment traditional medical treatment plans. Combining management strategies such as diet, hygiene, and exercise can form a sound treatment approach and facilitates multidisciplinary involvement of care providers, which has been shown to be beneficial in the treatment of sexual pain. Although more robust research is warranted, evidence suggests that complementary and alternative medicine strategies can be highly effective in the treatment of complicated, multifactorial, and often chronic sexual pain conditions in women.

REFERENCES

American College of Obstetricians and Gynecologists (2011). Vulvodynia. Frequently Asked Questions (FAQ127). Washington, DC. https://www.nichd.nih.gov/health/topics/vulvodynia/conditioninfo/Pages/treatments.aspx

American Urologic Association (2012). Guidelines. http://auanet.org/education/guidelines/icbladder-pain-syndrome-cfm.

Backman H, Widenbrant M, Bohm-Starke N, Dahlof LG (2008). Combined physical and psychosexual therapy for provoked vestibulodynia—an evaluation of a multidisciplinary treatment model. *J Sex Res.* 45(4):378–385.

Baggish MS, Sze EHM, Johnson R (1997). Urinary oxalate excretion and its role in vulvar pain syndrome. *Am J Obstet Gynecol.* 177(3):507–511.

Bassaly R, Downes K, Hart S (2011). Dietary consumption triggers interstitial cystitis/bladder pain syndrome patients. *Female Pelvic Med Reconstr Surg.* 17:36–39.

Bergeron S, Binik YM, Khalife S, et al. (2001). A randomized comparison group cognitive behavioral therapy, surface electromyographic biofeedback and vestibulectomy in the treatment of dyspareunia from vulvar vestibulitis. *Pain.* 91:297–306.

Bergeron S, Brown C, Lord MJ, Oala M, Binik YM, Kalife S (2002). Physical therapy for vulvar vestibulitis syndrome: a retrospective study. *J Marital Sex Ther.* 28:183–192.

Bergeron S, Lord MJ (2003). The integration of a pelvi-perineal re-education and cognitive behavioral therapy in the multidisciplinary treatment of sexual pain disorders. *Br Assoc Sex Relat Ther.* 18(2):135–141.

Bergeron S, Khalifé S, Dupuis M-J, McDuff P (2016). A randomized clinical trial comparing group cognitive-behavioral therapy and a topical steroid for women with dyspareunia. *J Consult Clin Psychol.* 84:259–268.

Bergeron S, Khalifé S, Glazer, HI, Binik YM (2008). Surgical and behavioral treatments for vestibulodynia: two-and-one-half year follow-up and predictors of outcome. *Obstet Gynecol.* 111(1):159–166.

Bergeron S, Morin M, Lord MJ (2010). Integrating pelvic floor rehabilitation and cognitive behavioural therapy for sexual pain: What have we learned and where do we go from here? *J Sex Relat Ther.* 25:304–313.

Bo K, Berghmans B, Morkved S, Van Kampen M (2007). *Evidence based physical therapy for the pelvic floor,* 1st ed. Philadelphia: Elsevier.

Brotto LA, Basson R, Smith KB, Driscoll M, Sadownik L (2015). Mindfulness-based group therapy for women with provoked vestibulodynia. *Mindfulness.* 6:417–432.

Corsini-Munt S, Bergeron S, Rosen NO, Mayrand MH, Delisle I (2014). Feasibility and preliminary effectiveness of a novel cognitive-behavioral couple therapy for provoked vestibulodynia: a pilot study. *J Sex Med.* Oct;11(10):2515–2527.

Curran S, Brotto LA, Fisher H, Knudson G. Cohen T (2010). The ACTIV study: acupuncture treatment in provoked vestibulodynia. *J Sex Med.* 7(2):981–995.

Danielsson I, Torstensson T, Brodda-Jansen G, Bohm-Starke N (2006). EMG biofeedback versus topical lidocaine gel: a randomized study for the treatment of women with vulvar vestibulitis. *Acta Obstet Gynecol Scand.* 85:1360–1367.

Dionisi B, Angianna F, Inghirami P, Lippa P, Senatori R (2008). Use of transcutaneous electrical stimulation and biofeedback for the treatment of vulvodynia: results of three years of experience. *Minerva Ginecol.* Jan;60:485–491.

Fariello JY, Whitmore KE, Moldwin RM (2013). Bladder pain syndrome and sexuality. In J Nordling, JJ Wyndaele, JP van de Merwe, P Bouchelouche, M Cervigni, M Fall (Eds.), *Bladder pain syndrome: a guide for clinicians* (pp.163–173). New York: Springer.

FitzGerald MP, Anderson RU, Potts J, et al. (2009). Randomized multicenter feasibility trial of myofascial physical therapy for the treatment of urological chronic pelvic pain syndromes. *J Urol.* 182(2):570–580.

FitzGerald MP, Kotarinos R (2003). Rehabilitation of the short pelvic floor II: treatment of the patient with the short pelvic floor. *Int J Pelvic Floor Dysfunction.* 14(4):269–275.

FitzGerald MP, Payne CK, Lukacz ES et al. (2012). Randomized multicenter clinical trial of myofascial physical therapy in women with interstitial cystitis /painful bladder syndrome and pelvic floor tenderness. *J Urol.* 187(6):2113–2218.

Friedlander JI, Shorter B, Moldwin RM (2012). Diet and its role in interstitial cystitis/ bladder pain syndrome (IC/BPS) and comorbid conditions. *BJU Int.* 109:1584–1591.

Friedrich EG. (1988). Therapeutic studies in vulvar vestibulitis. *J Repro Med.* 33:514–518.

Gentilcore-Saulnier E, McLean L, Golfinger C, Pukall CF, Chamberlain S (2010). Pelvic floor muscle assessment outcomes in women with and without provoked vestibultodynia and the impact of a physical therapy program. *J Sex Med.* Feb;7(2):1003–1022.

Glazer Hi, Rodke G, Swencionis C, Hertz R, Young AW (1995). Treatment of vulvar vestibulitis syndrome with electromyographic feedback of the pelvic floor musculature. *J Repro Med.* 40:283–290.

Goetsch MF (2007). Surgery combined with muscle therapy for dyspareunia from vulvar vestibulitis: an observational study. *J Repro Med.* 52:597–603.

Haefner HK, Collins E, Davis GD, et al. (2005). The vulvodynia guideline. *J Low Genital Tract Dis.* 9:40–51.

Harlow BL, Abenhaim HA, Vitonis AF, Harnack L (2008). Influence of dietary oxalates on the risk of adult-onset vulvodynia. *J Reprod Med.* 53(3):171–178.

Hartmann D (2001). The perceived effectiveness of physical therapy treatment on women complain of chronic vulvar pain and diagnosed with either vulvar vestibuitis syndrome or dysesthesic vulvodynia. *J Womens Health Phy Ther.* 25:13–18.

Hartmann D (2010). Chronic vulvar pain from a physical therapy perspective. *Dermato Ther.* Sept-Oct;23(5):505–513.

Hartmann D, Nelson CA (2008). Ten years after physical therapy for vulvodynia: Are there lasting benefits? *J Repro Med.* 53(11).

Hilton S, Vandyken C (2011) The puzzle of pelvic pain—a rehabilitation framework for balancing tissue dysfunction and central sensitization. *J Womens Phys Ther.* 35(3):103–113.

Kellogg-Spadt S, Iorio J, Fariello J, Whirmore KE (2012) Vaginal dilation: when its indicated and tips for teaching it. *OBG Manag.* 24:1–5.

Koziol, JA. (1994). Epidemiology of interstitial cystitis. *Urol Clin North Am.* 21(1):7–20.

Leclerc B, Bergeron S, Brassard A, Belanger C, Steben M, Lambert, B (2014). Attachment, sexual assertiveness, and sexual outcomes in women with provoked vestibulodynia and their partners: a mediation model. *Arch Sex Behav.* Aug;44(6):1561–1572.

Masheb RM, Kerns RD. Lozano C, Minkin MJ, Richman S (2009). A randomized clinical trial for women with vulvodynia: cognitive-behavioral therapy vs. supportive therapy. *Pain.* 141(1–2):31–40.

McKay E, Kaufman RH, Doctor U, Berkova Z, Galzer H, Redko V (2001). Treating vulvar vestibulitis with electromyographic feedback of pelvic floor musculature. *J Repro Med.* 46:337–342.

Munday P, Buchan A, Ravenhill G, Wiggs A, Brooks F (2007). A qualitative study of women with vulvodynia: II. Response to a multidisciplinary approach to management. *J Reprod Med.* 52(1):1922.

Murina F, Radicic G, Biano V (2004). Capsaicin and the treatment of vulvar vestibulitis syndrome: a valuable alternative. *Med Gen Med.* Dec:8;6(4):48.

Nahin RL, Barnes PM, Stussman BJ, Bloom B (2009). *Costs of complementary and alternative medicine (CAM) and frequency of visits to CAM practitioners: United States, 2007* (National Health Statistics Report 18). Hyattsville, MD: National Center for Health Statistics.

Nickel JC (2004). Interstitial cystitis: the paradigm shifts: international consultations on interstitial cystitis. *Rev Urol.* 6(4):200–202.

Oyama IA, Rejba A, Lukban JC, Fletcher E, Kellogg-Spadt S, Holzberg AS, Whitmore K (2004) Modified Thiele massage as a therapeutic intervention for female patients with interstitial cystitis and high tone pelvic floor dysfunction. *Urology.* 64:862–865.

Pazmany E, Bergeron S, Verhaeghe J, Van Oudenhove L, Enzlin P (2014). Sexual communication, dyadic adjustment, and psychosexual well-being in premenopausal women with self reported dyspareunia and their partners: a controlled study. *J Sex Med.* 11(7):1786–1797.

Prendergast S, Weiss J (2003). Screening for musculoskeletal causes of pelvic pain. *Clinical Obstet and Gynecol.* 46(4):773–782.

Pukall CF, Kandyba K, Amsel R, Khalifé S, Binik YM (2007). Effectiveness of hypnosis for the treatment of vulvar vestibulitis syndrome: a preliminary investigation. *J Sex Med.* 4(2):417–425.

Reissing ED, Binik YM, Khalife S, Cohen D, Amsel R (2004). Vaginal spasm, pain and behavior: an empirical investigation of the diagnosis of vaginismus. *Arch Sex Behav.* 33:5–17.

Rosen NO, Bergeron S, Sadikaj G, Glowacka M, Delisle I, Baxter ML (2014). Impact of male partner responses on sexual function in women with vulvodynia and their partners: a dyadic daily experience study. *Health Psychol.* 33(8):823–831.

Sadownik LA, Seal BN, Brotto LA (2012). Provoked vestibulodynia: women's experience of participating in a multidisciplinary vulvodynia program. *J Sex Med.* 9(4):1086–1093.

Sarma AV, Foxman B, Bayirli B, Haefner H, Sobel JD (1999). Epidemiology of vulvar vestibulitis syndrome: an exploratory case-control study. *Sex Transm Infect.* 75(5):320–326.

Shorter B, Lesser M, Moldwin RM, Kushner L (2007). Effect on comestibles on symptoms of interstitial cystitis. *J Urol.* 178:145–152.

Solomons CC, Melmed MH, Heitler SM (1991). Calcium citrate for vulvar vestibulitis. *J Reprod Med.* 36:879–882.

Spoelstra SK, Dijkstra, JR, van Driel MF, Weijmar Schultz WC (2011). Long-term results of an individualized, multifaceted, and multidisciplinary therapeutic approach to provoked vestibulodynia. *J Sex Med. 8*(2):489–496.

Steinberg AC, Oyama IA, Rejba A, Kellogg-Spadt S, Whitmore K (2005). Capsaicin for the treatment of vulvar vetsibulitis. *Am J Obstet Gynecol.* 192:1549–1553.

Travell J Simons D (1992). *The trigger point manual.* Vol. 2. Baltimore: Williams & Wilkins.

van Lankveld JJ, Granot M, Weijmar Schultz WC, et al. (2010). Women's sexual pain disorders. *J Sex Med. 7*(1):615–631.

Weiss JM (2000). Chronic pelvic pain and myofascial trigger points. *Pain Clinic.* 2(6):13–18.

Weiss JM (2001). Pelvic floor myofascial trigger points: manual therapy for interstitial cystitis and urgency frequency syndrome. *J Urol.* 166:2226–2231.

Zyczynski HM, Culbertson S, Gruss J, DeGroat WC (1997). Substance P and the pathophysiology of vulvar vestibultitis [abstract]. *J Soc Gynecol Invest.* 4(Suppl):107A.

SECTION IV

Traditional Eastern and Western Treatment Approaches to Healing Sexual Problems

17

Healing Sexual Problems in the Ayurvedic Tradition of India

KEESHA EWERS

Introduction

The ancient traditions from India offer many ways of individualizing the practice of sexual medicine for our patients. The three more well-known frameworks that incorporate sexual health into their vocabulary and protocols are Ayurveda, Yoga, and Tantra. Given that Tantra and Yoga are actually spiritual paths, with sexuality only being a tiny slice of the body of knowledge they encompass, this chapter focuses on Ayurvedic medicine. In fact, earlier Tantric (religious) texts did not mention specific anatomy relevant to sexual functioning, but these topics gradually found their way into Ayurveda (Rao, 2004).

A Short History of Ayurveda

Ayurveda is an ancient framework of preventative medicine and healthcare that is the sister science of Yoga, Vastu, and Jyotish. All of these sciences were compiled into the same body of texts known as the Vedas. They were written by ancient scholars who were intent on explaining the human condition as it relates to the cosmos, the earth and all beings we share it with, the seasons, the elements, and ourselves.

Much of the ancient texts of Ayurveda have been lost, but the two well-known texts often referenced today are the *Caraka Samhita* (Compendium of Caraka; also spelled Charaka) and the *Susruta Samhita* (Compendium of Susruta), both estimated to have been written around 100 AD. The *Susruta Samhita* deals primarily with surgery and the *Charaka Samhita* with what today is taught in the West as Ayurveda, or "the Science of Life." Banchoff (2009, 284) states that "an early Hindu text, the Caraka Samhita, vividly describes the beginning of life." Ayurveda is dated as far back as 5,000 years as a compendium of written texts, and even older as an oral tradition.

Overview of Ayurvedic Medicine

The following is a basic summary of what is a complex and multilayered framework of medicine that cannot be adequately taught in one chapter or even in one book. This does, however, offer a rudimentary understanding of how shifting our paradigm of symptom relief and polypharmacy to that of the whole person and whole foods/plant medicine can aid in integrating what is best from the East and the West, ancient and modern, to create a framework of medicine that addresses the root cause of sexual health issues—what this author calls Functional Sexology™.

One aspect of Ayurveda that until recently would be considered revolutionary by today's standards of medical care and research is that we are not all the same person. Only now through the power of modern technological medicine is Western medicine acknowledging something Ayurveda has always known: often there is no one-size-fits-all solution to many medical, sexual, emotional, or mental imbalances and issues. We are each constitutionally unique, and many diverse factors may influence our health status. In Ayurveda, we are thought of as microcosms of the macrocosm of the universe and therefore all of the elements that occur in our universe are present in us, including ether, air, fire, water, and earth. The proportions and relation of these elements comprising our bodies account for our uniqueness, physiologically and psychologically. The study of Tantra takes this one step further to embrace the essential nature of the masculine and feminine within each of us, recognizing that one cannot exist without the other.

To those of us steeped in the Western scientific model, a system of medicine based on the elements may at first seem quaint to say the least. Yet Ayurveda is a complex, nuanced approach to health that has yielded startling insights thousands of years in advance of our Western medicine. For example, Ayurveda recognized that obesity was a risk factor for sexual dysfunctions in men such as diminished erectile function, orgasmic function, sexual desire, and overall satisfaction (Geetha, 2012). See other chapters in this volume that discuss the growing modern understanding of obesity as a source of many types of sexual difficulties for men and women.

As in traditional Chinese medicine, homeopathy, naturopathy, and even the origins of Western medicine, in Ayurveda digestion is known to be the source of both wellness and illness. According to Ayurveda, we all digest our foods, thoughts, emotions—and even our spiritual and sexual experiences—differently.

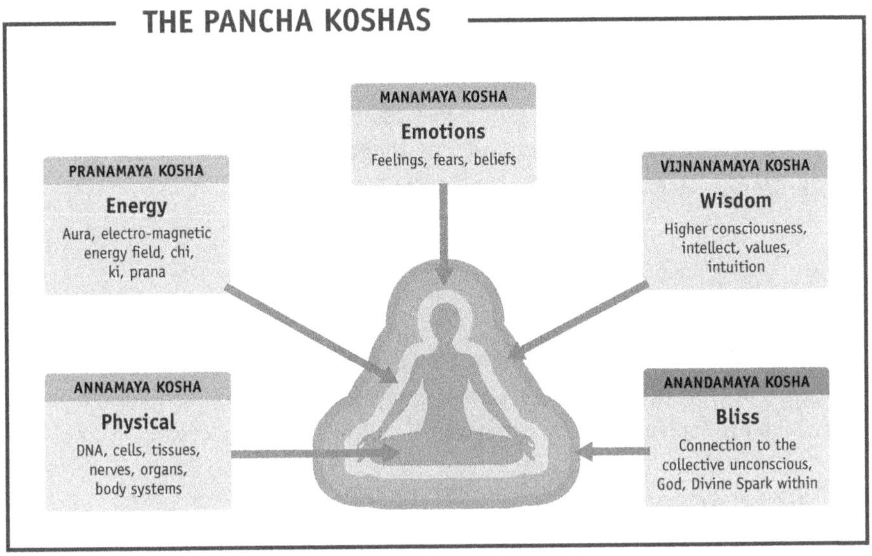

FIGURE 17.1. The pancha koshas.

We react to the seasons, exercise, sexual patterns, and sleep in ways that are unique to our constitutional type or what is called our *dosha*.

The way disease progresses through the body is recognized as an imbalance that has disrupted and blocked the channels (*srotas and nadis*) that nourish each of the body's layers known as *dhatus*. These blockages to nourishment are usually caused by toxins or *ama*. Another interesting feature of Ayurveda is that these channels—and even the toxic buildup that can create an imbalance with our doshas that can lead to disease—are thought of as intimately related to, yet beyond, the physical body.

That is, Ayurveda teaches that we are more than our physical body. We have, in fact, five layers or "sheaths" to our being. These sheaths are known as the *pancha koshas*, or five sheaths. When the physical sheath (or food body) is toxic, this will impact the subtler sheaths, which include the energetic or pranic body, the mental sheath, the sheath of higher wisdom, and the bliss body. The pancha koshas are depicted in Figure 17.1.

The Three Humors or Doshas

The doshas, or three humors, are psycho-physiological constitutional types known as vata, pitta, and kapha. As mentioned, Ayurveda is an elemental framework of medicine. Every person is comprised of a ratio of all three doshas that is determined at the time of conception. When the doshas are in balance, they support the normal functioning of the body; when out of balance, they create disease as they manifest in the layers of the body or *dhatus*. In addition to the three doshas, there are also 15 subdoshas. However, for the purposes of understanding integrative sexual medicine, this chapter focuses only on the three primary doshas.

VATA DOSHA

Vata is known as "king of the doshas," because when imbalanced, it can quickly imbalance the rest of the humors. Vata is comprised of ether and air, and a person with a vata predominate dosha can have the following characteristics:

- Thin body frame
- Thin lips and hair
- Dry skin and brittle nails
- Cold hands and feet
- Creaky joints
- Learns quickly and forgets quickly
- Walks and talks fast
- Restless and scattered in mind and body
- Imaginative and creative
- Anxious and fearful
- Trouble with sleep
- Constipation
- Difficulty "staying the course" and being consistent
- Subject to mood swings

PITTA DOSHA

A pitta person is the typical "type A personality" because of characteristics engendered by the predominance of fire and water elements. Some common characteristics of a predominately pitta type person can be:

- Medium body frame
- Fair skin, often with freckles or moles
- Strong appetite and digestive fire
- Warm constitution
- Early greying or baldness

- Sharp intelligence and understanding and a curious mind
- Prone to acne, rashes, sensitive teeth, heartburn, acidity, strong body odor
- Prone to inflammation, irritability (especially in the heat or when meals are skipped), and judgment of self, which can leak out onto others

KAPHA DOSHA

The kapha predominate person could be called the "cookie-baking mom or dad." Kapha is comprised of water and earth, which lends to the loyal stability they offer in relationships. Other characteristics can be:

- Larger body frame with smooth, oily skin and hair
- Grounded with slower speech and gait
- Cold skin and slow digestion
- Lubricated joints
- Learns slowly but never forgets
- Slow to get going but strong endurance
- Thick hair, nails, and skin
- Gentle and caring with a tendency toward excess weight and fat
- Copious mucus and congestion
- Prone to boils and cysts
- Sugar and salt cravings, poor carbohydrate metabolism, and frequent edema
- Sedentary lifestyle
- Attachments to love and relationships

Each person is born with their unique mix of the three doshas known as their *prakruti*. Any imbalance of this birth constitution is called their *vikruti*. It is in the state of an imbalanced vikruti in which disease and subsequently sexual health imbalances can occur.

The Seven Dhatus

As stated before, the human body is viewed from an Ayurvedic perspective in seven different tissue layers called dhatus. The word *dhatu* can be translated to mean the tissue that holds the body together, from atomic to molecular and from the cells to the gross matter of the organs. Table 17.1 lists eight dhatus, but male and female reproductive tissues are classified as one layer, so the actual number of dhatus is only seven.

Table 17.1. The Dhatus

Dhatus	*Corresponding Tissue in English*
Rasa (means juice of life)	Plasma, including white blood cells, lymph, and serum
Rakta (means red)	Red blood cells and oxygenation
Mamsa	Muscle, movement, support, and strength
Meda	Adipose tissue, steroids like cholesterol, and other lipids/insulation
Asthi	Bones and cartilage for protection, shape, and longevity
Majja	Bone marrow, nerve tissue, and connective tissue (associated with the endocrine system)
Artava	Female reproductive tissue and hormones
Shukra	Male reproductive tissue and the orgasmic fluid produced by women during sexual intercourse. Found in every layer of the body because every cell is a sex cell.

When we eat food, we digest it according to our level of digestive fire or *agni*. Each dhatu has its own digestive fire or *agni* that metabolizes nourishment to create new tissue and to pass on raw materials to the next tissue layer. If there is a disruption or blockage in this process, each dhatu will be affected, and disease can ensue due to an imbalanced dosha entering a tissue layer. As we can see, the reproductive or sexual tissue is the last dhatu, which explains why an imbalance of any sort in the other parts of the body can interrupt sexual desire and function.

Toxicity and Ama

As the HURT (Healing Un-Resolved Trauma) Study revealed in 2013, old emotional wounds can be triggered in women decades later and felt as if they are current, thereby adversely affecting their sexual desire (Ewers, 2014).

From an Ayurvedic perspective, this makes perfect sense. Any toxin, be it physical, such as poorly digested food, mental, such as an incomplete thought, or emotional, such as a held-onto trauma, can impair mental and bodily function. These result in the toxin called ama, which is a result of mental, emotional, or physical indigestion.

Remember that this toxic substance can now open the door to incomplete nourishment of the bodily tissues and then cause disease. See Figure 17.2 for an example of how ama can affect sexual desire, enjoyment, and function.

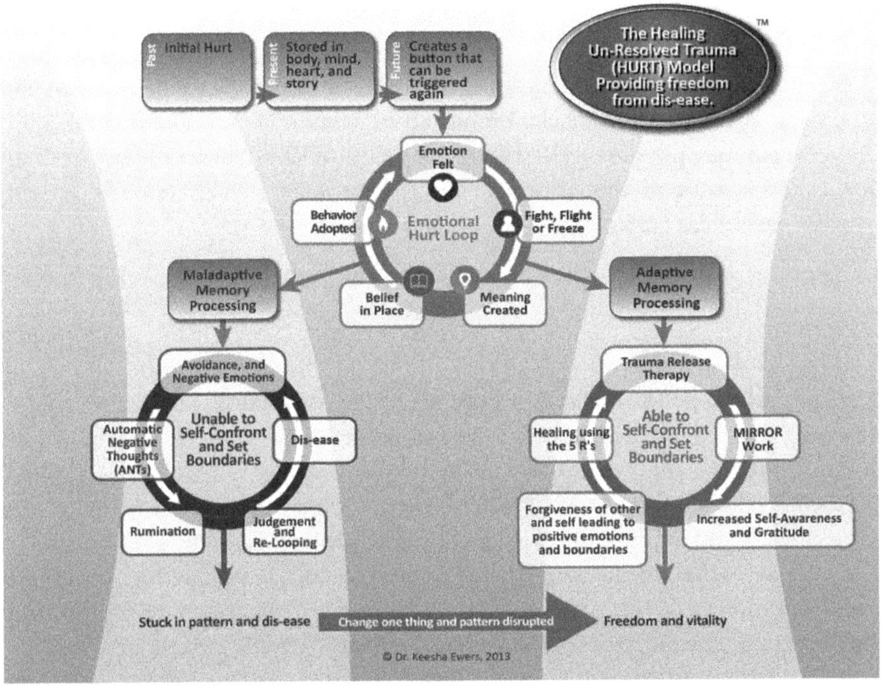

FIGURE 17.2. The Healing Un-Resolved Trauma (HURT) Model, simplified.

In this model, the person experiences a transgression and reacts with emotions, which arouse the sympathetic nervous system, resulting in a physiological response that will then store the memory of the hurt in the body and the mind. If the emotions and beliefs that emerge from this event are not fully and properly digested, they will get stuck in the body and mind as ama. This causes a maladaptive memory loop or muscle memory that can then be triggered in the future if a similar hurt is experienced, which can result in illness and sexual disorder.

Alternatively, if the emotions and beliefs are digested fully and processed completely through the body and mind, there will be no lingering toxicity. If a future transgression of a similar nature is experienced, the person or experiencer will move into an adaptive model of processing that includes trauma release methods if needed, an increase in self-awareness, a muscle memory of what it means to let go of hurt and to forgive, and a willingness to experience sexuality in a healthy way once again.

Within the Ayurvedic framework, this adaptive and complete digestion of a traumatic experience requires healthy digestive fire or *agni*. The resulting nourishment and balance of dhatus, all the way into the reproductive tissue, results in an increase in vitality or *ojas*.

Agni and Ojas

Agni allows us to fully experience life, and in this full experience lies passion and desire. So agni is a principle closely related to passion for life and sexual health. Also, on a purely physical level, it is by the power of agni that we digest our food, life experiences, and thoughts and absorb nutrients—which imbues us with vitality—as well as burns off waste products (detoxification).

OJAS

One product of good digestion is ojas, a subtle life-sustaining substance that is the very "juice" of life. In fact, it is what makes life "juicy" and even blissful—both necessary for the enjoyment of sexual activity and sexual function. In Ayurvedic medicine, ojas is referred to as a vital nectar and a powerful tonic to reinvigorate the senses, mind, body, and sexual health. The production of ojas is essential to life, because without vital energy there can be no cellular reproduction. Some have proposed understanding ojas from the Western biomedical framework, positing that ojas might be mapped onto stress and steroid physiology (Walton, 1995).

Ojas is not only dependent on healthy digestive fire or agni but also lifestyle choices, the way stress is handled, trauma, and the quality of relationships. An unhappy relationship is the number one cause of low female sexual desire and low ojas or life vitality.

Ways to increase ojas include the following.

Lifestyle

- Keep a regular daily routine; be in bed by 10 and wake up at sunrise.
- Meditate daily.
- Pranayama (conscious breathing techniques) to burn off accumulated ama.
- Exercise according to dosha type and balance.
- Do not smoke or drink alcohol.

Diet

- Eat pure, fresh, organic foods.
- Avoid foods that cause sensitivities or allergies.
- Avoid canned and processed foods.
- Do not overeat.
- Eat only when relaxed and in a calm environment.

- Eat only while seated and focus on eating.
- Eat only when hungry.

Detoxification

- Detox regularly to eliminate accumulated ama or toxicity.
- Do Pancha Karma (the Ayurvedic therapy for detoxification and rejuvenation) regularly.
- Make sure any detox regimen is overseen by a medical provider who can individualize the program to Ayurvedic constitution through pulse diagnosis.
- Work with a therapist if thoughts are toxic and of a victim nature.

Relationships

- Make time to communicate with significant others daily.
- Learn functional ways of conflict management.
- Find common interests and engage in them regularly.
- Make sure that for every negative comment directed to a partner there are five positive ones.
- Practice saying "yes" and notice how often "no" is said.

Applying Ayurveda to Sexual Health

Allowing Ayurveda to inform an integrative medical practice will expand the paradigm in which sexual health is approached and managed. No two people will be seen as the same. There will be no "one-size-fits-all" intervention. The patient will feel like a part of the team as he or she learns to take responsibility for his or her own health through dosha-specific lifestyle choices, stress-reducing techniques, and herbal remedies. Circling back to the five sheaths of the body, or pancha koshas, this would mean that a client in an integrated sexual medicine model will learn that detoxification of the five sheaths, balancing the doshas, and leading a life that ensures strong agni and ojas will result in bliss, a primary goal of sexual activity.

Herbal Remedies for Sexual Health

Herbal remedies for "sexual issues" include herbal remedies for detoxification, stoking digestive fire, enhancing ojas, and supporting reproductive tissue. The word "aphrodisiac" is defined differently in Ayurveda than it is in the West. In the West, an aphrodisiac is used to promote sexual desire. In Ayurvedic medicine, it is believed that one of the most important aspects of life for humans is sex and sexual relations. The ancient Ayurvedic system felt that this aspect of life was important enough to devote a whole branch of medicine to its study. This special branch of medicine is called Vajikarana (Dalal, 2013). Vajikarana can be loosely translated as "aphrodisiac" or "that which gives power to a horse." Fertility, the birth of healthy children, and sexual enjoyment are the goals of Vajikarana.

Some recent animal research supports the usefulness of Ayurvedic botanical medicine for a range of sexual and reproductive problems, showing for example improved fertility, greater frequency and endurance in sexual behaviors, and enhanced nitric oxide availability via diverse proposed mechanisms of action (Bansode, 2015; Goswami, 2013; Sharma, 2012; Thakur, 2009, 2011, 2012). More preclinical research will be needed for these traditional botanicals to be studied in human clinical trials (Pratap, 2012). Any possible future widespread use would be most appropriate and effective if prescribed and used traditionally as in Ayurveda, not as in Western medicine as a pill.

Recall that Ayurveda places great importance on the status of digestion, holding that the majority of diseases start in the intestines. Again, Western medicine is catching up: see other chapters in this book that discuss our growing understanding of the importance to health of the gut microbiome and optimum gut function. Ayurveda asserts that food improperly digested becomes toxic or poisonous in the intestines, which triggers an imbalance of the doshas, which then moves through the tissue layers to cause a block in the channels for nourishing the rest of the body and subsequently results in disease. Remember too that people are the sum total of what they eat, think, and drink. Properly digested food and thoughts will eventually become reproductive fluids and ojas in the final dhatu or tissue layer.

According to Ayurveda, aphrodisiacs are only used after a medically supervised detoxification program is undertaken. The Charaka Samhita, for instance, states that aphrodisiacs will not work if taken while the body is in a toxic state any more than a dirty cloth can be properly colored by dye (Caraka Samhita).

The most potent aphrodisiac mentioned in Ayurvedic texts is a beautiful, excited woman and a partnership between two people who love one another. The mention of women as aphrodisiacs highlights that Ayurvedic texts were written by men. An emotional connection, loving partner, and skilled lover might be said to be the most powerful aphrodisiac for women.

Common Sexual Issues from an Ayurvedic Perspective

Ayurveda teaches that sexual frequency is dosha specific and gives guidelines for sexual activity for each dosha. It is thought that ejaculating too often for men will reduce ojas and can lead to disease (Money, 1991). It is also advised to engage in sexual intercourse between 10 and 11 PM and not during the daytime. For the vata-predominate dosha types, sex is recommended only once or twice a month. Pitta individuals can engage sexually every two weeks and kapha types two or three times a week. Remember, sexual activity is usually between two people who are likely different dosha types. No one is easily fit into neat little doshic-specific boxes. Yet there is always a solution in Ayurveda. For example, sexual intercourse is not depleting if done with the intention of connection and not orgasm. For this reason, exercises are taught in the Yogic, Tantric, and Ayurvedic traditions for circulating sexual energy rather than releasing it (Money, 1991). This allows sexual intercourse to be enjoyed daily if desired. None of these teachings should be attempted without the guidance of a well-trained teacher.

The following are some common sexual health issues encountered by medical providers with basic recommendations. All Ayurvedic interventions are determined by doing pulse diagnosis, listening to the client's story, and doing a physical examination. None of these herbal formulas are meant to be taken as a one-size-fits-all solution.

LOW LIBIDO

Low libido arises from weak shukra and artava dhatus (male and female reproductive tissues), high stress, emotional issues, relationship dissatisfaction, cultural and spiritual barriers, and cognitive beliefs. As in functional medicine and functional sexology, treating the root cause within the context of the whole story is essential. If low sexual desire is due to circumstantial stressful events, those situations must be addressed and stress-reducing techniques taught. If there are relationship factors, which can cause low libido or be the result of low libido, these must be addressed in therapy.

For strengthening reproductive tissue, the following herbal recipes are helpful for enhancing libido:

For women: Warm 1 tsp shatavari (*Asparagus racemosus*) and ½ tsp vidari kanda (*Ipomoea digitata/paniculata)* in one cup of warm milk[1] and drink before bedtime.

For men: Warm 1 tsp ashwagandha (*Withania somnifera*) and ½ tsp vidari kanda in one cup of warm milk and drink before bedtime.

1. A note about milk: In India, thousands of years ago, the milk supplied by cows was much different than that available in grocery stores today. If you do not digest milk easily, try raw, whole, unpasteurized milk. Warm it slowly and add the spices. An alternative is to take your herbs with a small amount of raw honey or ghee (clarified butter).

Ayurveda places great importance on strengthening ojas after male ejaculation. There are many recipes taught for men to use before and after sexual activity. Probably the easiest is to peel 10 raw almonds that have soaked overnight in water. Pour off the water and add to a blender with 1 cup of warm milk. Blend with 1 pitted date and a pinch of ginger, nutmeg, and saffron. Drink after any sexual activity that ends in ejaculation.

PREMATURE EJACULATION

Premature ejaculation (PE) is usually caused by a vata disorder. Typical for a man with PE will be performance anxiety, fear, and nervous tension, all of which are signs of vata imbalance. Always screen men with this issue for high cholesterol, diabetes, prostate problems, and poor circulatory function.

Teach the male client with PE the following exercise:

1. Stimulate the penis until erect.
2. Sit in a comfortable seated position and pull the anal muscles inward 10 times.
3. Practice daily and before engaging with a partner sexually.

A man with PE must follow a vata-reducing diet and daily routine. Additionally, the patient's partner needs to be part of the therapeutic plan of care. A state of anxiety is often common in both partners if PE has been a sustained issue in the relationship.

PE Formula: Mix 1 oz of each of the following:

- Cinnamon
- Long pepper
- Cardamom
- Nutmeg
- Mace
- Pine nuts
- Dried ginger
- Cloves
- Dried coconut flakes
- Saffron

Use a mortar and pestle to grind all of the ingredients into a powder. Mix into enough raw honey to make a paste. Take ¼ tsp daily with warm milk or herbal tea. Results will be evident within three weeks, and the routine should be continued for several months.

ERECTILE DYSFUNCTION

Erectile dysfunction (ED) can be caused by a vata imbalance such as listed previously or as a side effect to a medication, alcohol, drugs, disordered sleep, a poor diet, constipation, or a circulatory issue.

Massaging the penis from the anus up the toward the base of the penis and the root of the penis with an herbalized Ayurvedic formula called mahanarayan oil can be helpful to improve circulation.

As with all disorders, examine the root cause and address it within the context of the client's story. If stress, sleep, diet, alcohol, medications, or drugs are factors, they will need to be addressed in any plan of care.

ED Herbal Formula: Warm 1 tsp ashwagandha, ½ tsp bala (*Sida cordifolia),* ½ tsp chopped garlic, and ½ tsp vidari kanda in one cup of warm milk. Drink twice a day for at least three months.

VAGINAL DRYNESS

Kama Shakti means "pleasure power." When a woman is experiencing vaginal dryness, her pleasure power diminishes rapidly due to the experience of pain. When we touch a hot stove, we are wired to stay away from hot stoves. Memory is a big factor in recurring pain. It is important to rectify vaginal dryness as soon as possible so a woman will not begin to equate sex with pain.

The cause of vaginal dryness can be a vata imbalance, adrenal fatigue, stress, emotional upset, cognitive beliefs, or a decrease in hormones secondary to menopause. Again, it is important to find the root cause and begin to balance at that level. Observing a vata-pacifying diet is essential for a menopausal woman. Menopause heralds the vata stage of life regardless of birth constitution.

Formula for shifting female hormones: Mix ½ tsp wild yam and ½ tsp shatavari into ½ cup aloe vera juice and warm water. Take twice daily after lunch and dinner throughout the menopausal stage of life.

Vaginal lubricant: Place a drop of water onto a hard surface and rub vigorously with your finger. You will soon feel heat under your finger and likely hear a high squeaking sound with the friction. This is what happens with water-based lubricants. Recommend instead an oil-based lubricant to your clients with vaginal dryness, remembering that whatever is put on the skin or any other bodily surface will enter the body to become digested. Use organic, cold-pressed sesame oil, organic coconut oil, or organic olive oil for the best lubricants. Whatever goes into the vagina needs to be safe to eat.

An alternative is to make a moisturizing tampon by soaking a large piece of sterile cotton in sesame oil, tying a piece of thread to one end, and inserting it into the vagina at night. A sanitary pad can be used to catch any leaked oil. These are only symptom managers. The most important long-term treatment will be balancing the hormones.

ANORGASMIA

Ayurveda teaches that everything we do is 95% mental with the exception of love and meditation. Where your mind goes, so goes your energy. Studies have shown women do not think erotic thoughts as often as men do. If a woman's mind is on work, stressful events, anxiety about pain, or anything other than the moment she is in with her partner, her likelihood of having an orgasm is lowered. For anyone who has "tried" to have an orgasm, they will know it's much like "trying" to go to sleep. Orgasms and sleep are both activities that must be preceded by relaxation and surrender. Once again, a vata-aggravated mind will have difficulty staying in the moment with the sensations of pleasure that accompany positive sexual relationships.

Treatment begins with pacifying the vata dosha by eating a vata-balancing diet, beginning mindfulness training through meditation on the body and senses, and homework activities for both partners to do together without the "goal" of an orgasm. The goal of sexual encounters must be pleasure and connection rather than climax. A woman's most potent block to eroticism is her mind. She needs to get into her body.

If there is a history of trauma, therapy using a trauma-release method such as eye movement desensitization reprogramming or brain spotting is essential. Both of these techniques teach clients to stay in the body while healing past hurts.

PARTNER PLAY

This simple exercise can be helpful in healing past hurt and establishing a sense of safety in touching and sexual interaction.

1. Lie on a comfortable surface on your back and close your eyes. Clothes can be on.
2. Allow your partner to touch all areas of your body except breasts and genitals.
3. Have your partner use different levels of touch from light to firm, from long strokes to short tickles.
4. Provide feedback to your partner about what you do and do not like.
5. Continue this using different objects, such as a feather, scarf, shower brush, fingers, etc.

6. When you are finished, the exercise is done.
7. The next time, if you feel comfortable with it, remove some clothing. Provide feedback to your partner again about what you like. Still no touching genitals or breasts.
8. When you are ready, repeat the exercise with no clothing and use sesame oil with a few drops of essential oil such as ylang ylang, patchouli, lavender, or geranium.
9. Only when you are fully aware of your body sensations and are comfortable being touched, allow your partner to touch your breasts and genitals. Again, give feedback to your partner about what you like and dislike. Stop when you are finished.

Solo play is also important for a woman who has not had an orgasm or is no longer having them. Massaging the vulva and clitoris with sesame oil and using a sex toy are all very helpful for self-discovery. If a woman does not know the map of her own body, it is difficult to give directions to a partner.

Letting go of orgasm as a measurement of a satisfying sexual encounter is essential. Explore motives for engaging sexually with your client. Find other reasons for her to be excited to relate in this way with her partner and herself. There is much talk about clitoral and vaginal and g-spot orgasms now in the media. Women who do not experience these sensations can begin to believe they are broken. In Tantra these sorts of orgasms are considered the same as a sneeze, and lovers are encouraged to instead unite energetically so fully that they have full body internal orgasms. Sex is considered a sacred activity and is never conceived of as dirty or bad. It is the natural outgrowth of spiritual practice both solo and with a partner. This is Kama Shakti.

In conclusion, viewing our clients as whole beings will allow for a more complete and thorough plan of care for any imbalance. Ayurveda offers a framework for incorporating the physical, emotional, mental, spiritual, and environmental aspects of life into our clinical process. Sexual medicine in the Western tradition has not offered robust treatment options for complex sexual issues that go beyond physical or mechanical sexual disorders, such as low libido. This is because libido is an outgrowth of strong ojas. Ojas is a product of healthy digestion within all of the body tissues, or dhatus. In our search for the right medication to "cure" low libido, we are not considering the multilayered complexity of sexual health. Ayurveda offers us a viable framework for coming up with simple solutions to complex sexual issues that continue to plague the patients we see in our offices. Modern scientific and clinical evidence, as presented in other chapters in this book, supports the wisdom of Ayurveda in recognizing that successful and rewarding sexual expression depends on supporting the health of the whole body and mind.

Recommended Reading on Ayurveda

- Arnold CH (1936). Historical gynecology. *Cal West Med.* 44(1): 40–3.
- Frawley D (2000). *Ayurvedic healing: a comprehensive guide,* 2nd ed., Vol. 1. Twin Lakes, WI: Lotus Press.
- Frawley D. (1997). *Ayurveda and the mind: the healing of consciousness.* Twin Lakes, WI: Lotus Press.
- Kaviratna AC, Sharma P (1913). *The Charaka Samhita.* 5 Vols. New Delhi, India: Sri Satguru Publications.
- Lad V (2006). *Textbook of Ayurveda.* Albuquerque, NM: Ayurvedic Press.
- Tierra M. (1998). *The way of herbs: Michael Tierra.* New York: Pocket Books.

REFERENCES

Banchoff T (2009). *Religious pluralism, globalization, and world politics.* New York: Oxford University Press.

Bansode FW, Arya KR, Singh RK, Narender T (2015). Dose-dependent effects of Asparagus adscendens root (AARR) extract on the anabolic, reproductive, and sexual behavioral activity in rats. *Pharm Biol.* Feb;53(2):192–200.

Caraka Samhita, Cikitsasthanam, II, (1), 50–51.

Dalal PK, Tripathi A, Gupta SK (2013). Vajikarana: treatment of sexual dysfunctions based on Indian concepts. *Indian J Psychiatry.* 55(Suppl 2):S273–S276.

Ewers K (2014). An integrative medicine approach to the treatment of HSDD: introducing the HURT Model. *Sex Relation Ther.* 29(1):42–55.

Geetha P, Aravind BS, Pallavi G, Rajendra V, Rao R, Akhtar N (2012). Sexual dysfunction (Kṛcchra Vyavāya) in obesity (Sthaulya): validation by an observational study. *Anc Sci Life.* 32(2):76–81.

Goswami SK, Inamdar MN, Jamwal R, Dethe S (2013). Efficacy of *Cinnamomum cassia Blume* in age induced sexual dysfunction of rats. *J Young Pharm.* Dec;5(4):148–153.

Money J, Prakasam KS, Joshi VN (1991). Semen-conservation doctrine from ancient Ayurvedic to modern sexological theory. *Am J Psychother.* Jan;45(1):9–13.

Pratap SA, Rajender S (2012). Potent natural aphrodisiacs for the management of erectile dysfunction and male sexual debilities. *Front Biosci (Scholar Ed).* Jan 1;4:167–180. Review.

Rao BR (2004). Influence of Tantra on ayurveda and Kamasutra. *Bull Indian Inst Hist Med Hyderabad.* Jan-Jun;34(1):41–49.

Sharma V, Thakur M, Dixit VK (2012). A comparative study of ethanolic extracts of Pedalium murex Linn. fruits and sildenafil citrate on sexual behaviors and serum testosterone level in male rats during and after treatment. *J Ethnopharmacol.* Aug 30;143(1):201–206.

Thakur M, Chauhan NS, Bhargava S, Dixit VK (2009). A comparative study on aphrodisiac activity of some ayurvedic herbs in male albino rats. *Arch Sex Behav.* Dec;38(6):1009–1015.

Thakur M, Chauhan NS, Sharma V, Dixit VK, Bhargava S (2012). Effect of Curculigo orchioides on hyperglycemia-induced oligospermia and sexual dysfunction in male rats. *Int J Impot Res.* Jan-Feb;24(1):31–37.

Thakur M, Thompson D, Connellan P, Deseo MA, Morris C, Dixit VK (2011). Improvement of penile erection, sperm count and seminal fructose levels in vivo and nitric oxide release in vitro by ayurvedic herbs. *Andrologia.* Aug;43(4):273–277.

Walton KG, Pugh ND (1995). Stress, steroids, and "ojas": neuroendocrine mechanisms and current promise of ancient approaches to disease prevention. *Indian J Physiol Pharmacol.* Jan;39(1):3–36.

18

Healing Male Sexual Function Concerns in Traditional Asian Medicine

JILLIAN CAPODICE

Background

The tenets of traditional Chinese medicine (TCM) are based on its fundamentals of anatomy and physiology including yin and yang, qi, blood, fluids, organs, meridians. The other fundamental concepts include the eight principles, five elements, and organ-bowel (*zang fu*) theory. For the TCM practitioner, advanced knowledge of current biomedicine is imperative; on the other hand, for the allopathically trained clinician, recognition of the concepts of the holistic based system as well as the understanding of the mechanisms and utility of some of its most prominent modalities, that is acupuncture, moxibustion, and herbal medicine, is also important. A whole or holistic medical system theory often takes into account fluctuating parameters of bodily status that are highly interrelated. Whatever one's clinical specialty, knowledge of the tenets of TCM as well as the biomedical conditions are critical when evaluating a patient for a chief complaint of a disorder of sexual function.

Starting with the pillars of TCM, the key introductory concepts that must be understood are the basics of yin and yang theory and the nature of *qi* (energy), *xue* (blood), and *jing* (fluid essence). The Chinese character for *qi* contains a rice grain and vapor. This refers to both the physical and energetic aspects of the concept of *qi* or energy. The Chinese character for *jing* is also the rice grain but the character also contains a part of the Chinese character symbol that represents a blue-green color and fluid. This helps us to understand the importance of the nature of *jing*, as in TCM *jing* corresponds to the essence of a person. It has also been characterized as related to the reproductive system and has been thought to characterize sperm in traditional texts. *Jing* is also closely related to *shen* or spirit, and it is a traditional concept that the spirit of a person and its seminal essence are closely related (regardless of one's sex or gender). *Jing* and *shen* also represent spirit, vitality, growth, and renewal. Understanding the complex relationships between the flow of *qi* and *xue* and their relationship with *jing* and *shen* establishes a foundation of these terms that will be introduced with regard to the pathophysiology later in this chapter.

The other traditional medicine basics that need to be understood are the fundamentals of the organ system (*zang fu*) and meridian system. These two systems help distinguish the whole medical system principles of TCM. The main *zang* organs are the liver (LIV), heart (HT), spleen (SP) kidney (KD), lung (LU), and pericardium (PC). The main *fu* organs are the gallbladder (GB), small intestine (SI), stomach (ST), urinary bladder (UB), large intestine (LI), and triple burner (*san jiao*) (TB). Interestingly, neither the genital organs nor the brain and spinal cord are considered as one of the 12 main *zang fu*. The male genital organs are introduced later in the chapter. Understanding the myriad functions of the *zang fu* organs is also critical to understanding the pathophysiology of a condition according to TCM. Briefly the methodology of *zang fu* system pattern identification utilizes the four traditional methods of examination; inspection, palpation, listening, and smelling and inquiry. This combined with the theory of meridians and the other tenets of traditional medicine are used to establish the proper traditional medicine diagnosis.

The meridian system is based on 12 main meridians as well as eight extraordinary meridians and collaterals. The 12 main meridians' names are also named in relation to the 12 *zang fu* organs. The meridians can also be characterized based on their location on the body as yin (medial) versus yang (lateral). These fundamentals are central in understanding TCM and classical Chinese medicine (CCM) pattern diagnoses for male sexual disorders. This is further discussed after introduction of the biomedical physiology, pathophysiology, and common treatments of male sexual disorders.

Male Sexual Dysfunction

BIOMEDICAL PATHOPHYSIOLOGY AND COMMON PATTERNS OF DIAGNOSIS IN TRADITIONAL ASIAN MEDICINE

Disorders of male sexual function are characterized as disorders of arousal, ejaculation, and/or libido. It is also important to know that male sexual disorders and in particular erectile dysfunction (ED) may be a harbinger of systemic diseases such as cardiovascular disease, vascular insufficiency, diabetes mellitus, and other comorbidities such as depression. A thorough history and physical exam are crucially important when evaluating a new patient who comes to a practitioner's office with a chief complaint of ED. It also is essential to obtain information on all medications the patient is taking, as many medications have sexual and related side -effects and can cause or contribute to ED or other sexual problems.

Surprisingly, much of our current understanding of the physiology of the human male erection has only been gained since the 1980s. The physiology of the penile erection is a complex event that requires four main components: normal blood flow, normal innervation, normal hormonal factors, and functional and healthy penile tissue. Arterial blood is supplied to the penis' corporal bodies by the pudendal arteries and venous drainage of blood from the penis is possible due to a complex network of penile and pelvic veins. Innervation to the penis receives autonomic input from the pelvic plexus that travels in very close proximity to the prostate gland and it's neurovascular bundles which are also responsible for normal penile erection. All of these functions are mediated by the male sex hormone testosterone, so normal testosterone and endocrine function are main contributors to healthy sexual function.

In general terms, most people think of male sexual dysfunction as ED or difficulty obtaining an erect penis; however, disorders have been further characterized to better describe ED as disorders of desire or libido, arousal, penile erectile function or ejaculatory or orgasmic disorders. There can be overlapping conditions that involve genital or sexual pain whether or not that is anatomical such as in the case of Peyronie's Disease or penile trauma or sexual trauma. For example, disorders of libido and desire are commonly linked with other sexual and medical disorders such as depression, sexual addiction, and sexual impulsivity. Disorders of hypoactive desire are commonly related to hypogonadal and endocrine disorders. For example, low serum testosterone may be a common cause of hypoactive desire but can also be related to aging or may be a side effect of medications.

ED is the chief disorder of sexual arousal in men. ED is defined as the inability to achieve or maintain an erection sufficient for satisfactory sexual intercourse or performance. The incidence of ED has been shown to increase with age. ED is also characterized based on organic or psychogenic causes and can be a combination of both. Organic causes of ED include neurogenic, hormonal, arterial, cavernosal, or drug-induced ED. Psychogenic causes include depression, stress, performance anxiety, relationship problems, and difficulties with sexual arousal. ED may also be an indicator of a serious medical condition such as uncontrolled hyperlipidemia, diabetes, and/or hypertension, and it is now well known that ED can play an important role in helping identify and manage cardiovascular and metabolic disease. ED may also be a side effect of some common medications such as selective serotonin reuptake inhibitors, thiazide diuretics, ketoconazole, and sympathetic blockers such as clonidine. First-line therapies for ED include lifestyle modifications, drug (pharmacologic) therapy modification, psychosocial counseling, patient education, androgen replacement therapy (oral, transdermal, intramuscular), and oral pharmacologic therapy with a class of medications called phosphodiesterase type 5 (PDE-5) inhibitors. If a comorbid condition is also found to be a cause of ED, such as in uncontrolled diabetes, then treatment of that condition such as medication and/or lifestyle interventions targeting glycemic control are an additional part of the treatment plan.

Disorders of ejaculation and orgasm include rapid or premature ejaculation (PE), anejaculation, anorgasmia, and retrograde ejaculation. PE is an ejaculatory dysfunction that is defined by perception of the individual as a deviation from an ill-defined "normal." Ejaculatory function disorders commonly occur along a continuum of time, and it is well known that individuals may experience variation in their perception of ejaculatory function over time or it may differ depending on specific sexual encounters or partners. At the very least, the distress to the patient and/or his partner must be present. Since the definition of PE can vary from person to person, PE ultimately refers to shorter ejaculatory latency, lack of control of ejaculation, and/or lack of sexual satisfaction to one or both partners. Common treatments include topical creams, PDE-5 inhibitors, antidepressants, and psychological treatment including mind body recognition and other cognitive behavioral therapies. The other ejaculatory disorders, anorgasmia and anejaculation, are not discussed in this chapter.

Finally, for sexual disorders associated with low desire or libido, hypogonadism is the most common cause however metabolic disorders and depression are common co-morbidities. Men may have symptoms of low desire with a normal testosterone level although low testosterone is a common cause of reduced desire. Low desire is also a potentially normal process of aging so this must be taken in context with the patient and his particular lifestyle and/or potential bother from low sexual desire.

Testosterone is the key androgenic hormone responsible for normal growth and development of male sexual organs. Testosterone helps men to maintain muscle mass and strength, is responsible for even distribution of fat, supports healthy bone mass and erythropoiesis, and affects male hair pattern, libido, and spermatogenesis. Testosterone production is controlled by the hypothalamic-pituitary-testicular axis. Testosterone is produced in the Leydig cells of the testes. These cells manufacture about 95%, of testosterone with the adrenal glands making up the rest. The other two steroidal hormones secreted by the testes are dihydrotestosterone (DHT) and estradiol-2.

The production and maintenance of normal testosterone levels are regulated by the pituitary gland and hypothalamus. Gonadotropin-releasing hormone is synthesized in the hypothalamus and stimulates the release of luteinizing hormone and follicle-stimulating hormone, which both play important roles in testosterone regulation. Luteinizing hormone binds to Leydig cells' receptors and initiates the release of androgens. Follicle-stimulating hormone is necessary for spermatogenesis and androgen-binding proteins. Testosterone then acts on most body tissues. The active intracellular hormone depends on the tissue response, and testosterone binds to and activates receptors in the reproductive, immune, hematopoietic, and central nervous systems. Testosterone acts directly on target cells and can be converted to DHT by the 5α-reductase enzymes or to estradiol by the aromatase enzyme complex. The 5α-reductase enzymes are most abundant in prostate, skin, and reproductive tissues. The main organs and tissues influenced by testosterone are the pituitary, liver, kidneys, muscles, bones, adipose, and skin.

Low testosterone may result from a number of disorders, including primary testicular failure of a genetic or developmental nature such as Klinefelter's syndrome, or of an acquired cause like viral orchitis. Testosterone deficiency also may occur with hypogonadotropic hypogonadism. This has a congenital basis such as in Kallmann's syndrome or results from an acquired condition such as a pituitary adenoma. Low testosterone can occur before or after puberty and then later on at any time in a man's life. Once the cause of low testosterone has been established the appropriate treatment can be put into place; testosterone replacement therapy is the most common primary treatment. Testosterone replacement therapy can help to restore libido and erectile function as well as mood, lean body mass, strength, and stamina. In men over the age of 50, testosterone therapy is less frequently and more controversially used due to reported risk of cardiovascular disease; however, this risk has not yet been definitively established.

Taking what we know about the basic biomedical pathophysiology of the common male sexual disorders, I now outline some of the common syndromes that are related to the male sexual disorders in the context of both TCM and CCM. Regardless of the theory that informs one's practice, traditional medicine fundamentals have long referenced the concept of balance in sexual function and in life. Classic texts commonly state that overindulgence in sexual activity is damaging to the kidney and *jing*. On the other hand, long-term abstinence can also diminish the *jing* and *shen*. The next portion of the chapter summarizes some common patterns for treatment of male sexual dysfunction in traditional Asian medical theory.

COMMON TRADITIONAL MEDICINE PATTERNS FOR MALE SEXUAL DYSFUNCTION DISORDERS

According to TCM, common causes of ED and low libido include physical problems such as diabetes, endocrine conditions, prostate disease, lifestyle issues such as over indulgence in alcohol consumption or obesity, drug interactions (i.e., hypertensive drugs or other pharmaceuticals), and psychological disorders. In TCM, the penis is considered the root of the liver, and most TCM patterns for complaints of male sexual function disorders have liver and kidney pathologies. The names for the male urogenital organs and their pin yin translations are listed in Table 18.1.

Table 18.1. Pinyin of Male Urogenital System

Part and Names	*Pinyin Name*
Scrotum (sac) Kidney sac Yin sac	Nang Shen nang Yin nang
Testes (often grouped with epididymis and called kidney seed)	Shen zi
Vas deferens and seminal vesicles (seminal pathway)	Jing dao
Prostate (essence chamber or life gate)	Jing shi or ming men
Spermatic cord (seminal tie)	Jing zi xi
Penis (stem) Penis (ancestral sinew) Penis (Jade stem)	Jing Zong jing Yu jing

It is interesting to note the different organ/feature names and the relationship to the traditional Chinese characters. When the TCM liver is out of balance with regard to male sexual function, it not surprisingly leads to liver qi stagnation and interrupted free flow of qi that can lead to ED. Liver qi stagnation is often combined with pathologies of the external evils such as dampness, heat, or cold. Lifestyle factors are common culprits that may contribute to the etiology of these pathologies. For example in TCM the overconsumption of greasy and spicy foods and excessive alcohol consumption is a cause of excessive heat and/or dampness. On the other hand, excessive exposure to cold can exacerbate cold and or cold/damp pathologies. Organ pathologies commonly affect any of the main *zang fu;* the most common organ pathologies associated with ED patterns include the liver, kidney, heart, lung, spleen, and *san jiao* (triple burner). The kidney organ and meridian is almost always a factor in a TCM pattern differentiation of ED. As previously mentioned, the kidney is related to the *jing* essence and this, coupled with kidney yin or yang deficiency, is a pattern diagnosis seen in men presenting with ED. Lifestyle factors also contribute to kidney yang deficiency over time and include overwork (both physical and mental taxation), excessive consumption of cold and raw foods, excessive exposure to cold, and overindulgence in sexual activity. Lifestyle factors related to kidney yin deficiency also include overwork, night shift work, and insomnia. These factors all deplete the *jing.*

Heart pathologies are more frequently associated with symptoms of PE and combined pathologies of the lung. The heart is also related to the kidney via its relationship to the connection between kidney essence. Moreover stagnation in the triple burner is frequently involved in patterns of both ED and PE. The concept of the *san jiao*, or triple burner, is debated in many texts. The *san jiao* is considered one of the yang organs; however, it is not a physical organ. Rather, its function is thought to be as a conduit for qi flow as well as dividing the upper, middle, and lower parts of the body. The *san jiao* is intimately related with its yin counterpart the pericardium, and there is a strong relationship between many of the acupuncture points on these two meridians and their use in clinical practice.

The three most common TCM patterns in male sexual dysfunction are liver qi stagnation, damp heat in the lower jiao, and kidney yang deficiency. In patients with liver qi stagnation predominance, common symptoms may include low libido with difficulty maintaining an erection with or without dysuria, lower abdominal or groin pain, and mental/emotional stress. In those with damp heat predominant pathologies, symptoms include difficulty obtaining an erection, PE due to blocked qi, and possible low sperm count or testosterone levels; poor lifestyle factors are often a contributing cause. In men with kidney yang deficiency predominance, patients often present with an inability to obtain an erection, low or no libido and fatigue.

Recognition and treatment principles aimed at external factors such as cold, heat, and dampness also need to be addressed and are amenable to treatment plans that focus on multimodal therapy including acupuncture, moxibustion, and lifestyle assessment modifications. In addition, traditional herbal medicine can be used and formulas should be tailored to individual symptoms and the proper diagnosis. For example, cold causes contraction and pain. This often manifests as stagnant qi and blood flow in the liver meridian. Cold- and heat-based patterns are common factors in pathologies including pelvic, prostate, perineum, and other male pelvic pain disorders. Treatment patterns aimed at warming the liver meridian and dispelling cold or coursing the liver meridian and draining damp heat are underlying treatment principles. Heat-predominant patterns also often present with common lower urinary tract symptoms such as dysuria, dribbling urine, and pain. In classic TCM texts priapism (an abnormally lasting penile erection with failure of detumescence) was listed as a problem caused by excess heat but priapism is not a modern-day chief complaint, it is a urologic emergency that can cause long-term damage to the penile corpora. Finally, dampness exacerbates both heat and cold pathologies and commonly settles into the middle and/or lower jiao. Symptoms of dampness related to the urogenital system include many common lower urinary tract symptoms such as weak stream, incomplete sensation of emptying, urinary frequency, urinary urgency, dysuria, nocturia, and PE. Dampness typically blocks the flow of qui and blood.

Both biomedicine and TCM recognize the aspect of the mind/psychology and *shen*/spirit/heart with respect to sexual function, libido, and desire. In urology, psychogenic ED is a common diagnosis as is the recognition of the role of the mind and body in TCM. Treatment based on the concept of soothing the relationship between the liver qi and its associated emotion of anger with that of the kidney essence and its relationship to fear in addition to heart/*shen* imbalances is a common pattern seen in men with either ED or PE. Patients with a biomedical diagnosis of psychogenic ED frequently present with TCM liver/kidney/heart pathologies. Treatment strategies aimed at soothing the *shen* and boosting the kidney are often the most effective. Talking about the condition, including education the patient on the physiology of erection, and presenting other non-invasive options for relaxation, also are important components of a successful treatment plan. Therapists (psychologists and psychiatrists) specializing in sex therapy are also very effective and underscore that a multidisciplinary approach is often warranted.

The three most common TCM patterns see in patients with PE are liver qi stagnation with heat or damp heat, heart and spleen vacuity, and *shen* or heart disturbance with kidney qi depletion. For the first pattern, additional hallmark symptoms may include emotional excitability, quickly obtained erections, and sometimes a hyperactive libido. For heart and spleen vacuity patterns, the common symptoms may include insomnia and fatigue. In patients with *shen* disturbance predominance patterns, performance anxiety is a common symptom. If it is coupled with kidney deficiency symptoms may also include low back pain and overall weakness. Again, thorough assessment leading to the proper diagnosis is the key to creating an effective treatment plan. For treatment of both ED and PE, addressing and creating treatment plans that engender a healthy lifestyle and qi-strengthening activities will also lead to successful clinical results.

COMMON CLASSICAL CHINESE MEDICINE PATTERNS

CCM theories can also be used to treat patients with ED and PE. CCM differs from TCM in a variety of ways. The fundamentals of CCM patterns are rooted in taking a holistic view of the patient and understanding and correctly applying the six-syndrome system and often, an appropriate herbal formulation. The six syndromes are Tai Yang, Yang Ming, Shao Yang, Tai Yin, Jue Yin, and Shao Yin. Practice according to this system is based on symptom presentation and relies heavily on accurate pulse and tongue diagnosis. Table 18.2 lists the common characteristics of each syndrome.

As with all cases, patients never present with one textbook syndrome; they usually present with combinations of two or three patterns. A common example of a CCM pattern in a patient presenting with ED is a patient with shao yin and tai yin syndromes. This patient often presents with symptoms such as inability to maintain a satisfactory erection, PE, and low back pain. This index patient can have a slow, wiry, and thin pulse and a white tongue coating. Other symptoms could included tinnitus or penile discharge. These symptoms commonly point to a yang ming pathology. If this were the case, a patient with these symptoms might present with signs such as a rapid and wiry pulse and a thicker tongue that could be yellow and/or greasy. When applying treatments for patients utilizing CCM diagnoses

Table 18.2. Classical Chinese Medicine Six Syndromes

Syndrome	*Common Symptoms*	*Pulse*
Tai Yang	Fever with or without sweating, aversion to cold, neck and joint pain, pruritis	Floating
Yang Ming	Fever and night sweating, insomnia, restlessness, fullness in the chest and stomach, gastrointestinal upset, swelling or joint pain	Slippery Rapid
Shao Yang	Fever, sore throat, restlessness, chest pressure	Wiry
Tai Yin	Cold, fatigue, bloating or weak digestion, insomnia, joint pain	Slow Deep
Jue Yin	Heat, skin disorders and rashes, fatigue, lower body coldness, joint pain	Slow Deep
Shao Yin	Cold, fatigue, diarrhea, increased urinary frequency	Weak Deep

traditional herbal medicine is integral as classic CCM texts do not commonly refer to acupuncture. In modern practice the clinician can choose to use a variety of treatment modalities.

HYPOGONADISM AND DISORDERS OF LOW LIBIDO

In both TCM and CCM, the concept of male hypogonadism is not characterized in the same way that it is biomedically. If we integrate the systems, meridian theory and applying TCM principles based on modern physiology/pathophysiology is compelling and useful. Recognizing the role of the extraordinary meridians also is useful when taking an integrative approach; understanding pathologies affecting the chong, ren, and du meridians is very useful when treating disorders of hypogonadism and low libido. In addition, thinking of the concepts of the kidney yin and yang and how they relate to the hormones and normal physiologic function of both the urogenital and endocrine systems can be helpful. Key concepts helpful in understanding the TCM kidney are listed in Table 18.3.

For example, patients presenting with hypogonadism can be seen as related to kidney yang depletion plus or minus metabolic disorders/dysfunction that may also be related to liver and spleen pathologies. Many metabolic functions that are thought of as neuroendocrine disorders in biomedicine are governed by kidney energies in the eyes of traditional medicine practitioners. Hormones affected in these cases include cortisol or thyroid hormones. For example, heart, spleen, and liver dysfunction coupled with kidney yin or yang pathology is a frequent TCM pattern in men with hypogonadal symptoms. A treatment principle based on

Table 18.3. Critical Concepts on the Role of the Traditional Chinese Medicine Kidney

Concept	*Description*
Life gate (*Ming men*)	The kidneys are the essence of the *ming men* or gate of vitality. The *ming men* is closely related to kidney yang. From a biomedical perspective this can relate to functional male sex hormones such as testosterone
Kidney yang	The foundation of all yang energies Critical in managing functions such as growth and development Often related to libido, erectile function, urination, and energy
Growth and maturation	Growth and maturation are critically related to kidney function. This includes both congenital and acquired qi
Jing	Jing is usually translated as essence. Jing is related to basic constitution, postnatal essence, and kidney essence, which helps to determine growth, reproduction, and development
Hearing	The kidney is closely related to the ears and hearing. Tinnitus is a common symptom of kidney pathologies
Relationship with the lung	The kidney and lung are related with primary regard to ascending and descending qi and fluids
Relationship with the heart	The kidney and heart represent the primary relationship between water and fire (kidney and heart, respectively) Governs root of essence and mind
Kidney yin	The foundation of all yin energies of the body Fundamental for reproduction and growth Often related to fertility, supporting the spleen and stomach, harmonizing sexual function and *jing*

nourishing the organs and replenishing qi can result in improvement in men with common symptoms of hypogonadism such as fatigue and listlessness. Functional medicine concepts that integrate various holistic theories can also be useful to the clinician as functional medicine practitioners also utilized concepts from TCM with regard to treatment principles aimed at balance and restoration of the patient and achieving optimal health.

The differentiation and individualistic nature of both TCM and CCM diagnosis and treatment varies greatly from the biomedical approach and it's differential diagnosis and treatment. Continued research of modalities such as acupuncture, herbal medicine, and mediation will add to the evidence based and advance acceptance and use in an integrated multidisciplinary clinical practice. Ultimately, medicine is integrative and we should utilize safe and effective treatments from all disciplines in order to provide the best care and outcomes for patients.

RESEARCH

Despite the complexity and differences in TCM and CCM diagnosis and treatment strategies, research in TCM is usually limited to the study of modalities such as the study of acupuncture needle mechanisms of action or a single herb or one of its components. The gold standard for clinical research is still based on the pharmacologic clinical trial study design of randomized, placebo controlled clinical trials although recognition of non-specific effects are also noted by researchers studying clinical trial design for complementary medicine practices. There are numerous peer-reviewed publications of clinical trials evaluating acupuncture for both ED and PE. There have been two randomized controlled studies and a handful of pilot studies examining acupuncture for ED.

The first study (Engelhardt, 2003) was a randomized controlled trial performed in 22 subjects. Subjects diagnosed with psychogenic ED were randomized to real acupuncture or a sham control. The real acupuncture procedure tested was an acupuncture point protocol designed to treat ED versus a sham control procedure which was an acupuncture point protocol designed to treat headache (point prescriptions were not available for review). Outcomes included scores on the well validated quantitative questionnaire, the International Index of Erectile Function (IIEF), tumescence testing, and serum hormone levels. Results demonstrated that there was a 68.4% improvement in the acupuncture group versus a 9% improvement in the sham group ($P = 0.0017$) as shown by scores on the IIEF questionnaire.

The second study (Aydin 1997) does not have data available for review but the abstract suggests that it was a randomized controlled study demonstrating that acupuncture and hypnosis improved ED symptoms versus placebo. This provides weak evidence from a clinical trial standpoint. For PE there has been one well-designed clinical trial testing acupuncture versus paroxetine versus placebo (sham acupuncture) (Sunay, 2011). Ninety heterosexual and sexually active men between the ages of 28 and 50 were randomized to one of three arms. The treatment groups were paroxetine 20 mg/day, true acupuncture, or sham acupuncture twice weekly for four weeks. The true acupuncture points were ST36, LI4, KI3, LIV3, EX-HN3 (yin tang), and CV3. The sham acupuncture used in this study was retractable needles. Results showed that the median Premature Ejaculation Diagnostic Tool scores (another quantitative questionnaire), for the paroxetine, acupuncture, and placebo groups, respectively, were 17.0, 16.0, and 15.5 at baseline and 10.5, 11.0, and 16.0 after treatment ($p = 0.001$, $p = 0.001$, $p = 0.314$, respectively). Extent of ejaculation delay was higher in the paroxetine versus the acupuncture group ($p = 0.001$) although acupuncture had a significantly stronger ejaculation-delaying effect than placebo. This is an interesting study that suggests that acupuncture may be useful in treating PE, but it is limited in its short duration and lack of follow-up.

No studies have looked at the effectiveness of acupuncture in treating hypogonadal symptoms. There have also been no randomized controlled studies testing any TCM herbal formulas for male sexual dysfunction. However, as mentioned earlier, a holistic approach utilizing lifestyle changes, physical activity, acupuncture, moxibustion, and herbal medicine are viable complementary medicine treatment strategies that can be utilized to treat a variety of symptoms related to male sexual function disorders.

Conclusion

Male sexual dysfunction including ED, PE, and hypogonadism are common conditions that affect the quality of life of men. These conditions may be due to a variety of factors including side effects of common medications. They can also be an indicator of another systemic disease such as depression or metabolic disorder. A variety of conventional biomedical treatment strategies including oral medications, injection therapies, hormone therapies, and surgery are available to treat these conditions, but traditional medicine for the treatment of male sexual function disorders may be useful for the refractory patient or in the patient that is seeking a non-pharmacologic approach to treatment. While there are distinct differences between the diagnosis and treatment strategies of biomedicine versus traditional Asian medicine, a complementary and integrated approach to treating any patient utilizing safe and effective treatment strategies is the future of multidisciplinary teams.

REFERENCES

Aydin S, Ercan M, Caşkurlu T, et al. (1997). Acupuncture and hypnotic suggestions in the treatment of non-organic male sexual dysfunction. *Scand J Urol Nephrol.* Jun;31(3):271–274.

Cheng X (2010). *Chinese acupuncture and moxibustion.* Beijing: Foreign Languages Press.

Damone B (2008). *Principles of Chinese medical andrology: an integrated approach to male reproductive and urological health.* Boulder, CO: Blue Poppy Press.

Engelhardt PF, Daha LK, Zils T, Simak R, König K, Pflüger H (2003). Acupuncture in the treatment of psychogenic erectile dysfunction: first results of a prospective randomized placebo-controlled study. *Int J Impot Res.* Oct;15(5):343–346.

Flaws B, Lake J (2001). *Chinese medical psychiatry: a textbook and clinical manual.* Boulder, CO: Blue Poppy Press.

Maciocia G (2011). *Obstetrics and gynecology in Chinese medicine.* New York: Churchill Livingstone.

Sionneau P (2002). *The treatment of modern Western medical diseases with Chinese medicine.* Beijing: Foreign Languages Press.

Sunay D, Sunay M, Aydoğmuş Y, et al. (2011). Acupuncture versus paroxetine for the treatment of premature ejaculation: a randomized, placebo-controlled clinical trial. *Eur Urol.* May;59(5):765–771.

Wu Y, Fischer W; Fratkin J, eds. (1997). *Practical therapeutics of Chinese medicine.* Brookline MA: Paradigm.

Zhang E, Hu X (1990). *Zhong yi ji chu li lun—Basic theory of traditional Chinese medicine.* Beijing: Ren min wei sheng chu ban she.

19

Homeopathy in the Prevention and Treatment of Sexual Problems

DANA ULLMAN AND KAREN ALLEN

Framework of Homeopathy

Homeopathic medicine is a specialized system of medicine and pharmacology that primarily utilizes specially prepared nanodoses of medicinal agents from the plant, mineral, animal, or chemical kingdom. The determination for usage of a homeopathic medicine is based on its toxicology, that is, whatever a substance has been found to cause in healthy people can and will elicit a healing response when given in nanodoses to sick people who have syndromes that are similar to the substance's toxicology.

The primary principle of homeopathy is called "the principle of similars," which is an ancient concept mentioned by Hippocrates, and by the Oracle at Delphi in ancient Greece, and was utilized in crude form by select primitive cultures. However, it was not until the early 1800s when a German physician and chemist Samuel Hahnemann, MD, developed a specific experimental basis for testing the toxicology of a medicinal agent in order to determine what it will heal (Hahnemann, 1842/1982).

The logic and rational for using a medicinal agent that mimics a person's pathology derives from theories of evolution that recognize that symptoms of illness are adaptations and defensive responses of an organism to infection and/or stress. By using a medicinal agent that mimics the person's symptoms, the medicine is thought to help the organism's immune and defense response. It is therefore not surprising that select well-respected and commonly practiced conventional medical treatments, such as vaccination and allergy desensitization, utilize this "like to treat like" strategy to either prevent or treat disease.

Homeopathy has remained highly controversial from its origin. Initially, it was attacked because Hahnemann and his colleagues provided such a strong critique of medicine of their day, at a time when physicians rarely criticized each other. As homeopathy developed further, Hahnemann and fellow homeopaths consistently found that the specialized method of making a homeopathic medicine through a "potentization" process of dilution (1:10 or 1:100), with vigorous shaking in between each dilution, resulted in sustained and even increased therapeutic results whenever a medicinal agent had an increased number of potentizations.

Recently, a compelling explanation for how homeopathic medicines work was developed by scientists and appeared in a leading journal, *Langmuir*, published by the American Chemistry Society (Chikramane,2012). Using market samples of six different metal-derived medicines from reputable manufacturers, scientists at the Department of Engineering at the Indian Institute of Technology have demonstrated, using three different types of spectroscopy (transmission electron microscopy, electron diffraction by selected area electron diffraction, and chemical analysis by inductively coupled plasma-atomic emission spectroscopy), that nanoparticles of the original medicinal agents persisted in solution even when they were potentized 1:100 six times, 30 times, and even 200 times. These modern technologies found a similar number of nanoparticles of each of the six metals tested at the similar amount in *all* of the potencies.

The metals tested in the 6C, 30C, and 200C were Aurum met. (gold), Cuprum met. (copper), Argentum met. (silver), Zincum met. (zinc), Stannum met.(tin), and Platina met. (platinum).

Skeptics of homeopathy have very commonly asserted that homeopathic medicines are so dilute that they are "beyond Avogadro's number" (dilutions beyond 1:10 24 times, dilution beyond 1:100 12 times, is the approximately mathematic determination) and therefore there should not be any molecules of the original medicinal agent in the solution. However, the making of a homeopathic medicine requires the usage of a double-distilled water, which is vigorously shaken once a medicinal agent is dropped into the solution. Homeopathic pharmacies have always made their medicines in glass bottles, assuming that glass was inert. However, it is not.

In fact, conventional material sciences research has found that vigorous shaking of a double-distilled water leads to silicate nanoparticles falling off the glass walls to go into the water. Further, the vigorous shaking of the glass vial creates bubbles and "nanobubbles" that bring oxygen into the water and that increase substantially the water pressure (William Tiller, PhD, the former head of Stanford's Department of Material Science, estimated this pressure to be 10,000 atmospheres).

Due to the high water pressure, every medicinal agent that the homeopathic pharmacy places in the water is literally pushed into the "silica chips." Then, when 90% or 99% of the water is poured out, the chips stick to the walls of the glass vial.

The increased process of dilution and vigorous shaking simply makes the chips smaller and smaller.

Conventional pharmacology and nanopharmacology research has confirmed that nanodoses of medicinal agents are able to sift through cell membranes and blood-brain barriers with much greater ease than larger doses (Bell, 2013, 2014). Further, modern research on many of our body's hormones and cell signaling agents confirm that they operate at nanodose levels similar to that which remain in the solutions found in homeopathic medicines (Eskinazi, 1999).

Homeopathic nanodoses have been found to have significant effects on gene expression. The famed *Nature* magazine (India's edition) has recently reported about this research influencing genes expression in cervical cancer cells (Saha, 2015). To date, there are now several dozen studies showing how different homeopathic medicines and in varying potencies have had up and down regulatory effects on genes (Khuda-Bukhsh, 2014) A review of this research concluded, "These findings support the hypothesis that homeopathic remedies could turn some important genes on or off, initiating a cascade of gene actions to correct the gene expression that has gone wrong and produced the disorder of disease" (Bellavite, 2015, 141).

The precise mechanism of action by which homeopathic medicines influence sex hormones and overall genitourinary health has not yet been elucidated, though the growing body of evidence showing that homeopathic medicines influence genetic expression lays the beginning groundwork for this understanding in light of the fact that certain genes are linked to expression of hormones and various cell signal agents that influence sexual health issues.

The most comprehensive review of basic sciences and clinical research ever conducted was funded by the government of Switzerland (Bornhöft, 2006). Not only did this report carefully and comprehensively review the body of evidence from randomized double-blind and placebo-controlled clinical trials testing homeopathic medicines, but it also evaluated the "real-world effectiveness" as well as safety and cost-effectiveness. The report also conducted a highly comprehensive review of the wide body of preclinical research (fundamental physio-chemical research, botanical studies, animal studies, and in vitro studies with human cells). And still further, this report evaluated systematic reviews and meta-analyses, outcome studies, and epidemiological research.

After assessing preclinical basic research and the high-quality clinical studies, the Swiss report affirmed that homeopathic high potencies seem to induce regulatory effects (e.g., balancing or normalizing effects) and specific changes in cells or living organisms. The report also reported that 20 of the 22 systematic reviews of clinical research testing homeopathic medicines detected at least a trend in favor of homeopathy.

The government of Switzerland also funded a review of clinical trials in homeopathy conducted by skeptics of homeopathy that found that the effects of

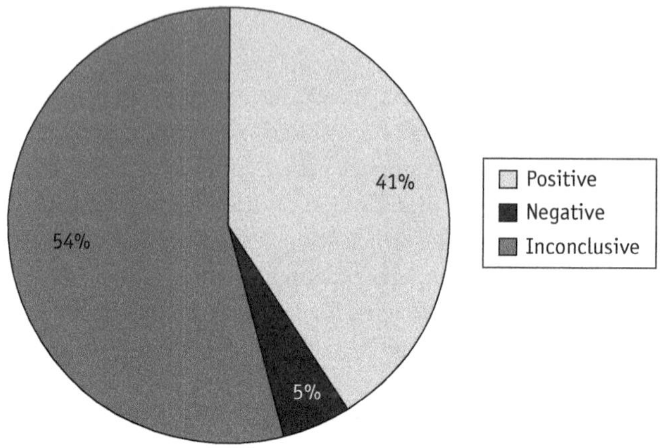

Figure 19.1. 189 RCTs pf homeopathy (end of 2014).
Source: https://www.hri-research.org/resources/homeopathy-faqs/there-is-no-scientific-evidence-homeopathy-works/
(Graphic developed by the Homeopathic Research Institute, used with permission).

homeopathic medicines are not different from that of a placebo (Shang, 2005). However, this review was deemed highly biased, and, using the same data set, reviewers who published their analysis in a leading journal known to evaluate research methodology found that the evidence provided also shows that homeopathic medicines provide distinctive benefits above and beyond that of a placebo (Lüdtke, 2008).

According to the Faculty of Homeopathy in the United Kingdom, which was incorporated by an Act of Parliament in 1950 and which regulates the education and practice of medical doctors who practice homeopathy in the UK, by the end of 2014, there were 189 randomized controlled trials of homeopathy on 100 different medical conditions that had been published in peer-reviewed journals. Of these, 104 papers were placebo-controlled and were eligible for detailed review:

- 41% were positive (43 trials)—finding that homeopathy was effective
- 5% were negative (5 trials)—finding that homeopathy was ineffective
- 54% were inconclusive (56 trials) (Figure 19.1)

Sexuality and the Genitourinary Microbiome

Great sexual function depends on health at every level, from the intimate couple to the smallest microbe. Healthy microflora supports normal function, just as dysbiosis disrupts it. Microbes leverage the human reproductive urge; they are readily communicated between partners in the genitourinary tract (De Francesco, 2011; Zozaya, 2016). They impact sexual function. The communal microbiome of the intimate couple, cross-pollinated through sexual activity, is a crucial factor. With each new sexual partner, the microbiome of both participants negotiates a new equilibrium. Not all flora play well in the sandbox together, so to speak; a client may report frequent urinary tract infections during an era with one partner that are completely absent with another. Multiple intimate partners increase the complexity of this communal ecosystem, increasing the risk of more frequent and varied genitourinary complaints (Nilsson, 1997) that interfere with enjoyable sexual activity. Discussion with clients about this intimate microflora ecosystem, along with more traditional coaching about safe sexual practices, empowers clients to make choices that engender health and a better quality of sexual experience.

Conventional medicine attends to overt sexually transmitted infections and pathogens in acute or chronic stages. It is less well equipped to address chronic genitourinary dysbiosis, especially in subclinical or latent conditions. Disturbances in sensation and function may appear without clinical lab findings. Symptoms may remain after multiple cycles of conventional therapeutics. Relapse is common. More than half of bacterial vaginosis clients have a recurrence of symptoms within 12 months after oral metronidazole therapy (Bradshaw, 2006). Research indicates that antimicrobial therapies do not benefit in asymptomatic bacteriuria and have some risk of harm (Nicolle, 2016). Negative urine lab cultures in just over one-quarter of women reporting urinary symptoms leave conventional providers without a specific therapeutic target (Hooton, 2013).

Genitourinary dysbiosis underlies a host of chronic complaints (Nibali, 2014) that disrupt pleasurable sexual activity, including vaginitis, bacterial vaginosis (van de Wijgert, 2014), cervicitis (Mattson, 2016), cervical cancer (Sheflin, 2014), pelvic inflammatory disease (Larsen, 2001), cystitis and interstitial cyctitis (Siddiqui, 2012), bladder pain syndrome, nonspecific urethritis (Bukharin, 2001), Bartholin's cyst, epididymitis, orchitis, prostatitis, and prostate cancer (Rosenblatt, 2001). Through a conventional medical lens, these are separate disease conditions. From an integrative viewpoint, they share the common terrain of disrupted microflora ecology. Most often, these clients present with a recent history of repeated but unsuccessful conventional interventions; the condition only partially relieves or returns. The integrative healthcare disciplines are better positioned to address the impaired local ecology in these chronic subacute or subclinical infections and in latent disease that cycles through active and inactive periods. When the microflora ecosystem is repaired, healthy sexual function reestablishes.

Homeopathy supports healthy immune engagement to optimize the genitourinary microbiome, supporting the body in clearing subclinical and latent infection that can interfere with normal sexual function. Each client's adaptation and immune response will create a characteristic way of managing the internal microbial ecosystem. He or she will describe symptoms in a slightly different way regarding the etiology, nature of discomfort, bodily sensation, modalities of aggravation, and amelioration. These details are important in selecting a well-matched homeopathic that is individualized to the client. In addition, it may be necessary to work with both partners of the intimate couple to achieve long-lasting improvement. With any genitourinary complaint, it is relevant to ask about recent change of sexual partner. Of the many homeopathic options available, the most frequently indicated are covered in Table 19.1. Homeopathy is a valuable contributor in an overall strategy of nutrition, probiotics, lifestyle balance, and other adjuncts.

Table 19.1. Genitourinary Dysbiosis: Leading Homeopathic Indicators

Calcarea iodatum	Best indicated with involuntary urination or dribbling with induration of prostate. Especially indicated for **seniors with enlarged prostate, frequent urination, and poor sleep**. Genital odors sour, foul, or fishy. Posology: 6c or 12c given 2–3 doses daily for 3–4 weeks, then reevaluate.
Conium	**Well indicated in senior men.** Enlarged, indurated prostate causes retained urine. Cutting pain in the scrotum. Burning and leaking of urine. Discharge of prostatic fluids while passing stool. Impotence, incomplete erections. Posology: 6c or 12c given 2–3 doses daily for 3–4 weeks, then reevaluate.
Fabiana imbricata	**Chronic inflammation of the prostate**. Subacute infection with recurrent cystitis. Interstitial cystitis. Chronic uric acid imbalance, frequent urination with irritation. Posology: 12x twice daily for 3–4 weeks, then reevaluate.
Nitricum acidum	Strong option when there is a known **history of genital warts, chlamydia, gonorrhea**. Chronic prostatitis. Splinter-like pains. Strong odor to urine. Hematuria. Genital or rectal fissures. Bartholin's cyst. Posology: 30c or 200c given in 1 or 2 doses, then reevaluate at 2–3 weeks.
Pulsatilla nigricans	Equally **well indicated in male and female clients**, with broad action in the genitourinary sphere. Nonirritating yellow or green discharges. Chronic vaginitis, uterine infection with changeable pains. Uterine displacement. Bartholin's cyst. Chronic prostate enlargement or inflammation. Spermatocele. Inflammation of testes, orchitis. Posology: 30c or 200c given in 1 or 2 doses, then reevaluate at 2–3 weeks.
Rhododendron	Consider in orchitis, epididymitis, spermatocele. **Strong affinity to testes and epididymus.** Atrophy, inflammation, swelling, or pain of the testes. Throbbing or cutting pain, sensitive to touch. Retraction of testes. Posology: 30c or 200c given in 1 or 2 doses, then reevaluate after 2–3 weeks.

Table 19.1. Continued

Sanicula aqua	Equally well indicated in male and female clients, with broad action in the genitourinary sphere. **Genital odor as of fish brine.** Urination with increased frequency, pain, or leaking. **Genital warts. Tenderness of reproductive organs.** Burning sensation. Increased perspiration. Posology: 12c or 30c given daily for 2–3 weeks, then reevaluate.
Spongia tosta	Equally well indicated in male and female clients, with broad action in the genitourinary sphere. **Chronic inflammation or enlargement of ovaries, testes, epididymitis following sexually transmitted infection.** Induration of any glands. Stitching pain in the testes. Posology: 12x twice daily for 3–4 weeks, then reevaluate.
Thuja	**Genital warts are indicative. History of gonorrhea, chlamydia.** Aching, squeezing, bruised pain in testes or ovaries. Ovarian pain extends down to the thighs. Ovarian cysts. Uterine polyps. Uterine displacement. Bartholin's cyst. Posology: 12c or 30c given daily for 2–3 weeks, then reevaluate.

Note: Posology: from the Greek *posos* (how much) and *logos* (science); the science or doctrine of dosage.

Sexuality and the Mental/Emotional Sphere

Perhaps more than any other physiological function, sexuality intertwines biological processes with the mental, emotional experience of the individual. Just as anxiety or fear can disrupt digestion or quicken respiration, it can disturb or halt a healthy sexual cycle of arousal, release, and relaxation. Homeopathy is uniquely well positioned to support the client with nonphysical sexual disruption because of the mind-body integration inherent in its use.

Trauma from past sexual abuse, molestation, or rape is a regrettably common finding in the consulting room. The Centers for Disease Control and Prevention's National Intimate Partner and Sexual Violence Survey (Black, 2010) reported that nearly one in five women have been raped. Approximately 45% of women and 22% of men have experienced sexual coercion or unwanted experience that did not include rape. While research shows that there has been a 49% drop in sexual assault since 1993 (Truman, 2015), this recent improvement underscores the greater problem in the larger cohort of clients over 25. These figures suggest that complex emotions about sexuality, even when consensual, may be likely in up to half of our female clients and a quarter of male clients. Additionally, some adult clients experience lingering fears from childhood or misunderstandings about the nature and process of sexual interaction, sometimes aggravated by negative or shaming messages from parents or caregivers. Fear, shame, or psychosexual conflict may be intense when even thinking of sex.

Homeopathy is an effective therapeutic intervention for sexual trauma and psychosexual conflict, aiding the client to release past experience and engage sexually in a healthy present. For each client, there will be a somewhat individual adaptation to the sexual trauma or conflict, and this is matched to a homeopathic that exhibits the same pattern. For example, after a traumatic sexual event, one person may become promiscuous, another may become very religious, and a third may develop characteristic pelvic pain that cannot be explained by any clinical findings. Each of these would indicate a different homeopathic that matches the exhibited pattern. In Table 19.2 frequently

Table 19.2. Sexual Trauma and Psychosexual Conflict: Leading Homeopathic Indicators

Staphysagria	Sequelae of rape or molestation, especially with presentation of suppressed anger and urinary irritation (cystitis after sex) and extreme sensitivity to touch. Promiscuity after sexual trauma, with intrusive sexual thoughts. Frequent swallowing, especially after oral sexual trauma.
Ignatia amara	Strong option immediately after a rape, useful for intense emotional turmoil, shock, and trauma. Weeping, sighing; mood swings are characteristic. Spasmodic symptoms are confirmatory. Useful in vestibulitis after trauma. Best used in higher potencies such as 200c and 1M.
Causticum	Internal sufferings, anger, especially with sense of injustice, sympathetic for sufferings of others. Strong affinity to respiratory tract (cough, hoarseness), urinary tract (involuntary urination), and central nervous system (ptosis of eyes, local paralytic conditions). Complaints with sensation of rawness and burning.
Aconitum napellus	Most useful when the client's adaptation to the sexual event is fright and shock. Client may be repeatedly reliving the event in her mind. Sensation of numbness, extreme restlessness, panic, palpitations. Worse at night, especially around midnight. Retention of urine caused by shock.
Platina	Indicated when the client's adaptation to the sexual abuse includes arrogance and narcissism, often with strong aversion to men. May experience hypersexuality with high libido. May use sexuality in revenge against men.
Sepia	Chronic stagnation, heaviness, or pain in pelvis after sexual trauma. Strong aversion to men. Relief from intense exercise, generally better from motion. Consider in chronic fatigue that follows sexual trauma. Also useful for libido disruption after oral contraceptives.

indicated homeopathics for sexual trauma are listed with their indications for use. Posology in these cases is individualized to the needs of the client, and higher potencies are more likely to be required.

Sexuality, Pain, and Functional Limitation

Even with healthy desire and a well-matched partner, a satisfying sexual relationship can be disrupted by functional limitation such as vaginal dryness or premature ejaculation. A wide variety of conditions can also create a painful experience of sexual activity. Dyspareunia, vaginismus, vulvodynia, vestibulitis, and chronic pelvic pain (CPP) are broad categories that cover many underlying causes and individual presentations. CPP is defined as pain in the pelvis that affects quality of life, lasting more than six months, not exclusively associated with menses, pregnancy, or intercourse. For 9 out of every 10 women with CPP, sexual intercourse is painful (Abercrombie, 2012; Zondervan, 1999).

Multifactorial presentations are common, and evidence suggests that multidisciplinary approaches are the most fruitful (Jarrell, 2005). The pelvic bowl is physiologically complex, and pain can arise from musculoskeletal, neurological, gynecological, urinary, and/or gastrointestinal factors. The experience of these women often includes depression and/or anxiety, which should be considered as part of the client's state. Often a woman with CPP has been evaluated for all of these factors, without a clear diagnosis, and is finally offered antidepressants or anti-anxiety medication and a referral for counseling. The unspoken message she receives is "It's all in your head." The homeopathic interview is well suited to bring out the full mind-body symptomatology for these clients, allowing a match to be made to a homeopathic medicine that addresses the full organism across all body systems.

A dominant etiology of sexual pelvic pain is endometriosis; it is found in an estimated one-quarter of women investigated for CPP (Schenken, 1996). Endometrial implants and subsequent formation of adhesions may limit motion of tissues, making penetration of intercourse painful. Long, painful, and frequent menses also interfere with sex; some women experience pain from ovulation through menses, 10 to 14 days of each 20- to 25-day cycle. Conventional therapeutics for endometrial pain include analgesics, surgery to remove implants, and medications to alter/downregulate reproductive cycles. These approaches, while providing short-term and near-term relief of pain, incur client detriments (postsurgical adhesions, bone loss, liver toxicity, increased cancer risk). Women using medications to downregulate the reproductive cycle can experience loss of libido, vaginal dryness, and loss of orgasm, so that even if pelvic pain is relieved, their sexual experience is not improved.

Endometrial pain, and its limitation to sexual function, has been an inconsistent finding across the collective population within the diagnosis. Some women with evidence of extensive implants experience very little or no pelvic pain, while others with very limited clinical findings have intense pain. Acknowledging that there were changes in the eutopic endometrium of women with endometriosis, researchers examined the presence of nerve fibers in the endometrial tissue. They found nerve fibers present in greater density in those with endometriosis and pain and less density or not present in those without endometriosis and pain (Al-Jefout, 2009). Homeopathy can address the characteristic endometrial triad of dysmenorrhea, neurological affinity, and formation of adhesions.

For men experiencing premature ejaculation, homeopathy can improve management of the sexual stimulation to allow normal function. In clients with erectile dysfunction, likely underlying limitation in cardiovascular function must be addressed. For these clients, homeopathy can support a gradual return to healthy cardiovascular function as part of an integrative protocol that gradually moves the client toward healthy sexual function.

In Table 19.3 frequently indicated homeopathics for sexual pain and functional limitation are listed with their indications for use. Posology in these cases is individualized to the needs of the client.

Table 19.3. Sexual Pain and Functional Limitation: Leading Homeopathic Indicators

Argentum nitricum	Premature ejaculation with overexcitement. Anticipation of future premajure ejaculation events worsens the condition.
Conium maculatum	Erectile dysfunction with weakness of the heart in older men. Best indicated with glandular complaints such as enlarged prostate.
Lillium tigrinum	Bearing down pains in the pelvis, worse from standing or walking. Concomitants include intense premenstrual emotional disturbances, psychosexual conflict, diarrhea with menses.
Lycopodium	Premature ejaculation, with marked performance anxiety. Erectile dysfunction.
Natrum muriaticum	Atrophic vaginitis. Vaginal dryness in perimenopause or after oral contraceptives or other hormonal therapeutics.
Nux vomica	Erectile dysfunction in overworked men with chronic heart disease.
Sabina officinalis	Abdominal pain felt in the sacrum, or extending toward the pubis or down thighs. Concomitants include fibroids, genital warts, spontaneous abortion. Tendency to gushing menses, with dark clots.

Table 19.3. Continued

Spigelia anthelmia	Neuralgic pains, especially with a left-sided predominance. Client may describe sensation as burning, like hot needles, squeezing or pressing outward. Pain tends to be worse night and tends to radiate from one location to another. Concomitants include head pains, especially around the eyes, heart palpitations.
Spongia tosta	Erectile dysfunction in clients with heart disease and atherosclerosis.
Sulphur lotum	Premature ejaculation, especially before intercourse can take place.
Viburnum opulus	History of early pregnancy loss. Abdominal pains that surround the pelvis, traveling from uterus to sacrum or vice versa. Colicky, spasmodic pain, worse from pressure at the umbilicus. Membranous dysmenorrhea. Concomitants include strong sensory awareness of the reproductive organs.

In Box 19.1 example cases demonstrate the homeopathic resolutions matched to the individualized experience of clients for the various conditions presented in this chapter. These cases demonstrate the integrated mind-body-spirit action of homeopathy, allowing constraint for the entire being to be lifted and normal function and sensation to be reestablished.

Box 19.1 Example Cases

Dysbiosis: Client report: 34-year-old female first noticed a fishy vaginal odor after she began a new sexual relationship six months prior. For the past two months, she has had painful intercourse with vaginal dryness, uterine tenderness on deep penetration, and occasional postcoital bleeding. There was no prior history of vaginal odor, dryness, sexual discomfort, or bleeding.

Clinical findings: All sexually transmitted infection tests reported negative and recent Pap test result was normal, but the clinician noted that cervical tissue appeared fragile.

Diagnosis and treatment: Bacterial vaginosis. Metronidazole three times over the prior five months, with temporary relief followed by relapse.

Individual client state: Upon inquiry about any other changes since these symptoms began, she volunteered that her feet had begun to sweat a lot, and the soles of her feet often had a sensation of burning, making it difficult to wear shoes.

Resolution: The client was given the homeopathic Sanicula aqua 30c, once daily for 2 weeks. She reported 75% improvement in symptoms at two-week follow-up. The foot sweat and burning were gone, as was the fishy genital odor. She still experienced some tenderness with penetration during sex, but there was no longer any vaginal dryness or postcoital bleeding. She was directed to continue Sanicula 30c daily for another 10 days. After that time, client reported that all symptoms were gone. At an annual follow-up, she reported no further recurrence of genitourinary symptoms.

Dysbiosis: Client report: 63-year-old male has experienced chronic bladder pain and urinary irritation for 10 years. Symptoms began after a new sexual partner; he contracted chlamydia and was treated with doxycycline. Although his physician retested and assured him the infection was cleared, he continued to have increasing urinary discomfort over the past decade.

Clinical findings: Urinalysis normal; culture is negative. Physical exam indicates that his prostate is enlarged.

Diagnosis (variously) and treatment: Urethritis, dysuria, prostatitis, benign prostatic hyperplasia, interstitial cystitis and CPP syndrome. 12 separate courses of antibiotics (azithromycin + ceftriaxone, doxycycline, erythromycin, minocycline, etc.) in past six years; symptoms unchanged.

Individual client state: He has pricking pain in the bladder when it is full. He has a cutting pain in the right side of the urethra during urination. He is pain-free only in the hour after urination. The urine has a very strong smell as if concentrated, although he drinks 10 to 12 glasses of water daily. Twice, small fissures have opened at the urethral meatus, with severe cutting pain. Over the past two years, he has marked increase in frequency of urination. Any sexual activity now creates sharp pricks of pain in the prostate, bladder, and urethra.

Resolution: The client was given the homeopathic Nitricum acidum 200c, three doses over 24 hours (Non rep.) and asked to stay well hydrated. At three-week follow-up, he reported very frequent urination with dark, cloudy urine the first night after he took the dose. Since then symptoms have improved. Bladder and urethral pain with increased urinary frequency was now present only in the first two to three hours after waking. For the remainder of the day he had no pain, normal urination, and no strong odor to urine. He also reported that a sense of fullness and heaviness in the prostate (not mentioned in the intake) was also relieved. Discomfort felt in the prostate with sex was now 50% less intense, more transient. He was asked to repeat the Nitricum acidum 200c one time if his symptoms began to worsen again; otherwise wait. At a second follow-up seven weeks later, he reported that he had repeated the homeopathic twice, and now all symptoms were gone. He experienced a relapse five months later after a stressful business reversal. He again repeated one dose of the Nitricum acidum 200c, and his symptoms disappeared within two days. At a one-year follow-up he reported no relapse.

Sexual trauma: Client report: A 24-year-old female reports strong reaction to gentle touch: overwhelming anger and stinging abdominal pain, onset two years prior when she was accosted on a train. Men removed some of her clothing, fondled her aggressively. She had no history of abdominal pain and had previously enjoyed a healthy sexual relationship with a former boyfriend.

Clinical findings: All physical exam and lab results normal.

Diagnosis and treatment: Posttraumatic stress disorder. Zoloft then Paxil, each for six months along with counseling; no change in symptoms.

Individual client state: With the anger, she experienced concomitant stinging pain in the lower right quadrant of her abdomen, as if from many paper cuts. The pain was intense (5 to 7 on a scale of 1 to 10) and lasted for two to three hours after the event. Any gentle touch was now unendurable, sometimes triggering visceral memory of the event on the train. She felt depressed and hopeless about having a normal loving relationship.

Resolution: She was given the homeopathic Staphysagria in 200c potency, three doses over 24 hours (Non rep.). At three-week follow-up she reported that she had slept very deeply for the first week after taking the dose. She dreamed through the next 10 days about the train event. In the first dream, she was tied and gagged by the men accosting her. Unable to defend herself, she felt trapped and enraged. She awoke with the same intense abdominal stinging. Later in the week, she dreamed that she was able to fight some of the men and awoke without pain. In a final third dream, she punched the men who were grabbing her, knocking them away and shouting for help. Others came to her aid, and the men were thrown off the train. These dreams gave insight to the healing process unfolding within. A few days after the last dream, she was gently stroked and realized there was no anger or stinging sensation. Over the next six months, she reported no recurrence and was enjoying a positive intimate relationship with her new boyfriend.

Psychosexual conflict: Client report: A 37-year-old male presented with involuntary urination during sexual arousal. Six months previously, his wife was raped while she was traveling out of town. She was unable to describe the assailant; no arrest could be made. The client felt intense anger, guilt, and outrage that he was not there to protect his wife. In the past two months, she had initiated sex with the client and was experiencing it in a healthy way. He wanted to be a good sexual partner but felt conflicted. With his erection, he had begun to involuntarily pass spurts of urine, which embarrassed him. This symptom was worsening over the past month. He had no history of urinary incontinence.

Clinical findings: Physical exam and urinalysis normal.

Diagnosis and treatment: Urinary incontinence. Duloxetine. Client experienced hives and nausea, stopped medication.

Individual client state: When asked if there had been any other changes to his health during this time, he reported that he had developed a hacking cough, triggered several times daily by a tickling in the throat pit. He had experienced this symptom in the past during times of sustained stress. Usually it began as a head cold; this time the cough started without any prior symptoms.

Resolution: He was given the homeopathic Causticum 30c once daily for two weeks. He was encouraged to seek counseling with a therapist experienced in sexual violence. At three-week return consult, he reported that the cough had worsened a little the day after he began the protocol and he had developed a raw sensation in his chest for 24 hours. The following day both cough and chest pain were gone. A week later, his wife initiated sex, and the client's sexual function was normal without loss of urine and without conflict or troubling thoughts about the rape. Afterwards he told her of his conflict and experience. He felt great relief, as if something had cleared. He continued well without relapse during the time the case was followed over the next year.

Premature ejaculation: Client report: A 31-year-old male presented with premature ejaculation throughout his adolescent and adult life. He and his wife want a pregnancy. He is not able to delay ejaculation long enough to allow penetration, even on repeated sexual activity.

Clinical findings: Physical exam and urinalysis normal.

Diagnosis and treatment: Premature ejaculation. Kegel exercises. Breathing techniques to delay/interrupt arousal.

Individual client state: When asked if there had been any other symptoms, he reports that during arousal he feels a strong sense of heat throughout his genitals.

Resolution: He was given the homeopathic Sulphur 1M, single dose (Non rep.) at the consult, and a vial of Sulphur 12c to take before each sexual activity. At a follow-up three weeks later, he reported that during the first week, there was no change. By the beginning of the second week, he felt less heat in the genitals with arousal. The arousal stimulation was more pleasurable and less compelling. He was able to maintain an erection long enough to allow for penetration before ejaculation once in the third week. Over the next six months, he repeated the Sulphur 1M 4 times when the symptoms became worse again. He continued to use Sulphur 12c before sexual activity. Gradual improvement through these months showed more frequent events of normal sexual function with penetration and slowly increasing duration of erection.

REFERENCES

Abercrombie PD (2012). Caring for women with chronic pelvic pain. *J Obstet Gynecol Neonatal Nurs.* 41:666–667.

Al-Jefout M, Dezarnaulds G, Cooper M, Tokushige N, Luscombe GM, Markham R, Fraser IS (2009). Diagnosis of endometriosis by detection of nerve fibres in an endometrial biopsy: a double blind study. *Hum Reprod.* 4(12):3019–3024.

Bell IR, Ives JA, Jonas WB (2014). Nonlinear effects of manoparticles: biological variability from hormetic doses, small particle sizes, and dynamic adaptive interactions. *Dose Response*. May;12(2):202–232.

Bell IR, Schwartz GE, Boyer NN, Koithan M, Brooks AJ (2013). Advances in integrative nanomedicine for improving infectious disease treatment in public health. *Eur J Integrat Med*. 5(2):126–140.

Bellavite P, Signorini A, Marzotto M, et al. (2015). Cell sensitivity, non-linearity and inverse effects. *Homeopathy*. 104(2):139–160.

Black MC, Basile KC, Breiding MJ, et al. (2010). *The National Intimate Partner and Sexual Violence Survey (NISVS): 2010 summary report*. Atlanta, GA: National Center for Injury Prevention and Control, Centers for Disease Control and Prevention.

Bornhöft G, Wolf U, von Ammon K, Righetti M, et al. (2006). Effectiveness, safety and cost-effectiveness of homeopathy in general practice—summarized health technology assessment. *Forschende Komplementärmedizin*. 13(Suppl 2):19–29.

Bradshaw CS, Tabrizi SN, Fairley CK, Morton AN, Rudland E, Garland SM (2006). The association of Atopobium vaginae and Gardnerella vaginalis with bacterial vaginosis and recurrence after oral metronidazole therapy. *J Infect Dis*. Sep 15; 194(6):828–836.

Bukharin OV, Ivanov IuB, Kuz'min MD, Mikhaĭlenko SV (2001). Characteristics of changes in microbiocenosis in patients with chronic nonspecific urethritis. *Zh Mikrobiol Epidemiol Immunobiol*. Jul-Aug;(4):86–89. [Article in Russian]

Chikramane PS, Kalita D, Suresh AK, Kane SG, Bellare JR (2012). Why extreme dilutions reach non-zero asymptotes: a nanoparticulate hypothesis based on froth flotation. *Langmuir*. 28(45):12864–15875.

De Francesco MA, Negrini R, Ravizzola G, Galli P, Manca N (2011). Bacterial species present in the lower male genital tract: a five-year retrospective study. *Eur J Contracept Reprod Health Care*. Feb;16(1):47–53.

Eskinazi D (1999). Homeopathy re-revisited: Is homeopathy compatible with biomedical observations? *Archives Intern Med*. Sept 27;159(17):1981–1987.

Hahnemann S (1842). *The organon of medicine*. 6th ed. [German: Organon der Heilkunst]. Republished as Kunzli J, Naude A, Pendleton P (1982). *The first integral English translation of the definitive sixth edition of the original work on homoeopathic medicine: a new translation*. Los Angeles: J. P. Tarcher. https://www.thehomeopathiccollege.org/wp-content/uploads/2015/11/Organon-of-Medicine-6th-edition.pdf.

Homeopathic drugs modify gene expression in cancer cells (2015). *Nature India*. Available at: http://www.natureasia.com/en/nindia/article/10.1038/nindia.2015.154.

Hooton TM, Roberts PL, Cox ME, Stapleton AE (2013). Voided midstream urine culture and acute cystitis in premenopausal women. *N Engl J Med*. 369:1883–1891.

Jarrell JF, Vilos GA, Allaire V, et al. (2005). Consensus guidelines for the management of chronic pelvic pain. *Br J Obstet Gynaecol*. Sep;106(11):1149–1155.

Khuda-Bukhsh AR (2014). Current trends in high dilution research with particular reference to gene regulatory hypothesis. *Nucleus (India)*. April;57(1):3–17.

Larsen B, Monif GRG (2001). Understanding the bacterial flora of the female genital tract. *Clin Infect Dis.* 32 (4):e69–e77.

Lüdtke R, Rutten ALB (2008). The conclusions on the effectiveness of homeopathy highly depend on the set of analyzed trials. *J Clin Epidemiol.* Dec;61(12):1197–1204.

Mattson SK, Polk JP, Nyirjesy P (2016). Chronic cervicitis: presenting features and response to therapy. *J Low Genit Tract Dis.* Dec;61(12):1197–1204.

Nibali L, Henderson B, Sadiq ST, Donos N (2014). Genetic dysbiosis: the role of microbial insults in chronic inflammatory diseases. *J Oral Microbiol.* 6:22962.

Nicolle LE (2016). The paradigm shift to non-treatment of asymptomatic bacteriuria. *Pathogens.* Apr;19;5(2).

Nilsson U, Hellberg D, Shoubnikova M, Nilsson S, Mårdh PA (1997). Sexual behavior risk factors associated with bacterial vaginosis and Chlamydia trachomatis infection. *Sex Transm Dis.* May;24(5):241–246.

Rosenblatt KA, Wicklund KG, Stanford JL (2001). Sexual factors and the risk of prostate cancer. *Am J Epidemiol.* Jun 15;153(12):1152–1158.

Saha, SK Roy S, Khuda-Bukhsh AR (2015). Ultra-highly diluted plant extracts of *Hydrastis canadensis* and *Marsdenia condurango* induce epigenetic modifications and alter gene expression profiles in HeLa cells *in vitro*. *J Integr Med.* 13:400–411.

Schenken RS (1996). Treatment of human infertility: the special case of endometriosis. In EY Adashi, JA Rock, Z Rosenwaks (Eds.), *Reproductive endocrinology, surgery, and technology,* Vol. 2 (pp. 2121–2140). Philadelphia: Lippincott-Raven.

Shang A, Huwiler-Müntener K, Nartey L, et al. (2005). Are the clinical effects of homoeopathy placebo effects? Comparative study of placebo-controlled trials of homoeopathy and allopathy. *Lancet.* Sep 2;366(9487):726–732.

Sheflin AM, Whitney AK, Weir TL (2014). Cancer-promoting effects of microbial dysbiosis. *Curr Oncol Rep.* Oct;16(10):406.

Siddiqui H, Lagesen K, Nederbragt AJ, Jeansson SL, Jakobsen KS (2012) Alterations of microbiota in urine from women with interstitial cystitis. *BMC Microbiol.* Sep 13;12:205.

Truman JL, Langton L (2015). *National Crime Victimization Survey.* Washington, DC: Bureau of Justice Statistics, US Department of Justice. http://www.bjs.gov/index.cfm?ty=pbdetail&iid=5366

van de Wijgert JHHM, Borgdorff H, Verhelst R, et al. (2014). The vaginal microbiota: What have we learned after a decade of molecular characterization? *PLoS One.* 9(8):e1059998.

Zondervan KT, Yudkin PL, Vessey MP, Dawes MG, Barlow DH, Kennedy SH (1999). Prevalence and incidence of chronic pelvic pain in primary care: evidence from a national general practice database. *Br J Obstet Gynaecol.* 106:1149–1155.

Zozaya M, Ferris MJ, Siren JD, et al. (2016). Bacterial communities in penile skin, male urethra, and vaginas of heterosexual couples with and without bacterial vaginosis. *Microbiome.* 4:16.

20

Chiropractic: How Musculoskeletal Involvement Impacts Sexuality

SUSAN R. EISEN

Chiropractic Fundamentals

A meaningful discussion of sexuality and chiropractic cannot be approached without an adequate understanding of what chiropractic is. The National Institutes of Health (NIH) defines chiropractic as a healthcare profession that focuses on the relationship between the body's structure, mainly the spine, and its functioning, and although practitioners may use a variety of treatment approaches, they primarily perform adjustments (manipulations) to the spine and other parts of the body with the goal of correcting alignment problems, alleviating pain, improving function, and supporting the body's natural ability to heal itself (NIH, 2012, 2016a, b).

It is notable that chiropractic is based on the understanding that the nervous system is responsible for all bodily functions, chiefly the central nervous system comprised of the brain and the spinal cord, which exits at and through our flexible vertebral spinal columns where it becomes the peripheral nervous system that ultimately innervates every muscle and organ in our bodies. The autonomic nervous system is responsible for bodily functions such as respiration, cardiovascular regulation, digestive processes, urination, and pupillary function and acts without conscious control, meaning one has does not will or think the heart to beat, the arteries to expand or contract, or the lungs to inhale air. The autonomic nervous system is a division of the peripheral nervous system. It is with this intrinsic understanding that chiropractors believe it is imperative that we all have healthy functioning spines. The Latin etymology for chiropractic literally means "working with the hands"; chiropractors do work with their hands to realign mechanical disorders, specifically vertebral subluxations, which are displacements of the facet joints, which are analogous to the hands and feet of our vertebrae.

Chiropractors must know in depth and work with clinical subtlety on the anatomy of the spinal vertebrae, especially those in the lumbar spine. All vertebrae throughout the cervical, thoracic, and lumbar spine are separated by our spinal discs that act as shock absorbers in the spinal column. One major function of the bony spinal column is to allow for support and flexibility, but the most important function is to protect the medulla spinae: the spinal cord. The spinal column consists of 33 bony segments that are partitioned into four separate groups: The cervical spine, consisting of 7 vertebrae, the thoracic spine with 12 vertebrae, the lumbar spine made up of 5 vertebrae, and the pelvic spine consisting of the sacrum and coccyx, which make up the remaining 5 and 3 segments, respectively (Figure 20.1). The five vertebral segments of the sacrum fuse into a triangular bone that sits at the base of the lumbar spine. The sacrum is the foundation of the spinal column and also an integral part of the bony pelvis articulating with the left and right iliac pelvises. The three remaining coccygeal vertebral segments fuse together to the sacrum and are our rudimentary tails. Complete fusion of the sacral and coccygeal segments typically occurs between the ages 18 and 25.

Each vertebra houses its respective portion of the spinal cord, protecting it from outside trauma and subsequent damage. This is accomplished by a posterior ring-like formation of bone called the neural foramen, which is attached to the anterior vertebral body. A bony prominence called the pedicle makes up the anterior portion of this ring. The posterior portion is called the lamina, which attaches to the spinous process and transverse process of each vertebra, completing the formation. Together the pedicle and lamina form the vertebral foramen through which the spinal cord passes. There are bony prominences formed at the base of each transverse process of every vertebra to form the superior and posterior articular processes, which form the opening called the intervertebral foramen and arch, where the root of the peripheral spinal nerves exit bilaterally, forming the beginning of the peripheral nervous system.

Other anatomy passing through the intervertebral foramen include the dorsal root ganglion, the spinal artery of the segmental artery, associated veins communicating with the internal and external plexuses, and the recurrent meningeal nerves (Figure 20.2). There are also transforaminal ligaments within the intervertebral foramen. The intervertebral foramen is an obviously crowded housing, and adequate spacing is essential to proper function of all nerves and circulation. Facet joints are also the surface stabilizing joints along with the vertebral body between each vertebra and the discs that separate each vertebra to give us spinal mobility. The facet joints, for the reasons stated previously, should optimally be in perfect alignment. Any misalignment of the vertebral complex between two juxapositioned vertebrae can result in pain, inflammatory response, and, most importantly, a change in the ability of the nerve root to function correctly. Facet joint

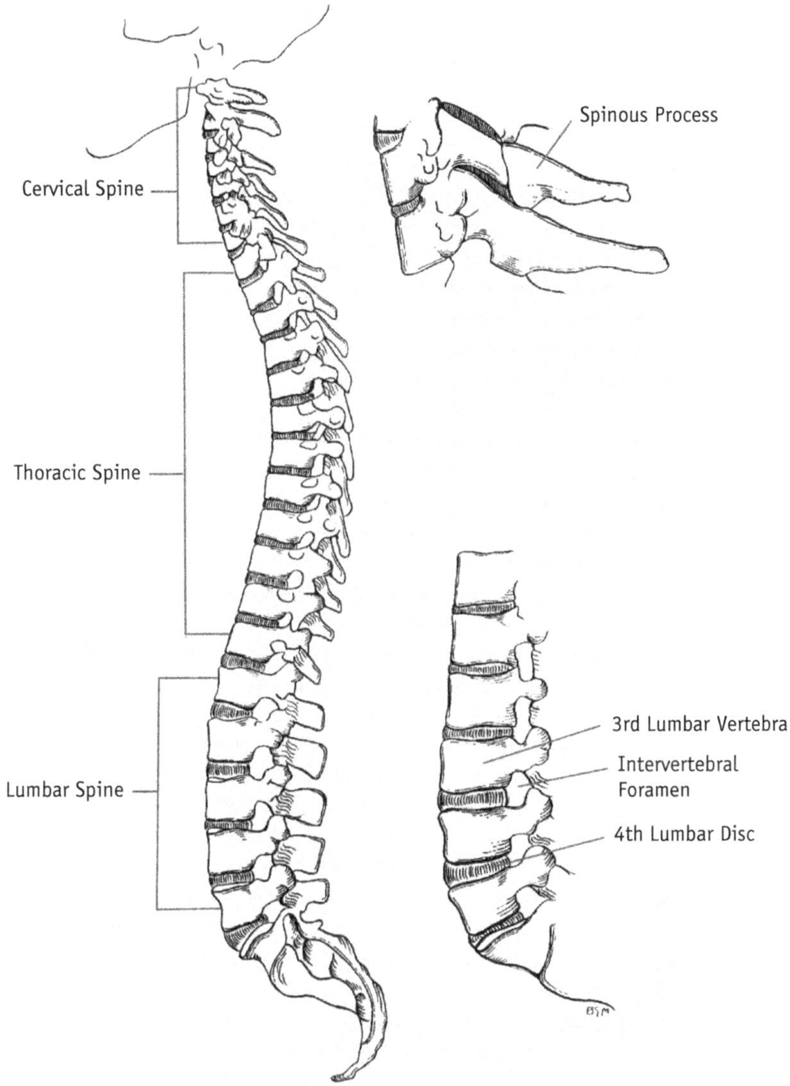

FIGURE 20.1. Spinal column.
Credits: © Barry Mordin, illustrator, Barbara Eisen, labeling and retouching.

sclerosis with arterial and nerve root occlusion can result in patients with osteoarthritis as well as in patients with degenerative disc disease including protruding and herniated discs.

Chiropractors believe that such mechanical disorders can effectively compromise spinal nerve roots to the extent of their corresponding bifurcating distal nerve structures and to the dermatomes, an area that can be best described as the

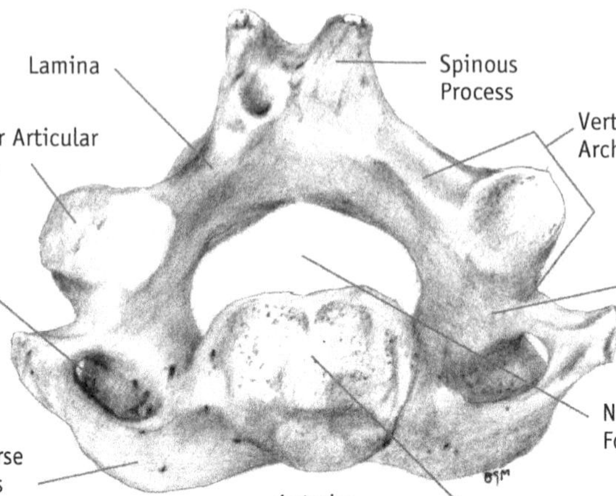

FIGURE 20.2. Vertebral diagram.
Credits: © Barry Mordin, illustrator, Barbara Eisen, labeling and retouching.

land mass of the body that the specific spinal nerve root innervates. Each spinal nerve root that exits through the intervertebral foramen extends to the various corresponding parts of the body. Proof of this can be determined when there is a frank interruption of the proper nerve root innervation such as that seen in injury to the spinal cord with resulting paralysis. Organs may continue to function due to an intact autonomic nervous system, but activity of muscles and other structures and functions will be compromised or cease altogether.

While modern-day chiropractic received its name and had its humble origins in Davenport, Iowa, in the 19th century, through the efforts of the humble Daniel D. Palmer, manipulation of the spine, like brain surgery, dates back to the time of the pharaohs in ancient Egypt. It was only "rediscovered" in 1895.

How Does Chiropractic Apply to Sexual Disorders?

How does this knowledge of nerve roots and of clinical chiropractic apply to sexual disorders such as sexual dysfunction (impotence, premature ejaculation, anorgasmia) amenorrhea, dysmenorrhea, and other conditions that affect our sexuality? The answer lies in the spinal nerve anatomy. The spinal nerves that innervate our sex organs reside in the lumbar spine and pelvis, and this, in part, includes our adrenals, where we manufacture our sex hormones. The adrenal glands, which sit atop the kidneys, are most known for the adrenal hormones that bring about the "fight or flight" instinctual response. They are, in some ways, largely responsible for the continuation of all species of the animal kingdom, as when presented with danger, adrenocortical hormones kick in and we make haste to get away from danger or to fight with superhuman strength if needed.

The adrenals also are an important part of the endocrine system and autonomic nervous system. The adrenal gland is innervated by the celiac plexus and the renal plexus of the sympathetic branch of the autonomic nervous system (Figure 20.3).

The renal plexus innervating the kidneys is formed by nerve filaments from the celiac plexus, aorticorenal ganglia, the lower thoracic splanchnic nerves, and the first lumbar splanchnic nerve. As mentioned earlier, the sex hormones are manufactured in the adrenal cortex; we would not be having much sex without

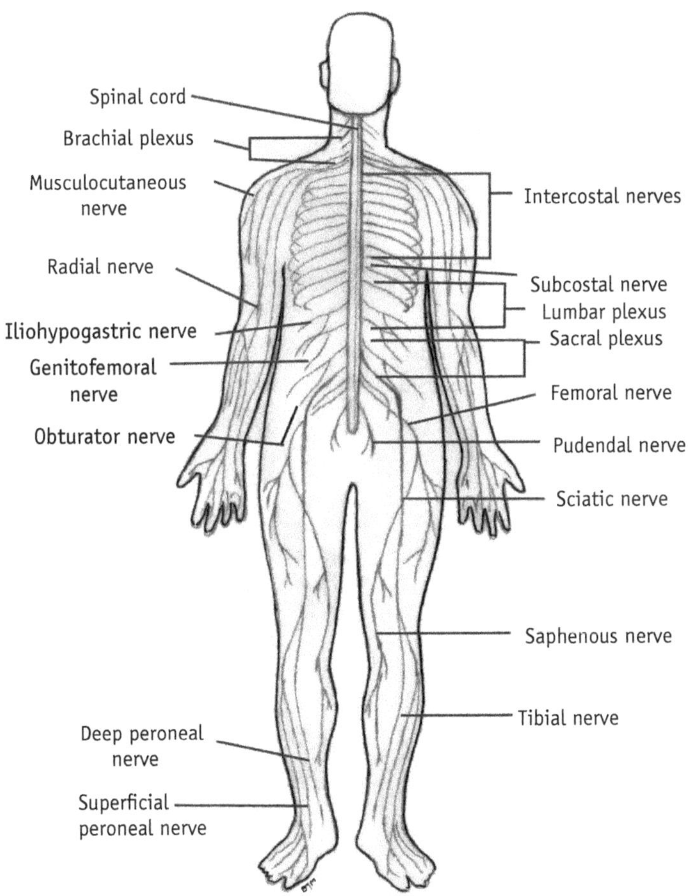

FIGURE 20.3. The nervous system.
Credits: © Barry Mordin, illustrator, Barbara Eisen, labeling and retouching.

them. Sexual response is not just all about estrogen and testosterone. The activity of the adrenals, commonly referred to as our stress organ, is much affected by the chronic work and life stressors of modern day. Weakened adrenals can affect us in a variety of ways, so it is always a good idea to get enough sleep and eat foods rich in vitamin B, which help to support a healthy adrenal gland.

Another important consideration with respect to the functioning of the adrenals is that cholesterol is essential for all hormone production, and that includes the sex hormones. For this reason, it is essential that we ingest a proper amount of good fats so that the necessary hormones can be manufactured in our bodies (Figure 20.4).

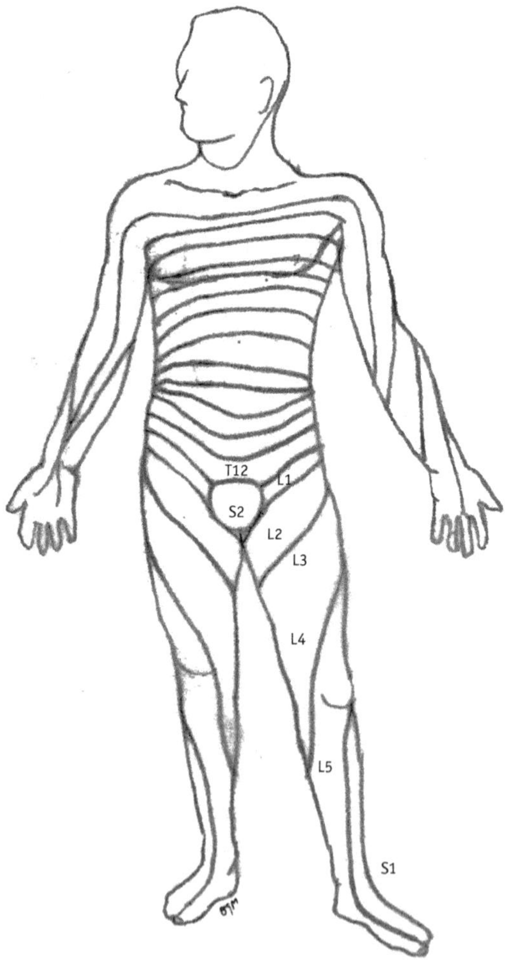

FIGURE 20.4. Dermatome chart.
Credits: © Barry Mordin, illustrator, Barbara Eisen, labeling and retouching.

The Genitofemoral nerve is also important to functions related to sexuality. The Genitofemoral nerve has its roots in the L1 and L2 spinal nerves and innervates the skin of the anterior scrotum in males and the skin over the mons pubis (pudendum) and labia majora in females. It passes directly through the psoas major muscle. The sciatic nerve, distinctive as the largest nerve in the human body, has its roots from the L4 through S3 spinal nerve roots and bifurcates along its S3 and S4 filaments to form the deep pelvic nerves. The pudendal nerve, which is part of the sacral plexus, has its roots in the S2, S3, and S4 spinal nerve roots and is a primary nerve of the perineum and pelvic floor, innervating and carrying sensation to the external genitalia of both sexes. The pudendal nerve leaves the pelvis through the greater sciatic foramen bilaterally and enters back into the pelvis through the lesser sciatic foramina along the sacroiliac joints bilaterally. The pudendal nerve (the word "pudenda" literally means "shame" in its Latin etymology) also innervates the urethra, anus, and perineum, all essential to healthy sexual function. Its motor nerve is responsible for function of the skeletal muscles in the perineum, external urethral sphincter, and levator ani. Its sensory function is to innervate the penis and clitoris, and as such it is the nerve primarily responsible for sexual pleasure and orgasm. If this nerve becomes damaged, irritated, or entrapped, no matter how mild or seemingly insignificant, sex can lose most of its pleasure. It is also possible that nerve root compromise of the sciatic nerve could create an environment where reduced function of a distal nerve of the plexus could result in compromised sexual function.

These are very important considerations because the sacrum as described earlier acts as the keystone of the skeletal spinal anatomy in that it articulates with both sides of the pelvis as well as being the foundation of the last lumbar vertebra, the weight-bearing fifth lumbar. It should be understood that, biomechanically, whatever the sacrum does the rest of the spinal column and pelvis follow. A mechanically displaced pelvis and sacrum can result in a variety of issues, one of them being nerve root irritation of the deep perineal nerve and the pudendal nerve. As we have become a sedentary society, we are now spending more time than ever before sitting on our pudendal nerves, which may have consequences for sexual function (Figure 20.5).

It now may be more evident why and how chiropractic can indirectly impact and improve sex life. Lumbosacral complaints are the leading driver of patients seeking a chiropractor. Sacral and iliac misalignments can produce spasm and torque pulling on all ligaments and musculature of the pelvic floor. It is also possible that chronic spasm of the psoas muscle can contribute to pelvic floor spasm and weakness when that muscle exerts constant pressure on spinal and nerve structures in the lumbar spine, pelvis, and hips. The Poupart's ligament, also known as the inguinal ligament, lines the pelvis from the symphysis pubis (pubic bone) to the anterior iliac spine. The psoas muscle runs deep to this ligament and can have an impact on it.

Pudendal nerve involvement including rare cases of pudendal nerve entrapment can present as deep pelvic pain, mild or significant pain while sitting,

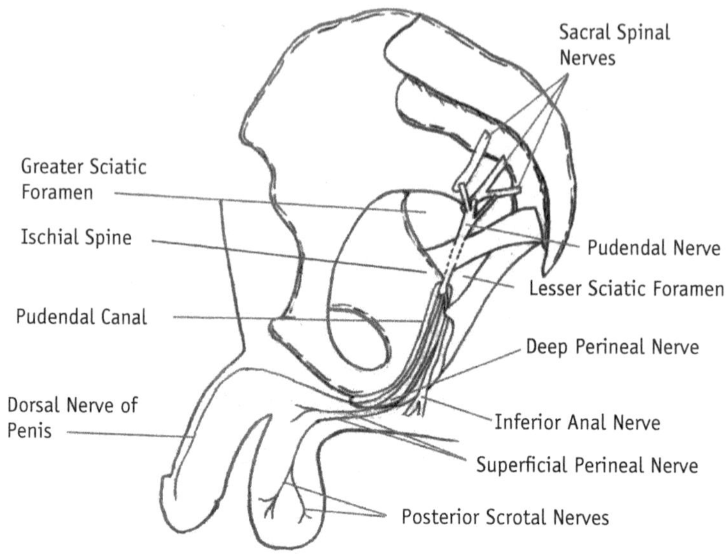

FIGURE 20.5. Male pelvis and pudendal nerves.
Credits: © Barry Mordin, illustrator, Barbara Eisen, labeling and retouching.

intermittent numbness in the anal or perineal regions, and the inability to attain orgasm.

Pudendal neuralgia can present in both men and women, though women are most affected, making up 65% of cases. Other organic causes of pudendal neuritis can include Lyme disease, sacral radiculopathies, and hypertrophy of the piriformis muscle seen in athletes when flexion and abduction of the thigh are repetitive motions. The piriformis exits through the greater sciatic notch where the pudendal nerve exits inferior to the muscle. A hypertrophic piriformis can cause irritation to the pudendal nerve. It is also not uncommon to see these cases in cyclists (Goodson, 1981; Leibovitch, 2005), rowing enthusiasts, and people doing increased numbers of sit-ups. It is notable from the chiropractic perspective that both rowing and sit-ups can affect the lumbar and pelvic nerve structures when improper postural position is maintained during these exercises. Like cycling, long-term effects of repetitive pressure or motion to affiliated nerve and muscular structures can cause irritation and potential involvement of deep pelvic structures, especially the pudendal nerves (Figure 20.6).

Chiropractic Training and Schools of Technique

Chiropractic school, not unlike medical school, includes an advanced focus and training in anatomy, physiology, and diagnosis of all organ systems. Structure is

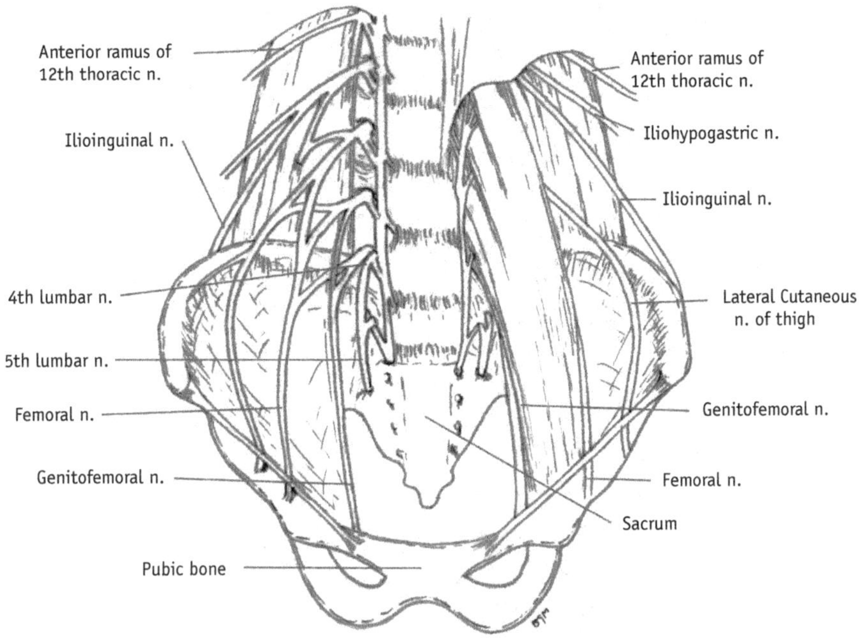

FIGURE 20.6. Female peritoneal structures.
Credits: © Barry Mordin, illustrator, Barbara Eisen, labeling and retouching.

function, as is drilled into chiropractic students. This author paid a lot of attention to the pelvis and low back and how to effectively adjust them; as chiropractors engage in physical manipulation, they must learn to work effectively with their own bodies, whether petite or large, to achieve optimum and safe outcomes. The author intended to become a sacro-occipital practitioner, given a petite physique and having been a patient of both the standard diversified style of chiropractic and the sacro-occipital technique (SOT) of chiropractic. However, when introduced to the hydraulic and electric air-operated table—quite revolutionary for chiropractic equipment during its inception in the late 1940s—this allowed for a gentler and more specific manipulation of a wider range of spinal areas. The table and technique associated with its use were developed by Dr. Clay Thompson, a wonderful and humble practitioner who taught many, including this author. Since its introduction, the Thompson table is the only chiropractic table exhibited at the Smithsonian Institute and Museum in Washington, D.C. It may also have been the inspiration for many other advanced technology chiropractic tables and techniques such as the Cox Technique with a table that focuses on disc ruptures and herniations.

SOT is a blended technique of eastern meridians from traditional Chinese medicine, developed by an osteopath and eventual chiropractor Dr. Major Bertrand

DeJarnette. He spent 25 years refining this technique, doing research on medical students in his native Nebraska, and continuing his research for some 60 years until his death in 1992.

Armed with the advanced hydraulic dropping table and the eastern meridian therapy of SOT called chiropractic manipulative reflex technique, the author started treating patients in clinic internship and discovered a great ability to regulate women's menses. Before long, female patients were seeking out this noninvasive means to regulate their menstrual cycles.

Case Study

FEMALE: RESTORATION OF NORMAL MENSES

In one such early case, a 26-year-old female with a long-standing history of amenorrhea presented in clinic suffering with diffuse low back pain and having her periods two to three times yearly. She also presented with a sacroiliac mechanical disorder and a thoracolumbar scoliosis. The convexities remarkable with scoliosis often create a rotational presentation of spinal column and, in her case, L3 and L4 were posteriorly rotated to the left. As the lumbar spinal nerves primarily innervate the reproductive organs, the uterus and ovaries, that was the focus of treatment. However, being more innovative and comprehensive required not ignoring any of the lumbar vertebrae or the sacrum that they rest upon. A thorough examination of the sacral notch trigger points was done with release of active trigger points. After four to five spinal manipulations, the patient got her period and then continued to do so monthly, much to her surprise and annoyance. She was reminded that a normally functioning reproductive system was her objective as well as an important part of her healthy sexuality lifestyle. Another such patient presented in the clinic one day, stating that she was having irregular periods. After two weeks of treatment and inability to start having her monthly menstrual cycle, she discovered that she was pregnant!

Use of the Thompson Terminal Points Drop Table became a clinical focus, along with a continuing respect for SOT, rich as it is with acupressure and meridian soft tissue work not unlike that used in acupuncture and Shiatsu massage. A skilled chiropractor should look not only for skeletal structural imbalances but also soft tissue involvement that may be primary or secondary to the mechanical disorders found. Patients often present with lumbopelvic imbalances associated with psoas muscle and hamstring muscle involvement. Simply manipulating the spine without restoring proper muscular tone and function will only allow the disorder to continue without abatement.

In another case, a 15-year-old female presented with no menses though she was otherwise well developed and well nourished. She was already being treated in the office for familial scoliosis. Her mother agreed to allow chiropractic treatment. Many therapeutic elements were used: L3–L5 manipulation, sacrum, and all of the associated acupressure trigger points. Three days after her second treatment, her menses began and has continued at regular monthly intervals without interruption.

These patient histories may be anecdotal to one practice but they do provide empiric evidence that chiropractic treatment can be effective when manipulation is utilized in conjunction with chiropractic manipulative reflex technique, specific to SOT.

Published papers linking pelvic pain and organic dysfunction in patients with low back pain appeared in the *Journal of Manipulative and Physiological Therapeutics* in July 1987 (Bower, 1987), and June 1996 in the *Journal of the Neuromusculoskeletal System* (Browning, 1996); these are also anecdotal in nature as no large long-term studies were presented in support of these cases.

Some common causes of sexual dysfunction are pelvic trauma, stress and anxiety, the side effects of medications such as diuretics and antidepressants, and the use of birth control pills in females. As all of these causes have been discussed in other chapters, so they are not detailed here.

Chiropractic is a complementary and alternative treatment modality that normally includes advocating for lifestyle changes to benefit the patient holistically, dietary and nutritional advice, and exercise, and the use of botanical and dietary supplements may be included.

MALE: WEIGHT LOSS AND RESTORING LOW LIBIDO

One such case involved the presentation of a 47-year-old male who wished to lose 80 to 100 pounds and improve his overall health. His chief complaints included sleep apnea and decreased libido and sexual dysfunction. The patient had been on a variety of psychotropic drugs since his early 20s. His related psychiatric diagnosis prescribed strict adherence to his medication. The patient was examined, and weak abdominals and a sacroiliac mechanical disorder with hypermobility of the sacroiliac joints were diagnosed. After reviewing his nutritional profile, all blood work, and laboratory data, it was determined that he likely suffered with a Candida albicans overgrowth. He was placed on an antifungal diet along with recommendations for whole foods and herbal supplementation to restore proper gut and digestive function.

After six months of the antifungal diet and whole foods and supplements, the patient had lost 80 pounds and was able to restore proper digestive health. He was then placed on an optimum diet eliminating sugars and wheat and including organic proteins and good fats, supplemented with daily probiotics and an herbal protocol including tribulus, B12 vitamins, and other herbs. This regimen restored his libido and elevated his associated low testosterone count from years of taking Haldol, which had increased his prolactin levels to 43. After changing to Abilify, his prolactin levels were returned to normal range, though he was recently put on Vraylar due to side effects with Abilify. His treatment continues for complete optimal effect regarding his sexual dysfunction and weight management, without being placed on a starvation diet and with improvement in his prolactin (lower) and testosterone (higher) levels. As a result of treatment, the patient has been able to reduce his medications to the lowest dosage with good result.

Research in Chiropractic

Research to try to establish a quality evidence base for any complementary and alternative medicine discipline is important to the field, to healthcare, and to patients. The National Center for Complementary and Integrative Health, a branch of the National NIH, funds research on chiropractic care. As of now, this has focused primarily on research for spinal manipulation of back pain, neck pain, and headache, as well as temporomandibular joint disorders. Chiropractic for sexual and reproductive disorders has not been a funding priority.

Randomized clinical trials to investigate chiropractic efficacy have been performed and reported on a greater scale in Great Britain with one such study performed at Northwick Park Hospital by the MRC Epidemiology and Medical Care Unit in Harrow, Middlesex. The study used the Oswetry Pain Scale for low back pain patients who had never sought medical or chiropractic care in the past. The results as reported by the *British Medical Journal* in June of 1990 revealed that chiropractic treatment was more effective than hospital outpatient management mainly for patients with chronic or severe low back pain (Meade, 1990). Other studies have been published regarding the effectiveness of manual treatment for the management of a variety of musculoskeletal and nonmusculoskeletal conditions based on widely accepted evidence-based guidelines in the United Kingdom and the United States (Bronfort, 2010).

When performed properly by well-trained practitioners, chiropractic is safe. In 2007, a study of treatment outcomes for 19,722 chiropractic patients in the United Kingdom concluded that the risks of a serious adverse event occurring as a result of chiropractic treatment were low to very low (Thiel, 2007).

The National Health Interview Survey, which has monitored the health of the United States since 1957, found that chiropractic was the most commonly used complementary health approach for back pain, based on data from 2002 (Kanodia, 2010). Among those who had used chiropractic for back pain, 66% perceived "great benefit" from their treatment. Further, according to the NIH, low back pain is one of the most common health complaints with 8 out of 10 people affected at some point during their lifetimes. Many people suffer with chronic low back pain, which can last for months or years. The annual costs for low back pain in the United States—including lost wages and reduced productivity—are more than $100 billion (NIH, 2012, 2016a, b).

How many of these chronic low back pain patients also suffer with sexual dysfunction? The answer is difficult to ascertain as no studies have been published. The relevance of studies in support of chiropractic for lower back pain to addressing sexual difficulties is self-evident; sexual life will be very compromised and inhibited when low back pain restricts normal and also intimate movement and causes acute or persistent discomfort.

Conclusion

In conclusion, chiropractic manipulation by a skilled and caring practitioner should be considered for referral by conventional physicians and by patients with types of sexual dysfunction, other related disorders, and reproductive complaints. The proposed benefits of expert chiropractic manipulation also should be investigated within the complementary and alternative medicine research community as a safe, minimally invasive, and holistic form of treatment for those suffering with sexual and related difficulties.

REFERENCES

Bower LD (1987). Treatment of lumbar motor unit instability with a custom orthosis: clinical procedure. *J Manipulative Physiol Ther.* Dec;10(6):310–315.

Bronfort G, Haas M, Evans R, et al. (2010). Effectiveness of manual therapies: the UK Evidence Report. *Chirop Osteopat.* 18(3):1–33.

Browning JE (1996). The mechanically induced pelvic pain and organic dysfunction syndrome: an often overlooked cause of bladder, bowel, gynecological, and sexual dysfunction. *JNMS.* June;4(2):52–66.

Goodson JD (1981). Pudendal neuritis from biking. *N Engl J Med.* Feb 5;304(6):365.

Kanodia AK, Legedza AT, Davis RB, et al. (2010). Perceived benefit of complementary and alternative medicine (CAM) for back pain: a national survey. *J Am Board Fam Med.* 23(3):354–362.

Leibovitch I, Mor Y (2005). The vicious cycling: bicycling related urogenital disorders. *Eur Urol.* Mar;47(3):277–286; discussion 286–287.

Meade, TW, Dyer S, Browne W, Townsend J, Frank AO (1990). Low back pain of mechanical origin: randomised comparison of chiropractic and hospital outpatient treatment. *BMJ.* June 2;300(6737):1431–1437.

National Institutes of Health, National Center for Complementary and Integrative Health. (2012). Chiropractic, spinal manipulation, and osteopathic manipulation. https://nccih.nih.gov/health/chiropractic.

National Institutes of Health, National Center for Complementary and Integrative Health. (2016a). Chiropractic: in depth https://nccih.nih.gov/health/chiropractic/introduction.htm.

National Institutes of Health, National Center for Complementary and Integrative Health (2016b). Spinal manipulation for low-back pain. https://nccih.nih.gov/health/pain/spinemanipulation.htm.

Thiel HW, Bolton JE, Docherty S, Portlock JC (2007). Safety of chiropractic manipulation of the cervical spine: a prospective national survey. *Spine*. Oct 1;32(21):2375–2378; discussion 2379.

OTHER RESOURCES

Agency for Health Care Policy and Research (1997). Publication no. 98-N002. Rockdale, MD: Agency for Health Care Policy and Research.

Gray H (1918). Gray's Anatomy Philadelphia, PA: Lea & Febiger; 20th Edition, revised and re-edited by Warren H. Lewis, New York: Bartleby.com, 2000.

McMennell J (1964). *Joint pain*. New York: Little Brown.

Stedman TL (1990). *Stedman's medical dictionary*, 25th ed. New York: Macmillan.

21

Aromatherapy in the Treatment of Sexual Dysfunction and Relationship Problems

MINDY GREEN

Introduction

Aromatherapy, the use of pure plant essential oils to enhance health and psychological well-being, is a pleasant, simple, and high-compliance modality that can be added to many other therapies to beneficially synergize therapeutic outcomes. These fragrant volatile (evaporative) oils can be incorporated into a variety of treatment options at any time of the day. The extremely small molecular size and lipid solubility of essential oils allow them to be easily absorbed via the skin, lungs, and brain. Applied externally or inhaled, the proper use of diluted essential oils can positively affect the psyche and emotions, helping to unravel the obstacles to a happy and healthy sex life, whether the issue is poor body image, past abuse, stress, or vaginal dryness. Scientific research has not thoroughly explored the extent to which essential oils can help in the realm of sexual dysfunction, but a 2007 article in *Psychology Today* showed that scent can lift the mood, provide relaxation, and connect us emotionally (PT staff, 2007).

There are, however, hundreds of published articles on mood enhancement, relaxation, and stress reduction with essential oils. Ushering in a sense of peace and tranquility is often the basis for feeling safe and emotionally supported, a great precursor to deepening a relationship and enhancing a successful sexual encounter. A 2015 study showed marked effects in stress reduction with simple inhalation, outperforming other relaxation techniques such as progressive muscle relaxation and other modalities (Schneider, 2015).

Herz (2016), in a review of the literature, concludes that odors evoking positive autobiographical memories potentially can increase positive emotions, decrease negative mood states, disrupt cravings, and reduce physiological indices of stress, including systemic markers of inflammation. The associated context of perceived smells is highly influenced by the memories these odors evoke, often imprinted in childhood or by an experience that evokes emotions (Herz, 2016). For these reasons, creating and then re-experiencing pleasant or compelling olfactory memories and their positive associations, for example in the context of interpersonal connection, can stimulate the desire for and enhance the positive experience of intimacy.

Beautiful and enhancing natural fragrances of botanicals are a deep pleasure, far superior, for example, to the cloying and synthetic scents we are bombarded with everywhere. They are a pleasure that beneficially stimulates many aspects of brain, body, and emotions.

Sense of Smell/Olfaction Overview

Olfaction is the process of smelling. Odor and emotions are processed in the same area of the brain—the limbic system, which is often referred to as the lizard brain or the primitive brain. This most ancient of senses is therefore anatomically connected to the basic emotional pathways that also are involved in the fight-or-flight response and other fundamental emotional responses and states.

The limbic system is defined as a complex system of structures and networks in the brain, involving several anatomical areas near the edge of the cortex governing feelings, motivation, and mood. The limbic system controls basic emotions (fear, pleasure, anger) and drives (hunger, sex, dominance, and care of offspring). It is comprised primarily of the hypothalamus, hippocampus, and amygdala.

The hypothalamus receives messages from the olfactory nerves, which similarly provide information that helps regulate appetite and sexuality. The limbic system influences the endocrine system and the autonomic nervous system and is highly interconnected with the brain's emotion, behavior, and pleasure centers, all of which play a role in sexual function. The hypothalamus also controls the pituitary (our master endocrine gland) both neutrally and chemically, pumping hormones into the bloodstream, based on a variety of mental, emotional, chemical, and neurological messages received. Scent can affect us emotionally, which can stimulate the production of neurotransmitters.

Many essential oils have the historic reputation, backed by anecdotal experience, of balancing hormones and alleviating various female complaints such as menopausal hot flashes and mood-related premenstrual syndrome (PMS). Some recent papers have contributed new experimental models and data to shed light on traditional use, and on possible mechanisms of action, although high-quality research still is insufficient and at times results are contradictory (Kiecolt-Glaser, 2008a, 2008b; Mahboubi, 2016; Matsumoto, 2013, 2016; Mizuno, 2015).

A randomized, double-blind, clinical trial detailed in the *Journal of Obstetrics and Gynaecology Research* (Ou, 2012) demonstrated that an essential oil cream dramatically reduced the pain associated with menstrual cramping (dysmenorrhea), via an external massage to the abdomen. It is well known that many plants (herbs and foods) contain phytosterols, dilute chemicals that occupy hormone receptor sites in the body. Some essential oil constituents also exert direct influence on brain chemistry and may affect the pituitary gland to release various chemical messengers into the body, providing a physiological effect on our mood and possibly influencing our hormones.

It is interesting to note that we need not be consciously aware of an odor to have it affect us emotionally. This is well known by many ad agencies and marketers. For example, low levels of fragrance are often piped into shopping malls, because marketing studies suggest they may unconsciously influence our purchasing decisions (Mitchell, 1995).

The olfactory nerve is the first cranial nerve and olfaction is the only sense to be fully developed at birth. In fact, babies can smell in-utero; the extremely small size of odor molecules allows them to cross the placenta. In humans as well as in animals, babies bond with their mothers through the sense of smell, and newborns can locate their mothers' nipples by smell.

Compounds from lactating women "have the potential to function as pheromones," according to a study by Jacob (2004). Exposure to breastfeeding compounds affects hormonal levels, which influence the menstrual cycle of other women who are in close proximity. Apocrine glands, which are believed to secrete pheromones, are also present on women's nipples, another possible pathway of mother-infant communication (Jacob, 2004).

These tiny odor molecules are also fat soluble, allowing them to pass through the lipid blood/brain barrier (Azimova, 2012.), encouraging relaxation and release of stress, or uplift and stimulation, depending on what chemical constituents are being inhaled. In fact, essential oils are such good penetrating agents that they are often used to enhance trans drug delivery (Herman, 2015).

Our ability to detect individual odors can be gene specific, explaining why different individuals may have differing perceptions of particular odors (Logan, 2014).

Each person has an odor print as unique as their fingerprint. It is influenced by diet, gender, heredity, health, medication, occupation, emotional state, and mood. Eccrine, apocrine, and sebaceous glands, as well as hair follicles and other sources of fatty secretions of the body, contribute to our own personal scent. Odor is a communication system; a statement about who we are. Diane Ackerman notes in her book, *A Natural History of the Senses*, "Because females have often been responsible for initiating mating, smell has been their weapon, lure and clue." Perfumes unconsciously reveal what they consciously aim to hide: sex appeal. Perfume was long used to mask and modulate human body odors, while today perfumes typically seem intended to be directly alluring, some more boldly than others. Ackerman also notes that most perfumes contain some flowers, harvested at the peak of their own sexual readiness (Ackerman, 1990, 44).

Smell, Sex, and the Role of Human Pheromones

History bears out our tenacious interest in which scents attract people to each other. The rise and fall of empires were often influenced by scent and seduction. In 800 BC the Queen of Sheba brought fragrance to seduce Solomon; and when Cleopatra visited Marc Antony the sails on her barge were scented to announce her arrival; scent historically has been a form of erotic foreplay. Today science has begun investigating not only fragrance, but specifically pheromones, and how they may affect our attraction to the opposite sex and our own sexual chemistry.

Entomologists discovered pheromones in the animal kingdom in 1959 and have studied them ever since. Animals, insects, and plants communicate through scent. Cats have scent glands in their face and rub their cheeks on people to mark them with their scent; they often wash themselves afterwards to taste human scent. Bees follow a scent trail to pollinate the most fertile flowers, salmon smell their way to their original spawning grounds, ants follow a fragrant trail to food and mates, and messages of danger are communicated through scent by numerous species of animals, insects, and plants. It is even suggested that sperm smell their way to the egg. There has long been a debate among scientists about whether or not humans create or are affected by pheromones and whether they influence our sexual behavior, puberty, menstruation, menopause, and more (Doty, 2014; Savic, 2014). Scientists have known about animal pheromones for a long time but have only recently acknowledged human pheromones. In 1986 *The Washington Post* reported on the first acknowledgment by scientists that human pheromones exist, based on a publication of that same year (Johns, 1986).

Many scientists assert that pheromones do not necessarily act as sex attractants, but acknowledge their link between sexual behavior and the health of a woman's reproductive system (whose menstrual cycles provide an easy research target, whereas men's reproductive physiology does not afford a comparable research model) (Grammer, 2005).

Some research suggests that the use of contraceptive pills could disrupt our mating preferences by having an effect on olfaction-driven behavioral choices. Genes in the major histocompatibility complex (MHC) govern our personal odor (Santos, 2005; Wedekind, 2000). Studies have shown that females often prefer the odor of MHC-dissimilar males, perhaps an evolutionary adaptation to increase genetic mixing. Women using oral hormonal contraceptives have been reported to have the opposite preference, choosing MHC-similar males, a result that was borne out in a study by Roberts (2008), suggesting that human modification of hormones and personal odor may directly affect our unconscious reproductive choices.

There is little information on how scent and pheromones affect men. Though there are fewer research studies on men, a 2004 publication acknowledged that chemosensory cues stimulate male sexual arousal and behavior, though it found no support for a link to pheromones (Keverne, 2004).

A study published in *Psychological Science* (Miller, 2010) reported that men's testosterone is influenced by olfactory cues in women during ovulation. Gildersleeve (2012), publishing in *Hormones and Behavior*, stated that their study builds on a growing body of evidence indicating that men are attracted to cues of impending ovulation in women, raising the intriguing question of whether women's cycling hormones influence men's attraction and sexual approach behavior. Hirsch (2014) researched what scents males found arousing, as measured by increased penile blood flow. He found that lavender and pumpkin pie odor were the most effective.

Pheromones are defined as "chemical compounds produced by one individual that affect the sexual physiology of another." Some researchers believe they are nonaromatic. The word *pheromone* is from the Greek "phero," to carry, and "hormon," to excite. Pheromones are odoriferous and nonodoriferous substances chemically similar to hormones (which are secreted by endocrine glands) but are manufactured by the apocrine glands (sweat glands). Apocrine glands occur in the armpits, face, chest (nipples), anal, and genital regions and are activated at puberty. Before puberty, a child's perspiration has no odor. This makes perfect sense; there is no need to "signal readiness" before we are fertile. It is interesting that girls who are separated from boys in adolescence (as in a same-gender boarding school situation) generally go through puberty later than do girls in a coed setting. The highest density of apocrine glands occur in the axillae and the perineum; they release their secretions in response to a wide variety of emotions including anxiety, fear, pain, or sexual arousal (Wilke, 2007).

A 2002 paper by Dalton validated the theory that females have a keener sense of smell than males. This may account for the greater prevalence of odor-related complaints from females, for example concerning sick-building or chemical intolerance syndromes.

A report by Paddock (2014) in *Medical News Today* concurs and credits women with having up to 50% more olfactory neurons (present in females at birth; Oliveira-Pinto, 2014; Paddock, 2014).

Adult male underarm scent has been used to regulate female cycles. In a 1985 study, seven women who had irregular menstrual cycles used an extract of male underarm odor placed on their upper lips three times a week for three months. As a result, more than 50% experienced menstrual regularity. The researchers also noted that preference for mate-related traits varies across different phases of the menstrual cycle, a result that has been demonstrated repeatedly (Havlicek, 2005).

Despite the finding of McClintock's (1971) famous paper, not all scientists are in agreement that being in the company of women may cause other women's menstrual cycles to shift into synchronicity. However, there is evidence that women who live or work together often begin to ovulate at the same time. The odor signal is released in sweat, and the women's cycles can take three to four months to regulate.

It has been reported that the ability to detect odors is lowest at menstruation and peaks at ovulation when olfactory sensitivity increases. This makes biological sense when we consider that peak fertility occurs when women are most aware of sexual attractants and when they most signal fertile readiness. *Scientific American* reported on results of two papers, from *Physiology and Behavior* and *Hormones and Behavior* (March, 2013), suggesting that as fertility peaks, a woman's sense of smell sharpens (McNeil, 2013; Rodriguez, 2013). In addition, use of contraceptives may blunt olfactory acuity. Rodriguez (2013) tested naturally cycling woman and women on hormonal contraception; they sniffed odors of lemon, peppermint, rose, musk, and the male pheromones androstenone and androsterone. Naturally cycling women near ovulation were more sensitive to musk and the pheromones than the women on contraceptives. However, some studies have found conflicting results, and the physiological mechanisms causing these effects remain unclear. For example, McNeil's (2013) research did not find a relationship between levels of specific hormones and olfactory sensitivity.

Sobel (1999) found that an air-borne fragrant pheromone (oestra- 1,3,5(10),16-tetraen-3yl acetate) would activate brain centers even when present at concentrations below a threshold of conscious detection. The researchers tested with functional magnetic resonance imaging and electroencephalogram to prove that exposure to pheromones at undetectable levels nonetheless activates brain centers (Sobel, 1999).

Pheromones are produced in sweat, blood, saliva, and vaginal secretions. Both volatile and nonvolatile compounds (proteins, steroids, cholesterols) can act as pheromones (Karlson, 1959).

After menopause, studies show that women's ability to detect musky odors greatly declines. Animal musk is very close to human testosterone and can be detected in minute quantities. Although there is no *biological* reason for a post-menopausal woman to be attracted to the male scent after she is no longer fertile, it is interesting to note that the ability to detect musk odors is restored when hormone replacement therapy is given. It was unclear whether the musk used in this study was synthetic or natural (derived from the testes of the musk deer). Also, fertile women who were exposed to a musk odor had shorter menstrual cycles, ovulated more often, and conceived more easily. Finally, it has been found that people with olfactory disorders often lose interest in sex (Keville, 2009, 19).

Experiments conducted in the early 1970s and replicated in 1994 revealed that children do not develop sensitivity to certain odors until they reach puberty. In these studies, nine-year-olds showed a pronounced insensitivity to two musk odors, although their ability to detect other odors was the same as that of postpubescents and adults (Fox, n.d.).

As previously noted, research on the MHC suggests that we sniff out a mate with a genetic complex different from our own to insure that our offsprings' MHC genes are more diverse and therefore will promote stronger immunity in them (Boehm, 2006; Roberts, 2008). Although this and other human chemical communication research is considered controversial, there exist numerous studies indicating that human odors play a role in sexual behavior.

The Social Issues Research Center noted in its recent publication on smell that "Schizophrenics, depressives, migraine sufferers and very-low-weight anorexics often experience olfactory deficits or dysfunctions. One group of researchers claims that certain psychiatric disorders are so closely linked to specific olfactory deficits that smell-tests should be part of diagnostic procedures. Zinc supplements have been shown to be successful in treating some smell and taste disorders" (Fox, n.d.). Impaired olfaction is common in schizophrenics (Turetsky, 2009) and may also be an early sign of Alzheimer's disease (Wilson, 2010).

Happiness can be conveyed by scent. *Psychological Science* (de Groot, 2015) reported on findings suggesting that not only negative affect but also a positive state (happiness) can be transferred by human odors. The sweat of a happy person can induce happiness in one who inhales that odor. Our sweat odor is caused by our diverse secretions—greatly influenced by our age, gender, physiological state, activity level, diet—and how microbes act on them. Transfer of mood via odor is broadly analogous to another recent finding that fecal matter from depressed humans can induce rodent depressive behavior in mice fed this particular microbiome contents (i.e., that bacteria and their associated products can transfer emotions from one animal source to another; Kelly, 2016). We are entering an era of discovery of the subtle influences animals and plants can have on each other, and on human and animal emotions and behavior, via transfer of scents and other molecules and microscopic vectors.

Cultural Odor Differences, from Armpits to Love Apples

> "The purest union that can exist between a man and a woman is that created by the sense of smell and sanctioned by the brain's normal assimilation of the animate molecules emitted by the secretions produced by two bodies in contact and sympathy, and in their subsequent evaporation."
>
> —Auguste Galopin, French writer

There is no doubt that the culture we live in influences our perception of what we consider to be acceptable, normal, or pleasant smelling. Climate, environment, global placement, local plants, regional customs, ceremonies, and foods all contribute to the smells we learn to associate from an early age with our culture and family.

Our olfactory practices and preferences are often culturally determined. There are many examples of odors deemed too strong or offensive in one culture being accepted and even preferred in another, such as certain food odors, and of course body odors. Members of a tribe in New Guinea say good-bye by putting a hand in each other's armpit and rubbing themselves with it, coating each with the other's scent. In Elizabethan times lovers exchanged "love apples." Peeled apples were kept in a woman's armpit until saturated with sweat, then given to her lover to inhale her fragrance when they were apart. It is reported that Napoleon sent a message to Josephine, "Home in three days, don't wash." The word "kiss" means "smell" in some languages. A kiss is a prolonged smelling of a loved one—"a desire to linger where one's personal scent glows," according to Diane Ackerman's (1990) book *The Natural History of the Senses*. Saliva also carries one's personal scent and gives off chemical clues as to our health.

In conclusion, human beings—in common with all other animals—possess a complex olfactory signaling system, even if those signals are somewhat attenuated for our species by evolutionarily blunted sensory capacities and the overriding influence of many other stimuli and behaviors. There is abundant evidence that in healthy postpubertal humans, humans remain hard-wired for types of sexual and erotic olfactory stimulation and communication.

Might our fundamental sensory pathways, culturally familiar olfactory habits and preferences, and enjoyable natural products all together offer a means to promote positive mood and experiences and mitigate types of sexual dysfunction by marshaling the power of scent molecules and pheromones?

Defining Essential Oils and Aromatherapy

Essential oils are the volatile fraction of fragrant plants. It is estimated that only about 20% of all plants create fragrance, though nearly all have some medicinal value. In many cases the medicinal properties of an herbal remedy are due to the volatile compounds found in the plant; digestive herbs such as culinary spices and seeds are good examples. These volatile compounds, as the name indicates, are inherently likely to evaporate and degrade; all essential oils should be stored, tightly sealed, away from heat and light. Stored properly, most essential oils will last for several years. Refrigeration will extend their shelf life. Personal use is best served by purchasing small quantities; they often come in 5 ml bottles, which is ideal. Depending on plant chemistry, some essential oils are considered better as they age, but from a therapeutic standpoint, the chemistry of the essential oils is in a constant state of flux and metabolic transition. Two to five years is ideal for using up an essential oil, depending on storage conditions throughout that time.

Aromatherapy can be defined as the skilled use of pure plant essential oils to enhance health and psychological well-being. Essential oils derived from fresh and dried botanicals through the primary extraction of distillation are the tools of clinical aromatherapy. True aromatherapy does not include compounded aromatics, synthetics, or reconstructed fragrances, often mislabeled as "aromatherapy" products, even though one might find these in the drug store aisle.

Developing Intimacy

There is no doubt that a body massage feels good, may help us relax, and may help us develop intimacy, but the addition of essential oils can enrich the emotional experience through the inhalation of pleasant fragrance. Like a song that brings one back to an important romance, the aroma of a specific fragrance that is used during a positive sexual experience can cement the memory of that odor and help one to emotionally revive a positive feeling simply by inhaling that scent. One can enhance and create a haven of serenity, safety, and support with the proper use of pleasant-smelling essential oils derived from plants (Herz, 2008, 2016).

Reducing Stress

Sexual problems such as impotence, infertility, and failure to reach orgasm are often related to psychological problems that stem from stress-related conditions (Graziottin, 2000). There is even some research suggesting that genetic odor variants in the perception of smell may influence one's association with depression and anxiety (Sookoian, 2011). Reducing stress is one of the premier actions of properly chosen essential oils. This may be a factor of the chemistry of a specific essential oil or a pleasant memory association with the oil. We are also influenced by scents that are perceived as desirable and pleasing, whether more universally or more personally, and whether simply pleasant or more specifically erotic. When we enjoy the aroma of a certain fragrance, we naturally breathe more deeply. Deep breathing has been a technique of relaxation for eons; it helps slow heart rate, calm the nervous system, and lower high blood pressure, among a host of other positive responses. Deep breathing is taught in traditional meditation practices, yoga, and relaxation techniques since their development many centuries ago, apart from any recent "mystical" or "new age" practices or associations.

A recent research study employed an essential oil blend of lavender and bergamot applied topically to the abdomen, resulting in downregulation of autonomic arousal, with significant decreases of pulse rate and blood pressure (Hongratanaworakit, 2011). Subjects rated themselves as "more calm" and "more relaxed" than those in the control group, suggesting a decrease of subjective behavioral arousal. The researcher concluded that there was evidence for using this blend in treating depression or anxiety in humans (Hongratanaworakit, 2011).

An aromatic bath or a room diffuser can also deepen relaxation. A scented massage can be a sensual experience and an enjoyable part of foreplay, opening the doors of subtle communication between partners, often breaking down tensions and psychological barriers that may hold us back from the potential of deeply satisfying relationships. Scents have long been known for their ability to create immediacy and dimensionality of memory (think of Proust's madeleine) and to bypass mental chatter to play directly on emotions and aesthetic pleasure.

Aromatherapy and Women's Health

Essential oils have been used in women's healthcare for decades and have been shown to be safe and acceptable and to be a pleasing and simple treatment modality for a wide range of female complaints and physical conditions, including childbirth, menopause, menstrual cramps, anxiety, depression, and palliative care for women dealing with breast cancer and its treatment.

PREGNANCY, LABOR AND DELIVERY, POSTPARTUM

Positive results are seen with aromatherapy tested in laboring women or employed by nurses in labor and delivery settings (Kaviani, 2014; Tillett, 2010), for healing episiotomy (Sheikhan, 2011), and for the postpartum period (Conrad, 2012). For example, Kaviani studied 156 nulliparous women in labor, randomly divided into salvia, jasmine, and control groups (52 in each group). The first two groups had aromatherapy via an incense mask for 15 minutes during the labor period; distilled water inhalation was used for the control group. Pain severity was measured before and 30 and 60 minutes after the incense aromatherapy; also measured were the duration of the first and second stages of labor and the newborn's APGAR score. Kaviani (2014) found that pain severity 30 minutes after the intervention was significantly lower, and duration of the first and second stages of labor was shorter, in the aromatherapy group inhaling salvia (both $p < 0.001$). No significant difference was found among the three groups regarding pain severity 60 minutes after the aromatherapy. Aromatherapy with salvia was found to have short-term beneficial effects on pain relief; in addition, it shortened the labor stages and had no negative impact on the baby's APGAR score. As this is a probable safe intervention for mother and baby, and cost-effective, it could be repeated during labor.

MENOPAUSE

A study population of 90 menopausal women was divided equally into an aromatherapy massage group, a placebo massage group, or a control group. Each participant in the aromatherapy massage group received 30-minute aromatherapy treatment sessions twice a week for four weeks with aromatherapy oil; participants in the placebo massage group received the same treatment with plain oil. No treatment was provided to the last group. The results demonstrated that aromatherapy massage was the most effective for menopausal symptoms, as measured by scores on the Menopause Rating Scale, for vegetative symptoms (Darsareh, 2012). A related paper by the same group, based on the same trial of 90 menopausal Iranian women, looked at outcome measures for psychological symptoms, as obtained through the psychological subscale of the Menopause Rating Scale. Again, they found that aromatherapy massage was more effective at alleviating psychological symptoms of menopause, compared to massage with plain oil and no intervention (Taavoni, 2013).

During menopause, uncomfortable symptoms such as hot flashes can cause poor sleep, and vaginal dryness or other genital changes may cause discomfort in intercourse and a reluctance to be intimate. As aromatherapy has been shown to be able to reduce some of these symptoms, it may indirectly help to reduce intimacy avoidance during menopause.

DYSMENORRHOEA

The *Journal of Obstetrics and Gynaecology* published a clinical trial on 75 women students and found that massage with rose oil reduced the severity of dysmenorrhoea, in comparison with massage alone (Sadeghi, 2015).

A study by Bakhtshirin (2015) showed effectiveness of lavender essential oil for primary dysmenorrhea. Pain was measured in two consecutive menstrual cycles of 80 students, before or 30 minutes after a massage with plain oil or with lavender-scented oil. Both forms of massage significantly reduced rated pain, but the lavender oil massage was significantly more effective.

A randomized, double-blind clinical trial of 48 outpatients with primary dysmenorrhea compared synthetic fragrance and essential oil blend of lavender (*Lavandula officinalis*), clary sage (*Salvia sclarea*), and marjoram (*Origanum majorana*) in a 2:1:1 ratio diluted in unscented cream at 3% concentration. After daily use on the abdomen the duration of pain was significantly reduced from 2.4 to 1.8 days, leading the researchers to conclude that four key analgesic components (linalyl acetate, linalool, eucalyptol, and β-caryophyllene) were significant in the outcome (Ou, 2012).

Another study concurred: essential oil massage was helpful for women with painful menstruation (Apay, 2012).

DEPRESSION, ANXIETY, AND STRESS

A number of studies have been conducted on the use of essential oils to reduce stress and anxiety. Reviewing studies from 2000 to 2008, Yim (2009) examined the clinical effects of using essential oils on depressed patients; the authors recommended that aromatherapy be used for patients with depression and secondary depressive symptoms arising from various chronic medical conditions. Rose essential oil has had anecdotal evidence of relaxation, stress reduction, and relief from depression; a recent study concurs (Hongratanaworakit, 2009).

The results of a study of 12 breast cancer patients receiving a 30-minute aromatherapy massage twice a week for four weeks showed that it significantly reduced anxiety and improved patients' immunologic status, as measured for example by increased numbers of leukocytes and lymphocytes over the study period (Imanishi, 2009). However, the aromatherapy massage did not significantly reduce depression scores. Given such a small sample size, possibly this study was underpowered to find effects for depression; compare this negative result for example to the conclusion of Yim (2009) and to the psychological benefits to 90 menopausal women also studied for four weeks (Taavoni, 2013). Nonetheless, positive results were obtained for both anxiolytic effects and immune upregulation, with an *N* of 12.

Another study also found benefits of aromatherapy and massage in 280 women being treated for breast cancer, reducing symptom severity and increasing quality of life (QOL). Assessment after the four interventions—fragrance, massage, aromatherapy massage, and control groups—showed that the first three study conditions had higher scores than the control group on total QOL and subdomain scores; improvement was greatest among patients in the aromatherapy massage group. Further, psychological and physical symptoms were experienced more intensely in the control group, while severity of these symptoms in the fragrance, massage, and aromatherapy massage groups decreased at weeks 6 and 10, compared with baseline, particularly in the massage with aromatherapy group (Ovayolu, 2014).

In a pilot study, 28 postpartum women at high risk for depression and anxiety were given aromatherapy or a control intervention. An essential oil blend of rose otto (distilled oil) and *Lavandula angustifolia* at 2% dilution was used in all treatments. All participants completed the Edinburgh Postnatal Depression Scale (EPDS) and Generalized Anxiety Disorder Scale (GAD-7) preintervention. Ratings on the EPDS and GAD-7 were then obtained at two weeks and at the end of all treatments at four weeks. Scores on the anxiety and depression scales were reduced for the aromatherapy group only. No adverse events were reported, and the study authors concluded that aromatherapy may be a useful, minimal-risk complementary therapy for this population (Conrad, 2012).

Given possible maternal nursing and therefore the transmission of pharmaceuticals in breast milk, there is a great need to find safe and effective alternatives to mainstream medications to treat depression and anxiety during pregnancy and fetal development and during the postpartum period in nursing mothers.

Among several essential oils tested, clary sage proved the most antidepressant (Seol, 2010) via modulation of the dopaminergic pathway; the authors suggest further research into therapeutic drug development of this essential oil. From the integrative medicine perspective, many might ask: Why create a pill—why not prescribe clary sage essential oil to be used in a traditional, nonconcentrated protocol?

Study of clary sage, historically used to treat postpartum depression, also has some encouraging research results for menopausal women (Lee, 2014). After inhalation of clary sage oil, cortisol levels were significantly decreased while 5-hydroxytryptamine concentration (5-HT, or serotonin) was significantly increased.

A 2013 pilot study using a mixture of essential oils (peppermint, basil, and helichrysum) found that their inhalation may reduce perceived level of mental exhaustion and moderate psychological "burnout" (Varney, 2013).

As noted, a clinical research study of an essential oil blend of lavender and bergamot applied topically to the abdomen of 40 healthy volunteers found that this protocol decreased autonomic arousal (significant decrease in pulse rate, blood pressure). Aromatherapy participants rated themselves as "more calm" and "more relaxed" than those in the control group; at a minimum, subjective perceptions of stress were reduced. It was concluded that there was sufficient evidence for this aromatic blend to treat depression or anxiety in humans (Hongratanaworakit, 2011).

Evidence for the antidepressant, antistress impact of aromatic botanicals also has been found in an animal model, where placebo responding presumably is not a factor (Park, 2015). Psychological stressors may cause reactive affective disorders such as depression and anxiety by altering expression of corticotropin releasing factor (CRF), serotonin (5-HT), and tyrosine hydroxylase (TH) in the brain. Mice challenged with stress were studied for the effects of an inhaled essential oil from *Asarum heterotropoides* (EOAH). *A. heterotropoides* is a Chinese botanical use to treat, pain, fever, and other ailments; the oil can be toxic. (See http://practicalplants.org/wiki/Asarum_heterotropoides#cite_note-PFAFimport-176-3.) Depression-like behaviors were assessed by subjecting the mice to the forced swimming test (FST) and tail suspension test (TST); duration of immobility in response to the FST and TST was measured as a standard mouse experimental analogue for anxiety/depression. Brain expression of CRF, 5-HT, and TH in the mice also was assessed directly in their brain tissue postsacrifice. The selective serotonin reuptake inhibitor fluoxetine was used as the positive control. Results showed that EOAH at higher doses (1.0 and 2.0 g) reduced immobility time in the FST and TST. EOAH at a dose of 1.0 g significantly reduced expected increases in expression of CRF positive neurons in the paraventricular nucleus of the hypothalamus, which contains groups of neurons activated by stressful and/or physiological changes; in the expression of TH positive neurons in the locus coeruleus, a nucleus in the pons (an area of the brainstem) involved in physiological responses to stress and panic; and the expected decreases of the 5-HT positive neurons in the dorsal raphe nucleus, the largest serotonergic nucleus of the brain, with projections to the amygdala, caudate, putamen, and olfactory bulb. These results provide strong support for the ability of this botanical to effectively inhibit formation of the substrate for depression-like behavioral responses, by mitigating the immediate negative brain responses to stress.

This author is not aware of research on possible benefits of aromatherapy as an adjunct treatment for stress, anxiety, or depression. Given the positive results set out here, studying standard antidepressants and anxiolytics plus appropriate aromatherapy could be productive. It is logical to suggest that patients also using low-invasive, safe, low-cost aromatherapy might then be able to do better on a lower dose of standard medications for affective conditions. As discussed in other chapters, safely reducing daily doses of needed medications is an important therapeutic goal in order to reduce sexual side effects.

Essential Oils to Inspire Passion

There are many natural fragrances with a history of use as aphrodisiacs. In nature the fragrance of a fully blossomed flower signals insects to pollinate at the height of a plant's sexual maturity. It is no wonder that the most famous essential oils employed in perfumes to attract a sexual partner comprise those derived from flowers. Distilled essential oils from flowers include rose, ylang ylang, lavender, neroli, chamomile, and others. Jasmine, another well-known and beautifully fragrant flower used in aromatherapy, is made by solvent extraction, creating an "absolute"; it is not made by distillation, as the subtle and delicate scent is destroyed by the high heat of distillation.

This chapter next reviews the uses of 15 essential oils, though many more could be considered useful and are listed for reference. Some of this information may be based on empirical and anecdotal evidence and historic usage. Essential oils used in aromatherapy to balance or promote hormonal activity include clary sage, rose, and geranium. Intense, deep, and earthy notes, especially from woods, roots, and resins, are very helpful for inducing relaxation and reducing stress. These include vetiver, atlas cedarwood, spikenard, patchouli, frankincense, and juniper. Essential oils that help stimulate blood flow and circulation such as black pepper, rosemary, cinnamon, ginger, and nutmeg are often used in aphrodisiac blends. Other oils can help to work on underlying emotional issues such as fear, depression, and performance anxiety. Consider adding accent oils such as cardamom, coriander, vanilla, basil, myrrh, sandalwood, citrus, and mint oils, which for many are exceptionally pleasant and evocative scents.

A Tantric sexual ritual from India uses scent to arouse the woman. She is anointed with several different oils on various areas of her body to lift her spirits so she may manifest as a goddess and be worshiped as the embodiment of the creative force, Shakti. Her mate adorns her hands with jasmine *(Jasminum officinale)* and her neck and cheeks with patchouli *(Pogostemon cablin)*; caresses the breasts with chamomile (*Chamaemelum nobile*); and uses sandalwood *(Santalum album; S spicatum)* to stroke the thighs. (Please refer to the dilutions and applications section before using any oils.) No single fragrance will be loved by everyone. Aroma preference is highly individual, as is choosing a scent that is sexually exciting. Scents also can have strong associative memories, including some that discourage a sexual connotation; for example, most people are not consciously likely to associate sexual attraction with a scent worn by a close family member. If associated with a passionate relationship, one's own unique natural scent, perfume, or essential oil blend will attract a lover and will always be remembered.

Materia Medica

There is very little scientific information on how specific essential oils may affect us sexually, beyond what is offered in historical literature and anecdotal or empirical evidence. The occasional research paper is included when available. The essential oils chosen are all nontoxic but may irritate the skin when not properly diluted. Follow safety guidelines and note when an essential oil is particularly strongly scented, unpleasantly overpowering a blend.

The following are traditional aromatherapy profiles for key essential oils (* = essential oil with high odor intensity; use sparingly).

1. **Cedarwood** (*Cedrus atlantica*). True cedarwood oil from the Himalayas is a beautiful sweet scent of root, earth, and spicy incense. It is a grounding, relaxing blissful odor reminiscent of deep woods. This ancient scent can arouse emotions and calm the body, encouraging sexual readiness.

   Cedrus atlantica essential oil may contain analgesic, anti-inflammatory, and sedative properties according to Martins (2015), who investigated whether it alleviates acute postoperative pain in an animal-model. Their findings suggest it does, via activating the descending pain modulation pathways on the opioidergic, serotonergic, noradrenergic and dopaminergic (dopamine D1 and D2 receptors) pathways (Martins, 2015).
2. **Chamomile** (*Chameamelum nobile*).* Roman chamomile is fruity, playful, and sweet. It has a long-documented history of relaxing effects; most of the research is on the herb, not the essential oil.

3. **Clary sage** (*Salvia sclarea)**. This relative of garden sage is spicy, pungent, and sharp with tea-like undertones. It is one of the few essential oils with a constituent (sclariol) that mimics estrogen. It is especially useful for PMS, cramps, and postnatal depression and is often used to revive sexual desire through regulating hormone levels.

   Researchers tested essential oils thought to have antidepressant effects in a rat model. Of all the essential oils in the 5% dilution blend (roman chamomile, clary sage, rosemary, and lavender), their findings indicate that clary sage demonstrated the strongest antistressor effect and could be developed as a therapeutic agent for patients with depression; the proposed mechanism of action involves modulation of the dopaminergic pathway (Seol, 2010).

   A small pilot study combined clary sage with lavender to investigate possible benefits of its topical use for work related stress in intensive care unit nurses. Decreased perception of stress during three 12-hour shifts was found (Pemberton, 2008).
4. **Frankincense** (*Boswellia carterii*). A famous resin of the Middle East, frankincense has been used as incense for countless centuries and as a dietary supplement for decades for its beneficial effects in many health ailments. The essential oil is best known as a meditation aid and calming agent. Its fixative qualities in a blend help to extend aromas, creating a fragrance that can linger for hours. Many species of Boswellia have been used in folk medicine and religious and cultural ceremony for thousands of years. It is well researched for its effective antiinflammatory activity (Siddiqui, 2011). Inflammation as an underlying, often chronic condition can affect mood and possibly normal sexual activity and function, directly and indirectly.
5. **Geranium** (*Pelargonium graveolens*). Scented geraniums distilled for their beautiful aromas should not be confused with the potted geraniums with showy pink, red, or white flowers. These smaller flowered relatives are quite different, with long-standing historical references to hormone balancing for women and mood-uplifting benefits. The complex odor is a perfume unto itself: spicy, stimulating, slightly citrusy, green, and sharp.

   A study carried out on 100 nulliparous women showed that the inhalation of geranium essential oil can effectively reduce anxiety during labor and can be recommended as a safe and noninvasive anti-anxiety aid during childbirth (Rashidi Fakari, 2015).
6. **Ginger** (*Zingiber officinale*).* Spicy food or stimulant blood mover? Ginger is both. The pungent essential oil derived from the rhizome of this common Asian spice also acts to increase circulation in general and bring warmth locally, potentially helping to heat up the libido and arouse emotions.

Chronic low back pain can be a serious deterrent to a healthy sex life. A randomized controlled trial investigated the use of massage with ginger essential oil for chronic low back pain and disability in older adults. The findings suggest that this therapy could be considered as an additional option by health professionals to provide holistic care to those sufferers (Sritoomma, 2014).

7. **Jasmine** (*Jasminum officinale*).* Often referred to as "king of the flowers," Jasmine absolute can be used with those who are timid or hesitant to create trust and open up emotionally. The eastern version of Cupid, *Kama* is the Hindu god of love; jasmine is used on the tip of his arrows.

   A recent clinical study found Jasmine to be physically stimulating, causing participants to be emotionally more alert with an increase of subjective behavioral arousal. Compared with placebo, jasmine oil caused significant increases of breathing rate, blood oxygen saturation, and systolic and diastolic blood pressure, indicating an increase of autonomic (sympathetic) arousal. These results support the stimulating/activating effect of jasmine oil. They provide evidence for its use in aromatherapy, for the relief of depression and to lift mood. (Hongratanaworakit, 2010).

8. **Lavender** (*Lavandula angustifolia*). The most used essential oil throughout all aromatherapy markets, lavender is the relaxant-extraordinaire. It has a wide range of uses from muscle relief to depression and can uplift mood while being an effective anxiolytic.

   A clinical study from 2012 showed evidence of the relaxing effects of inhaled lavender oil. Subjects showed significant decreases of blood pressure, heart rate, and skin temperature, indicating a decrease of autonomic arousal. Subjects reported being more active, fresher, and more relaxed than subjects inhaling unscented base oil. Lavender oil also increased the power of theta (4–8 Hz) and alpha (8–13 Hz) brainwave activity, particularly in bilateral temporal and central areas (Sayorwan, 2012). Another contemporary research study showed effectiveness of lavender essential oil for primary dysmenorrhea (Bakhtshirin, 2015).

9. **Neroli** (*Citrus aurantium*). The blossom of bitter orange yields a magnificent floral odor without the irritation of most citrus peel oils. It is famous for its soothing aroma and its penchant for promoting relaxation and is the main ingredient in the popular perfume *Poison*. Royal in its origins, neroli flowers are used to counter emotional shock, fear, and lack of confidence. It has been used as an aphrodisiac since ancient times; Spanish prostitutes used it to seduce their customers. The tree is in fruit and flower at the same time, considered a symbol of open communication in a perfect union.

   A randomized clinical trial was conducted on two groups of 63 pregnant women in each, to determine the efficacy of aromatherapy in reducing anxiety during the first stage of labor. One group was exposed to inhalation of a gauze pad with neroli essential oil; the other group had no aroma on their gauze pads. The results confirmed that neroli oil is a simple, inexpensive, noninvasive, safe, and effective intervention to reduce anxiety during labor (Namazi, 2014).

Neroli may also be effective in reducing preoperative anxiety before minor operations (Akhlaghi, 2011). The Spielberger State-Trait Anxiety Inventory (STAI-State) and the Amsterdam Preoperative Anxiety and Information Scale (APAIS) were administered before a minor operation. Patients' STAI-State and APAIS scale responses were significantly decreased in the neroli- preoperative treated group ($p < 0.05$), compared to no significant response change in the saline solution control group.

10. **Orange** (*Citrus sinensis*). Sweet and fruity orange oil is the least photosensitizing of all the citrus fruits and should be preferred if exposure to sunlight or UV light will occur after oil application. Also, use with care in the bath or in any water dilution as it can be irritating in large proportions in a blend or used alone. Citrus oils have the unique effect of being uplifting to the spirit while relaxing the body. An effective sedative, orange essential oil is a useful accompaniment in blends for anxiety (Goes, 2012; Lehrner, 2000).
11. **Patchouli** (*Pogostemon cablin*) * An unusual member of the mint family, this essential oil smells more like a root than the leaf from which it is derived: earthy, musty, and distinctively heavy. In India, where it is native, it is associated with prosperity and is said to balance libido. Who can forget the scent of patchouli as the hallmark of the hippie generation? (A possible turn-on for those who remember those uninhibited times.) It is helpful in blends that address nervousness, oversensitivity, and anxiety and has been used in aphrodisiac blends and well-known earthy perfumes such as *Tabu*. Recently, a comprehensive literature review of patchouli (Swamy, 2015) was conducted. Besides the numerous therapeutic properties noted, it listed biological activities including aphrodisiac and antidepressant.
12. **Rose** (*Rosa damascena*). The most famous of blooms, rose is the undeniable "queen of the flowers," with centuries of quotes attesting to its enhancement of love and sensuality, from Cleopatra to Shakespeare to the present. Rose oil is said to soothe the senses by reducing anxiety connected to sex. Rose carries a message of an open and receptive heart, helping unify spiritual and physical love. Employed in the realm of relationships to heal conflict and a waning sexual desire, rose oil is reputed to be able to help lift depression and reduce envy, anger, sorrow, intolerance, and lack of confidence. It is said to ease loss and sadness, rebuilding trust and connection.

A clinical study of 40 healthy volunteers investigated the effect of dermally applied rose oil on human autonomic parameters and emotional responses. Subjects rated themselves as more calm, more relaxed, and less alert than subjects in the control group. The researcher concluded a relaxing effect of the rose oil and evidence of relief of depression and stress in humans. (Hongratanaworakit, 2009). An overview of *Rosa damascena* and its applications can be found in Mahboubi (2016).

13. **Sandalwood** (*Santalum spicatum*). This more sustainable Australian sandalwood has replaced East Indian oils in recent years. The comparable sweet, earthy, and woody note brings a soothing component to blends, grounding them and offering an earthy reminder to slow down and appreciate the moment. Sandalwood promotes spiritual devotion, peaceful relaxation, and comfort to those who have undergone loss. Avicenna, the famous Arabian physician of the 10th century, used it to exhilarate and to heal passions of the heart. The *Kama Sutra*, India's ancient sexual text, suggests that a man "give her your jasmine garland, lie her back gently, and massage her body with sweet sandalwood oil."

    The skin application of sandalwood oil revealed some interesting physiological effects when several well-known essential oil scientists participated in a research project (Hongratanaworakit, 2004.) Alpha-santalol (a sesquiterpene major component of sandalwood oil) caused significant physiological changes indicative of a relaxing/sedative effect, yet application of sandalwood oil provoked both physiological deactivation and behavioral activation. These findings may suggest an uncoupling of physiological and behavioral arousal processes by sandalwood oil.

14. **Vetiver** (*Vetiveria zizanioides*).* Deep, balsamic, heavy grounding is offered by this dark and thick essential oil derived from the root of a tropical grass. It is deceptive in its above-ground appearance; the thin roots grow many feet down into the earth stabilizing hillsides and preventing erosion. So too does it help to stabilize feelings of uncertainty, release deep fear, and uplift the body and mind while it centers the emotions and comforts the spirit. Vetiver essential oil has anecdotal evidence in aromatherapy for relaxation.

    An animal study investigated its effects and found that anxiolytic properties may be associated with altering neuronal activation in a subarea of the amygdala, the brain region most associated with fear responding (Saiyudthong, 2015).

15. **Ylang ylang** (*Cananga odorata*).* Heady, floral, and often cloying, this tropical flower is sedative while releasing emotional fatigue. The euphoric scent can be overwhelming and is sometimes considered intoxicating. It is traditionally used treat impotency and frigidity. In Indonesia, where the tree is endemic, ylang ylang flowers are scattered on the beds of

newlyweds to promote desire and bless the union with children. The dilutions should be kept low to prevent headache and nausea from the overly sweet scent.

In a clinical study of 40 healthy volunteers, the effects of ylang ylang oil were evaluated after transdermal absorption; parameters included skin temperature, pulse rate, breathing rate, and blood pressure. Results showed a significant decrease of blood pressure and increase of skin temperature, and self-reporting of relaxation and relief from depression and stress (Hongratanaworakit, 2006).

Dilutions, Safety, and Applications

Essential oils are very concentrated forms of plant distillates. They can be irritating to the skin and should not be used undiluted. Following these suggestions for dilutions will ensure safe and effective results. Do not use essential oils on pregnant women during the first trimester; thereafter use in 0.5% to 1% dilutions of only the safest essential oils: lavender, chamomile, rose, orange, frankincense, or neroli.

A carrier oil is the fatty oil used to dilute the essential oil to a safe preparation for direct application to the skin or bath. There are many organic seed and nut oils readily available; some well-tolerated favorites are coconut, almond, jojoba, sunflower, or olive oil.

General dilution guidelines:

- 2% dilution: 10 drops of essential oils per one ounce of carrier oil for general application.
- 1% dilution: 5 drops of essential oils per one ounce of carrier oil for milder application.
- 0.5% dilution: 2 to 3 drops of essential oils per one ounce of carrier oil for those with extremely sensitive skin.

Note: The following sample formulas provided are in the general dilution of 2%; dilute further as needed and noted previously.

For the treatments addressed here, essential oils should *never* be ingested. Follow the guidelines for dilution and keep out of the reach of children and away from the eyes. They can be inhaled through a variety of means explained here and used externally on the skin or in the bath. The following are suggestions for a variety of applications.

All of the following examples are for general application. Add the suggested drops given in each recipe into a one-ounce bottle of carrier oil or lotion. Follow specific instructions for massage, bath, inhalation, and so on. Reduce the number of drops for sensitive skin according to the previous guidelines.

Massage oil—in addition to the obvious relaxation benefits to the nervous system and muscles, self-massage or partner massage is a wonderful prelude to an intimate encounter.

Example: 4 drops orange, 4 drops lavender, 2 drops chamomile in one ounce of carrier oil

Bath oil—a relaxing bath will help reduce tension, relax muscles, and calm mind chatter.

Example: 6 drops sandalwood, 3 drops neroli, 1 drop ylang ylang in one ounce of carrier oil. Add one teaspoon to a full bathtub.

Body lotion—a great option for a nongreasy massage or after-shower skin moisturizer. Deeply relaxing, this grounding combination can break down emotional and psychological barriers, calm the body, and uplift the spirit.

Example: 5 drops rose; 4 drops orange; 1 drop vetiver in one ounce of unscented lotion

Diffusers/inhalers—an efficient way to micronize one's essential oil molecules into tiny droplets to cover a wide area (usually up to 300 square feet), diffusers are the best choice to create a sensual atmosphere. Individual inhalers can be made by adding 10 drops of essential oil (single note or blend) to a small vial of rock salt; inhale as needed. This and other blends can also be dropped on a handkerchief and inhaled as desired. Another low-tech option is a simple mixture of water and essential oil in a receptacle over a burning candle. Do not use essential oils on a lightbulb due to combustion possibilities.

Example: 5 drops lavender, 3 drops sandalwood, 2 drops orange in one ounce of carrier oil will help to normalize mood, emotions, and hormones, according to anecdotal information.

Body mist/room/linen spray—A low-tech option for enhancing air quality and creating inviting bedding and sensually scented skin (avoid the face). Mix 10 drops of essential oil into one ounce of water and shake well before spraying.

Example: 6 drops frankincense, 3 drops patchouli, 1 drop ylang ylang will meld into a grounding and sensual combination.

Anointing oil—can be incorporated into the most intimate of moments; apply to neck, thighs, breasts, or most anywhere (avoid mucus membranes).

Example: 5 drops neroli, 3 drops rose, 2 drops jasmine in one ounce of carrier oil.

Blending Essential Oils

A number of formulas are provided here, but once users gain a little confidence, they are encouraged to explore their creativity and develop blends that delight them and their loved one. The suggested instructions are limited to three essential oils per formula to simplify the process. Readers can expand their options as they gain experience, but for safety's sake, do not exceed the 2% dilution of 10 drops per ounce of carrier. Use a disposable pipette for accurate drops.

Fun and Flirty Blends:

- **Relax**: 7 drops lavender, 2 drops chamomile, 1 drop vetiver
- **Release**: 5 drops lavender, 3 drops neroli, 2 drops chamomile
- **Restore**: 4 drops geranium, 4 drops clary sage, 3 drops rose
- **Enrich**: 5 drops orange, 3 drops lavender, 2 drops ginger
- **Femme Fatale**: 5 drops neroli, 3 drops rose, 2 drops ylang ylang
- **Hormone Harmonizer**: 4 drops lavender, 3 drops geranium, 2 drops clary sage
- **Invite**: 5 drops cedarwood, 4 drops frankincense, 1 drop ylang ylang
- **Long and Strong**: 5 drops sandalwood, 3 drops jasmine, 2 drops patchouli
- **Lover Boy**: 6 drops cedarwood, 3 drops neroli, 1 drop jasmine
- **Pain Relief:** 4 drops frankincense, 3 drops lavender, 3 drops ginger

Blend for Vaginal Dryness:

- 2 oz. sea buckthorn or pomegranate as a carrier oil
- 1 drop each: clary sage, rose, geranium
- Blend this low-dilution formula together and apply a few drops to the vulva and/or vagina daily for chronic dryness. Dilute further if irritation occurs or discontinue if it persists. Take sea buckthorn oil capsule orally; it is the only plant source of omega 3, 6, 7, 9, an important naturally occurring blend of essential fatty acids. Follow dosages on product label.

Conclusion

When properly employed, aromatherapy, used for centuries and providing a deep connection to medical and cultural traditions and to nature, has proven itself to be a safe, pleasant, and simple addition to every treatment option for improving intimacy and emotional and sexual health. A growing body of human and animal research supports this assertion. Whatever blend or single note one uses to improve mood, reduce stress, and enhance one's most cherished relationships, the use of pure plant essential oils will help to build self-confidence, reduce stress, encourage affectionate feelings, and benefit individuals and their partners in developing a deeper sense of connection and trust, all sure to lead to a happier and healthier sex life.

REFERENCES

Ackerman D (1990). *A natural history of the senses.* New York: Vintage Books.

Akhlaghi M, Shabanian G, Rafieian-Kopaei M, Parvin N, Saadat M, Akhlaghi M (2011). Citrus aurantium blossom and preoperative anxiety. *Rev Bras Anestesiol.* Nov–Dec;61(6):702–712.

Apay SE, Arslan S, Akpinar RB, Celebioglu A (2012). Effect of aromatherapy massage on dysmenorrhea in Turkish students. *Pain Manag Nurs.* Dec;13(4):236–240.

Azimova SS, Glushenkova IA, Vinogradova IV (2012). *Lipids, lipophilic components and essential oils from plant sources.* New York: Springer.

Bakhtshirin F, Abedi S, YusefiZoj P, Razmjooee D (2015). The effect of aromatherapy massage with lavender oil on severity of primary dysmenorrhea in Arsanjan students. *Iran J Nurs Midwifery Res.* Jan-Feb;20(1):156–160.

Boehm T, Zufall F (2006). MHC peptides and the sensory evaluation of genotype. *Trends Neurosci.* 29(2):100–107.

Conrad P, Adams C (2012). The effects of clinical aromatherapy for anxiety and depression in the high risk postpartum woman—a pilot study. *Complement Ther Clin Pract.* Aug;18(3):164–168.

Dalton P, Doolittle N, Breslin PA (2002). Gender specific induction of enhanced sensitivity to odors. *Nature Neurosci.* Mar;5(5):199–200.

Darsareh F, Taavoni S, Joolaee S, Haghani H (2012). Effect of aromatherapy massage on menopausal symptoms: a randomized placebo-controlled clinical trial. *Menopause.* Sep;19(9):995–999.

de Groot JH, Smeets MA, Rowson MJ, et al. (2015). A sniff of happiness. *Psychol Sci.* 26(6):684–700.

Doty LR (2014). Human pheromones: Do they exist? In C Mucignat-Caretta (Ed.), *Neurobiology of chemical communication* (Frontiers in Neuroscience). Boca Raton, FL: CRC Press/Taylor & Francis.

Fox K (n.d.). *The smell report: an overview of facts and findings.* Oxford: Social Issues Research Centre. http://www.sirc.org/publik/smell.pdf.

Gildersleeve AK, Haselton GM, Larson MC, Pillsworth GE (2012). Body odor attractiveness as a cue of impending ovulation in women: evidence from a study using hormone-confirmed ovulation. *Horm Behav.* 61(2):157–166.

Goes TC, Antunes FD, Alves PB, Teixeira-Silva F (2012). Effect of sweet orange aroma on experimental anxiety in humans. *J Altern Complement Med.* Aug;18(8):798–804.

Grammer K, Fink B, Neave N (2005). Human pheromones and sexual attraction. *Eur J Obstet Gynecol Reprod Biol.* Feb 1;118(2):135–142.

Graziottin A (2000). Libido: the biologic scenario. *Maturitas.* Jan;34(1):S9–S16.

Havlicek J, Roberts SC, Flegr J (2005). Women's preference for dominant male odour: effects of menstrual cycle and relationship status. *Biol Lett.* Sep 22;1(3):256–259.

Herman A, Herman AP (2015). Essential oils and their constituents as skin penetration enhancer for transdermal drug delivery: a review. *J Pharm Pharmacol.* Apr;67(4):473–485.

Herz RS (2008). *The scent of desire: discovering our enigmatic sense of smell.* New York: HarperCollins.

Herz RS (2016). The role of odor-evoked memory in psychological and physiological health. *Brain Sci.* Jul 19;6(3).

Hirsch RA, Gruss JJ (2014). Human male sexual response to olfactory stimuli. Chicago: American Academy of Neurological and Orthopaedic Surgeons. Smell and Taste Treatment Research Foundation. http://aanos.org/human-male-sexual-response-to-olfactory-stimuli/.

Hongratanaworakit T (2009). Relaxing effect of rose oil on humans. *Nat Prod Commun.* Feb;4(2):291–296.

Hongratanaworakit T (2010). Stimulating effect of aromatherapy massage with jasmine oil. *Nat Prod Commun.* Jan;5(1):157–162.

Hongratanaworakit T (2011). Aroma-therapeutic effects of massage blended essential oils on humans. *Nat Prod Commun.* Aug;6(8):1199–1204.

Hongratanaworakit T, Buchbauer G (2006). Relaxing effect of ylang ylang oil on humans after transdermal absorption. *Phytother Res.* Sep;20(9):758–763.

Hongratanaworakit T, Heuberger E, Buchbauer G (2004). Evaluation of the effects of East Indian sandalwood oil and alpha-santalol on humans after transdermal absorption. *Planta Med.* Jan;70(1):3–7.

Imanishi J, Kuriyama H, Shigemori I, et. al. (2009). Anxiolytic effect of aromatherapy massage in patients with breast cancer. *Evid Based Complement Alternat Med.* Mar;6(1):123–128.

Jacob S, Spencer NA, Bullivant SB, Sellergren SA, Mennella JA, McClintock MK (2004). Effects of chemosignals on the human menstrual cycle. *Hum Reprod.* Feb 19(2):422–429.

Johns MA (1986). The role of the vomeronasal organ in behavioural control of reproduction. *Ann NY Acad Sci.* Dec 474:148–157.

Karlson P, Luscher M (1959). "Pheromones": a new term for a class of biologically active substances. *Nature.* 183:55–56. http://www.sirc.org/publik/smell_human.html

Kaviani M, Maghbool S, Azima S, Tabaei MH (2014). Comparison of the effect of aromatherapy with *Jasminum officinale* and *Salvia officinale* on pain severity and labor outcome in nulliparous women. *Iran J Nurs Midwifery Res.* Nov-Dec;19(6):666–672.

Kelly JR, Borre Y, O'Brien C, et al. (2016). Transferring the blues: depression-associated gut microbiota induces neurobehavioral changes in the rat. *J Psychiatr Res.* 82:109–118.

Keverne EB (2004). Importance of olfactory and vomeronasal systems for male sexual function. *Physiol Behav.* Nov 15;83(2):177–187.

Keville K, Green M (2009). *Aromatherapy: a complete guide to the healing art,* 2nd ed. New York: Crossing Press.

Kiecolt-Glaser JK, Graham JE, Malarkey WB, Porter K, Lemeshow S, Glaser R (2008a). Olfactory influences on mood and autonomic, endocrine, and immune function. *Psychoneuroendocrinology.* Apr; 33(3):328–333.

Kiecolt-Glaser JK, Graham JE, Malarkey WB, Porter K, Lemeshow S, Glaser R (2008b). Response to letter to the editor, Oct; 33(9):1303, regarding "Olfactory influences on mood and autonomic, endocrine, and immune function." *Psychoneuroendocrinology.* Apr; 33(3):328–339.

Lee KB, Cho E, Kang YS (2014). Changes in 5-hydroxytryptamine and cortisol plasma levels in menopausal women after inhalation of clary sage oil. *Phytother Res.* Nov;28(11):1599–1605.

Lehrner J, Eckersberger C, Walla P, Pötsch G, Deecke L (2000). Ambient odor of orange in a dental office reduces anxiety and improves mood in female patients. *Physiol Behav.* Oct 1–15;71(1–2):83–86.

Logan DW (2014). Do you smell what I smell? Genetic variation in olfactory perception. *Biochem Soc Trans.* Aug;42(4):861–865.

Mahboubi M (2016). Rosa damascena as holy ancient herb with novel applications. *J Tradit Complement Med.* Jan; 6(1):10–16.

Martins DF, Emer AA, Batisti AP (2015). Inhalation of *Cedrus atlantica* essential oil alleviates pain behavior through activation of descending pain modulation pathways in a mouse model of postoperative pain. *J Ethnopharmacol.* 175:30–38.

Matsumoto T, Asakura H, Hayashi T (2013). Does lavender aromatherapy alleviate premenstrual emotional symptoms? A randomized crossover trial. *Biopsychosoc Med.* 7:12.

Matsumoto T, Kimura T, Hayashi T (2016). Aromatic effects of a Japanese citrus fruit—yuzu (Citrus junos Sieb. ex Tanaka)—on psychoemotional states and autonomic nervous system activity during the menstrual cycle: a single-blind randomized controlled crossover study. *Biopsychosoc Med.* 10:11.

McClintock KM (1971). Menstrual synchrony and suppression. *Nature.* 229:244–245.

McNeil J, Cameron JD, Finlayson G, Blundell JE, Doucet É (2013). Greater overall olfactory performance, explicit wanting for high fat foods and lipid intake during the mid-luteal phase of the menstrual cycle. *Physiol Behav.* Mar 15;112–113:84–89.

Miller SL,. Maner JK (2010). Scent of a woman: men's testosterone responses to olfactory ovulation cues. *Psychol Sci.* Feb;21(2):276–283.

Mitchell DJ, Kahn BE, Knasko SC (1995). There's something in the air: effects of congruent or incongruent ambient odor on consumer decision making. *J Cons Res.* 22(2):229–238.

Mizuno D, Konoha-Mizuno K, Mori M, Yamazaki K, Haneda T, Koyama H, Kawahara M (2015). An in vitro system comprising immortalized hypothalamic neuronal cells (GT1–7 cells) for evaluation of the neuroendocrine effects of essential oils. *Evid Based Complement Alternat Med.* 343942.

Namazi M, Akbari SAA, Mojab F, Talebi A, Alavi Majd H, Jannesari S (2014). Aromatherapy with citrus aurantium oil and anxiety during the first stage of labor. *Iran Red Crescent Med J.* Jun;16(6):e18371.

Oliveira-Pinto AV, Santos RM, Coutinho RA, et al. (2014). Sexual dimorphism in the human olfactory bulb: females have more neurons and glial cells than males. *PLoS One.* Nov 5;9(11):e111733.

Ou MC, Hsu TF, Lai AC, Lin YT, Lin CC (2012). Pain relief assessment by aromatic essential oil massage on outpatients with primary dysmenorrhea: a randomized, double-blind clinical trial. *J Obstet Gynaecol Res.* May;38(5):817–822.

Ovayolu O, Sevi? U, Ovayolu N, Sevinç A (2014). The effect of aromatherapy and massage administered in different ways to women with breast cancer on their symptoms and quality of life. *Int J Nurs Pract.* Aug;20(4):408–417.

Paddock C (2014). Women's finer sense of smell may be due to more brain cells. *Medical News Today.* http://www.medicalnewstoday.com/articles/284991.php

Park HJ, Lim EJ, Zhao RJ, et al. (2015). Effect of the fragrance inhalation of essential oil from Asarum heterotropoides on depression-like behaviors in mice. *BMC Complement Altern Med.* Mar 6;15:43.

Pemberton E, Turpin PG (2008). The effect of essential oils on work-related stress in intensive care unit nurses. *Holist Nurs Pract.* Mar-Apr;22(2):97–102.

PT Staff (2007). The hidden force of fragrance. *Psychology Today.* November 1. https://www.psychologytoday.com/articles/200711/the-hidden-force-fragrance

Rashidi Fakari F, Tabatabaeichehr M, Kamali H, et al. (2015). Effect of inhalation of aroma of geranium essence on anxiety and physiological parameters during first stage of labor in nulliparous women: a randomized clinical trial. *J Caring Sci.* Jun 1;4(2):135–141.

Roberts SC, Gosling LM, Carter V, Petrie M (2008). MHC-correlated odour preferences in humans and the use of oral contraceptives. *Proc Biol Sci.* Dec 7;275(1652):2715–2722.

Rodriguez T (2013). Fertile women have a heightened sense of smell. *Scientific American.* September 1. http://www.scientificamerican.com/article/fertile-women-heightened-sense-smell/.

Sadeghi Aval Shahr H, Saadat M, Kheirkhah M, Saadat E (2015). The effect of self-aromatherapy massage of the abdomen on the primary dysmenorrhoea. *J Obstet Gynaecol.* May;35(4):382–385.

Saiyudthong S, Pongmayteegul S, Marsden CA, Phansuwan-Pujito P (2015). Anxiety-like behaviour and c-fos expression in rats that inhaled vetiver essential oil. *Nat Prod Res.* Jan 2:1–4.

Santos PS, Schinemann JA, Gabardo J, Bicalho Mda G (2005). New evidence that the MHC influences odor perception in humans: a study with 58 Southern Brazilian students. *Horm Behav.* April 47(4):384–8.

Savic I (2014). Pheromone processing in relation to sex and sexual orientation. In C Mucignat-Caretta (Ed.), *Neurobiology of chemical communication* (Frontiers in Neuroscience). Boca Raton, FL: CRC Press/Taylor & Francis.

Sayorwan W, Siripornpanich V, Piriyapunyaporn T, Hongratanaworakit T, Kotchabhakdi N, Ruangrungsi N (2012). The effects of lavender oil inhalation on emotional states, autonomic nervous system, and brain electrical activity. *N J Med Assoc Thai.* Apr; 95(4):598–606.

Schneider R (2015). There is something in the air: testing the efficacy of a new olfactory stress relief method (AromaStick®). *Stress Health.* Oct;32(4):411–426.

Sheikhan F, Jahdi F, Khoei EM, Shamsalizadeh N, Sheikhan M, Haghani H (2012). Episiotomy pain relief: use of lavender oil essence in primiparous Iranian women. *Complement Ther Clin Pract.* Feb;18(1):66–70.

Seol GH, Shim HS, Kim PJ, et al. (2010). Antidepressant-like effect of *Salvia sclarea* is explained by modulation of dopamine activities in rats. *J Ethnopharmacol.* Jul 6;130(1):187–190.

Siddiqui MZ (2011). Boswellia serrata, a potential antiinflammatory agent: an overview. *Indian J Pharm Sci.* May;73(3):255–261.

Sobel N, Prabhakaran V, Hartley CA, et al. (1999). Blind smell: brain activation induced by an undetected air-borne chemical. *Brain.* Feb;122 (Pt 2):209–217.

Sookoian S, Burgueño A, Gianotti TF, Marillet G, Pirola CJ (2011). Odor perception between heterosexual partners: its association with depression, anxiety, and genetic variation in odorant receptor OR7D4. *Biol Psychol.* Mar;86(3):153–157.

Sritoomma N, Moyle W, Cooke M, O'Dwyer S (2014). The effectiveness of Swedish massage with aromatic ginger oil in treating chronic low back pain in older adults: a randomized controlled trial. *Complement Ther Med.* Feb;22(1):26–33.

Swamy MK, Sinniah UR (2015). A comprehensive review on the phytochemical constituents and pharmacological activities of pogostemon cablin benth: an aromatic medicinal plant of industrial importance. *Molecules.* May 12;20(5):8521–8547.

Taavoni S, Darsareh F, Joolaee S, Haghani H (2013). The effect of aromatherapy massage on the psychological symptoms of postmenopausal Iranian women. *Complement Ther Med.* Jun;21(3):158–163.

Tillett J, Ames D (2010). The uses of aromatherapy in women's health. *J Perinat Neonatal Nurs.* Jul-Sep;24(3):238–245.

Turetsky IB, Hahn G-C, Borgmann-Winter K, Moberg JP (2009). Scents and nonsense: olfactory dysfunction in schizophrenia. *Schizophr Bull.* Nov;35(6):1117–1131.

Varney E., Buckle J (2013). Effect of inhaled essential oils on mental exhaustion and moderate burnout: a small pilot study. *J Altern Complement Med.* Jan;19(1):69–71.

Wedekind C, Penn D (2000). MHC genes, body odours, and odour preferences. *Nephrol Dial Transplant.* Sep;15(9):1269–1271.

Wilke K, Martin A, Terstegen L, Biel SS (2007). A short history of sweat gland biology. *Int J Cosmet Sci.* 29:169–179.

Wilson RS, Arnold SE, Schneider JA, Boyle PA, Buchman AS, Bennett DA (2009). Olfactory impairment in presymptomatic Alzheimer's disease. *Ann N Y Acad Sci.* Jul; 1170:730–735.

Yim VW, Ng AK, Tsang HW, AY Leung AY (2009). A review on the effects of aromatherapy for patients with depressive symptoms. *J Altern Complement Med.* Feb;15(2):187–195.

22

The Benefits of Water Therapy for Sexual and Pelvic Problems

ALEXA FLECKENSTEIN

Introduction

Water is the most erotic and playful of elements—without it, the swelling of wombs, of male and female pleasure organs would not happen. Water makes us silly, youthful, and exuberant—as any beach scene proves. Water and sexuality—the two have been going together since the dawn of mankind. Water already surrounds us in utero—like Botticelli's *Venus*, we are all ocean-born creatures.

Still, for someone not familiar with hydrotherapy, its touted benefits sound doubtful. Why would water even work? After all, water hits only the outside skin—how can it exert an effect on such far-reaching places as cells in different organs, including the brain?

No Life without Water: A Brief History

All ancient cultures held water sacred. We lost that deep water wisdom at about the time water began arriving from taps—effortless, cheap, mundane. Now water is a given and neglected and ignored. But our prehistoric ancestors stood under a refreshing waterfall, walked barefoot in the early morning dew over pebbles or soft low growth, swam in the ocean, washed themselves in a cold rivulet or a stream warmed by the sun, inhaled steam over hot stones, built sweat lodges. Renowned primatologist Jane Goodall, observing our close relations, the chimpanzees, near waterfalls, concluded that water inspires awe and spiritual feelings in the animals (Goodall, n.d.).

Nearly every culture has had a water culture—think of Turkish hammam (Benouis, 2008), Finnish sauna (Sutkowy, 2014), Russian steam bath (Levina, 1993), Roman thermae (Mulec, 2015), Jewish mikvot (Obladen, 2017), Native American sweat lodge (Mehl-Madrona, 2010), highland Mayas' steam-bathing (Groark, 2005), Japanese hot springs (Hayasaka, 2013), and Guatemalan chuj or temazcal (Radoff, 2013). Since earliest times, people have sought healing by drinking or bathing in mineral springs. To our forbears (and still in many tribal cultures), the use of sacred water always meant healing too. Intuitively, it makes sense that water can be used to treat sexual and pelvic problems.

This chapter presents the simple water modalities that can be used in treating disease and discusses the different sexual and pelvic ailments to which they can be applied. Finally, we discuss the complicated biological mode of water working in the body. After all, for someone not familiar with hydrotherapy, its touted benefits sound doubtful. Why would water even work? Water hits only the outside skin—how can it exert an effect on such far-reaching places as cells in different organs, including the brain (Furnari, 2014)? How can cold water on the skin, for instance, douse inflammation deep in the body?

SEBASTIAN KNEIPP (1821–1897)

The Bavarian priest Sebastian Kneipp was the first to publicize water therapy worldwide and in a systematic way. He created a comprehensive system of natural healthcare by tying hydrotherapy to the four other "Pillars of Health": nutrition, herbs, movement, and what he names *order therapy* (Kneipp, 1892). For him, order means *natural order,* encouraging regularity of daily and seasonal rhythms. His notions go against modern "freedoms"—such as working, shopping, and pursuing fun activities during the wee hours of the night. New chronobiological research supports the health benefits of a lifestyle that conforms more closely to the natural circadian rhythms of sleep and wakefulness (Musiek, 2015) and to the cycles of the seasons. Even on a molecular level, science has found a circadian rhythm (Hirota, 2015), substantiating Kneipp's view. *Order*—as it is used in *European Natural Medicine* (Fleckenstein, 2006)—covers sufficient sleep, social needs (including sexual relationships), breathing techniques, and a spiritual life, among other aspects.

Sebastian Kneipp's *My Water Cure* (Kneipp, 1886) became the first worldwide self-help bestseller. The book sold in the millions, was translated into most major languages (Lust, 1901), and is still in print. The then-new transatlantic telegraph cables might have played a role in the rapid spreading of Kneipp's ideas. The time was ripe: for several generations, people had been leaving their rural homes to work in the big industrial cities and, increasingly, they were falling victims to so-called *civilization diseases*—chronic diseases and epidemics, tuberculosis foremost.

In America, the Great Water Craze (Kemble, 1967) and its cold-water spas sprouted from Vancouver to Louisiana. Kneipp had cured himself from consumption (tuberculosis) by jumping into the wintry gray Danube River. As tuberculosis was still a major health problem in the United States at the beginning of the 20th century, Kneipp's self-cure was great advertisement for the spa and cold-water movement. But with World War I Germany lost its general credibility, compounded by World War II and the Holocaust. The Flexner Report (1910), in its zeal to end all quackery, threw out the baby with the bathwater—so to speak. As a result, the large sanatoriums and water spas fell into neglect and oblivion, at least in the Americas. In Europe, they survived and are thriving again with the modern trend to natural health.

Owing in large part to the paucity of modern water research, for most of the 20th-century conventional medicine ignored or ridiculed hydrotherapy—or *hydropathy,* as it was also called at times, awkwardly (Trall, 1880/1996). Most of the sparse science done on water for health is buried in old, obscure European journals, as the original hydrotherapy studies were mostly published in German or eastern European languages. In any case, those old studies would not satisfy our present-day research requirements for multicenter, double-blind, randomized, placebo-controlled studies.

JOHN FLOYER (1649–1734)

Nearly 150 years before Kneipp, the Englishman John Floyer had written his *Psychrolousia* (Floyer, 1732/2010), an entertaining history of cold bathing, from the Egyptians, Jews, Greeks, and Romans to more modern times in England; Floyer also describes different spa waters that gain their specific health benefits through minerals washed out of the ground, including tin waters, chalk waters, and nitrous purging waters (Floyer, 1732/2010). John Floyer is important mostly because he introduces the brilliant term "inflammatory lassitude" (Floyer, 1732/2010) for a constellation of rheumatic diagnoses we nowadays might call chronic fatigue syndrome, fibromyalgia, polymyalgia rheumatica, and so forth. Today we know that inflammation underlies the pathology of many chronic diseases, such as diabetes, arthritis, and depression (Fleckenstein, 2014). We will see that cold water has anti-inflammatory effects (Lee, 2012), the most important of its many benefits, and as such can be used to cure inflammatory processes, including those of the sexual and reproductive apparatus.

VINZENZ PRIESSNITZ (1799–1851)

Vinzenz Priessnit is the third "water doctor" that merits short mentioning. Before Kneipp, Priessnitz invented wraps: a hot-water wrap around the neck against sore throat, a hot wrap on the abdomen for a case of belly ache, and so on (Claridge, 1842–1843/2015). But it was Sebastian Kneipp's ingenious mind that observed the distant effect of water applications: a hot footbath influences the head and pelvis, a cold armbath lowers blood pressure and improves breathing, and so on.

Cold and Warm Hydrotherapy in the Home

There are two kinds of water treatment: cold and warm. Cold water works mainly in two ways, as discussed previously: it is anti-inflammatory and invigorating. Invigoration happens by creating daily mini-stresses, so that one is better able to withstand the real stresses of life. Warm water provides comfort for pain and relief of cramps. The following are the most common forms of water cure that patients can do at home:

COLD SHOWER

Given that most people nowadays will not have access to a trained hydrotherapist equipped with the precise diameter of watering hose Kneipp prescribed (about half an inch), the modern cold shower (Fleckenstein, 2007) must serve as an approximation of a Kneipp *gush* (Kneipp, 1886). So water therapy in the privacy of the home must use what is readily at hand: a sink, a tub, a shower.

After a hot shower, step out of the water spray and turn the handle on cold (really cold). Briefly rinse feet with cold water, then hands, then some splashes into the face. That is enough for the first time. One always starts at the feet, away from the heart. Next time, let cold water run over the feet up to the knees, then hands and face. In about a month, one can reach the goal: a cold shower to the whole body for about 20 to 30 seconds. Soap/shampoo use should be kept to a minimum as they break down the healthy skin barriers of fatty acids and benign bacteria.

A randomized crossover albeit extremely short duration study (each arm a week only), with very limited cold-water exposure (only applied to face and neck), showed a measurable difference in cognitive response in the elderly (Doering, 2001). One could extrapolate that a cold shower should have at least as much benefit for the brain. The benefits go beyond the brain though: Cold water—or any cold exposure—increases immunity and raises immune parameters (Castellani, 2002). Short, daily cold exposures will fend off frequent infections like the common cold and even cancer by improving immune function (Shevchuk, 2007).

Contraindications: Cold-water treatments are generally well tolerated in younger people unless overdone. The frail and elderly should start with cold washings (see later discussion) and advance cautiously, if at all. Uncontrolled hypertension, arterial disease (including Raynaud's), and any acute diseases are contraindications; blood pressure reasonably controlled with medications is not. In susceptible asthmatics, cold exposure might trigger bronchospasm (Chen, 1978). Overt coronary disease might preclude the cold-water shock treatment; a knowledgeable physician should be consulted. Cold showering is not winter swimming (even if we use some research results from winter swimming); at no point is dunking the body in ice water for prolonged periods recommended. Some studies have found benefits in winter swimming—mostly regarding mood (Huttunen, 2004)—but such extremes of cold can probably be endured only by larger people and less so by people at their ideal body mass index (Knechtle, 2009). There is little research on the impact of cold-water application on people of African descent, but they react to cold-water stressors with higher blood pressure and higher heart rates, which means they might also be more susceptible to other stressors of modern life (Calhoun, 1993). When taking a bath or a shower (or doing any water application), the surrounding room temperature does not so much affect outcomes, it seems, but it does influence subjective well-being (Hashiguchi, 2002).

COLD WASH

A more easily tolerated variation of a cold shower is the cold wash: The patient stands in front of the sink, or in the tub, and quickly washes down with a cold, slightly wrung face cloth, head to toe. This is the way elderly and frail people should start. If one comes out of the shower and still feels cold after a minute or two, the exposure was too long. One can reduce exposure time or use another time of the day: As much as a shower belongs to the morning routine, people in reduced health often tolerate cold better in the early afternoon hours—this is another example of observing natural circadian rhythms. For gradual adaptation to the cold, never turn the handle to warmer water temperatures—that is counterproductive: The gradient between temperatures is pivotal for inducing the neuro-endocrine-immune cascade of internal events that will lead to invigoration. Instead, shorter exposure times or smaller body exposure areas will do the trick.

Kneipp, late in his life, admitted that in his youth he had advocated longer cold-water exposure but had learned since then that overwhelming stressors could be detrimental. The severe self-imposed treatment of his youth—jumping into the wintry Danube River—should be discouraged, especially in the frail and the elderly: "The shorter the [cold] bath, the better the effect" (Kneipp, 1886).

WALKING BAREFOOT IN COLD WATER, AT THE OCEAN, OVER DEWY PEBBLES

Done regularly, this has balancing and invigorating properties. Again, if the feet are still cold afterwards for a prolonged time, the exposure was too much. A study on children preliminarily showed an invigorating effect in walking barefoot, even during winter, measured again by elevated norepinephrine levels (Park, 2004). But overdoing walking without shoes might hurt the foot anatomy (Vaseenon, 2015). Yet another study showed benefits for the knees. Walking barefoot in the house, however, makes falls more likely (McCarthy, 2014). A review article came to ambivalent conclusions about barefoot walking (Perkins, 2014). In sum, walking barefoot in cold water, in dewy grass, and at the beach has always been touted as healthful, especially for the pelvic organs, but decisive studies for walking barefoot for a longer time (or always) are lacking (Tam, 2014).

WARM FOOTBATH

Because of the reflex relationship between the feet and the pelvic area, a warm to hot (increasing temperature) footbath is useful in cases of sexual tension, pelvic pain, cramps (Emmanuel, 2000), and acute urinary tract infections (Uehlecke, 1997). A warm footbath induces relaxation in a twofold way: by decreasing autonomous sympathetic firing and increasing parasympathetic activity (Yamamoto, 2011). Therefore, a warm footbath lends itself as an adjuvant treatment in many pain and anxiety states (Kheirkhah, 2014)—which often accompany sexual dysfunction. A footbath comes in handy for insomnia patients as warm feet induce sleepiness reliably (Liao, 2013). Even here the rule applies: every warm application needs a very short cold gush at the end.

SITZBATH

For a sitzbath, fill a tub or a commercial basin with 1 to 3 inches of water. The sitzbath can be warm to hot, mostly for comfort and relaxation, or cold, for reducing swelling and edema and promoting healing (LaFoy, 1989). The cold sitzbath should last not much longer than half a minute. Always be careful of slipping in the tub and of burns if hot water is involved.

In a safer variant, temperature will be raised slowly from warm to hot by adding hotter water. An Australian study on sitzbaths for anal fissures showed only moderate effects (Gupta, 2006), but then the sitzbaths were done with plain lukewarm water, without the extra benefits of antibacterial (e.g., tea tree oil, native to Australia) or emollient herbs (mallow). Another study showed that a plain warm sitzbath does not add benefits to hemorrhoid treatment (Gupta, 2008), yet astringents were not added, and the water was warm. European Natural Medicine recommends short cold sitzbaths for hemorrhoids (Beer, 2011). Warm sitzbaths after transurethral resection of the prostate seem to prevent complications and rehospitalizations (Park, 2010). Anal pain was relieved by warm to hot sitzbaths—the higher the temperature, the better the pain relief (Shafik, 1993b). Micturition after anal surgery could be induced in most patients by sitting in a warm to hot sitzbath; the higher the temperature, the more relief patients found (Shafik, 1993b).

VAGINAL DOUCHES ARE DISCOURAGED

A large review of research on douching linked it to adverse reproductive and gynecologic outcomes, including bacterial vaginosis, preterm birth, low-birth-weight infants, pelvic inflammatory disease, chlamydial infection, tubal pregnancy, higher rates of HIV transmission, and cervical cancer (Cottrell, 2003). A study on hundreds of women with pelvic inflammatory disease showed more cases of this disease and kidney infections in women who used frequent douching (Ness, 2001).

FULL BATH

A full bath is the most familiar of water applications. Essential oil extracts or herbal teas turn the bathwater into even more enjoyable aromatherapy (Fleckenstein, 2007). Warm water soothes via the parasympathetic system, whereas cold water excites via the sympathetic system. So it seems that all we need is a warm bath to counteract modern stress. Unfortunately, it is not that easy, because the parasympathetic mode used alone and for prolonged time leads to sluggishness. Our bodies need the instigation of the sympathetic branch. Further, the effects of a warm bath do not last and do not invigorate: A study of warm-water immersion showed subjective feelings of well-being and relaxation, but the change in physical markers measured—blood pressure, body temperature, respiratory rate, muscle tension, and pupil size—did not show lasting effects; neither did psychological/psychiatric parameters afterwards (Robiner, 1990).

As we have stressed, every warm bath should end with a short cold wash, gush, or shower—"to close the pores" as Germans say, and to keep the warmth inside. Psychoanalytically, the similarity of luxuriating in a warm bath with floating in the womb cannot be overlooked. Any kind of warm treatment might be a positive emotional throwback to the times of feelings of security and being cared for in the womb or early childhood (Sorri, 1988). Physiologically however, a warm bath is a burden on an already weakened heart, owing to the heat, and also to the water pressure on the chest, which can cause/exacerbate congestive heart failure (CHF). In those cases, a ¾ warm bath (up to the umbilicus) gives warmth but little extra pressure on the heart.

WARM WATER BOTTLE

A simple variation on Kneipp's hot compresses and *wrappings*—which are complicated to do—is placing a warm water bottle on a painful body area, often the abdomen. Like a full bath, it can soothe pains (Masuda, 2005) and cramps. The highland Mayas have an ancient concept of health as *vital warmth* (Groark, 2005), and as much as cold exposure is important for invigoration, warmth is of the utmost importance for certain diseases, including fatigue, lassitude, loss of appetite, pain, and cramps (Graz, 2014). Never use boiling water in a hot water bottle; it should always be a little less hot. Every hot bottle should be wrapped in a towel: especially infants, the elderly, and postoperative patients are at risk for burns (Graz, 2014).

In European natural medicine, using hot or cold water often boils down to the questions: Is the patient in a state of *fullness*? in a state of *emptiness*? *Fullness* pertains to excess—stroke risk, high blood pressure, overweight, abscess, and so on—and needs to be treated with cold. *Emptiness* is seen in asthenia, low blood pressure, lassitude, weakness, exhaustion, and so on—and needs pampering with warmth.

SAUNA ENDING IN COLD-WATER IMMERSION

Because of the high temperature gradient between the hot, dry sauna and the cold water, sauna makes the strongest thermotherapy. Regular sauna visits have been linked to a reduction in all-cause mortality (Laukkanen, 2015).

The Finnish sauna is a small, heated room; ideal temperature is between 80° to 100°C (Helamaa, 1988). People sit or lie naked on towels on wooden planks until they sweat. Afterward they briefly dunk into a vat with cold water, take a cold shower, or go into a lake or the ocean for cooling, followed by rest. Sauna is thought to detoxify (Crinnion, 2007) and invigorate (Sobajima, 2013). Sauna is contraindicated in breast cancer after axillary surgery, to avoid lymphedema (Showalter, 2013), and in unstable heart disease (Basford, 2009; Imai, 1998). Sickle cell disease might also be a contraindication (Eisenbach, 2005), but there are too few cases reported in the literature. Sweat lodges function similarly to sauna (Livingston, 2010) but will raise core temperature considerably higher.

Dry sauna, contrary to popular belief—and contrary to whirlpool (Jacuzzi) and steam bath—does not raise core body temperature (Pilch, 2014) and is therefore not contraindicated in healthy pregnancy, but sauna can temporarily decrease sperm count (Yoon, 2011) and sperm motility (Guo, 2006) by heating up scrotum contents (Saikhun, 1998); long term, however, sauna does not cause fertility problems (Wähä-Eskeli, 1988). Sauna might reduce the likelihood of coming down with the common cold (Ernst, 1990b). Sauna bathers' exposure to heat and cold improves blood circulation (Vuori, 1988); better organ perfusion leads to better organ function (Kauppinen, 1989). Sauna can reverse even microcirculation problems in diabetes (Huang, 2012; Shui, 2015). In sauna, naturally occurring opiates are elevated, adding to the feeling of well-being (Ernst, 1990b). Sauna stress increases prolactin levels to a similar extent as exercise (Sirvio, 1987). Other sex hormones—testosterone, follicle stimulating hormone and luteinizing hormone—are not elevated in regular sauna visits (Leppaluoto, 1986). Sauna has been found an effective treatment for many medical conditions, including hypertension, congestive heart failure, and postmyocardial infarction care. It has helped with chronic obstructive pulmonary disease, chronic fatigue, chronic pain, and addictions (Crinnion, 2011). An overview seems to refute that infrared sauna lowers cholesterol (Beever, 2009). Several heart parameters, especial CHF, were improved with regular *waon* therapy—a modified low-temperature, no water-added Japanese sauna (Fujita, 2011; Ohori, 2012). Again, sauna makes a case for better general health leading to better sexual health—but we have no studies for proof.

As to negative impact on health, public saunas may be a source of infections in men who have sex with men and female sex workers (Holtz, 2012; Kang, 2011).

The Romans had a well-known bath culture. Their thermae were public meeting places (males only) to attend to business and enjoy entertainment with food, massage, and music. Roman citizens might enjoy their baths several times a week, which made ancient Rome probably the cleanest place in history—at least among the upper classes. Originally, as the name *thermae* implies, they used hot water mostly. But after the Roman Emperor Augustus (63 BC–14 AD) had been cured of a liver abscess by his physician, Musa, applying cold water (Floyer, 1732/2010), hot water went out of fashion and the first cold-water craze swept the antique world.

Hydrotherapy for Sexual/Pelvic Problems

Water therapy acts on/in the entire body, effecting overall well-being and general health. General health translates into sexual health—even if we are still missing major studies to support this view. Some preliminary research has been done: a recent review of different water applications concludes that hydrotherapy has good evidence behind it to recommend a wider application in medical practice (Mooventhan, 2014).

DIFFERENTIAL DIAGNOSIS

We cannot blindly assume that any pain *down there* (as Kneipp bashfully referred to the pelvic area) is caused by a sexual problem; pain thereabouts can have many roots, sexual and nonsexual. Any complaints of pelvic pain/discomfort could signal many diagnoses, some of them serious. Box 22.1 gives some of the differential diagnoses we have to consider before applying hydrotherapy.

The list of pelvic ailments in Box 22.1 is by no means exhaustive; it is included solely as a warning. But once we have ruled out more dire diagnoses, the afflictions left are exactly the kind of conditions that will be ideally suited for hydrotherapy and other lifestyle measures (Stiefelhagen, 2005). If no cause for a complaint can be found, problems might be called *functional,* which used to imply an elusive health disturbance, often equated with *psychogenic.* Now we reinterpret *functional* as an off-kilter autonomic nervous system (ANS). Surely there is a bodily equivalent to

Box 22.1 Differential Diagnosis of Complaints of the Lower Abdomen

Abdominal hernias
Adnexal tumors
Appendicitis
Benign lesions of the ovaries
Bladder cancer
Cervicitis
Chronic pelvic pain syndrome
Colon cancer
Colonic obstruction
Constipation
Depression
Diverticulitis
Dysmenorrhea
Ectopic pregnancy
Endometrial carcinoma
Endometriosis
Endometritis
Fistula
Herpes zoster
Infertility
Inflammatory bowel disease
Interstitial cystitis (cystitis, nonbacterial)
Intestinal motility disorders
Irritable bowel syndrome
Levator ani syndrome
Lumbar disc disease
Lumbar spondylosis
Megacolon
Neurogenic bladder
Ovarian cancer
Ovarian cysts
Ovarian torsion
Pelvic inflammatory disease
Perianal abscess
Perianal cysts
Perianal fissures
Polycystic ovarian syndrome
Porphyria
Prostatic abscess
Prostatitis, acute and chronic
Prostatodynia
Pyelonephritis, chronic
Salpingitis
Sexually transmitted disease
Ulcerative colitis
Urethral cancer
Urethral diverticula
Urethral strictures
Urethral syndrome
Urinary tract infection
Uterine cancer
Uterine prolapse
Vaginitis
Vulvar lesions
Vulvovaginitis

dysmenorrhea—it is not "all in her head." But mostly, disturbed internal balances are not as easily observed from the outside as a high fever, a boil, or a broken bone. The disturbances happen in cells, in glands, in interstitial tissues, along the autonomic axis—leading to long years of vague complaints before they manifest themselves in cancer (Fernandes, 2015), for instance. Hydrotherapy can bring such a stressed system back into physiological balance (Brenke, 2006).

Dysmenorrhea. Dysmenorrhea—the myriad of problems around the monthly female bleeding, to a great part triggered by higher levels of oxytocin (the lovey-lovey hormone; Fan, 2013; Guo, 2013)—can be approached with herbal remedies, improved nutrition (a healthier diet and weight loss lead to less premenstrual syndrome and dysmenorrhea symptoms [Barnard, 2000]), regular, moderate movement therapy—more intensive before the period, gentler during the period. But an important part of the therapy should be hydrotherapy (Beer, 2011).

In the acute phase, treatment will be with warm water. Warm footbath (Uehlecke, 1997), armbaths, and/or a simple warm water bottle (Graz, 2014) on the cramping belly or hurting lower back (Hosono, 2010), together with rest/early bedtime, are an old-fashioned recommendation. Chamomile (Sharifi, 2013), rosemary, or the pine needle extract pycnogenol (Stanislavov, 2003) are soothing additions for the footbath, working as adjuvant aromatherapy. Rosemary has analgesic (Raskovic, 2015) and relaxing (Diego, 1998) effects without sedation (Diego, 1998). An alternative would be adding hay flowers (Szekely, 1973) to the water.

In between monthly periods, a program for invigoration and balancing with cold water (cold shower) is slowly introduced and then increased (Uehlecke, 1997). For more severe cases, a Kur-like program in a spa is recommended with daily three to four different water applications: early morning, morning, afternoon, and before bed.—This course toward invigoration could begin even without water—by brushing the upper and the lower body, alternately, and always in the direction of the heart. Floyer already recommended dry brushing of the skin with a strigil (Floyer, 1732/2010)—a dry sponge, brush, or blade—to improve blood circulation. After dry brushing, the first cold treatment recommended is to tread barefoot in a cold basin or tub (or walk in early morning dewy grass, in a river, or in the ocean) for a few minutes just before bedtime, for relaxation and normalizing peristalsis of the bowels as gastrointestinal complaints often accompany menses (Parandavar, 2014).

Pelvic pain syndromes. Hydrotherapy works similarly in other functional pelvic pain syndromes.

Functional infertility in either sex, where no physical cause can be found, may respond to hydrotherapy. Regarding female infertility, alternating cold and hot baths and showers (always ending with a cold one), including short sitzbaths, is recommended. Aromatherapy will add sensual pleasure. In Germany, a whole Kneipp cure is often recommended with hydrotherapy, movement, massage, herbs, and general lifestyle advice (Uehlecke, 1999). The Kur is especially effective as it takes the woman out of her usual environment and daily stresses. However, not many women have the luxury and the knowledge to care only for themselves, even for a limited time. Women—like men—have a higher likelihood of developing diabetes if they are infertile (Tobias, 2015). Male infertility is often related to lifestyle indiscretions (smoking, drinking, eating, not moving), confirmed by a study that showed that infertile men are at higher risk of developing diabetes than fertile men (Kumar, 2014). A daily cold shower or a short tepid sitzbath is a mood enhancer (Shevchuk, 2008) and might thus increase motivation to change one's lifestyle in other ways too. Hot-water applications should be avoided because they lower sperm count (Rao, 2015), as already mentioned. This suggests that hot baths might be a form of contraception (Lissner, 1992)—though not foolproof. Historically, they have been used for that purpose.

Pregnancy. If pregnancy is achieved, cold-water treading and cold leg gushes are recommended against heavy legs and venous stasis problems (Beer, 2013; Uehlecke, 1999). Regular cold washings might prevent pregnancy striae, or stretch marks (Uehlecke, 1999). Water birth is increasingly and successfully used in obstetric centers (Cluett, 2009), but that topic is only loosely related to our subject of sexuality and hydrotherapy.

Menopause. Change-of-life problems like hot flashes can be ameliorated with cold washings, cold showers, and regular sauna with cold-water immersion (Uehlecke, 1999). In one study, simple cold splashes to the face immediately alleviated hot flashes (Molnar, 1975). A yearlong barefoot walking program decreased menopausal complaints in women (Monteiro, 2011).

Decreased or absent sexual pleasure—male or female—is high on the list of sexual complaints. After serious marital, physical, psychological, and lifestyle impedances have been ruled out, the couple should be *forbidden to* consummate the sexual act (Master, 1976) and instead play in warm water, use the sauna, take warm and cold showers, explore warm foot- and armbaths, and so on, without the pressure to perform sexually.

Sexual dysfunction might be accompanied by milder or more severe forms of depression (Thangadurai, 2014). Cold showers have been suggested even against endogenous depression (Shevchuk, 2008). The proposed mechanism is balancing of the hypothalamic-pituitary-adrenal (HPA) axis, that may be unhinged by the demands of modern life. Also, together with invigoration, some general analgesia seems to be effected by cold showers (Shevchuk, 2008). Rats exposed daily to several hours of standing in cold water showed greater longevity and lower weight, despite taking in more calories (Holloszy, 1986). Cold stress in rats increased beta-endorphins and adrenocorticotropic hormone immunoreactivity (Giagnoni, 1983). These are all signs that regular cold exposure furthers well-being.

Breast tenderness (mastalgia), especially during menses, can be relieved with regular cold washings of the breast (Uehlecke, 1999) and elimination of dairy products (Abraham, 1983) and possible other allergens. Research on treatment compared hot and cold compresses for painful breast engorgement while nursing. Both were effective. Using traditional cabbage leaves as compresses also worked, only slightly less so (Arora, 2008).

Sexuality after menopause, and in older age, in both sexes, should be approached playfully in the same way as sexual anhedonia in younger couples. Water, with its cleansing properties, has the additional advantage of lessening body odors, which in themselves can become impediments to intimate relationships in the elderly.

Erectile dysfunction. Floyer addresses male impotence as one of ailments readily helped by cold water. He calls the state of unwilling detumescence *weakness of the penis,* and *lost erection* (Floyer, 1732/2010)—brilliantly mocked by Laurence Sterne in the opening scene of his novel *Tristram Shandy* (Sterne, 1750–1767/2003), when Madame Shandy interrupts the procreating process with the untimely question if her husband had wound up the clock too. Fittingly, Floyer claims that his cold-water treatment helped 20 men with erectile dysfunction (ED): "the cold Water . . . has wound up their Watch, and set their Pendulum in statu quo" (Floyer, 1732/2010). His regimen was to immerse the bottom of such a man in cold water for an hour at a time (Floyer, 1732/2010), a practice that should be discouraged as shorter exposure times (half a minute to utmost three minutes) lead to better outcomes (Kneipp, 1886). Admittedly, the scientific literature for using potency-enhancing herbs in ED is more convincing than that for cold water. While direct hydromassage (water jet) has been recommended for ED (Karpenko, 1979), addressing potential underlying chronic disease causes such as diabetes, being overweight, and heart problems might be the better approach in the long run. Starting invigoration with a daily cold wash or cold shower and going for a walk every day is an auspicious beginning.

Frequent urinary tract infections/irritated bladder. During an acute urinary tract infection, no sitzbaths/baths should be taken—especially no cold applications. Warm footbaths, warm compressions, or a warm water bottle to the lower abdomen are advisable. If a high white blood cell count but no bacteria in the urine rule out infection, suggesting the diagnosis of irritable (allergic) bladder, a warm sitzbath is allowed and might be soothing, if possible with hay flowers. When symptoms are gone, cold-water therapy can be started cautiously for invigoration, starting with short gushes to the lower legs, ending with daily short (20 to 30 seconds at most) whole-body cold showers. Supporting measures are hydration with warm herbal teas for bladder and nerve health, regular moderate exercises like walking, lubrication during sexual activity with virgin coconut oil (its antibacterial properties are well documented), and prophylaxis with cranberry extract capsules after intercourse (Thormar, 2011; Uehlecke, 1997).

Nightly emission, later called *nocturnal pollutions,* involves involuntary nightly semen emissions, a natural occurrence, against which Kneipp himself—the "father of the cold water cure"—had recommended cold-water applications.

During Kneipp's time, many still believed that masturbation would harm body, mind, and soul. As a Catholic priest who could not bring himself to name sexual organs Kneipp, ironically, died of something *down there*—probably colon or prostatic cancer. And, as records show, his self-prescribed cold sitzbaths did not prevent his demise. A warning we should heed is that water therapy does not replace the good care of a physician.

The Science Behind the Water Cure

Our understanding of the effects of cold water therapy comes mostly from military studies of submerging volunteers in ice water for prolonged periods of time and from studies on winter swimming (doing the same thing as a hobby). Unfortunately, the Nazis misused prolonged ice baths in concentration camps, and they also have been forced onto psychiatry patients (Lifton, 1988, Whitaker, 2010). Cruel as a one-hour ice-water bath is, those studies have illuminated some of the molecular mechanisms of how cold water works in the body, which is helpful in a funding situation unlikely to provide grand-scale population studies on hydrotherapy in the near future, as there is no money to be made on the results. Water is available to (nearly) everyone in the developed world. Actually, access to water and sanitation is the main yardstick by which we judge the term "developed."

But some studies have been done. Even though hydrotherapy originally derives from folk practice, increasingly it is finding its place back among other tools in the doctor's bag, if judged by recent publications: Water has been used against fibromyalgia, multiple sclerosis, pressure ulcers, sports-related muscle damage, in critically ill intensive care unit patients, spondyloarthritis, and depression (Donze, 1983/2015; Felten-Barentsz, 2015; Fernandez Garcia, 2015; Latorre-Roman, 2015; Velez-Diaz-Pallares, 2015; Zamuner, 2015). While we focus on the healing properties of water for sexual afflictions, it still requires that we look at the entire body.

FROM THE SKIN TO THE THALAMIC-PITUITARY-ADRENAL AXIS

Touching the skin is sexual in grown-ups and might be necessary for infant survival (Abdallah, 2013). The specific thermal sensation of feeling hot or cold water starts with nerve endings in the skin (Campero, 2009, Noel, 2009). Those temperature sensors are necessary because humans are *homeotherms*—animals that keep their core temperature close to 37°C, plus or minus half a degree. Hypo- and hyperthermia both threaten health and survival. Responsible for this temperature *homeostasis* is the hypothalamus (under the thalamus), an almond-sized structure deep in the middle of the brain that also keeps other functions in balance: water and electrolyte concentration, hormone levels, metabolic parameters, circadian rhythms, and more. The hypothalamus also increases temperature settings—the result of which is called *fever*—to kill off invading germs. As one of the old brain structures preserved since long before the dinosaurs, the hypothalamus likewise gives rise to the most basic of feelings of hunger and thirst, hate and love, aggression and rage, and sexual urges and behaviors and sends out hormones that govern all the glands along the HPA axis (Kioukia-Fougia, 2002). The HPA axis is only an abbreviated way to speak of that axis. In reality, it comprises the hypothalamus-pituitary-thyroid-adrenal-gonadal glands. As the strong emotions and urges governed by the hypothalamus obviously can create stress when clashing with the real world, it is not surprising that, physiologically, the HPA axis controls the body's stress response. The skin and its complicated system of neurotransmitters, receptors, and ion channels directly relays to the hypothalamus and can therefore influence its hormonal output and stress responses. This is the basis of water therapy for sexual dysfunction.

The thalamus, the larger brain structure on top of the hypothalamus, consists of two halves, suggesting that the thalamus might relate to the two brain hemispheres in general. Similarly, the thalamus is the great communication center between the lower brain—the brain stem—and the cortex, the gray *thinking* matter. The thalamus brings sensations and feelings into consciousness. Already in the thalamus basic decisions about behavior are made (Davis, 1999).

GLANDS

In contrast to the conscious decisions of the cortex of the brain—the gray matter, the HPA axis governs the unconscious workings of hormones and glands. Incoming thermal skin stimuli affect all the endocrine glands in the body—from the pineal gland to the gonads: pineal, pituitary, thyroid, thymus, pancreas, adrenals, ovaries, and testes (Barra, 2014; Bodnar, 1979; Brennan, 2006; Nikolopoulos, 2013; Quintanar-Stephano, 1991; Rakonczay, 2001; Seggie, 1985; Smith, 1990). The pineal gland—sometimes called the *third eye*—nestles close to the thalamus and produces melatonin when it is dark. The pineal gland is the highest authority to regulate the HPA axis. For instance, hamsters that had undergone pinealectomy could not achieve the cold adaptations necessary to overwinter (Larkin, 2001). The gonads—our main subject here—are located on the other end of the axis; we discuss them later.

THE AUTONOMIC NERVOUS SYSTEM

Any cold or hot exposure also influences the autonomic nervous system, or ANS (Jansky, 2006). *Autonomic* is synonymous with *vegetative, visceral,* or *involuntary*—all terms expressing that we have no conscious influence over that part of the nerve system. However, newer science shows that meditation can voluntarily influence even the unconscious processes of the vegetative system (Kox, 2014).

SYMPATHETIC AND PARASYMPATHETIC BRANCHES OF THE AUTONOMIC NERVOUS SYSTEM

The ANS has two parts: the sympathetic fight-or-flight branch and the parasympathetic rest-and-digest branch. The main neurotransmitters in this system are dopamine, epinephrine, and norepinephrine. For our purposes here, it is important to know that the parasympathetic branch incites sexual arousal; the sympathetic branch incites orgasm. But it is more complicated than that: Both catecholamines/neurotransmitters epinephrine and norepinephrine are involved in the whole sexual cascade of hormones: if skin water sensations begin the activation of that cascade, it follows that everything that pertains to the skin might also pertain to sexuality.

Erogenous zones—around the mouth, the nipples, and the genital area—are well known. They very directly link cutaneous nervous sensation to sexual response. But those areas are by no means universal; considerable individual differences exist.

NOREPINEPHRINE

The catecholamine norepinephrine is part of the fight-or-flight mechanism of the ANS. It causes the heart to beat faster and furnishes the excitement and energy one needs in a difficult or dangerous situation (Mooventhan, 2014). In our stressful modern times, sympathetic and parasympathetic are out of balance: We overuse the sympathetic part. This lopsidedly overexcited flight-or-fight mechanism swamps the body with epinephrine and norepinephrine hormones and triggers the typical sympathetic response: piloerection (goose bumps) (which in furry animals is a great way to save core heat by plumping up the pelt and make them look more ferocious but is pretty useless in nearly hairless humans), increasing of metabolic rate (burning more cell fuel to stoke the internal fires), muscle shivering, skin vasoconstriction to keep the core temperature inside instead of letting it evaporate, and brown fat oxidation creating nonshivering thermogenesis.

Interestingly, norepinephrine is the signaling agent that is consistently elevated in cold-water studies (Leppaluoto, 2008; Smith, 1990). Because norepinephrine levels stay up in cold-exposed people, it is postulated that it may play a role in the cold *adaptation* process (Marino, 1998). Cold activation of the ANS, which in prehistoric times came naturally by inevitable weather exposure—among other triggers—now is blunted by civilization with winter heating and summer air conditioning and a sedentary lifestyle.

BETA-ENDORPHINS

Beta-endorphins (Gerra, 1992) are another group of neurotransmitters produced in the hypothalamus and pituitary gland. They are feel-good hormones, generated by cold-water exposure, exercise, and sex (Bodnar, 2014; Giagnoni, 1983; Hildebrandt, 2014). Women produce more feel-good hormones (Gerra, 1992). Age does not dampen the generation of beta-endorphins (Romero, 1987).

CYTOKINES

The direct access of the hypothalamus and the neuroendocrine axis is but one way in which cold and hot exposures influence the body. In and under the skin, temperature perception sets in motion small signaling molecules, also called *messenger molecules*. These include *cytokines*, mostly, as well as substance P, which helps decrease pain perception. The different kinds of cytokines can act as anti-inflammatory or proinflammatory (Dugue, 2000) agents. In a single cold exposure, the proinflammatory effects dominate, but with repeated exposures—through adaptation, habituation, and invigoration (Siems, 1994; Tipton, 2013; Wakabayashi, 2012)—the anti-inflammatory cytokines prevail (Castellani, 2002). Tiny cytokines travel abundantly and quickly throughout the body, functioning as an alarm and first-aid system against invading germs, announcing wounds and burns, cancer and inflammation, cranking up hormones in the glands (Smith, 1990) and immune markers (Brenner, 1999). These small molecules influence all organs, from brain to fat cells (Koska, 2002). And, not surprising, cytokines are involved in all aspects of male and female parts of reproduction and sexuality (Gaafar, 2014, Heng, 2014; Shi, 2014). Which means that the immune system has to function in order for sexuality to function.

THE IMMUNE SYSTEM

On many levels, cold and warm water have an effect on the immune system: Cold water seems to activate cellular immunity (phagocytes, T-lymphocytes, plus cytokines) and warm water humoral immunity (antibodies; Kalenova, 2005). The two parts are also called *slow* and *fast* immunity, or *innate* and *acquired* immunity.—Again, the ANS and the immune system work hand in hand: if sympathetic nerves are cut, immune function is impaired (Livnat, 1987). In a chemically ablated sympathetic nerve system the T-cell response—an important test for the function of the immune system—is hampered in mice (Madden, 1989). When norepinephrine—the substance consistently increased by cold-water immersion—was added to a T-cell culture, it sped up maturation of T cells (Swanson, 2001). After the first cold exposure, immune markers go down; however, habituation happens quickly, and adaptation shows by rising immune markers (Dugue, 2000). Exercise also boosts norepinephrine—that would account for why people in good physical shape have better immune systems (Kappel, 1998). Immune and autonomic nerve functions are closely interrelated but not yet understood in every detail (Kenney, 2014).

GUT BRAIN

Adaptation to our surroundings happens via the immune system, the brain, and the gut brain (Elenkov, 2000). The three parts, of course, are not independent of each other—they exchange information constantly and maintain *homeostasis* on many physical levels, including temperature, blood pressure, electrolyte balance, pH regulation, blood glucose stability, water level, and oxygen/carbon dioxide.

Interactions of the brain and the immune system are well documented (Heneka, 2015). The third participant in this interaction of body and world, the gut, assimilates or discards food material, with the help of beneficial bacteria. The bacterial sum living in and on our bodies is called the *microbiome* (Galland, 2014). Not surprisingly, there is a feedback loop to the brain (Luna, 2014) and to the immune system (Al-Nedawi, 2015). Water works on all three components: brain (Xiuwen, 2014), immune system (Kalenova, 2005), and gut brain (Hassan, 2014). This back-and-forth talk between brain, gut brain, and immune system again rides happily on the HPA axis, influencing the ANS (Elenkov, 2000).

HORMESIS INDUCES COLD ADAPTATION AND INVIGORATION

Contrary to expectations, repeated small noxious stimuli—called *hormesis*—will not harm but invigorate the body in the long run (Arumugam, 2006). Hormesis is how certain plant chemicals work: to ward off germs and critters, the plant develops polyphenols with antioxidant effects that are actually poisons. If we ingest those polyphenols in very small doses, they startle our immune system into quickly getting rid of those noxious substances, thereby activating self-defenses. Eventually they *invigorate* the body—or, in more modern language, *strengthen immune responses*. Cold water works exactly in the same way via hormesis: a regular small stress that over time invigorates. Cold exposure on the skin starts the whole balancing cascade of the *neuro-endocrine-immune system* (Slominski, 2012).

A daily cold shower will come to fruition after about six weeks or earlier (Jansky, 1996). Other stressors, working via hormesis, are exercise, fasting, and cognitive stimulation (Arumugam, 2006). The point here is that the stressors—as with cold-water exposure—must be not too strong and not too long-lasting to be effective.

Conclusion

While hydrotherapy seems to work for pelvic and sexual ills, not enough research has been done. But this volume shows that many other integrative modalities also induce healing: herbs, nutrition, psychiatric/psychological treatment, exercise, acupuncture, massage and bodywork, sufficient sleep, mindfulness, water, and many more—they all have been observed and documented to bring about better health and better sexual health. Contrastingly, in conventional—and often useful—medicine the paradigm is that one agent—a pill—will heal a specific ailment. The single-agent medication aspirin (acetylsalicylic acid) was discovered in 1853 and trademarked by Bayer in 1899. We are still finding new ways the drug works in our bodies (Kwiecien, 2015). The silver bullet theory may work in a small minority of diseases—mostly acute and of infectious and genetic origin. In the real world, diseases—especially chronic diseases—are multifactorial, developing from insufficient sleep, long hours of sitting, relationship problems and childhood trauma, bad food choices, stress at work and at home—stress even when having fun—pollution, and thwarted hopes. Even the simplest of medications is not simple, because the body is not.

Another way to look at water therapy is from the standpoint of genetics. Genes, in order to work, have to be switched on and off. Increasingly, we are learning that diseases such as atherosclerosis, obesity, heart attack, stroke, autoimmune diseases, and cancer have to do with *lifestyle*, which is a short form of saying genes are switched on and off. That can happen through unhealthy habits or through time-honored practices our ancient bodies respond to, because this comes naturally for them (Relle, 2015). The results differ.

We have seen that hydrotherapy affects the whole body, pushing it back into balance—beginning at the skin, going to the HPA axis, and to all of the glands (especially sexual glands like ovaries, testes, and prostate) and inner organs—not by an artificial plan that could go awry and yield unanticipated and unwanted side effects but by the simple way water and climate have always influenced the human body, and always helped to maintain internal homeostasis. It works in the same way other natural modalities work. Not much can go wrong if one takes a warm footbath or a cold shower; both will add to the inner balance of hormones and glands and neurotransmitters, affecting sexuality and sexual health too. We do not have to precisely know the way each molecule moves in the body and influences other molecules (even as we are trying to understand better the mechanisms involved). But we can observe the result: enhanced well-being, elevated mood, and improvement in health.

Acknowledgment

My thanks to Dorothée Berghaus and Roanne Weisman for helping to prepare this manuscript.

REFERENCES

Abdallah B, Badr LK, Hawwari M (2013). The efficacy of massage on short and long term outcomes in preterm infants. *Infant Behav Dev.* 36(4):662–669.

Abraham GE (1983). Nutritional factors in the etiology of the premenstrual tension syndromes. *J Reprod Med.* 28:446–464.

Al-Nedawi K, Mian MF, Hossain N, Karimi K, Mao YK, Forsythe P, Min KK, Stanisz AM, Kunze WA, Bienenstock J (2015). Gut commensal microvesicles reproduce parent bacterial signals to host immune and enteric nervous systems. *FASEB J.* 29:684–695.

Arora S, Vatsa M, Dadhwal V (2008). A comparison of cabbage leaves vs. hot and cold compresses in the treatment of breast engorgement. *Indian J Community Med.* 33:160–162.

Arumugam TV, Gleichmann M, Tang SC, Mattson MP (2006). Hormesis/preconditioning mechanisms, the nervous system and aging. *Ageing Res Rev.* 5:165–178.

Barnard ND, Scialli AR, Hurlock D, Bertron P (2000). Diet and sex-hormone binding globulin, dysmenorrhea, and premenstrual symptoms. *Obstet Gynecol.* 95:245–250.

Barra R, Cruz G, Mayerhofer A, Paredes A, Lara HE (2014). Maternal sympathetic stress impairs follicular development and puberty of the offspring. *Reproduction.* 148:137–145.

Basford JR, Oh JK, Allison TG, Sheffield CG, Manahan BG, Hodge DO, Tajik AJ, Rodeheffer RJ, Tei C (2009). Safety, acceptance, and physiologic effects of sauna bathing in people with chronic heart failure: a pilot report. *Arch Phys Med Rehab.* 90:173–177.

Beer AM (2011). [Naturopathy consultation. Dysmenorrhea]. *MMW Fortschr Med.* 153:23.

Beer AM (2013). [Naturopathy consultation. Venous insufficiency]. *MMW Fortschr Med.* 155:25.

Beever R (2009). Far-infrared saunas for treatment of cardiovascular risk factors: summary of published evidence. *Can Fam Physician*. 55(7):691–696.

Benouis K, Benabderrahmane M, Harrache-Chettouh D, Benabdeli K (2008). [Is Turkish bath water potable?: The baths of Sidi-Bel-Abbes]. *Santé* 18:63–66.

Bodnar RJ, Glusman M, Brutus M, Spiaggia A, Kelly DD (1979). Analgesia induced by cold-water stress: attenuation following hypophysectomy. *Physiol Behav*. 23:53–62.

Bodnar RJ (2014). Endogenous opiates and behavior: *Peptides*. 75:18–70.

Brenke R (2006). [Hydrotherapy and thermotherapy in the treatment of functional disorders]. *MMW Fortschr Med*. 148:26–28, 30.

Brennan CM, Breen EP, Porter RK (2006). Cold acclimation and oxygen consumption in the thymus. *Biochim Biophys Acta*. 1757:1463–1468.

Brenner IK, Castellani JW, Gabaree C, Young AJ, Zamecnik J, Shephard RJ, Shek PN (1999). Immune changes in humans during cold exposure: effects of prior heating and exercise. *J App Physiology*. 87:699–710.

Calhoun DA, Mutinga ML, Collins AS, Wyss JM, Oparil S (1993). Normotensive blacks have heightened sympathetic response to cold pressor test. *Hypertension*. 22:801–805.

Campero M, Baumann TK, Bostock H, Ochoa JL (2009). Human cutaneous C fibres activated by cooling, heating and menthol. *J Physiology*. 587:5633–5652.

Castellani JW, IK MB, Rhind SG (2002). Cold exposure: human immune responses and intracellular cytokine expression. *Med Sci Sports Exer.* 34:2013–2020.

Chen WY, Horton DJ (1978). Airways obstruction in asthmatics induced by body cooling. *Scand J Respiratory Dis*. 59:13–20.

Claridge RT (2015). Hydropathy or the cold water cure, as practiced by Vincent Priessnitz. ReInk Books/AbeBooks.com. (Original work published 1842–1843)

Cluett ER, Burns E (2009). Immersion in water in labour and birth. *Cochrane Database Sys Rev*. CD000111.

Cottrell BH (2003). Vaginal douching. *J Obstet Gynecol Neonatal Nurs*. 32:12–18.

Crinnion W (2007). Components of practical clinical detox programs: sauna as a therapeutic tool. *Altern Ther Health Med*. 13:S154–S156.

Crinnion WJ (2011). Sauna as a valuable clinical tool for cardiovascular, autoimmune, toxicant- induced and other chronic health problems. *Altern Med Rev.* 16:215–225.

Davis KD, Lozano RM, Manduch M, Tasker R., Kiss ZH, Dostrovsky JO (1999). Thalamic relay site for cold perception in humans. *J Neurophysiol*. 81:1970–1973.

Diego MA, Jones NA, Field T, Hernandez-Reif M, Schanberg S, Kuhn C, McAdam V, Galamaga R, Galamaga M (1998). Aromatherapy positively affects mood, EEG patterns of alertness and math computations. *Int. J Neurosci*. 96:217–224.

Doering TJ, Thiel J, Steuernagel B, Johannes B, Konitzer M, Niederstadt C, Schneider B, Fischer GC (2001). [Changes of cognitive brain functions in the elderly by Kneipp therapy]. *Forsch Komplementmed*. 8:80–84.

Donze C (2015). *Update on rehabilitation in multiple sclerosis*. Paris: Presse Medicale. (Original work published 1983)

Dugue B, Leppanen E (2000). Adaptation related to cytokines in man: effects of regular swimming in ice-cold water. *Clin Physiol.* 20:114–121.

Eisenbach C, Pohl J, Dikow R, Stremmel W, Encke J (2005). Severe rhabdomyolysis and renal failure triggered by a sauna visit in sickle cell trait: a case report. *Clin Nephrol.* 63:229–231.

Elenkov IJ, Wilder RL, Chrousos GP, Vizi ES (2000). The sympathetic nerve—an integrative interface between two supersystems: the brain and the immune system. *Pharmacol Rev.* 52:595–638.

Emmanuel AV, Kamm MA, Beard RW (2000). Reproducible assessment of vaginal and rectal mucosal and skin blood flow: laser doppler fluximetry of the pelvic microcirculation. *Clin Sci.* 98:201–207.

Ernst E (1990a). [Hardening against the common cold—is it possible?]. *Fortschr Med.* 108:586–588.

Ernst E, Pecho E, Wirz P, Saradeth T (1990b). Regular sauna bathing and the incidence of common colds. *Ann Med.* 22:225–227.

Fan YS, Miao FR, Liao AN, Xu F (2013). [Effect of drug-paste separated moxibustion on expression of estrogen, progestogen and their endometrial receptor mRNA in rats with primary dysmenorrhea]. *Zhen ci yan jiu = Acupuncture Res/Zhongguo yi xue ke xue yuan Yi xue qing bao yan jiu suo bian ji* 38:352–357.

Felten-Barentsz KM, Haans AJ, Slutsky AS, Heunks LM, van der Hoeven JG (2015). Feasibility and safety of hydrotherapy in critically ill ventilated patients. *Am J Respir Crit Care Med.* 191:476–477.

Fernandes JV, Cobucci RN, Jatoba CA, de Medeiros Fernandes TA, de Azevedo JW, de Araujo JM (2015). The role of the mediators of inflammation in cancer development. *Pathol Oncol Res.* 21(3):527–534.

Fernandez Garcia R, Sanchez Sanchez LC, Lopez Rodriguez MD, Sanchez Granados G (2015). [Effects of an exercise and relaxation aquatic program in patients with spondyloarthritis: A randomized trial.]. *Med clinica* 145(9):380–384.

Fleckenstein A, Weisman R (2006). *Healthy to 100—aging with vigor and grace.* Deerfield Beach, FL: Health Communications.

Fleckenstein A, Weisman R (2007). *Health2O—tap into the healing powers of water to fight disease, look younger, and feel your best.* New York: McGrawHill

Fleckenstein A (2014). *the diabetes cure: the 5-step plan to eliminate hunger, lose weight, and reverse diabetes—for good!* Emmaus, PA: Rodale.

Flexner A (1910) *Medical education in the United States and Canada: a report to the Carnegie Foundation for the Advancement of Teaching.* University of Michigan Library.

Floyer J (2010). *Psychrolousia, or, the history of cold bathing: both ancient and modern: in two parts (London).* Gale ECCO, Print Editions. (Original work published 1732)

Fujita S, Ikeda Y, Miyata M, et al. (2011). Effect of Waon therapy on oxidative stress in chronic heart failure. *Circ J.* 75(2):348–356.

Furnari A, Calabro RS, Gervasi G, La Fauci-Belponer F, Marzo A, Berbiglia F, Paladina G, De Cola MC, Bramanti P (2014). Is hydrokinesitherapy effective on gait and balance in patients with stroke? A clinical and baropodometric investigation. *Brain Inj.* 28:1109–1114.

Gaafar TM, Hanna MO, Hammady MR, Amr HM, Osman OM, Nasef A, Osman AM (2014). Evaluation of cytokines in follicular fluid and their effect on fertilization and pregnancy outcome. *Immunol Invest.* 43:572–584.

Galland L.(2014). The gut microbiome and the brain. *J Med Food.* 17(12):1261–1272.

Gerra G, Volpi R, Delsignore R, Maninetti L, Caccavari R, Vourna S, Maestri D, Chiodera P, Ugolotti G, Coiro V (1992). Sex-related responses of beta-endorphin, ACTH, GH and PRL to cold exposure in humans. *Acta Endocrinol.* 126:24–28.

Giagnoni G, Santagostino A, Senini R, Fumagalli P, Gori E (1983). Cold stress in the rat induces parallel changes in plasma and pituitary levels of endorphin and ACTH. *Pharm Res Comm.* 15:15–21.

Goodall, J (n.d.). Waterfall displays. YouTube. https://www.youtube.com/watch?v=jjQCZClpaaY.

Graz B, Savoy M, Buclin T, Bonvin E (2014). [Dysmenorrhea: patience, pills or hot-water bottle?]. *Rev Med Suisse.* 10:2285–2288.

Groark KP (2005). Vital warmth and well-being: steambathing as household therapy among the Tzeltal and Tzotzil Maya of highland Chiapas, Mexico. *Soc Sci Med.* 61:785–795.

Guo H, Zhang HG, Xue BG, Sha YW, Liu Y, Liu RZ (2006). [Effects of cigarette, alcohol consumption and sauna on sperm morphology]. *Zhonghua nan ke xue = Natl J Androl.* 12:215–217, 221.

Guo SW, Mao X, Ma Q, Liu X (2013). Dysmenorrhea and its severity are associated with increased uterine contractility and overexpression of oxytocin receptor (OTR) in women with symptomatic adenomyosis. *Fert Steril.* 99:231–240.

Gupta P (2006). Randomized, controlled study comparing sitz-bath and no-sitz-bath treatments in patients with acute anal fissures. *ANZ J Surg.* 76:718–721.

Gupta PJ (2008). Warm sitz bath does not reduce symptoms in posthaemorrhoidectomy period: a randomized, controlled study. *ANZ J Surg.* 78:398–401.

Hashiguchi N, Ni F, Tochihara Y (2002). Effects of room temperature on physiological and subjective responses during whole-body bathing, half-body bathing and showering. *J Physiol Anthro App Hum Sci.* 21:277–283.

Hassan AM, Jain P, Reichmann F, Mayerhofer R, Farzi A, Schuligoi R, Holzer P (2014). Repeated predictable stress causes resilience against colitis-induced behavioral changes in mice. *Front Behav Neurosci.* 8:386.

Hayasaka S, Goto Y, Maeda-Yamamoto M. (2013). The effects of bathing in hot springs on the absorption of green tea catechin: a pilot study. *Complement Ther Clin Pract.* 19:243–245.

Helamaa E, Aikas E (1988). The secret of good "löyly." *Ann Clin Res.* 20:224–229.

Heneka MT, Carson MJ, Khoury JE, et al. (2015). Neuroinflammation in Alzheimer's disease. *Lancet Neurol.* 14:388–405.

Heng YJ, Liong S, Permezel M, Rice GE, Di Quinzio MK, Georgiou HM (2014). The interplay of the interleukin 1 system in pregnancy and labor. *Reprod Sci.* 21:122–130.

Hildebrandt T, Shope S, Varangis E, Klein D, Pfaff DW, Yehuda R. (2014). Exercise reinforcement, stress, and beta-endorphins: an initial examination of exercise in anabolic-androgenic steroid dependence. *Drug Alcohol Depend.* 139:86–92.

Hirota T, Kay SA (2015). Identification of small-molecule modulators of the circadian clock. *Meth Enzymol.* 551:267–282.

Holloszy JO, Smith EK (1986). Longevity of cold-exposed rats: a reevaluation of the "rate-of-living theory." *J App Physiol.* 61:1656–1660.

Holtz TH, Thienkrua W, McNicholl JM, et al. (2012). Prevalence of Treponema pallidum seropositivity and herpes simplex virus type 2 infection in a cohort of men who have sex with men, Bangkok, Thailand, 2006–2010. *Int J STD AIDS.* 23(6):424–428. doi:10.1258/ijsa.2011.011256.

Hosono T, Takashima Y, Morita Y, Nishimura Y, Sugita Y, Isami C, Sakamoto I, Tagami K, Hidaka Y, Suzuki A (2010). Effects of a heat- and steam-generating sheet on relieving symptoms of primary dysmenorrhea in young women. *J Obstet Gynaec Res.* 36:818–824.

Huang PH, Chen JW, Lin CP, Chen YH, Wang CH, Leu HB, Lin SJ (2012). Far infrared therapy promotes ischemia-induced angiogenesis in diabetic mice and restores high glucose-suppressed endothelial progenitor cell functions. *Cardiovasc Diabetol.* 11:1–13.

Huttunen P, Kokko L, Ylijukuri V (2004). Winter swimming improves general well-being. *Int J Circumpolar Health.* 63:140–144.

Imai Y, Nobuoka S, Nagashima J, Awaya T, Aono J, Miyake F, Murayma M (1998). Acute myocardial infarction induced by alternating exposure to heat in a sauna and rapid cooling in cold water. *Cardiology.* 90:299–301.

Jansky L, Matouskova E, Vavra V, Vybiral S, Jansky P, Jandova D, Knizkova I, Kunc P (2006). Thermal, cardiac and adrenergic responses to repeated local cooling. *Physiol Res/Acad Sci Bohemoslovaca* 55:543–549.

Jansky L, Sramek,P, Savlikova J, Ulicny B, Janakova H, Horky K (1996). Change in sympathetic activity, cardiovascular functions and plasma hormone concentrations due to cold water immersion in men. *Eur J App Physiol Occup Physiol.* 74:148–152.

Kalenova LF, Sukhovei YG, Fisher TA (2005). Specific and nonspecific reactions of mouse immune system under the effect of short-term exposure in warm and/or cold water. *Bull Exp Biol Med.* 140:720–722.

Kang D, Liao M, Jiang Z, et al. (2011). Commercial sex venues, syphilis and methamphetamine use among female sex workers. *AIDS Care.* 23(Suppl 1):26–36.

Kappel M, Poulsen TD, Galbo H, Pedersen BK (1998). Effects of elevated plasma noradrenaline concentration on the immune system in humans. *Eur J App Physiol Occup Physiol.* 79:93–98.

Karpenko EI, Skripal' Iu A. (1979). [Hydromassage of the prostate gland in the treatment of sex disorders]. *Vrachebnoe Delo*, 89–92.

Kauppinen K (1989). Sauna, shower, and ice water immersion. Physiological responses to brief exposures to heat, cool, and cold. Part II. Circulation. *Arctic Med Res.* 48:64–74.

Kemble HB, Weiss HR (1967). *The great American water-cure craze: a history of hydropathy in the United States.* Trenton, NJ: Past Times Press.

Kenney MJ, Ganta CK (2014). Autonomic nervous system and immune system interactions. *Comp Physiol.* 4:1177–1200.

Kheirkhah M, Vali Pour NS, Nisani L, Haghani H (2014). Comparing the effects of aromatherapy with rose oils and warm foot bath on anxiety in the first stage of labor in nulliparous women. *Iranian Red Crescent Med J* 16:e14455.

Kioukia-Fougia N, Antoniou K, Bekris S, Liapi C, Christofidis I, Papadopoulou-Daifoti Z (2002). The effects of stress exposure on the hypothalamic-pituitary-adrenal axis, thymus, thyroid hormones and glucose levels. *Prog Neuropsychopharmacol Biol Psychiatry.* 26:823–830.

Knechtle B, Christinger N, Kohler G, Knechtle P, Rosemann T (2009). Swimming in ice cold water. *Irish J Med Sci.* 178:507–511.

Kneipp S (1886). *Meine Wasserkur.* (J. Kösel'sche Buchhandlung). [My water cure]. (English translation 1893)

Kneipp S (1892). *So sollt ihr leben! Winke und Rathschläge für Gesunde und Kranke zu einer einfachen, vernünftigen Lebensweise und einer naturgemässen Heilmethode.* (J. Kösel'sche Buchhandlung). [Thus shalt thou live: hints and advice for the healthy and the sick on a simple and rational mode of life and a natural method of cure].

Koska J, Ksinantova L, Sebokova E, Kvetnansky R, Klimes I, Chrousos G, Pacak K (2002). Endocrine regulation of subcutaneous fat metabolism during cold exposure in humans. *Ann N Y Acad Sci.* 967:500–505.

Kox M, van Eijk LT, Zwaag J, van den Wildenberg J, Sweep FC, van der Hoeven JG, Pickkers P (2014). Voluntary activation of the sympathetic nervous system and attenuation of the innate immune response in humans. *Proc Natl Acad Sci U S A* 111:7379–7384.

Kumar S, Murarka S, Mishra VV, Gautam AK (2014). Environmental & lifestyle factors in deterioration of male reproductive health. *Indian J Med Res.* 140(Suppl):S29–S35.

Kwiecien S, Magierowska K, Sliwowski Z, Wojcik D, Magierowski M, Brzozowski T (2015). New insight into the mechanisms of gastroduodenal injury induced by nonsteroidal anti-inflammatory drugs: practical implications. *Pol Arch Med Wewn.* 125(3):191–198.

LaFoy J, Geden EA (1989). Postepisiotomy pain: warm versus cold sitz bath. J *Obstet Gynecol Neonatal Nurs.* 18:399–403.

Larkin JE, Freeman DA, Zucker I (2001). Low ambient temperature accelerates short-day responses in Siberian hamsters by altering responsiveness to melatonin. *J Biol Rhythms.* 16:76–86.

Latorre-Roman PA, Rentero-Blanco M, Laredo-Aguilera JA, Garcia-Pinillos F (2015). Effect of a 12-day balneotherapy programme on pain, mood, sleep, and depression in healthy elderly people. *Psychogeriatrics.* 15:14–19.

Laukkanen T, Khan H, Zaccardi F, Laukkanen JA (2015). Association between sauna bathing and fatal cardiovascular and all-cause mortality events. *JAMA Intern Med.* 175(4):542–548.

Lee EC, Watson G, Casa D, Armstrong LE, Kraemer W, Vingren JL, Spiering BA, Maresh CM (2012). Interleukin-6 responses to water immersion therapy after acute exercise heat stress: a pilot investigation. *J Athlet Train.* 47:655–663.

Leppaluoto J, Huttunen P, Hirvonen J, Vaananen A, Tuominen M, Vuori J (1986). Endocrine effects of repeated sauna bathing. *Acta Physiol Scand.* 128:467–470.

Leppaluoto J, Westerlund T, Huttunen P, Oksa J, Smolander J, Dugue B, Mikkelsson M (2008). Effects of long-term whole-body cold exposures on plasma concentrations of ACTH, beta-endorphin, cortisol, catecholamines and cytokines in healthy females. *Scand J Clin Lab Invest.* 68:145–153.

Levina IL, Klepikov II, Mandrova RR (1993). [The therapeutic efficacy of the Russian bath]. *Vopr Kurortol Fizioter Lech Fiz Kult.* 5:54–57.

Liao WC, Wang L, Kuo CP, Lo C, Chiu MJ, Ting H (2013). Effect of a warm footbath before bedtime on body temperature and sleep in older adults with good and poor sleep: an experimental crossover trial. *Int J Nurs Stud.* 50(12):1607–1616.

Lifton RJ (1988). *The Nazi doctors: medical killing and the psychology of genocide.* New York: Basic Books.

Lissner E. (1992). New nonhormonal contraceptive methods for men. *Chang Men.* Summer-Fall:24–25.

Livingston R (2010). Medical risks and benefits of the sweat lodge. *J Altern Complement Med.* 16:617–619.

Livnat S, Madden KS, Felten DL, Felten, SY (1987). Regulation of the immune system by sympathetic neural mechanisms. *Prog Neuropsychopharmacol Biol Psychiatry.* 11:145–152.

Luna RA, Foster JA (2014). Gut brain axis: diet microbiota interactions and implications for modulation of anxiety and depression. *Curr Opin Biotech.* 32C:35–41.

Lust B (1901). [Frontispiece]. *The Kneipp Water Cure monthly copies and Herald of Health.* 2:171.

Madden KS, Felten SY, Felten DL, Sundaresan PR, Livnat S (1989). Sympathetic neural modulation of the immune system. I. Depression of T cell immunity in vivo and vitro following chemical sympathectomy. *Brain Behav Immun.* 3:72–89.

Marino F, Sockler JM, Fry JM (1998). Thermoregulatory, metabolic and sympathoadrenal responses to repeated brief exposure to cold. *Scand J Clin Lab Invest.* 58:537–545.

Master WH, Johnson VE (1976). Principles of the new sex therapy. *Am J Psychiatry.* 133:548–554.

Masuda A, Koga Y, Hattanmaru M, Minagoe S, Tei C (2005). The effects of repeated thermal therapy for patients with chronic pain. *Psychotherap Psychosom.*74:288–294.

McCarthy C, Fleming N, Donne B, Blanksby B (2014). Barefoot running and hip kinematics: good news for the knee? *Med Sci Sports Exer.* 47(5):1009–1016.

Mehl-Madrona L (2010). Sweat lodge. *J Altern Complement Med.* 16:609–610.

Molnar GW (1975). Body temperatures during menopausal hot flashes. *J App Physiol.* 38:499–503.

Monteiro MA, Gabriel RE, Sousa MF, Abrantes JM, Moreira MH (2011). Impact of a 12-month exercise program on the temporal parameters of the foot rollover during walking in postmenopausal women. *Menopause.* 18:771–777.

Mooventhan A, Nivethitha L (2014). Scientific evidence-based effects of hydrotherapy on various systems of the body. *N Am J Med Sci.* 6:199–209.

Mulec J, Kristufek V, Chronakova A, Oarga A, Scharfen J, Sestauberova M (2015). Microbiology of healing mud (fango) from roman thermae aquae iasae archaeological site (varazdinske toplice, croatia). *Microb Ecol.* 69:293–306.

Musiek ES, Xiong DD, Holtzman DM (2015). Sleep, circadian rhythms, and the pathogenesis of Alzheimer disease. *Exp Mol Med.* 13(47):e148.

Ness RB, Soper DE, Holley RL, et al. (2001). Douching and endometritis: results from the PID Evaluation and Clinical Health (PEACH) study. *Sex Trans Dis.* 28:240–245.

Nikolopoulos I, Osman W, Haoula Z, Jayaprakasan K, Atiomo W (2013). Scrotal cooling and its benefits to male fertility: a systematic review. *J Obstet Gynaec.* 33:338–342.

Noel J, Zimmermann K, Busserolles J, et al. (2009). The mechano-activated K+ channels TRAAK and TREK-1 control both warm and cold perception. *EMBO J.* 28:1308–1318.

Obladen M (2017). Social birth: rites of passage for the newborn. *Neonatology.* 28;112(4):317–323.

Ohori T, Nozawa T, Ihori H, Shida T, Sobajima M, Matsuki A, Yasumura S, Inoue H (2012). Effect of repeated sauna treatment on exercise tolerance and endothelial function in patients with chronic heart failure. *Am J Cardiol.* 109:100–104.

Parandavar N, Mosalanejad L, Ramezanli S, Ghavi F (2014). Menopause and crisis? Fake or real: comprehensive search to the depth of crisis experienced: a mixed-method study. *Glob J Health Sci.* 6:246–255.

Park SJ, Kikufuji N, Hyun KJ, Tokura H (2004). Effects of barefoot habituation in winter on thermal and hormonal responses in young children: a preliminary study. *J Hum Ergol.* 33:61–67.

Park SU, Lee SH, Chung YG, Park KK, Mah SY, Hong SJ, Chung BH (2010). Warm sitz bath: Are there benefits after transurethral resection of the prostate? *Korean J Urol.* 51:763–766.

Perkins KP, Hanney WJ, Rothschild CE (2014). The risks and benefits of running barefoot or in minimalist shoes: a systematic review. *Sports Health.* 6:475–480.

Pilch W, Szygula Z, Palka T, Pilch P, Cison T, Wiecha S, Tota L (2014). Comparison of physiological reactions and physiological strain in healthy men under heat stress in dry and steam heat saunas. *Biol Sport.* 31:145–149.

Quintanar-Stephano JL, Quintanar-Stephano A, Castillo-Hernandez L (1991). Effect of the exposure to chronic-intermittent cold on the thyrotropin and thyroid hormones in the rat. *Cryobiology.* 28:400–403.

Radoff KA, Thompson LM, Bly KC, Romero C (2013). Practices related to postpartum uterine involution in the Western Highlands of Guatemala. *Midwifery.* 29:225–232.

Rakonczay Z Jr, Takacs T, Mandi Y, Ivanyi B, Varga S, Papai G, Boros I, Lonovics J (2001). Water immersion pretreatment decreases pro-inflammatory cytokine production in cholecystokinin-octapeptide-induced acute pancreatitis in rats: possible role of HSP72. *Int J Hyperthermia.* 17:520–535.

Rao M, Zhao XL, Yang J, Hu SF, Lei H, Xia W, Zhu CH (2015). Effect of transient scrotal hyperthermia on sperm parameters, seminal plasma biochemical markers, and oxidative stress in men. *Asian J Androl.* 17(4):668–675.

Raskovic A, Milanovic I, Pavlovic N, Milijasevic B, Ubavic M, Mikov M (2015). Analgesic effects of rosemary essential oil and its interactions with codeine and paracetamol in mice. *Eur Rev Med Pharmacol Sci.* 19:165–172.

Relle M, Foehr B, Schwarting A (2015). Epigenetic aspects of systemic lupus erythematosus. *Rheumatol Ther.* 2(1):33–46.

Robiner WN (1990). Psychological and physical reactions to whirlpool baths. *J Behav Med.* 13:157–173.

Romero MT, Bodnar RJ (1987). Maintenance of beta-endorphin analgesia across age cohorts. *Neurobiol Aging.* 8:167–170.

Saikhun J, Kitiyanant Y, Vanadurongwan V, Pavasuthipaisit K (1998). Effects of sauna on sperm movement characteristics of normal men measured by computer-assisted sperm analysis. *Int J Andol.* 21:358–363.

Seggie J, Campbell L, Brown GM, Grota LJ. (1985). Melatonin and N-acetylserotonin stress responses: effects of type of stimulation and housing conditions. *J Pineal Res.* 2:39–49.

Shafik A (1993a). Role of warm-water bath in anorectal conditions: the "thermosphincteric reflex." *J Clin Gastroenterol.* 16:304–308.

Shafik A (1993b). Role of warm water bath in inducing micturition in postoperative urinary retention after anorectal operations. *Urol Int.* 50:213–217.

Sharifi F, Simbar M, Mojab F, Majd HA (2013). Comparison of the effects of Matricaria chamomila (Chamomile) extract and mefenamic acid on the intensity of premenstrual syndrome. *Complement Ther Clin Pract.* 20(1):81–88.

Shevchuk NA (2008). Adapted cold shower as a potential treatment for depression. *Med Hypoth.* 70:995–1001.

Shevchuk NA, Radoja S (2007). Possible stimulation of anti-tumor immunity using repeated cold stress: a hypothesis. *Infect Agents Canc.* 2:20.

Shi B, Wei RX, Cui Y, Zhou J, Zhang XX (2014). [IL-6 and sICAM-1 in seminal plasma relate to male immune infertility]. *Zhonghua nan ke xue = Natl J Androl.* 20:1098–1102.

Showalter SL, Brown JC, Cheville AL, Fisher CS, Sataloff D, Schmitz KH (2013). Lifestyle risk factors associated with arm swelling among women with breast cancer. *Ann Surg Oncol.* 20:842–849.

Shui S, Wang X, Chiang JY, Zheng L (2015). Far-infrared therapy for cardiovascular, autoimmune, and other chronic health problems: a systematic review. *Exp Biol Med.* 240(10):1257–1265.

Siems WG, van Kuijk FJ, Maass R, Brenke R (1994). Uric acid and glutathione levels during short-term whole body cold exposure. *Free Radical Biol Med.* 16:299–305.

Sirvio J, Jolkkonen J, Pitkanen A (1987). Adenohypophyseal hormone levels during hyperthermia. *Endocrinologie.* 25:21–23.

Slominski AT, Zmijewski MA, Skobowiat C, Zbytek B, Slominski RM, Steketee JD (2012). Sensing the environment: regulation of local and global homeostasis by the skin's neuroendocrine system. *Adv Anatomy Embryol Cell Biol* 212(v– vii):1–115.

Smith DJ, Deuster PA, Ryan CJ, Doubt TJ (1990). Prolonged whole body immersion in cold water: hormonal and metabolic changes. *Undersea Biomed Res.* 17:139–147.

Sobajima M, Nozawa T, Ihori H, Shida T, Ohori T, Suzuki T, Matsuki A, Yasumura S, Inoue H (2013). Repeated sauna therapy improves myocardial perfusion in patients with chronically occluded coronary artery-related ischemia. *Int J Cardiol.* 167:237–243.

Sorri P (1988). The sauna and sauna bathing habits: a psychoanalytic point of view. *Ann Clin Res.* 20:236–239.

Stanislavov R, Nikolova V (2003). Treatment of erectile dysfunction with pycnogenol and L-arginine. *J Sex Marital Ther.* 29:207–213.

Sterne L (2003). *The life and opinions of Tristram Shandy, gentleman (or Tristram Shandy).* (Original work published 1759–1767). New York: Oxford University Press, Oxford World's Classics.

Stiefelhagen P (2005). [Functional disorders call for total therapy—Kneipp's hydrotherapy instead of psychopharmaceuticals]. *MMW Fortschr Med.* 147:4–6, 8.

Sutkowy P, Wozniak A, Boraczynski T, Mila-Kierzenkowska C, Boraczynski M (2014). The effect of a single Finnish sauna bath after aerobic exercise on the oxidative status in healthy men. *Scand J Clin Lab Invest.* 74:89–94.

Swanson MA, Lee WT, Sanders VM (2001). IFN-gamma production by Th1 cells generated from naive CD4+ T cells exposed to norepinephrine. *J Immunol.* 166:232–240.

Szekely EB (1973). *Healing waters.* San Diego, CA: Academy Books.

Tam N, Astephen Wilson JL, Noakes TD, Tucker R (2014). Barefoot running: an evaluation of current hypothesis, future research and clinical applications. *Br J Sports Med.* 48:349–355.

Thangadurai P, Gopalakrishnan R, Kuruvilla A, Jacob KS, Abraham VJ, Prasad J (2014). Sexual dysfunction among men in secondary care in southern India: nature, prevalence, clinical features and explanatory models. *Natl Med. J India.* 27:198–201.

Thormar H, ed. (2011). *Lipids and essential oils as antimicrobial agents.* New York: John Wiley.

Tipton MJ, Wakabayashi H, Barwood MJ, Eglin CM, Mekjavic IB, Taylor NA (2013). Habituation of the metabolic and ventilatory responses to cold-water immersion in humans. *J Therm Biol.* 38:24–31.

Tobias DK, Gaskins AJ, Missmer SA, Hu FB, Manson JE, Buck Louis GM, Zhang C, Chavarro JE (2015). History of infertility and risk of type 2 diabetes mellitus: a prospective cohort study. *Diabetologia.* 58:707–715.

Trall RT (1996). *True healing art or hygienic vs. drug medication.* Washington, DC: Kessinger. (Original work published 1880)

Uehlecke B, Brüggemann W (1997). *Kneipp Vademecum pro medico—Hydrotherapie–Phytotherapie—Bewegungstherapie—Ernährung—Ordnungstherapie.* [Kneipp Vademecum pro medico—hydrotherapy—phytotherapy—movement therapy—nutrition—order therapy] Würzburg, Germany: "Sebastian-Kneipp"-Gesundheitsmittel-Verlag.

Uehlecke B, Hentschel H-D (1999). Gesund leben mit Kneipp—Vorbeugen und Heilen mit der Kneipp-Methode [Living healthily with Kneipp—Prevention and curing with the Kneipp system]. München: Ehrenwirth.

Vaseenon T, Wattanarojanaporn T, Intharasompan P, Theeraamphon N, Auephanviriyakul S, Phisitkul P (2015). Foot and ankle problems in Thai monks. *J Med Assoc Thailand = Chotmaihet Thangphaet.* 98:71–76.

Velez-Diaz-Pallares M, Lozano-Montoya I, Abraha I, Cherubini A, Soiza RL, O'Mahony D, Montero-Errasquin B, Cruz-Jentoft AJ (2015). Nonpharmacologic interventions to heal pressure ulcers in older patients: an overview of systematic reviews (The SENATOR-ONTOP Series). *J Am Med Dir Assoc.* 16(6):448–469.

Vuori I (1988). Sauna bather's circulation. *Ann Clin Res.* 20:249–256.

Wähä-Eskeli, K., and Erkkola, R. (1988). The sauna and pregnancy. *Ann Clin Res.* 20, 279–282.

Wakabayashi H, Wijayanto T, Kuroki H, Lee JY, Tochihara Y (2012). The effect of repeated mild cold water immersions on the adaptation of the vasomotor responses. *Int J Biometrol.* 56:631–637.

Whitaker R (2010). *Mad in America: bad science, bad medicine, and the enduring mistreatment of the mentally ill.* New York: Basic Books.

Xiuwen Y, Hongchen L, Ke L, Zhen J, Gang L. (2014). [Effects of noxious coldness and non-noxious warmth on the magnitud e of cerebral cortex activation during intraoral stimulation with water]. Hua xi kou qiang yi xue za zhi = Huaxi kouqiang yixue zazhi = *West China J Stomatology.* 32:552–555.

Yamamoto K, Nagata S (2011). Physiological and psychological evaluation of the wrapped warm footbath as a complementary nursing therapy to induce relaxation in hospitalized patients with incurable cancer: a pilot study. *Cancer Nurs.* 34:185–192.

Yoon SJ, Park JK, Oh S, Jeon DW, Yang JY, Hong SM, Kwak MS, Choi YS, Rim SJ, Youn HJ (2011). A warm footbath improves coronary flow reserve in patients with mild-to-moderate coronary artery disease. *Echocardiography.* 29(10):1119–1124.

Zamuner AR, Andrade CP, Forti M, Marchi A, Milan J, Avila MA, Catai AM, Porta A, Silva E (2015). Effects of a hydrotherapy programme on symbolic and complexity dynamics of heart rate variability and aerobic capacity in fibromyalgia patients. *Clin Exp Rheumatol.* 33:73–81.

23

Botanicals (Adaptogens and Others) that Support or Stimulate Desire: Traditions and Science

ERIC YARNELL

Numerous herbs were used traditionally as aphrodisiacs, meaning agents that enhance male and/or female sexual desire. Here the most historically important, most scientifically validated, and most clinically relevant herbs in this category from numerous cultures are discussed along with case studies highlighting their effective use (Table 23.1). Many trials have looked at the effects of herbs on libido in the context of erectile dysfunction (ED) in men, so the impact of these herbs on ED beyond sexual desire must also be mentioned.

Adulteration of Herbal Products with Drugs

Given the high demand for safe and effective aphrodisiacs and treatments for ED, the global marketplace is full of low-quality products claiming to be dietary supplements that are in fact intentionally spiked with approved and unapproved phosphodiesterase type 5 inhibitor (PDE5i) drugs, anabolic steroids, or other agents not claimed on the label. For example, 134 of 175 (77%) purported herbal sexual aids seized by authorities in Singapore contained sildenafil or other PDE5i drugs (Low, 2009). Some of these illegal drugs contained glyburide, and 10 people died due to unwitting overdose of this hypoglycemic agent. Among 91 purported herbal sexual performance-enhancing products taken from stores in two American cities or seized by US customs, 75 (81%) contained unclaimed PDE5i or analogs (Campbell, 2013).

Besides being completely unethical, adulteration of dietary supplements with these ingredients is illegal in most jurisdictions and quite unsafe as consumers do not know the drugs are present and thus do not know the dose or danger they are

Table 23.1. Botanicals Relevant to Sexual Function

Latin Name	*Common Name*	*Part Used*	*Main Action(s)*	*Dose Form(s) and Dose(s)*
Crocus sativus	Saffron	Stamen	Endothelial tonic, mild aphrodisiac	Extract: 200 mg qd-bid Tincture: 1–2 ml tid
Epimedium spp	Horny goat weed	Leaf and stem	Endothelial tonic, mild aphrodisiac	Tincture: 3 ml tid Infusion: 3 g tid
Panax ginseng	Asian ginseng, red ginseng	Steamed, dried root	Stimulating adaptogen, endothelial tonic, mild aphrodisiac	Powder: 1 g bid-tid Tincture: 2–3 ml tid
Pausinystalia johimbe	Yohimbe	Bark	Strong aphrodisiac, central α-2 adrenergic antagonist, stimulant	Tincture: 5 gtt bid Pure yohimbine: 5.4 mg bid-tid
Tribulus terrestris[a]	Caltrop vine	Seed capsule	Mild aphrodisiac, otherwise unclear	Tincture: 3 ml tid
Turnera diffusa[a]	Damiana	Leaf	Stimulating nervine, mild aphrodisiac	Tincture: 3–5 ml tid
Withania somnifera[a]	Ashwagandha	Fresh root	Calming adaptogen, mild aphrodisiac	Powder: 1–5 g bid-tid Tincture: 3–5 ml tid

[a] Clinical trial evidence is weak or lacking but has a strong history of use in traditional medicine.

exposing themselves to. The burden on typically understaffed and underfunded government regulatory agencies to detect and act against companies making, importing, and selling adulterated products is enormous and extremely difficult to combat, as companies can shut down and reopen in a nearby location with a new name and address far more quickly and easily than regulators can find and stop them. In many cases, even when regulators take actions such as demanding recalls, the companies simply do not comply and the regulators lack funding or staffing to enforce their demands (Cohen, 2013). This situation also casts a pall over companies that are selling authentic, nonadulterated products legitimately, making the entire dietary supplement industry look bad.

Patients should be educated about the problem of PDE5i adulteration and counseled to never purchase purported dietary supplements for sexual enhancement advertised on the Internet or from health food stores. Providers should recommend specific brands (in which they have no financial interest) from legitimate companies with a broad range of products and not just sexual enhancement supplements to avoid exposing patients to undeclared drugs in fake products. Hopefully funding and staffing problems at regulatory agencies will be corrected so that aggressive enforcement action can be taken, making the cost of violating the law too high and reducing the incentive to cheat.

Adaptogens as Sexual Aides

The concept of an herbal adaptogens was created by Soviet scientists, notably doctors Nicholai Lazarev and Israel Brekhman (Pawar, 2012). These were defined as nontoxic agents that increased resistance to stressors of all kinds, and in particular counteracted fatigue. While the effects of these agents on libido were not explicitly discussed by those who coined the term and developed the concepts, the fairly well-documented negative effects of stress on libido and sexual function suggest that adaptogens might have a role in counteracting stress-induced sexual dysfunction (Hamilton, 2014).

Panax ginseng (Asian ginseng or rén shēn in Mandarin Chinese), *P. quinquefolius* (American ginseng, xī yáng shēn), and *Eleutherococcus senticosus* (eleuthero, cì wǔ jiā) are three members of the Araliaceae or ginseng family that are among the best studied of adaptogenic herbs. Asian ginseng (also called Korean or Chinese ginseng) is native to eastern Asia and is one of the most well-known and prized herbs in the world. Its Latin genus name derives from the word "panacea," giving one a sense of its broad esteem and applicability. It is traditionally considered a "heating" or stimulating herb in Chinese traditional medicine, though the degree to which it is energizing versus calming (as a result of counteracting the harmful effects of stress) appears to have been lost in translation as Asian ginseng moved from traditional Chinese medicine to Western commodity (Dharmananda, 2002). Asian ginseng is the most associated with male health and is focused on here, also because neither American ginseng nor eleuthero has been subjected to human clinical trials to definitively determine their effects on libido.

A meta-analysis of seven clinical trials on steamed, dried roots of four-year-old Asian ginseng roots (known as red ginseng or hóng shēn) in men with ED of various types found that overall red ginseng was effective compared to placebo (Jang, 2008). The trials were generally considered to be of low quality, as most did not report methods of randomization or blinding, details of drop-outs or withdrawals, or efficacy of allocation concealment. Doses ranged from 1,000 mg daily to 600 mg tid. Men had a range of types of ED, but three trials dealt with men with psychogenic ED and three with men with mixed ED. A subgroup analysis of the men with psychogenic ED did find red ginseng significantly effective compared to placebo, supporting that red ginseng worked at least in part by stimulating libido. Most animal research suggests Asian ginseng has both a beneficial effect on endothelial functional as well as central-acting effects on libido (Choi, 1999).

Asian ginseng is extremely safe, despite a few questionable case studies in the literature (Paik, 2015). In 40 clinical trials on Asian ginseng that reported on adverse effects, no major events occurred (Shergis, 2013). The majority of clinical trials confirm that Asian ginseng has no adverse interaction with anticoagulants. A fairly typical dose is 1,000 mg of crude root powder two to three times per day taken with or without food. Unfortunately, given the high value of Asian ginseng, it is frequently adulterated.

Withania somnifera (ashwagandha) root, a member of the Solanaceae family, also has a long history of use as what would be considered an adaptogen. As its Latin binomial suggests ("somnifera" coming from the Latin word for sleep), it is more clearly a calming or relaxing agent. It has a strong history of use as a sexual tonic particularly for men. However, in a randomized, single-blind trial, crude root powder 2 g tid was not superior to placebo at improving psychogenic ED in 86 men (Mamidi, 2011).The short duration of this trial, just 60 days, was probably a major limiting factor as adaptogens typically take more than three months to start to have noticeable effects. Given that there are trials indicating an anxiolytic effect of ashwagandha, this herb should be studied in different extracts, for longer durations, and in men who particularly have anxiety-related sexual dysfunction to determine if it can be clinically helpful (Andrade, 2000).

Two Illustrative Aphrodisiacs

One of the best-studied and validated herbal treatments for low libido is *Pausinystalia johimbe* (yohimbe) bark (Rubiaceae family) and its indole alkaloid β-yohimbine. This evergreen tree is native to the rainforests of West Africa, particularly Nigeria, Cameroon, and the Democratic Republic of the Congo. This herb and its alkaloid are central-acting aphrodisiacs.

β-Yohimbine is a potent antagonist of the α-2 adrenergic receptors (Millan, 2000). These receptors, in the central nervous system, act as a reuptake mechanism for catecholamines. Blocking the receptor therefore leaves more of these neurotransmitters active in the synapses, resulting in a libido- and mood-enhancing effect (Bremner, 1997).

Meta-analysis of clinical trials of β-yohimbine have uniformly reported it helps improve libido in men compared to placebo (Ernst, 1998). The clinical trials have been criticized for being relatively small and having methodological problems. One trial in postmenopausal women found a combination of arginine 6 g and yohimbine 6 mg increased physiologic responses to erotic stimuli significantly better than placebo, with no difference between groups in subjective measures of arousal (Meston, 2002). A separate trial found yohimbine, mirtazapine, and olanzapine equally ineffective as placebo at offsetting sexual dysfunction due to selective serotonin reuptake inhibitors (SSRIs) in women (Michelson, 2002). The role of yohimbe and yohimbine in women is unclear but should probably only be considered only when first-line approaches to low libido have failed.

The author has used yohimbe as an informal test to help differentiate psychogenic and pathophysical ED, though this has not been strictly studied in a clinical trial. Patients with an unclear ED etiology but without major signs of arteriosclerosis are given yohimbe tincture for two weeks. If at the end of this period they report a significant improvement in their symptoms, it is highly likely they have mostly psychogenic problems as yohimbe does not appear to directly affect vascular or endothelial function to any significant degree. If it has minimal to no effect, vascular, neurogenic, and endocrine ED (though the latter two are much less common) have to be ruled out. Patients are clearly told this is a test only and not a long-term treatment of their problem.

For example, a 64-year-old very healthy white male presented with sudden onset of total loss of erectile function, nocturia associated with mild to moderate benign prostatic hyperplasia, and a fluctuating, benign, cystic mass in the left scrotum. He was happily married and had one grown son. He had not noted any change in his relationship with his wife and reported that their sex life was satisfactory. He was very physically active and fit, ate a healthful vegetarian diet, and had good-quality sleep. He agreed to try yohimbe tincture. Though he was told to take five drops twice per day and record what happened, and though the bottle was labeled with this dose and a hand-written treatment plan also stated this dose, what he actually ended up doing was taking five droppersful twice a day, or a 25-fold overdose as there are approximately 25 drops per dropperful. Within three days, he noticed a very strong increase in his libido, near continual thoughts about sex, and total restoration of erectile function, though this was accompanied by some chest and head pressure and flushing. Intriguingly, the force of his urine stream also dramatically improved. After two weeks of this accidental overdose, he took a dose after running for an hour and became extremely sleepy, slept for an hour, ate, then started to feel a worsening sensation of panic and a certainty he was having a heart attack. Emergency services were summoned and after it was determined he was not having a stroke or heart attack but it was determined he was taking yohimbe, he went to the hospital where further testing confirmed no cardiac damage. He stopped the yohimbe of course and angrily wanted to know what had happened. After a careful review it became clear he had made the error in overdosing. Nevertheless, it was pretty clear he had psychogenic ED as the yohimbe had such an immediate and strong effect that would not be possible with significant arterial, neurological, or endocrine dysfunction.

Pure β-yohimbine is not available in most jurisdictions, so only yohimbe tinctures or powders are an option. A typical dose of a 1:3 to 1:5 weight:volume (w:v) tincture of yohimbe bark is five drops (0.2 ml) twice a day. It is imperative that patients not misconstrue this dose as five droppersful (5 ml) twice a day, a massive overdose almost guaranteed to cause hypertension, anxiety, and/or a panic attack after a few days. Yohimbe needs to be taken regularly; it is not an on-demand treatment though it should take effect within two to three days.

Yohimbe and β-yohimbine are contraindicated in patients with anxiety disorders, obsessive-compulsive disorder, and active posttraumatic stress (Bremner, 1997; Charney, 1984). They can cause hypertension and anxiety in overdose, which if severe enough can feel like a heart attack (Goldberg, 1983). That being said, pioneering trials have now shown that under supervision of a trained therapist, the combination of yohimbine with anxiety triggers (so-called exposure therapy) actually helps desensitize people to their fears far more effectively than beta-blockers, which mitigate fear responses (Smits, 2014). They are not recommended for use during pregnancy or lactation.

Turnera diffusa (damiana, mizib-coc in Mayan), formerly known by the descriptive Latin name *T. aphrodisiaca*, is an interesting contrast to the comparatively well-researched yohimbe. This native northern and central Mexican plant is a member of the Passifloraceae family. The leaves have a history of use as an aphrodisiac, and this remains the major basis for its continued popularity, though damiana is also noted for helping with diarrhea and gastrointestinal ulcers, among other things (Hernández, 2003; Taha, 2012). Based on animal studies damiana has anxiolytic properties that support the traditional concept that it helps patients with performance anxiety or anxiety-induced low libido (Kumar, 2005).

Unlike yohimbine, there are very few clinical studies of damiana. The only one in patients with low libido involved a formula containing damiana along with arginine, Asian ginseng, *Ginkgo biloba* (ginkgo) leaf, and several vitamins and minerals (Ito, 2001). Among 77 women, those randomized to the formula had significant improvement in libido and sexual satisfaction compared to those taking placebo after four weeks. No adverse effects were observed. No studies were located on the effects of damiana on low libido in men, though it does appear to increase sexual desire in male rats (Arletti, 1999). The mechanisms of action of damiana are largely unknown, though one study in male rats suggested flavonoids in damiana leaf had similar, though milder, effects as yohimbine (Estrada-Reyes, 2009). Other flavonoids, notably apigenin, contribute an anxiolytic action though its exact molecular mechanism is not fully known (Kumar, 2008).

A typical dose of damiana tincture (1:2–1:3 w:v) is 0.5 to 1 tsp (3–5 ml) three times per day. It also needs to be taken regularly to be effective, often taking one or more months to begin to be noticeable. Damiana is extremely safe with no known drug interactions.

The following case illustrates a complicated patient and how he was managed in actual naturopathic practice utilizing multiple herbs in combination. A 73-year-old man with a serum total prostate-specific antigen (PSA) of 8.0 ng/ml underwent a prostate biopsy, which found a single core positive for low-grade cancer, Gleason 3 + 3, with no perineural invasion. He was still having hematuria though the biopsy had been two weeks prior. He had a very large prostate at 127 cc (measured by pelvic magnetic resonance imaging) and significant lower urinary symptoms. He had spinal stenosis and a herniated intervertebral disc between L3 and L4 causing a lot of pain; treatment to correct this had been delayed until the elevated PSA was resolved. He had type 2 diabetes mellitus fairly well controlled with morning blood glucose levels generally below 120 mg/dl and hemoglobin A1c levels <6.5%. He historically had a strong libido and good erections but in the past two years associated with his various health concerns and life stress, he began to lose his sex drive and had trouble getting and maintaining an erection. He was stable on metformin, pioglitazone, levothyroxine, lisinopril, and as-needed dextropropoxyphene for back pain. He was also taking saw palmetto/β-sitosterol/pumpkin seed oil, holy basil tea, ginkgo extract, lecithin, coQ10, vitamin E, glucosamine/chondroitin sulfate, garlic extract, and a multivitamin.

The patient was prescribed two herbal tincture formulas. The first was a combination of several herbs for stress, immune support (to help fight the cancer as well as infection prevention), cancer suppression, stopping bleeding, and further support of blood glucose regulation. It contained *Eleutherococcus senticosus* (eleuthero) root 12%, *Schisandra chinensis* (wu wei zi) fruit 10%, *Oplopanax horridus* (devil's club) root bar 10%, *Momordica charantia* (bitter melon) fruit 10%, *Vaccinium myrtillus* (bilberry) leaf and fruit 10%, *Mahona aquifolium* (Oregon grape) root 9%, *Catharanthus roseus* (Madagascar periwinkle) herb 9%, *Geranium maculatum* (cranesbill) root 8%, *Viscum album* (European mistletoe) herb 8%, *Fouquieria splendens* (ocotillo) bark 5%, *Rheum palmatum* (rhubarb) cooked root 5%, and *Taxus brevifolia* Pacific yew) leaf 4%. He was directed to take 1 teaspoon of this mixture three times daily.

The second formula directly addressed his sexual dysfunctions, which were important for him to correct. The formula was created to address what seemed to be a combination of stress- and anxiety-related libido reduction and metabolic syndrome-induced endothelial dysfunction. It contained damiana 19%, ginkgo leaf 14%, ashwagandha root 14%, muira puama bark 11%, Epimedium leaf 11%, *Astragalus membranaceus* (astragalus) root 9% (another adaptogen), *Polygonum multiflorum* (he shou wu) cooked root 9% (a tonic similar to Epimedium), *Mucuna pruriens* (velvet bean) seed 7% (a central-acting stimulant), yohimbe bark 4%, and *Capsicum frutescens* (cayenne) fruit 2% (to increase absorption of the formula ingredients). Again the dose was 1 teaspoon three times per day.

Within one month he had noted real improvement in his sexual desire and erectile quality with the formula. He had previously tried using all three available PDE5i drugs with little success. He now noted that if he took the herbal formula regularly, a dose of just 5 of tadalafil was now extremely helpful. He disliked the cayenne and so this was removed from the formula. His postbiopsy bleeding had stopped, and he had been declared ready for the back surgery. He took the two formulas for a total of three months with excellent results. Ultimately they were simplified and combined into one for ease, and he continued to take them for several more months with good sexual function and improvement in most of his concerns.

Overselling Horny Goat Weed and Caltrop Vine

The suggestively named Chinese herbal medicine yín yán huò literally translates as horny goat weed. Despite this name, horny goat weed should be viewed as more of a slow-acting tonic that has much more in common with adaptogens than aphrodisiacs: in Chinese medicine this herb is regarded as what is known as a kidney yang tonic and is used to increase libido, improve erectile function, and improve fertility among other actions (Munyangaju, 2014). Five species in the Berberidaceae family are generally recognized as interchangeable sources of horny goat weed: *Epimedium brevicornum, E. koreanum, E. pubescens, E. sagittatum,* and *E. wushanense* (Bensky, 2004). The leaves are used as medicine.

Horny goat weed and various flavonoids found in it have been repeatedly shown in preclinical trials to act as cyclic guanosine monophosphate (cGMP) phosphodiesterase 5-inhibiting, and nitric oxide synthase stimulating medicines (Chen, 2009; Chiu, 2006; Zhang, 2012). Icariin, a prenylated flavonoid that is the best-studied compound in horny goat weed, in particular is effective, with an IC_{50} in vitro of 0.688 μmol/L, very comparable to that of papaverine at 0.432 μmol/L (Xin, 2003). In another in vitro study icariin was one-tenth as potent as sildenafil at increasing cGMP levels, though its other effects not related directly to cGMP were not accounted for (such as rho-kinase inhibition; Jiang, 2006). Lest one think, however, that horny goat weed can be simply reduced to the activity of this one compound, icariin combined with the flavonoid scutellarin (also known as breviscapine), both found in horny goat weed, were significantly more effective when combined compared to when used separately to enhance erectile function in spontaneously hypertensive rats (Li, 2014).

Horny goat weed has only been assessed for enhancing erectile function in two clinical trials and in combination with other herbs and nutrients. In one open clinical trial, 49 men with mild to moderate ED were randomized to take a formula with undisclosed amounts of horny goat weed, deer antler velvet (lù róng), Asian ginseng, *Allium sativum* (garlic) aged bulb, *Cuscuta chinensis* (dodder, tù sī zǐ) seed, and cattle gallstone (niú huáng) or an eight ingredient traditional Chinese formula known as jiā wèi xiāo yáo sǎn (Augmented Rambling powder, kamishōyō-san in Japanese; Nishimatsu, 2014). Deer antler velvet and dodder have similar action as horny goat weed according to Chinese medicine, while garlic is a endothelial tonic and cattle gallstone is a digestive tonic. Erectile function improved significantly more in the group taking the horny goat weed–containing combination than the jiā wèi xiāo yáo sǎn group. Unfortunately, PDE5i drug contamination was not addressed in this trial. Note that the use of deer antler velvet is not recommended as it is not a clearly sustainable or humane medicine.

Another clinical trial, this one double-blind, compared placebo and a complex formula containing primarily deer antler velvet, horny goat weed, *Cynomorium songaricum* (cynomorium, suǒ yáng) stem, *Cistanche deserticola* (broomrape, ròu cōng róng) stem, and *Carthamus tinctorius* (safflower, hóng huā) with smaller amounts of *Dioscorea opposita* (Chinese yam, shān yào) tuber, *Lycium chinense* (Chinese wolfberry, gǒu qǐ zǐ) fruit, *Broussonetia papyrifera* (paper mulberry, chǔ shí zǐ) fruit, *Allium tuberosum* (Chinese leek, jiǔ cài zǐ) seed, *Morinda officinalis* (morinda, bā jǐ tiān) root, *Ilex cornuta* (Chinese holly, gǒu gǔ yè) leaf), *Cornus officinalis* (Asiatic cornelian cherry, shān zhū yú) fruit, and honey (Punyawudho, 2012). A group of 63 men with mild to moderate ED were randomized and took a single tablet one hour before sexual activity. This is highly unusual and not in alignment with the traditional use of these herbs on an ongoing basis and suggests the formula may have been adulterated with PDE5i (this was not explicitly addressed in the paper). The formula was said to improve erectile function significantly compared to placebo.

Horny goat weed should be considered a long-term tonic medicine for men with ED and low libido. It takes one to three months to start to take effect and gets stronger with long-term use in most cases. A typical dose of a 1:2 to 1:3 w:v tincture is 0.5 tsp (3 ml) three times daily. Up to 3 g of the crude herb can also be made into tea by infusing it into recently boiled water for 15 minutes, straining, then drinking one cup three times per day (total daily dose 9 g). Similarly, a granulated, spray-dried powder form is available of this as an instant tea, and typical doses would be

1 to 2 g three times daily. Horny goat weed is generally safe, particularly if taken with food to avoid any digestive upset. One rodent study found that large doses of horny goat weed interfered with absorption of sildenafil, but such studies rarely translate to human clinical reality (Hsueh, 2013). For the greatest safety, sildenafil should be taken 30 minutes before or two hours after taking horny goat weed.

Tribulus terrestris (caltrop vine, puncture vine, tribulus) in the Zygophyllaceae family is native to dry parts of southern Europe, southern Asia, and across Africa. It has become a global weed in desert areas, and because its fruit (the medicinal part) is a spiky capsule, it is usually considered a noxious weed. The fruit and leaves have been used traditionally, though this information is somewhat unclear. They do appear to have been used for sexual dysfunction in Ayurvedic medicine. Steroidal saponins are believed to be important to caltrop vine's medicinal activity (Dinchev, 2008).

Multiple trials support that caltrop vine can have some benefit on sexual function in men and women. A combination of caltrop vine 450 mg, the brown algae *Ecklonia cava* 300 mg, and glucosamine oligosaccharide 250 mg twice a day was compared to placebo in a single-blind, randomized trial (Sansalone, 2014). A total of 177 Italian men with mild to moderate ED participated in this three-month trial. The combination formula significantly improved erectile function compared to placebo and most effectively in men with arteriogenic eosinophil count.

In a randomized trial a similar combination formula (but with 150 mg *Ecklonia*, 396 mg tribulus, and 144 mg glucosamine per tablet) twice daily was compared to tadalafil 5 mg daily for two months in 70 men with low libido and variably erectile function (Iacono, 2012). Both products improved erectile function modestly, while only the herbal combination increased libido significantly. The herbal combination had fewer adverse effects, though all issues were minor in both groups.

One small (n = 30) double-blind, randomized trial in Spanish men with ED compared 400 mg caltrop vine to placebo for 1 month (Santos, 2014). There was no difference in erectile function between the groups at the end of the trial. This trial also found no effect of caltrop vine on serum testosterone (see later discussion for more about this). Results may have been different since a distinct extract was used, because of the short duration of the trial, or simply because the extract did not work.

A double-blind trial randomized 70 Iranian women with low libido to take either a 7.5 mg extract of caltrop vine or placebo daily for four weeks (Akhtari, 2014). Desire and sexual function improved significantly by several measures in the caltrop vine group compared to placebo. Adverse effects were mild and not significantly different between the groups. Unfortunately, none of these three studies appears to have assessed the product being tested for PDE5i drug adulteration.

Unfortunately, there is much hype selling caltrop vine as a way to raise testosterone levels, primarily aimed at athletes and young men who already have normal testosterone levels. At least two clinical trials have shown these claims are unfounded (Neychev, 2005; Rogerson, 2007). This mismarketing of what could be a useful herbal supplement for the wrong indication confuses the public and should be prevented and stopped more aggressively by regulatory authorities.

A typical dose of caltrop vine tincture (1:2–1:3 w:v) is 0.5 tsp (3 ml) three times daily. In capsules, extract doses are typically 400 mg two to three times daily. There are generally no adverse effects at these doses.

Saffron

Crocus sativus (saffron), of the Iridaceae family, has medicinal stamens (an unusual medicinal plant part) with effects both as a mood-elevating yet calming aphrodisiac and an agent that enhances endothelial function. In a meta-analysis of five clinical trials, two placebo and three antidepressant drug-controlled, saffron extracts are effective in patients with major depression (Hausenblas, 2013). Given the common co-occurrence of depression and low libido (Waldinger, 2015), this finding is suggestive that saffron might be helpful in some patients with sexual dysfunction.

Also importantly, saffron may help offset libido-lowering effects of antidepressants. A double-blind trial in 36 Iranian men with major depression all taking fluoxetine randomized them to saffron 15 mg twice daily or placebo (Modabbernia, 2012). After four weeks, saffron significantly improved erectile function and libido compared to placebo without any serious adverse effects.

In an initial open trial in 20 Iranian men with ED, all took 200 mg saffron extract daily for just 10 days (Shamsa, 2009). Compared to baseline, the subjects' sexual function and erectile firmness significantly improved. A follow-up randomized, open clinical trial failed to find a benefit for saffron 30 mg bid compared to sildenafil in 346 Iranian men with ED (Safarinejad, 2010). This 12-week study found sildenafil was dramatically effective at improving erectile function while saffron was basically inactive. It is still possible that saffron may help increase libido, which PDE5i drugs do not, and may have a place in clinical therapy, but its efficacy is still unclear. Robust, double-blind, randomized trials at higher doses are warranted.

Tropical Aphrodisiacs: Biopiracy and Clinical Reality

Numerous natural products from the Global South have periodically made their way into the Global North as potential treatments for sexual dysfunction. In various ways these tell a similar story of biopiracy (stealing resources and knowledge of indigenous people) and also of the drive by some unscrupulous marketers of natural products to exaggerate and distort the reality of these products in search of sales, not truth.

Lepidium meyenii (maca) in the Brassicaceae family is a well-documented example of this phenomenon, given that the biopiracy of this species occurred in the recent past. It is native to the central Andes and is related to *Raphanus sativus* (radish) and *Brassica rapa* (turnip). The medicinal portion of all of these plants is known as a hypocotyl, which is a tuberous root structure. Documentation is good that maca was used traditionally as a healing food to enhance fertility and sexual virility (Pulgar, 1978). In 2001, the American company Pure World Botanicals was granted American patents on the extraction and use of maca for sexual dysfunction (Smith, 2003). The government of Peru began to fight these patents soon after on the basis that the United States was using its patent system to steal the biological resources of another country and that the patents (for uses and extraction methods well established in traditional medicine) were not novel, that many of the claims were not substantiated, and that the raw materials were obtained illegally from Peru (Delegation of Peru, 2003; Landon, 2007). The company Naturex, a French company, bought Pure World then granted free licenses to wholly Peruvian owned maca companies and increased fees paid to maca growers in 2007, though this came at the price of tacit approval to the legitimacy of the patents (Douaud, 2007). The patents appear to still be in force today.

Maca does have libido-enhancing properties in men based on open and randomized trials. A double-blind trial in 50 Italian men with mild ED randomized them to take either maca 2,400 mg per day or placebo for 12 weeks (Zenico, 2009). Sexual function and desire improved significantly more in the maca group than the placebo group. A very similar trial in healthy Pervuian men also found gelatinized maca 1,500 or 3,000 mg daily superior to placebo for increasing libido (Gonzales, 2002). This effect was independent of effects on mood in general and, as shown in other trials, had no effects on serum levels of sex steroids (Gonzales, 2003). Another randomized, double-blind trial in 16 American adults, most of whom were women, suffering sexual dysfunction due to SSRIs took maca 1,500 or 3,000 mg daily or placebo (Dording, 2008). The higher dose was associated with significantly less sexual dysfunction than placebo.

So while maca may be a valuable tonic medicine for people with sexual dysfunction, it is not clear that it can be sourced in a way that does not support biopiracy or in a way that directly and primarily supports the native people of Peru. Instead, international corporations likely reap most of the profits from its sale. Until this situation can be remedied, the use of maca is not recommended except in the most recalcitrant cases.

Similar though less extreme examples of this phenomenon abound. *Gracilaria spp* (sea moss), a red algae from the Caribbean and South America, is a traditional aphrodisiac/virility enhancer that has been severely overharvested for the medical trade (Smith, 1984). *Ptychopetalum olacoides* (muira puama = "potency wood") root and bark, an Amazonian shrub, is sold in the Global North as an aphrodisiac based on traditional beliefs though there is little evidence the profits from its sale flow back to native people in any significant way. Caution is warranted as an unrelated Amazonian tree, *Liriosma ovata*, is also called by this same common name but does not appear to be an aphrodisiac. One open trial found a combination of muira puama (whether it was *Ptychopetalum* or *Liriosma* is unclear) and ginkgo found it could increase libido (Waynberg, 2000). It is unknown how sustainable this product is, and its efficacy remains unknown.

The Curious Case of Chocolate

Chocolate (*Theobroma cacao*) is widely reputed as an aphrodisiac, though there is actually no evidence supporting this idea. One observational study in women actually found it has no effect on libido once age was factored in (Salonia, 2006). It is not clear the origin of the connection between chocolate and libido, though certainly chocolate was revered by the native people in Mexico. The power of an idea may be strong though, and it seems unlikely chocolate will be replaced by yohimbe or Epimedium candy for Valentine's day any time soon. The connection between chocolate and Valentine's day was a marketing effort that began in the 1840s with the Victorian chocolatier Richard Cadbury, so it is hardly scientific or traditional.

REFERENCES

Akhtari E, Raisi F, Keshavarz M, et al (2014). *Tribulus terrestris* for treatment of sexual dysfunction in women: randomized double-blind placebo—controlled study. *Daru.* 22:40.

Andrade C, Aswath A, Chaturvedi SK, et al (2000). A double-blind, placebo-controlled evaluation of the anxiolytic efficacy of an ethanolic extract of *Withania somnifera*. *Indian J Psychiatry.* 42(3):295–301.

Arletti R, Benelli A, Cavazzuti E, et al (1999). Stimulating property of *Turnera diffusa* and *Pfaffia paniculata* on the sexual behavior of male rats. *Psychopharmacology.* 143(1):15–19.

Bensky D, Clavey S, Stöger E, Gamble A (2004). *Chinese herbal medicine materia medica*, 3rd ed. Seattle: Eastland Press.

Bremner JD, Innis RB, Ng CK, et al. (1997). Positron emission tomography measurement of cerebral metabolic correlates of yohimbine administration in combat-related posttraumatic stress disorder. *Arch Gen Psych.* 54:246–254.

Campbell N, Clark JP, Stecher VJ, et al. (2013). Adulteration of purported herbal and natural sexual performance enhancement dietary supplements with synthetic phosphodiesterase type 5 inhibitors. *J Sex Med.* 10(7):1842–1849.

Charney DS, Heninger GR, Breier A (1984). Noradrenergic function in panic anxiety: effects of yohimbine in healthy subjects and patients with agoraphobia and panic disorder. *Arch Gen Psychiatry*. 41:751–763.

Chen CY, Chang YH, Bau DT, et al. (2009). Discovery of potent inhibitors for phosphodiesterase 5 by virtual screening and pharmacophore analysis. *Acta Pharmacol Sin*. 30:1186–1194.

Chiu JH, Chen KK, Chien TM, et al. (2006). *Epimedium brevicornum* Maxim extract relaxes rabbit corpus cavernosum through multitargets on nitric oxide/cyclic guanosine monophosphate signaling pathway. *Int J Impot Res*. 18:335–342.

Choi YD, Rha KH, Choi HK (1999). In vitro and in vivo experimental effect of Korean red ginseng on erection. *J Urol*. 162(4):1508–1511.

Cohen PA, Venhuis BJ (2013). Adulterated sexual enhancement supplements: more than mojo. *JAMA Intern Med*. 173(13):1169–1170.

Delegation of Peru (2003). Patents referring to *Lepidium meyenii* (maca): response of Peru. Intergovernmental Committee on Intellectual Property and Genetic Resources, Traditional Knowledge and Folkore, Fifth Session. World Intellectual Property Organization, Geneva, Switzerland, July 7–15.

Dharmananda S (2002). The nature of ginseng: traditional use, modern research, and the question of dosage. *HerbalGram*. 54:34–51.

Dinchev D, Janda B, Evstatieva L, et al. (2008). Distribution of steroidal saponins in *Tribulus terrestris* from different geographical regions. *Phytochemistry*. 69:176–186.

Dording CM, Fisher L, Papakostas G, et al. (2008). A double-blind, randomized, pilot dose-finding study of maca root (*L. meyenii*) for the management of SSRI-induced sexual dysfunction. *CNS Neurosci Ther*. 14(3):182–191.

Douaud C (2007). Naturex eases maca patents for benefit of Peruvian growers. NutraIngredients-USA, June 20. http://www.nutraingredients-usa.com/Suppliers2/Naturex-eases-maca-patents-for-benefit-of-Peruvian-growers.

Ernst E, Pittler MH (1998). Yohimbine for erectile dysfunction: a systematic review and meta-analysis of randomized clinical trials. *J Urol*. 159:433–436.

Estrada-Reyes R, Ortiz-López R, Gutiérrez-Ortíz J, Martínez-Mota L (2009). *Turnera diffusa* Wild (Turneraceae) recovers sexual behavior in sexually exhausted males. *J Ethnopharmacol*. 123:423–429.

Goldberg MR, Hollister AS, Robertson D (1983). Influence of yohimbine on blood pressure, autonomic reflexes, and plasma catecholamines in humans. *Hypertension*. 5(5):772–778.

Gonzales GF, Cordova A, Vega K, et al. (2002). Effect of *Lepidium meyenii* (maca) on sexual desire and its absent relationship with serum testosterone levels in adult healthy men. *Andrologia*. 34(6):367–372.

Gonzales GF, Cordova A, Vega K, et al. (2003). Effect of *Lepidium meyenii* (maca), a root with aphrodisiac and fertility-enhancing properties, on serum reproductive hormone levels in adult healthy men. *J Endocrinol*. 176(1):163–168.

Hamilton LD, Julian AM (2014). The relationship between daily hassles and sexual function in men and women. *J Sex Marital Ther*. 40(5):379–395.

Hausenblas HA, Saha D, Dubyak PJ, Anton SD (2013). Saffron (*Crocus sativus* L) and major depressive disorder: a meta-analysis of randomized clinical trials. *J Integr Med.* 11(6):377–383.

Hernández T, Canales M, Avila JG, et al. (2003). Ethnobotany and antibacterial activity of some plants used in traditional medicine of Zapotitlán de las Salinas, Puebla (Mexico). *J Ethnopharmacol.* 88:181–188.

Hsueh TY, Wu YT, Lin LC, et al. (2013). Herb-drug interaction of *Epimedium sagittatum* (Sieb et Zucc) Maxim extract on the pharmacokinetics of sildenafil in rats. *Molecules.* 18(6):7323–7335.

Iacono F, Prezioso D, Illiano E, et al. (2012). Sexual asthenia: tradamixina versus tadalafil 5 mg daily. *BMC Surg.* 12(Suppl 1):S23.

Ito TY, Trant AS, Polan ML (2001). A double-blind placebo-controlled study of ArginMax, a nutritional supplement for enhancement of female sexual function. *J Sex Marital Ther.* 27(5):541–549.

Jang DJ, Lee MS, Shin BC, et al. (2008). Red ginseng for treating erectile dysfunction: a systematic review. *Br J Clin Pharmacol.* 66(4):444–450.

Jiang Z, Hu B, Wang J, et al. (2006). Effect of icariin on cyclic GMP levels and on the mRNA expression of cGMP-binding cGMP-specific phosphodiesterase (PDE5) in penile cavernosum. *J Huazhong Univ Sci Technolog Med Sci.* 26:460–462.

Kumar S, Madaan R, Sharma A (2008). Pharmacological evaluation of bioactive principle of *Turnera aphrodisiaca. Indian J Pharm Sci.* 70:740–744.

Kumar S, Sharma A (2005). Anti-anxiety activity studies of various extracts of *Turnera aphrodisiaca. Ward. J Herbal Pharmacother.* 5(4):13–21.

Landon AJ (2007). Bioprospecting and biopiracy in Latin America: The case of maca in Perú (Nebraska Antrhopologist Paper 32). http://digitalcommons.unl.edu/nebanthro/32.

Li Y, Jiang J, He Y, et al. (2014). Icariin combined with breviscapine improves the erectile function of spontaneously hypertensive rats. *J Sex Med.* 11:2143–2152.

Low MY, Zeng Y, Li L, et al. (2009). Safety and quality assessment of 175 illegal sexual enhancement products seized in red-light districts in Singapore. *Drug Saf.* 32(12):1141–1146.

Mamidi P, Thakar AB (2011). Efficacy of ashwagandha (*Withania somnifera* Dunal Linn) in the management of psychogenic erectile dysfunction. *Ayu.* 32(3):322–328.

Meston CM, Worcel M (2002). The effects of yohimbine plus L-arginine glutamate on sexual arousal in postmenopausal women with sexual arousal disorder. *Arch Sex Behav.* 31(4):323–332.

Michelson D, Kociban K, Tamura R, Morrison MF (2002). Mirtazapine, yohimbine or olanzapine augmentation therapy for serotonin reuptake-associated female sexual dysfunction: a randomized, placebo controlled trial. *J Psychiatr Res.* 36(3):147–152.

Millan MJ, Newman-Tancredi A, Audinot V, et al. (2000). Agonist and antagonist actions of yohimbine as compared to fluparoxan at alpha(2)-adrenergic receptors (AR)s, serotonin (5-HT)(1A), 5-HT(1B), 5-HT(1D) and dopamine D(2) and D(3) receptors: significance for the modulation of frontocortical monoaminergic transmission and depressive states. *Synapse.* 35(2):79–95.

Modabbernia A, Sohrabi H, Nasehi AA, et al. (2012). Effect of saffron on fluoxetine-induced sexual impairment in men: randomized double-blind placebo-controlled trial. *Psychopharmacology (Berl).* 223(4):381–388.

Munyangaju JE, Miao L, Gan GW, et al. (2014). Yang-tonifying traditional Chinese medicinal plants and their potential phytoandrogenic activity. *Chin J Nat Med.* 12(5):321–334.

Neychev VK, Mitev VI (2005). The aphrodisiac herb *Tribulus terrestris* does not influence the androgen production in young men. *J Ethnopharmacol.* 101(1–3):319–323.

Nishimatsu H, Kitamura T, Yamada D, et al. (2014). Improvement of symptoms of aging in males by a preparation LEOPIN ROYAL containing aged garlic extract and other five of natural medicines—comparison with traditional herbal medicines (Kampo). *Aging Male.* 17:112–116.

Paik DJ, Lee CH (2015). Review of cases of patient risk associated with ginseng abuse and misuse. *J Ginseng Res.* 39(2):89–93.

Pawar VS, Shivakumar H (2012). A current status of adaptogens: natural remedy to stress. *Asian Pac J Trop Dis.* 2(Suppl 1):S480–S490.

Pulgar J (1978). La Maca y el uso de la región Puna VIII. Periódico Expreso, July 4. Lima, Perú: 18.

Punyawudho B, Puttilerpong C, Wirotsaengthong S, Aramwit P (2012). A randomized, double-blind, placebo-controlled crossover study of Cappra® for the treatment of mild or mild to moderate erectile dysfunction in Thai male. *Afr J Tradit Complement Altern Med.* 10:310–315.

Rogerson S, Riches CJ, Jennings C, et al. (2007). The effect of five weeks of *Tribulus terrestris* supplementation on muscle strength and body composition during preseason training in elite rugby league players. *J Strength Cond Res.* 21(2):348–353.

Safarinejad MR, Shafiei N, Safarinejad S (2010). An open label, randomized, fixed-dose, crossover study comparing efficacy and safety of sildenafil citrate and saffron (*Crocus sativus* Linn) for treating erectile dysfunction in men naïve to treatment. *Int J Impot Res.* 22(4):240–250.

Salonia A, Fabbri F, Zanni G, et al. (2006). Chocolate and women's sexual health: an intriguing correlation. *J Sex Med.* 3(3):476–482.

Sansalone S, Leonardi R, Antonini G, et al. (2014). Alga *Ecklonia bicyclis, Tribulus terrestris,* and glucosamine oligosaccharide improve erectile function, sexual quality of life, and ejaculation function in patients with moderate mild-moderate erectile dysfunction: a prospective, randomized, placebo-controlled, single-blinded study. *Biomed Res Int.* 2014:121396.

Santos CA Jr, Reis LO, Destro-Saade R, et al. (2014). *Tribulus terrestris* versus placebo in the treatment of erectile dysfunction: A prospective, randomized, double blind study. *Actas Urol Esp.* 38(4):244–248.

Shamsa A, Hosseinzadeh H, Molaei M, et al. (2009). Evaluation of *Crocus sativus* L (saffron) on male erectile dysfunction: a pilot study. *Phytomedicine.* 16(8):690–693.

Shergis JL, Zhang AL, Zhou W, Xue CC (2013). *Panax ginseng* in randomised controlled trials: a systematic review. *Phytother Res.* 27(7):949–965.

Smith AH, Nichols K, McLachlan J (1984). Cultivation of seamoss (*Gracilaria*) in St. Lucia, West Indies. In CJ Bird, MA Ragan (Eds.), *Developments in hydrobiology: Eleventh International Seaweed Symposium,* vol. 22 (pp. 249–251). Amsterdam: Springer Netherlands.

Smith E (2003). Maca root: modern rediscovery of an ancient Andean fertility food. *J Amer Herbalists Guild.* 1:15–21.

Smits JA, Rosenfield D, Davis ML, et al. (2014). Yohimbine enhancement of exposure therapy for social anxiety disorder: a randomized controlled trial. *Biol Psychiatry.* 75(11):840–846.

Taha MME, Salga MS, Ali HM, et al. (2012). Gastroprotective activities of *Turnera diffusa* Willd. ex Schult. revisited: Role of arbutin. *J Ethnopharmacol.* 141:273–281.

Waldinger MD (2015). Psychiatric disorders and sexual dysfunction. *Handb Clin Neurol.* 130:469–489.

Waynberg J, Brewer S (2000). Effects of Herbal vX on libido and sexual activity in premenopausal and postmenopausal women. *Adv Ther.* 17(5):255–262.

Xin ZC, Kim EK, Lin CS, et al. (2003). Effects of icariin on cGMP-specific PDE5 and cAMP-specific PDE4 activities. *Asian J Androl.* 5(1):15–18.

Zenico T, Cicero AF, Valmorri L, et al. (2009). Subjective effects of *Lepidium meyenii* (maca) extract on well-being and sexual performances in patients with mild erectile dysfunction: a randomised, double-blind clinical trial. *Andrologia.* 41:95–99.

Zhang J, Wang YB, Ma CG, et al. (2012). Icarisid II, a PDE5 inhibitor from *Epimedium wanshanense,* increases cellular cGMP by enhancing NOS in diabetic ED rats corpus cavernosum tissue. *Andrologia.* 44(Suppl 1):87–93.

INDEX

Note: Page numbers followed by 'f' and 't', refers to figures and tables